Inorganic and Organometallic Polymers II

ACS SYMPOSIUM SERIES **572**

Inorganic and Organometallic Polymers II

Advanced Materials and Intermediates

Patty Wisian-Neilson, EDITOR
Southern Methodist University

Harry R. Allcock, EDITOR
Pennsylvania State University

Kenneth J. Wynne, EDITOR
Office of Naval Research

Developed from a symposium sponsored
by the Division of Polymer Chemistry, Inc.,
at the 205th National Meeting
of the American Chemical Society,
Denver, Colorado,
March 28–April 2, 1993

American Chemical Society, Washington, DC 1994

Library of Congress Cataloging-in-Publication Data

Inorganic and organometallic polymers II: advanced materials and intermediates: developed from a symposium sponsored by the Division of Polymer Chemistry at the 205th National Meeting of the American Chemical Society, Denver, Colorado, March 28–April 2, 1993 / Patty Wisian-Neilson, editor, Harry R. Allcock, editor, Kenneth J. Wynne, editor.

    p.   cm.—(ACS symposium series, ISSN 0097–6156; 572)

Includes bibliographical references and indexes.

ISBN 0–8412–3036–6

1. Inorganic polymers—Congresses. 2. Organometallic polymers—Congresses.

I. Wisian-Neilson, Patty, 1949– . II. Allcock, H. R. III. Wynne, Kenneth J., 1940– . IV. American Chemical Society. Division of Polymer Chemistry V. American Chemical Society. Meeting (205th: 1993: Denver, Colo.) VI. Title: Inorganic and organometallic polymers 2. VII. Series.

QD196.I53   1994
546—dc20   94–36598
                                                           CIP

PRINTED IN THE UNITED STATES OF AMERICA

Foreword

THE ACS SYMPOSIUM SERIES was first published in 1974 to provide a mechanism for publishing symposia quickly in book form. The purpose of this series is to publish comprehensive books developed from symposia, which are usually "snapshots in time" of the current research being done on a topic, plus some review material on the topic. For this reason, it is necessary that the papers be published as quickly as possible.

Before a symposium-based book is put under contract, the proposed table of contents is reviewed for appropriateness to the topic and for comprehensiveness of the collection. Some papers are excluded at this point, and others are added to round out the scope of the volume. In addition, a draft of each paper is peer-reviewed prior to final acceptance or rejection. This anonymous review process is supervised by the organizer(s) of the symposium, who become the editor(s) of the book. The authors then revise their papers according to the recommendations of both the reviewers and the editors, prepare camera-ready copy, and submit the final papers to the editors, who check that all necessary revisions have been made.

As a rule, only original research papers and original review papers are included in the volumes. Verbatim reproductions of previously published papers are not accepted.

M. Joan Comstock
Series Editor

Contents

MAIN GROUP ELEMENT POLYMERS

INDEXES

Preface

INORGANIC AND ORGANOMETALLIC POLYMERS were the subject of a second international symposium held last year. Seven years have passed since the first symposium and six years since publication of ACS Symposium Series, Volume 360, on this subject. In this period of time, good progress has been made on a number of research fronts, and work in this area remains at the frontier of materials science. The symposium was held to update this rapidly expanding field; this volume reviews recent advances in several areas, including new polymerization processes, new polymer systems, and numerous specialty applications.

The organization of this book is similar to the first symposium volume and addresses each of the following areas: silicon-containing polymers, oxopolymers, polyphosphazenes, other main group element polymers, and metal-containing polymers. Unlike the first volume, which contained both topical reviews of each area and specialist reports by symposium participants, this volume contains only the latter.

The ever-growing list of unique properties and ultimate applications of inorganic and organometallic polymers is a major driving force for continuing research, for this symposium, and for assembling these manuscripts. This volume, which brings together many of the most recent advances of this interdisciplinary field, should therefore be of interest to materials scientists, polymer chemists, ceramists, metallurgists, materials engineers, and inorganic and organometallic chemists.

We gratefully acknowledge the generous financial support from the Office of Naval Research, the American Chemical Society Division of Polymer Chemistry, Inc., the Petroleum Research Fund of the ACS, and the Dow Corning Corporation.

PATTY WISIAN-NEILSON
Southern Methodist University
Dallas, TX 75275

KENNETH J. WYNNE
Office of Naval Research
Arlington, VA 22217–5660

HARRY R. ALLCOCK
Pennsylvania State University
University Park, PA 16802

June 10, 1994

Chapter 1

Recent Advances in Inorganic and Organometallic Polymers

An Overview

Patty Wisian-Neilson

Department of Chemistry, Southern Methodist University,
Dallas, TX 75275

As in the first symposium volume (1), the term "inorganic and organometallic polymers" is primarily used to describe macromolecules that contain inorganic elements in the polymer backbone with either organic or inorganic side-groups. In addition to linear polymers with repeating monomer units, macromolecules consisting of complex rings or with highly branched, crosslinked structures (e.g., see oxopolymers) are included in this definition.

In addition to reviews cited in the first volume, several other reviews and textbooks have appeared in recent years. These include articles by Allcock in *Science* (2) *Advanced Materials* (3), an excellent textbook by Mark, Allcock, and West (4), and a volume based on a related ACS symposium edited by Sheats, Carraher, Pittman, Zeldin, and Currell (5). A brief overview of the field can also be found in a polymer textbook by Allcock and Lampe (6). The papers in this volume describe specific topics and are listed in the approximate order of presentation at the 1993 symposium.

The main driving force for the interest in inorganic and organometallic polymers is the quest for new specialty materials that meet the rigorous demands of high technology. As described by Allcock (reference 1 and Chapter 17, this volume), classical materials can be grouped into four major categories: ceramics, metals, semiconductors, and organic polymers. Inorganic macromolecules bridge these dissimilar areas and, with appropriate design, can be made to incorporate the useful properties of more than one of these systems. Thus, this volume includes discussions of inorganic polymers with a broad range of properties such as electrical and ionic conductivity, high temperature and radiation resistance, and biological compatibility. Potential applications include resist materials, gas permeable membranes, high temperature thermosets, precursors to important ceramics, etc.

The articles in this volume are grouped roughly by elements in the polymer backbone: (a) Silicon Containing Polymers, (b) Oxopolymers, (c) Poly(phosphazenes) (d) Other Main Group Element Polymers and (e) Metal Containing Polymers. As discussed in the following paragraphs, each of these groups consists of work that reflects the long-term goals in the field: (a) synthesis of entirely new polymers, (b)

0097–6156/94/0572–0001$08.00/0

new preparative routes to known systems, (c) modifications of familiar materials, and (d) new properties of inorganic polymers.

While a number of new polymer systems have been prepared in the past few years, several that have totally new backbones are particularly notable. One is the novel class of main group element polymers called poly(alkyl/aryloxothiazenes), $[N=S(O)R]_n$ (Roy, Chapter 26). These structural analogs of polyphosphazenes are prepared by the condensation polymerization of suitably designed new silicon-nitrogen-sulfur compounds. Another set of fascinating polymers is the polymetallocenyl-silanes, -germanes, and -phosphines which contain either ferrocene or ruthenocene groups as well as main group elements in the backbone (Manners, Chapter 33). Ring opening polymerization of strained metallocenophanes was the synthetic method used to prepare these well-defined polymers. Chujo (Chapter 30) reports the synthesis of another new class of polymers that may be considered as inorganic-organic hybrid polymers. These systems, which contain both boron atoms or boron-nitrogen rings and short chains of carbon in the polymer backbone, are prepared by hydroboration reactions. Basic structure-property correlations for each of these systems are reported. The synthesis of new polymer systems is not restricted to main group element polymers as demonstrated by novel photodegradable polymers with cyclopentadienyl metal carbonyl dimers in the backbone (Tyler, Chapter 36) and conjugated organometallic NLO polymers with fluorenyl-ferrocene side-groups (Wright, Chapter 34). Indeed, these inorganic polymer systems not only demonstrate the tantalizing diversity of inorganic polymeric materials yet to be synthesized, but also illustrate the broadening scope of the methods available for the synthesis of new materials.

A great deal of work has centered on modifications, improvements, and extensions of synthetic approaches to known inorganic and organometallic polymer systems. For example, several new synthetic pathways to polysilanes are reported here. These are the anionic ring opening polymerization of "masked disilenes" or disilane bridged cyclic compounds (Sakurai, Chapter 2); simple, inexpensive electrochemical coupling of dichlorosilanes which allows for some measure of control of chain length (Biran, Chapter 3); and ring opening polymerization of cyclotetrasilanes initiated by silyl cuprates which results in stereoregular polymers (Matyjaszewski, Chapter 4). Paine and Sneddon (Chapter 27) and Kimura (Chapter 28) expand on the synthesis of borazine based polymers and discuss various aspects of their use in the preparation and fabrication of boron nitride ceramic materials. Allen (Chapter 29) expands on the well-known addition polymerization of vinylic compounds by studying the polymerization rates of vinyl monomers substituted with inorganic rings.

Modifications in the preparation of polyphosphazenes by the condensation polymerization of N-silylphosphoranimines are also reported. The preparation of a series of new poly(alkyl/arylphosphazenes), relevant chemistry of the Si-N-P precursors, and catalysis by phenoxide anions are presented (Neilson, Chapter 18). Matyjaszewski (Chapter 24) discusses the use of fluoride catalysts to prepare polyphosphazene random and block copolymers with alkoxy and fluoroethoxy substituents and proposes mechanisms for their formation.

Reactions of preformed polymers have proven to be an excellent method for preparing new materials and for tailoring polymers to achieve desirable properties.

This is most clearly illustrated by a variety of work in polyphosphazene chemistry. Allcock (Chapter 17) provides an overview and gives specific examples of applications for new materials accessible by this approach. He discusses surface modification, incorporation of groups that impart liquid crystallinity or NLO behavior, the use of cross-linking for improved mechanical properties, and structure-property relationships (e.g., T_g). Wisian-Neilson (Chapter 19) discusses a variety of derivatization reactions of poly(alkyl/arylphosphazenes) that have been used to incorporate reactive functional groups attached by P-C bonds and for the synthesis of graft copolymers via anionic grafting reactions. Other inorganic-organic systems based on polyphosphazenes include graft copolymers formed by free radical reactions (Gleria, Chapter 22) and blends of polystyrene with polyphosphazenes (Chen-Yang, Chapter 23). Finally, Ferrar (p. Chapter 20) reports the preparation and optical and mechanical properties of novel phosphazene-ceramic composites formed by in situ sol-gel polymerization of metal alkoxides in poly(methoxyethoxyethoxyphosphazene), (MEEP).

Although less common, the modifications of several preformed silicon-containing polymers have also been used to prepare new inorganic polymers. A very notable example is provided by West (Chapter 9) who reports on the hydrosilyation of C_{60} with poly(methylsiloxane) giving "shrink-wrapped buckeyball." Additional studies of modification of polysilanes, which exploit the reactivity of the Si-H groups, are outlined by Waymouth (Chapter 6). A somewhat different approach to polymer modification is given by the preparation of copolymers of poly(dimethylsiloxane) and urea-urethane (Wynne, Chapter 7). This is designed to provide "inorganic" surface properties that minimize surface adhesion.

The use of inorganic polymers as processable ceramic precursors remains a major area of study. In the section on sol-gel chemistry, new systems, property studies, and methods for controlling particle size and porosity are discussed. In the paper by Schmidt (Chapter 15), some potential optical applications and applications as anti-soil or corrosion resistant coatings of several such systems are presented. Livage (Chapter 12) outlines the preparation of mixed metal oxopolymers, while Barron (Chapter 13) and Rees (Chapter 14) report on improved methods for the synthesis of aluminum oxopolymers. Brinker (Chapter 10) reports on the influence of kinetic effects on ceramic film structures formed by sol-gel methods.

A number of processable nitrogen-containing ceramic precursors are also discussed. These include polyalazanes (Jensen, Chapter 32), cross-linkable vinyl substituted polysilazanes (Schwark, Chapter 5), mixed sol-gels from aminolysis reactions (Gonsalves, Chapter 16), and polyborazines (Sneddon and Paine, Chapter 27 and Kimura, Chapter 28) which serve as precursors to AlN, Si_3N_4, BN/AlN composites, and BN, respectively. Arylene and alkylene bridged polysilsesquioxanes (Loy, Chapter 11) and carborane-polysiloxanes (Keller, Chapter 31) have also been employed to make modified silicas.

Finally, no discussion of inorganic polymers would be complete without examples of metal-containing polymers. In addition to organometallic polymers mentioned above (Manners, Wright, Tyler), polymers formed by classical metal coordination are an important area of investigation. A particularly novel system, which has been studied by Bohle (Chapter 37) is β-hematin, an inorganic biopolymer. In addition to its important biological functions, this system serves as a model for potentially large numbers of yet undiscovered biopolymers. Very different applications of coordination polymers are discussed in Hanack's paper (Chapter 35) on semiconducting metal phthalocyanines polymers.

Several of the papers in this volume focus not only on the synthesis of new materials but also on properties and applications. The applications of polysiloxanes as photoresists (van de Grampel, Chapter 8), polyphosphazenes for microencapsulation of biologically active species (Allcock, Chapter 17), and inorganic-organic hybrid oxopolymers with optical applications (Schmidt, Chapter 15) are examples. New information is also presented on the photochemistry and photophysics (Hoyle, Chapter 25), the oxygen permeability (Kajiwara, Chapter 21), and a variety of optical properties of polyphosphazenes (Allcock).

Because inorganic polymers range from linear polymers to highly crosslinked networks, include a large number of elements in the periodic table, and encompass almost limitless combinations of elements, those outside the field may find its diversity to be almost overwhelming. Indeed, those scientists concerned with making measurements frequently find the array of so many polymers to be staggering. This is, however, part of the mystery and challenge of the field. With so many options, there must be materials in the realm of inorganic polymers that have unique and extremely useful properties. Indeed, as described in this volume, the range of such properties has certainly expanded in recent years.

Although the current trend appears to be toward focusing on specific properties for well-defined applications, the importance of developing economical, environmentally sound syntheses must remain a priority. For example, the wide-scale commercialization of polysilanes and polysilazanes would be enhanced by more efficient organosilicon hydride intermediates (7). Similarly, the cost effectiveness of poly(alkyl/arylphosphazenes) might be improved by a "direct process" analogous to that used in the silicone industry for attaching alkyl and aryl groups to the phosphine or phosphoranimine precursors. Simple modifications of known preformed systems can also be a feasible means of optimizing known inorganic polymers. This is, perhaps, best exemplified by the surface modification of polyphosphazenes where the chemistry can be modeled on well-defined polymers or even on small molecule analogs. In this way, properties such as biocompatibility or adhesion can be enhanced while retaining useful bulk properties.

Several other factors will play a pivotal role in the future of the inorganic polymer systems. Continued progress in the field requires extensive, well-coordinated, collaborative efforts of a range of scientists and engineers as well-as a considerably broadened scope of work within individual labs. For example, synthetic chemists will expand into materials characterization and materials scientists and engineers will find chemical characterization necessary to ensure the compositional and structural integrity of the materials under investigation.

Finally, industry can play a vital role in commercial development of inorganic polymers, as demonstrated by the success of silicone polymers. In the late 1930's and early 1940's a great deal of industrial effort was directed toward the development of these polymers. Better than half of all corporate research in chemistry in at least one major company was directed toward this goal during that period. General Electric alone added over 30 scientists to this project in less than four years (8). While several industries delved into the polysilanes heavily in the mid 1980's, it does not appear that these efforts were anywhere near that of the polysiloxane projects. The same can be said for polyphosphazenes. However, the scope and scale of present day applications of polysiloxanes, which are far beyond anything anticipated by those working in the field almost 50 years ago, suggests that these newer polymer systems might ultimately be extremely beneficial. With the prevailing attitudes in large corporations, it seems that real progress toward commercialization is most likely to occur in small companies where a variety of administrative factors foster development of new materials. Low volume specialty applications particularly in the electronic and biomedical fields will provide markets that can presently be adequately pursued by small industries.

In summary, the field of inorganic and organometallic polymers has advanced significantly in the past six years. Moreover, the opportunities for future progress are almost unlimited when the diversity of synthetic methods, side-group variations, and the large numbers of inorganic elements and possible combinations of these, are considered. There is, however, much to be learned about polymerization mechanisms and structure-property relationships, both of which facilitate the intelligent, systematic design of new materials. With the increased ability of chemists to design new materials, additional interactions with materials scientists and engineers are mandatory to efficiently channel synthetic efforts toward the most useful inorganic polymers. Finally, the optimistic interest and support of industry are required to sustain the ongoing development of this field.

Literature Cited

1. Zeldin, M.; Wynne, K. J.; Allcock, H. R., Ed. *ACS Symposium Series* **1988**, *360*, 283.
2. Allcock, H. R. *Science* **1992**, *255*, 1106.
3. Allcock, H. R. *Advanced Materials* **1994**, *6(2)*, 106
4. Mark, J. E.; Allcock, H. R.; West, R. *Inorganic Polymers*; Prentice Hall: Englewood Cliffs, New Jersey, 1992.
5. *Inorganic and Metal-Containing Polymers*; Sheats, J. E.; Carraher, C. E.; Pittman, C. U.; Zeldin, M.; Currell, B., Eds; Plenum: New York, 1990.
6. Allcock, H. R.; Lampe, F. W. *Contemporary Polymer Chemistry*; Prentice Hall: Englewood Cliffs, New Jersey, 1990, Chapter 9.
7. Arkles, B. *Main Group Chemistry News* **1994**, *2(2)*, 4.
8. Liebhafsky, H. A. *Silicones Under the Monogram*; Wiley: New York, 1978.

RECEIVED August 1, 1994

SILICON-CONTAINING POLYMERS

Chapter 2

Anionic Polymerization of Masked Disilenes to Polysilanes

Mechanism and Applications

H. Sakurai, K. Sakamoto, Y. Funada, and M. Yoshida

Department of Chemistry, Faculty of Science, Tohoku University, Aoba-ku, Sendai 980, Japan

The mechanism of initiation and propagation of the anionic polymerization of masked disilenes (1–phenyl–7,8–disilabicyclo[2.2.2]octa–2,5–diene derivatives) is investigated in detail. In the initiation with alkyllithium, regiospecific attack of alkyllithium occurs to guarantee the regiospecificity of the polymerization. Phenyl(dimethyl)silyl anions with suitable cryptands were found as an excellent initiator in benzene at room temperature. Under the conditions, polymerization of masked disilenes proceeded with a living manner to give highly monodispersed polysilanes. Finally, preparation of new polysilanes such as various amino group–substituted polysilanes is described. The amino–substituted polysilanes showed unexpectedly large red shift in uv absorption at 363 nm with unprecedented thermochromism. Iodine doped amino–substituted polysilanes showed the highest conductivity among doped polysilanes examined.

Polysilanes have been widely investigated in the past decade because of their potential applications in the field of materials science (*1*). Polysilanes are usually prepared by the Wurtz–type reductive coupling of dichlorosilanes with alkali metals. Although many kinds of polysilanes have been prepared by this method, the structure of the polymers is difficult to control. Molecular weight and polydispersity are also unmanageable (*2*). Recently, we have reported the anionic polymerization of masked disilenes (*i.e.* 1–phenyl–7,8–disilabicyclo[2.2.2]octa–2,5–diene derivatives, **1**) as a novel promising method for preparing polysilane (*3*). The polymerization gave polysilanes with highly ordered structures (*4*).

In typical examples of preparative experiments, the polymerization was carried out in THF under the conditions of high vacuum or argon atmosphere with a catalytic amount of alkyllithium as an initiator. Thus, a monomer was placed in a 50–ML two–necked flask, equipped with a magnetic stirrer, a rubber septum, and

0097–6156/94/0572–0008$08.00/0

a three–way stopcock under dry argon. A solution of monomer in THF, freshly distilled from sodium, was placed in the flask. A hexane solution of n–BuLi was added to the solution at −78 °C. The color of the solution turned red immediately. The mixture was kept stirring after removing the cooling bath for 30 min. A few drops of ethanol were added to the mixture. The polymers were purified by removal of the solvent, precipitation, and further cycles of dissolving–precipitation followed by freeze–drying.

Preparation of Masked Disilenes

Polymerization of Masked Disilenes

Mechanism of Initiation

First, regioselectivity of the initiation was examined with a monomer **2**. Thus, the reaction of **2** with 1 equiv. of methyllithium, followed by quenching with ethanol, gave 1,1,1–trimethyl–2,2–di(2–methylpropyl)disilane **3** and 1,1,1,3,3–pentamethyl–2,2,4,4–tetra(2–methylpropyl)tetrasilane **4** as the two main products. These products contained no regioisomers. Deuteriosilanes **3**–d_1 and **4**–d_1 were also obtained from a similar reaction, but quenched with deuterium oxide. These products were derived by protonation of the disilanyl and tetrasilanyl anions **5** and **6**. The important structural characteristic of the products is the existence of trimethylsilyl and di(2–methylpropyl)hydrosilyl groups for both **3** and **4**. The structures of the products clearly indicate that both initiation and propagation steps of the polymerization proceed exclusively by the attack of anionic species on the sterically less hingered silicon atom at the 8–position of the monomer. This is the origin of the regio-selectivity of the anionic polymerization process.

7: R^1 = Bu; R^2 = R^3 = R^4 = Me
8: R^1 = R^3 = Me; R^2 = R^4 = Pr
9: R^1 = R^3 = Me; R^2 = R^4 = Hex
10: R^1 = R^3 = Me; R^2 = R^4 = Bu

Table I. Anionic Polymerization of 1–Phenyl–7,8–disilabicyclo[2.2.2]–octa–2,5–dienes in THF

Mono- mer (M)	Initia- tor (In)	[I]/[M]	Temp (°C)	Time (h)	Yield (%)	$\overline{M}n/10^{-4}$	$\overline{M}w/\overline{M}n$
7	MeLi	0.04	−110→rt	6	72	3.4	1.9
7	sec–BuLi	0.03	−110→rt	9	60	5.3	1.5
7	n–BuLi	0.03	−110→rt	18	79	11.0	1.5
7	n–BuLi	0.03	−110→rt	6	62	11.0	1.5
7	n–BuLi	0.03	rt	11	52	9.5	1.5
7	n–BuLi	0.04	rt	3	54	8.9	1.6
8	n–BuLi	0.02	−110→rt	2	33	7.2	1.5
8	PhLi	0.13	−78→rt	1	63	4.7	1.3
9	n–BuLi	0.02	−110→rt	15	68	14.0	1.3

Mechanism of Propagation

The polymerization was thought to involve living polysilanyl anions. The livingness was, however, not provided in a rigorous manner. If the polymerization is perfectly living, the degree of polymerization should be equal to the ratio of the monomer and the initiator at the initial stage of the polymerization. However, as seen in Table I, polysilanes, prepared earlier in this investigation, have far higher molecular weight than the theoretical one. This indicates that the initiation is incomplete due to some side reactions of the initiators (*3*).

Single–Electron–Transfer Process as a Side Reaction. First, we have examined possible side reactions to decrease the efficiency of the initiation process. A solution of butyllithium was added to a THF solution of **9** at −70 °C. The color of the solution turned to red immediately and a strong ESR signal was observed with a well separated hyperfine structure (Figure 1). The observed radical species was identified as the anion radical of silyl substituted biphenyl by computational simulation as well as by comparison with spectra of a model compound $Me_3Si-C_6H_4-C_6H_5^{-\bullet}$ (*5*).

The anion radical should be a product of a single electron transfer (SET) process from butyllithium to the monomer. Since no polymeric product was obtained under the above–mentioned conditions, the SET process is an undesired side reaction of the initiation. The SET process is one of the origin of too much higher molecular weight of the polymer than the expected as observed.

Table II. Anionic Polymerization of Masked Disilene 10 Initiated by Silyl Anions in the Presence of Cryptand in Benzene at Room Temperature

In^a	[In]/[**10**]	Time (min)	Yield (%)	$\overline{Mn}_{obs}^{\ b}$	$\overline{Mn}_{cal}^{\ c}$	$\overline{Mw}/\overline{Mn}$
K	0.01	3	71	19,000	20,000	1.2
K	0.02	1.2	79	11,000	10,000	1.2
Na	0.01	16	72	24,000	20,000	1.4
Na	0.02	10	63	13,000	10,000	1.3
Li	0.015	60	64	19,000	14,000	1.3

a) K: **11a**–cryptand[2.2.2], Na: **11b**–cryptand[2.2.1], Li: **11c**–cryptand[2.1.1]. b) From GPC elusion volume relative to polystyrene. c) Calculated for the formula of $Ph_2MeSi-(BuMeSi-SiBuMe)_n-H$ (n = [**10**]/[In]).

Silylmetal as an Initiator. In order to avoid the SET process, we chose diphenyl–methylsilyl anions (Ph_2MeSiM: **11a**, M = K; **11b**, M = Na; **11c**, M = Li) as initiators for **10** and benzene as a solvent (*6*) and the polymerization was carried out under strictly anhydrous conditions by using a vacuum line technique. The polymerization did not take place in benzene with silyl anions alone. However, in the presence of suitable equimolar cryptands, the silyl anions initiated the polymerization (*7*).

The results are summarized in Table II. The molecular weights of polysilanes thus obtained were in agreement with the calculated values within experimental errors. The rate of polymerization are exceedingly high, especially so when silylpotassium is used as an initiator.

Living Anion Polymerization. Figures 2 and 3 show GPC profiles during the polymerization of **10** in benzene with **11a**–cryptand[2.2.2] as the initiator and the relationship between molecular weights and degrees of monomer conversion, respectively. Clearly, in the propagation stage of the polymerization, the molecular weight of the polymer increased proportionally to the degree of conversion of the monomer. Thus, the polymerization proceeds in a living manner under the conditions. Similar relations were observed for polymerization of **10** with **11b**–cryptand–[2.2.1] and **11c**–cryptand[2.1.1], although the rates of polymerization were much slow (Figure 4).

The polymerization initiated by the silylpotassium with cryptand[2.2.2] was completed within a few minutes. Yield of the polymer reached to 98.9%. However, after standing the solution for 16 h, nearly all the polymer underwent degradation to cyclopentasilane **12** and cyclohexasilane **13**. For the silyl sodium and lithium, the rate of both polymerization and degradation were slow, no ready degradation being observed. Apparently, the counter cations influence both propagation and degradation processes of the polymerization.

$$R\text{-}(SiMeBuSiMeBu)_n^{\ominus} \ Li^{\oplus}$$

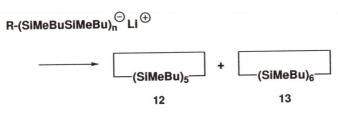

Degradation by Back–Biting Reaction. The degradation of the polysilanylpotassium in the presence of cryptand[2.2.2] and the formation of cyclic oligosilanes **12** and **13** were traced by means of their yields. The plots fit in with logarithmic curves indicating unimolecular reactions. During the degradation, the average molecular weights ($\overline{M}n$) of polymer changed slightly only from 24,000 to 19,000 (Fig. 5). Moreover, the dispersity ($\overline{M}w/\overline{M}n$) was almost kept constant at 1.3; no polymer of a medium molecular weight was formed. These results showed that the degradation of polysilanylpotassium occurred very rapidly after slow first–order initiation of degradation. The half–life of the first–order reaction is about 200 min at room temperature.

The result can be explained as follows. In a nonpolar solvent such as benzene, most of the polysilanylpotassium forms a tight ion pair which did not undergo degradation. The tight ion pair will dissociate into a solvent separated ion pair or free ions slowly by a unimolecular process, and the dissociated polysilanyl anions undergo degradation very rapidly to cyclic oligosilanes by a back–biting reaction. (8)

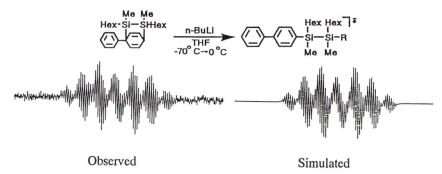

Observed   Simulated

Figure 1. ESR Spectra of Anionic Radicals Observed at Low Temperature.

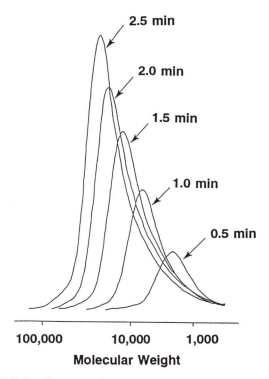

Figure 2. GPC Profiles of Polysilane in the Propagation Process of the Anionic Polymerization in Benzene Initiated by **11a**–Cryptand[2.2.2].

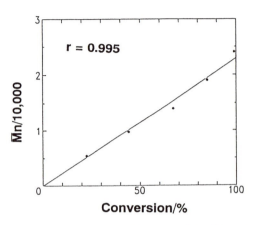

Figure 3. Anionic Polymerization of **10** Initiated by **11a**–Cryptand[2.2.2].

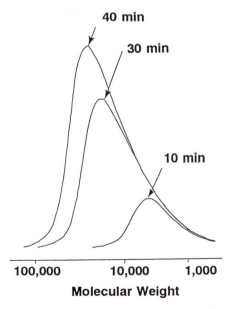

Figure 4. GPC Profiles of Polysilane in the Propagation Process of the Anionic Polymerization in Benzene Initiated by **11c**–Cryptand[2.1.1].

Dialkylamino–Substituted Polysilanes

Physical properties of polysilanes prepared by anionic polymerization are quite different to some extent as have been already seen in thermochromism of alternating copolymers (*4*).   Moreover, the anionic polymerization can afford polysilanes of special structures otherwise very difficult to prepare.   In this section, one of the most dramatic examples is shown.

The chemical and physical properties of polysilanes are strongly influenced by substituents attached to the polymer backbone.   In this respect, heteroatom–substituted polysilanes should be very much intriguing on their properties.   However, heteroatom–functional substituents such as amino and alkoxy groups on silicon cannot survive under the vigorous synthetic conditions of polysilanes by the conventional Wurtz–type condensation of dichlorosilanes.   Therefore, it is difficult to prepare heteroatom–functional polysilanes.   We have recently found that amino–substituted masked disilenes can be prepared and polymerized successfully to un–precedented amino–substituted polysilane of the completely alternative structure, poly[1,1,2–trimethyl–2–(dialkylamino)disilene].

The requisite masked disilenes **14a–c** were prepared as shown below.   Rather surprisingly, high regiospecificity of the reaction was observed; only one of the possible regioisomers was obtained in moderate yield.   Indeed, no regioisomer was detected at all.

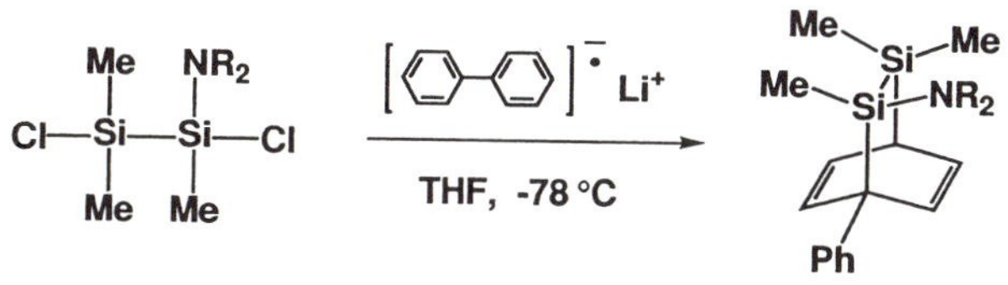

14a: R = Et
14b: R = Pr
14c: R = Bu

Our previous results showed that the isomer ratio depended on the steric bulkiness of the substituents on silicon, however, we have not observed such a complete regioselectivity before (*4*).   For the amino–substituted derivatives, the regioselectivity was amplified not only by the steric bulkiness but also by an electronic property of the amino groups.

The polymerization of these monomers in THF solutions with a catalytic amount of butyllithium gave the corresponding amino–substituted polysilanes **15a–d** in high yields.   Although polysilanes **15a** and **15b** were insoluble, **15c** was soluble in various organic solvents.   Since the solid state ^{29}Si NMR spectra of **15a** and **15b** were almost similar to that of **15c**, the insoluble polymer **15a** and **15b** are also pure amino–substituted polysilanes.   The molecular weight distribution of **15c** was determined by gel permeation chromatography calibrated by polystyrene standards with toluene as eluent; $\overline{M}n = 2.7 \times 10^4$, $\overline{M}w/\overline{M}n = 1.7$.

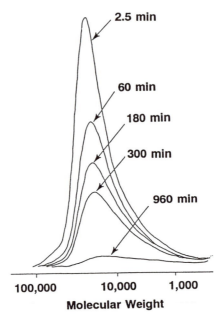

Figure 5. Degradation of Living Polysilanyl Anion (GPC Profiles).

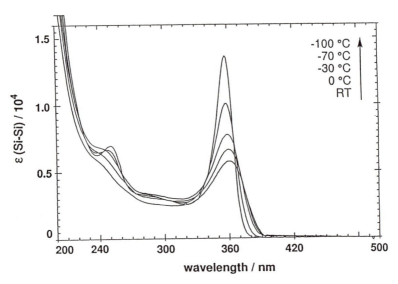

Figure 6. Temperature Dependent UV Spectra of **15c**.

Structure of **15c** was investigated by means of ^1H, ^{13}C and ^{29}Si NMR spectra. In the ^1H and ^{13}C NMR spectra of **15c**, the butyl moiety was observed to indicate that the nitrogen–silicon bond of the monomer was maintained during the polymerization. The ^{29}Si NMR spectra showed sharp signals indicating a highly ordered alternating structure of the polymer.

Figure 6 shows temperature dependent uv spectra of **15c**. The absorption maximum at 363.4 nm ($\varepsilon_{Si-Si} = 5,800$) is unusually located at the longer wavelength region as a polysilane without phenyl substituent. Moreover, with decreasing temperature, the maxima show continuous blue shift. In contrast with this thermochromic behavior, absorption maxima of typical poly(dialkylsilylene) shifted to longer wavelength at low temperatures corresponding to a conformational change of the polysilane backbone to an all–trans form (*9*). The origin of the unusually red–shifted absorption of **15c** is not merely a conformational effect (*10*) but an electronic perturbation of amino groups for the σ–conjugation of polysilanes. In this regard, it is interesting to note that another heteroatom–functional polysilane, poly(dimethoxysilylene) also exhibited a red–shifted absorption at 332 nm (*11*).

In relation to the substituent effect observed for the amino–substituted polysilanes, the conductivity of the doped samples was measured. Interestingly, it was found that the conductivity of **15c** doped by I_2 vapor in vacuo was 4×10^{-2} S cm^{-1}, that was the highest value among all the known polysilanes up to date. This result may be explained in terms of stabilization of the hole generated in the polysilane backbone by strong π–donation of amino substituents.

Acknowledgment

We are grateful for the financial support of the Ministry of Education, Science, and Culture of Japan (Specially Promoted Research No. 02102004).

Literature Cited

1. Miller, R. D.; Michl, J. *Chem. Rev.* **1989**, *89*, 1359.
2. Miller, R. D.; Thompson, D.; Sooriyakumaran, R.; Fickes, G. N. *J. Polym. Sci. Part A, Polym. Chem.* **1991**, *29*, 813.
3. Sakamoto, K.; Obata, K.; Hirata, H.; Nakajima, M.; Sakurai, H. *J. Am. Chem. Soc.* **1989**, *111*, 7641.
4. Sakamoto, K.; Yoshida, M.; Sakurai, H. *Macromolecules* **1990**, *23*, 4494.
5. Sipe, Jr., H. J.; West, R. *J. Org. Chem.* **1974**, *70*, 353.
6. Evans, A. G.; Jones, M. L.; Rees, N. H. *J. Chem. Soc.* (B), **1969**, 894.
7. Cypryk, M.; Gupta, Y.; Matyjaszewski, K. *J. Am. Chem. Soc.* **1991**, *113*, 1046.
8. Ruehl, K. E.; Davis, M. E.; Matyjaszewski, K. *Organometallics* **1992**, *11*, 788.
9. Harrah, L. A.; Zeigler, J. M. *J. Polym. Sci., Polym. Lett. Ed.* **1985**, *23*, 209.
10. Trefonas, III, P.; Damewood, Jr., J. R.; West, R.; Miller, R. D. *Organometallics* **1985**, *4*, 1318.
11. Gupta, Y.; Matyjaszewski, K. *Polym. Prep. (Am. Chem. Soc., Div. Polym. Sci.)* **1990**, *31*, 46.

RECEIVED July 11, 1994

Chapter 3

Electrochemical Access
to Di-, Tri-, and Polysilanes

M. Bordeau, C. Biran, M.-P. Léger-Lambert, F. Spirau, and D. Deffieux

Laboratoire de Chimie Organique et Organométallique, Unité de
Recherche Associée, 35 Centre National de la Recherche Scientifique,
University of Bordeaux I, F-33405 Talence, France

Di-, tri-, polysilane oligomers and polymers can be synthesized from
diorganodichlorosilanes and triorganochlorosilanes, using simple,
inexpensive and practical electrochemical techniques (undivided cell,
constant current density, sacrificial anode). The high selectivity of the
method allows the control of the formation of the polysilane backbone
since the monosilylation of dichlorosilanes can be performed in high
yield. In this way polydimethylsilane, similar to that commonly used
as a precursor of polycarbosilanes involved in the preparation of SiC-
based ceramics, can be easily obtained. A process in which dimethyldichlorosilane was used as the solvent was investigated and optimized
to afford polydimethylsilane in 90 % faradaic yield, using an
aluminum anode. Electro-copolycondensation of mixtures of
dimethyldichlorosilane and methylphenyldichlorosilane was also
investigated.

Although the first polysilanes may have been prepared more than 70 years ago by
Kipping (polydiphenylsilane) [1] and then by C. Burkhard (polydimethylsilane) [2], in
1949, they have attracted much scientific interest only since the end of the seventies
when the pioneering work of Yajima showed that poly(dimethylsilane) could be
converted through a series of pyrolytic steps into SiC fibers [3-6] and when West
found a useful preparative method for a soluble and fusible Me_2Si-PhMeSi copolymer.
  Interest in polysilane polymers has been reawakened because they have
appeared as new potential industrial raw materials for production of conducting and
semi-conducting electronic devices, photomemories, photoresists, UV-absorbing and
thermochromic materials, radical photoinitiators for polymerization, precursors for
silicon carbide ceramics and fibers, organic glasses and medical drugs [7-12].
  Among a variety of new preparative methods for creating the Si-Si bond such
as transition metal catalysed dehydrocoupling of hydrosilanes [12-14a], ring opening
polymerization [11, 12], polymerization of masked disilenes [14b], the most practical
one is still the Wurtz type condensation of organomonochloro- or dichlorosilanes
which requires a stoichiometric amount of an alkali metal and high reaction
temperatures in the cases of Na and K [3-9, 12] (Equations 1 and 2). Lower
temperature variants have yet recently been reported [14c].

0097–6156/94/0572–0018$08.00/0

$$2 \equiv SiCl + 2M \xrightarrow[-2M\,Cl]{} \equiv Si\text{-}Si \equiv \qquad (1)$$

$$n\ Cl\text{-}\underset{|}{\overset{|}{Si}}\text{-}Cl + 2nM \xrightarrow[-2n\,MCl]{} \left(Si \right)_n \qquad (2)$$

$$M = \text{alkali metal}$$

Considering that alkali metals are expensive and relatively dangerous to handle on an industrial scale, however, an electroreductive method for creating Si-Si bonds operating at room temperature has been proposed. Thus, around 1980, Hengge [15-17], Corriu [18-20] and Boudjouk [21] obtained several disilanes using "classical electrochemistry", that is to say using a divided cell equipped with Pt electrodes (and, curiously for a divided cell, with a sacrificial Hg or Cd anode) and under either controlled-potential or controlled-current conditions. The divided cell, however, presented disadvantages for industrial applications. For example, separators necessitate high cell voltages, membranes are often not sufficiently stable in organic medium and potentiostats are too expensive and not sufficiently powerful. Moreover, independently of the type of the cell, large quantities of siloxanes were formed because of residual traces of water in the medium.

For these reasons, we preferred a technique using an undivided cell with a sacrificial anode made of a readily oxidized metal. Anodic oxidation of metals has been previously shown to provide organometallic compounds from either preformed carbanions [22], or carbanions formed in situ by cathodic reduction of organic halides [23]. A ligand was often added to form stable unreactive organometallic compounds [24]. This process, first applied to organic electrosynthesis in the early 80's by Silvestri [25,26] and Matshiner [27], was still, however, a potentiostatic method. Finally, a more practical intensiostatic method has been utilized in France since 1985 by J. Perichon et al. in the field of organic chemistry [28,29] and by our group for the generation of C-Si [30-33] and Si-Si bonds [33,34].

Principle of the Sacrificial Anode Technique

The general scheme for forming Si-Si bonds involves first the generation of a $\equiv Si^-$ anion at the cathode by reduction of a chlorosilane ($\equiv SiCl$). This silicon anion is subsequently trapped in the solution either by another chlorosilane less readily reduced than the former or by the same chlorosilane, also acting as a good electrophile (Equations 3-6).

$$\text{Cathode}: \equiv SiCl + 2e^- \longrightarrow \equiv Si^- + Cl^- \qquad (3)$$

$$\text{in solution}: \equiv Si^- + \equiv Si'Cl \longrightarrow \equiv Si\text{-}Si' \equiv + Cl^- \qquad (4)$$

$$\text{Anode}: 2/n\ M - 2/n\ e^- \longrightarrow 2/n\ M^{n+} \qquad (5)$$

$$\text{overall reaction}: \equiv SiCl + 2/n\ M + \equiv Si'Cl \xrightarrow{\text{Electricity}} \equiv Si\text{-}Si' \equiv + 2/n\ MCl_n \qquad (6)$$

At the anode, the metal is oxidized in preference to the $\equiv Si^-$ and Cl^- anions, thereby avoiding secondary chlorination reactions (especially on aromatic rings) in an undivided cell. Normally, for the reaction to proceed, the reduction potentials of the different species present in solution must satisfy the order shown in Figure 1a.

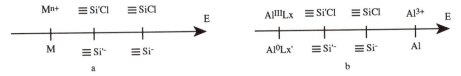

Figure 1. Potentials order

When M^{n+} is Mg^{2+}, this order is always observed but with Al, Zn, Cu cations it is not always the case, and a metallic deposition can subsequently occur at the cathode. To avoid this undesired reaction, we added a small quantity of a complexing agent like HMPA or tris (3,6-dioxaheptyl) amine (TDA-1) to the aprotic solvent, to shift the reduction potential of the cation towards more cathodic values (Figure 1b).

Experimental

Cell and Power Supply. An undivided cylindrical glass cell (100mL) fitted with a sacrificial cylindrical aluminum or magnesium bar (1 cm diameter) as the anode and a concentric cylindrical stainless steel grid (1.0 ± 0.2 dm^2, surface area of the wires) as the cathode was used. A constant current (0.1A, density 0.1 ± 0.05 A.dm^{-2}) was provided by a Sodilec EDL 36-0.7 regulated DC power supply which is very inexpensive compared to a potentiostat.

Solvent-Cosolvent. The solvent used was tetrahydrofuran (THF) containing a small amount of hexamethylphosphoramide (HMPA) or tris(3,6-dioxaheptyl)amine (TDA-1) [35] as complexing cosolvents. THF and HMPA respectively were dried over sodium benzophenone ketyl and CaH$_2$; TDA-1 was distilled under vacuum just before use.

Supporting Electrolytes. The electrochemical oxidation of the sacrificial anode forms stoichiometric metallic salts that permit the supporting electrolyte (Bu$_4$NBr, LiCl, AlCl$_3$ or Bu$_4$NBF$_4$) to be used at a very low concentration (0.02 M) instead of at 0.2-0.4 M (about that of the substrate), as usually used in classical methods. Supporting electrolytes were dried by heating for 24 h under vacuum at 50°C, and AlCl$_3$ was sublimed at 150°C under vacuum.

Drying the Medium. Residual traces of water were removed by adding a small amount of Me$_3$SiCl (about 0.2-0.5 mL), or the desired quantity if it was the reagent ; a fast hydrolysis afforded hexamethyldisiloxane, which is not electroactive, and HCl, which was eliminated by pre-electrolysis.

Chlorosilanes. (gifts from Rhône-Poulenc Co.). They were distilled under argon gas over Mg powder just before use.

Electrolytic Procedures. i) Crossed coupling of mono or dichlorosilanes with Me$_3$SiCl in a solvent medium : To the dried cell containing a magnetic spin bar was added 0.5 g (1.6 mmol) of Bu$_4$NBr (or another supporting electrolyte when specified). The cell was then pumped out and flushed with Ar two times. THF (40 mL), HMPA (7 mL) or TDA-1 (10 mL) and Me$_3$SiCl in large excess (25 to 30 mL) were introduced through a septum by means of a syringe. The solution was degassed by bubbling argon, for 10 min, under sonication. The substrate (30 mmol of a more easily reducible mono- or dichlorosilane) was then introduced. Pre-electrolysis was carried out at room temperature under a controlled current (0.1 A), to reduce the formed HCl into H$_2$. The end of the pre-electrolysis was detected by measuring the working potential of the cathode vs. an Ag wire as a reference ; thus, after initially remaining constant (-0.7 to -0.8 V) a sudden drop of the potential to about -2 V was observed. The electrolysis (0.1 A) was performed under stirring until the desired charge was consumed (18 h for 2.2 F. mol^{-1}). The progress of the reaction was monitored by gas chromatography (GC). After elimination of the salts from the resulting solution by adding anhydrous pentane, filtering and evaporating the solvents two times, the crude product was purified by distillation. The current efficiency (CE) was calculated on the basis of [yield (%) x theoretical number of electrons/charge supplied (F. mol^{-1})].

ii)**Electrosynthesis of Poly(dimethylsilane) (PDMS) "without solvent"** : A previously preelectrolyzed solution of Bu$_4$NBr (0.5-1 g) in 13 mL of HMPA, or in a mixture of 15 mL HMPA and 10 mL TDA-1, was added into the cell followed by 70 mL of Me$_2$SiCl$_2$. After the electrolysis, the crude PDMS was filtered off. THF (50 mL) was added to dissolve the salts and the crude powder filtered again. At this point, 150 mL of methanol were added and stirred overnight at room temperature. The PDMS was then filtered, washed and dried.

Results and Discussion

Reduction potential of chlorosilanes. The electrochemical reduction of chlorosilanes was shown to be an irreversible process taking place at low potentials for the monochlorosilanes (under -2.4 V vs SCE [18,19,36] and from -1.3 to -2.5 V for the dichlorosilanes [34, 37]. Nevertheless, the initial values reported in the literature are erroneous [38,39]. The main difficulty comes from the tendency of these compounds to hydrolyze, forming hydrogen chloride in the medium. The latter, as shown by Corriu [18-20, 36], was responsible for reduction peaks observed between -0.4 and -1 V vs. SCE by cyclic voltammetry [38, 39]. Actually, the values of the reduction potentials reported in the literature were determined under different conditions. For example, values (in V vs. SCE) of -2.4 [19,36], -2.5 [40], -2.85 [21] for Ph$_3$SiCl, -2.5 [19, 36] for Ph$_2$MeSiCl, -2.5 [19, 36] for PhMe$_2$SiCl, -2.28 [34] for Me$_2$SiCl$_2$, were derived from voltammetrical measurements on a hanging mercury drop in TBAP/DME ; -2.5 [37] for Me$_2$SiCl$_2$, -1.8 [37] for PhMeSiCl$_2$, -1.5 [37] and -1.3 [40] for Ph$_2$SiCl$_2$ resulted from steady-state voltammograms at a Pt cathode in DME. Me$_3$SiCl shows no reduction wave at more positive potentials than -3.0 V vs. SCE [41] ; the value -3.2 V has been proposed but without specifying the conditions [21].

Di- and Trisilanes as Models. We first studied Si-Si bond formation in di- and trisilanes from chlorosilanes, as models.

Crossed coupling of monochlorosilanes with Me$_3$SiCl. When phenylmethylmonochlorosilanes (30 mmol) were electrolyzed with 9 mol equiv of trimethylchlorosilane, which is more difficult to reduce than the former, until 2.2 F. mol^{-1} of charge was consumed, the only product formed was the unsymmetrical disilane (Equation 7) with yields similar to or better than those obtained by the alkali metal route (Table I). Both Al and Mg anodes gave similar results.

$$\text{Ph}_n\text{Me}_{3-n}\text{SiCl} + \text{Me}_3\text{SiCl} \xrightarrow[\substack{\text{Al or Mg anode} \\ \text{THF + HMPA or TDA-1} \\ \text{Room temperature}}]{\substack{2.2 \text{ F.mol}^{-1}}} \text{Ph}_n\text{Me}_{3-n}\text{Si-SiMe}_3 \qquad (7)$$

n = 1-3   excess

The absence of symmetrical dimers supports the silyl anion hypothesis, as represented in equations 3-5, which is more likely than a coupling mechanism of silyl radicals. It is suggested that the more easily formed phenylated silyl anions are then trapped by Me_3SiCl that is favored by the fact it is present in a large excess. This mechanism is corroborated by the kinetic curve for phenylpentamethyldisilane formation, obtained by GC analysis as a function of the charge passed (Equation 8, Figure 2).

$$PhMe_2SiCl \ + \ Me_3SiCl \xrightarrow[\text{THF + TDA-1}]{\text{Al, 2e}^-} PhMe_2Si\text{-}SiMe_3 \qquad (8)$$
$$\text{(5 equiv. excess)}$$

Table I.   Electrochemical Synthesis of Unsymmetrical Disilanes $Ph_nMe_{3-n}Si\text{-}SiMe_3$ - Comparison with the Classical Route

Substrate	Product						
	Electrosynthesis				Alkali Metal Route		
	Al anode			Mg anode			
	GC yield %	Ref	Isolated yield %	Isolated yield %	Intermediate	yield %	Ref
$PhMe_2SiCl$	90		80	83	≡SiLi	47	[42]
$Ph_2MeSiCl$	90	[34]	82	84	≡SiLi	74	[42]
Ph_3SiCl	90		74[a]	77[a]	≡SiLi/Na/K	75	[43]

[a] Yield after recrystallization

Homocoupling of monochlorosilanes. After 1.2 $F.mol^{-1}$ of electricity had been passed through either an Al or Mg anode, the symmetrical coupling of monochlorosilanes (65 mmol) led, as expected, to the symmetrical disilanes (equation 9) in good isolated yields which were similar to those of the metal route (Table II).

$$2 \ Ph_nMe_{3-n}SiCl \xrightarrow[\text{THF + HMPA or TDA-1}]{\text{Al or Mg anode}} Ph_nMe_{3-n}Si\text{-}SiMe_{3-n}Ph_n \qquad (9)$$
$$n = 0\text{-}3$$

In this reaction only one half of the engaged chlorosilane is reduced, the rest acting as the electrophile ; so the theoretical total current is 1 $F.\ mol^{-1}$.

In spite of its very negative reduction potential, Me_3SiCl could also be reduced to Me_6Si_2 in the above conditions ; but other procedural modifications were unsuccessful in conditions of anode/solvent/supporting electrolyte as follow : $Mg/DMF/MgCl_2$, Al/DMF + TDA-1 (40 mL/10 mL)/LiCl, Al/ACN/LiCl and Al/NMP/AlCl_3, Me_3SiCl probably being not reduced in DMF and the aluminum salts being less soluble in ACN or NMP than in THF.

In the case of homocoupling of monochlorosilanes, the kinetic curve again followed the theoretical straight line but deviated from it a little sooner because of the decrease, at the end of the electrolysis, of the amount of chlorosilane acting both as the substrate and the electrophile (Figure 3).

Table II. Electrochemical Synthesis of Symmetrical Disilanes $(Ph_n Me_{3-n} Si)_2$. Comparison with the Classical Route

Substrate	Product	Electrosynthesis Isolated yield (%)			Metal Route		
		Al anode	Ref	Mg anode	Reaction conditions	Yield %	Ref
Me_3SiCl	$(Me_3Si)_2$	74	[34]	78	Mg/THF	40	[44]
					Li/THF	76	[42]
						84	[45]
					Na/K/xylene	70	[46]
						85	[47]
					$MeMgCl/Et_2O$	72[a]	[48,49]
$PhMe_2SiCl$	$(PhMe_2Si)_2$	80		83			
$Ph_2MeSiCl$	$(Ph_2MeSi)_2$	60		68			
Ph_3SiCl	$(Ph_3Si)_2$	70		71	Li/THF ultra-sound	73	[50]

[a] s-residues as the substrate.

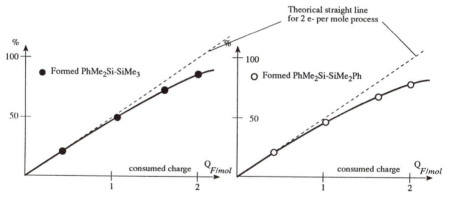

Figure 2. Kinetics of the cross-coupling of $PhMe_2SiCl$ and Me_3SiCl

Figure 3. Kinetics of the homocoupling of $PhMe_2SiCl$

Coupling of dichlorosilanes with excess Me_3SiCl. The coupling of dichlorosilanes (30 mmol) with excess Me_3SiCl (10 mol. equiv.) led, after the passage of up to 4.2 F. mol^{-1}, to the respective trisilanes (Equation 10) in isolated yields much higher than those obtained from chemical routes (Table III). Both Al and Mg anodes gave similar results.

$$Ph_nMe_{2-n}SiCl_2 + Me_3SiCl \xrightarrow[\substack{\text{Al or Mg anodes} \\ \text{THF + HMPA or TDA-1}}]{4.2\ F.mol^{-1}} Ph_nMe_{2-n}Si\text{-}(SiMe_3)_2 \quad (10)$$

$n = 0\text{-}2$ excess

Table III. Electrochemical Synthesis of Trisilanes $Ph_nMe_{2-n}Si(SiMe_3)_2$ (4.2 F. mol^{-1} passed).Comparison with the Classical Route

Substrate	Product						
	Electrosynthesis				Metal Route		
	Al anode			Mg anode			
	GC yield %	Isolated yield %	Ref	Isolated yield %	Reaction Conditions	yield %	Ref
$PhMeSiCl_2$	98	90		92	Na/K/ xylene reflux	32	[51]
Ph_2SiCl_2		77		79	Li/THF	14	[52]
Me_2SiCl_2		60	[34]	63	Li/THF	13	[45]
					3 steps[a]	42	[45,53,57]

[a] s-residues as the substrate.

This simple electrochemical method seems to be the most practical way to produce these trisilanes, since the use of alkali metals gives low yields and the best alternative route for $Me_3SiSiMe_2SiMe_3$, which requires 3 steps via $Me_3SiSiMe_2Cl$ [45, 53, 54], leads to only a 42 % yield ($Me_3SiSiMe_2Cl$ is obtained from $Me_3SiSiMe_3$ [55-57], itself prepared from the "s-residues" of the "direct synthesis" of dimethyldichlorosilane ($Me_3Si_2Cl_3 + Me_2Si_2Cl_4$).

Knowing that the reduction potentials of dichlorosilanes are much less cathodic than that of Me_3SiCl, we can assume that the dichlorosilanes are first reduced to monochlorosilyl anions which are then trapped by the Me_3SiCl present in large excess rather than by the dichlorosilane. Moreover, the kinetic curves (obtained from GC analysis) of the electrochemical reduction of $PhMeSiCl_2$ with an Al anode, as an example, show that the two silicon-chlorine bonds are reduced in two well-separated steps ; that is the monochlorinated disilane was formed initially until 2.1 F/mol of $PhMeSiCl_2$ had been consumed (Equations 11, Figure 4).

$$PhMeSiCl_2 \xrightarrow[\substack{excess\ Me_3SiCl \\ Al\ anode \\ THF + HMPA}]{2.1\ F.mol^{-1}} PhMeClSi\text{-}SiMe_3 \xrightarrow{+2.1\ F.mol^{-1}} PhMeSi\text{-}(SiMe_3)_2 \qquad (11)$$

83%(GC)   98% (GC)
63% (isolated)   90%(isolated)

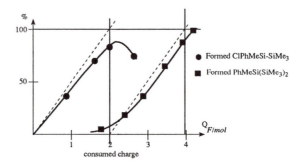

Figure 4.   Kinetics of the electrochemical synthesis of $ClPhMeSi\text{-}SiMe_3$ and $PhMeSi(SiMe_3)_2$ with an Al anode

This selectivity, not available by the chemical routes, allows a good synthesis of the 1-chloro-1-phenyltetramethyldisilane in 63 % isolated yield. But it must be pointed out that the optimization of the monosilylation procedure depends strongly on the total current passed because formation of the trisilane begins just before 2 F.mol^{-1}.

The influence of the nature of the anodic metal on the selectivity of the monosilylation was also investigated. As shown above, for 2.1 F. mol^{-1} of charge passed, the selectivity was good with an aluminum anode but, when we used a magnesium anode, the selectivity was lost (Equations 12, 13).

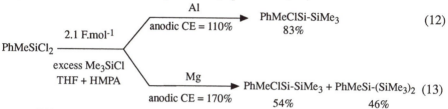

This can be understood by comparing the anodic current efficiencies (anodic C.E.) (percentage of consumed metal relative to the theoretical mass corresponding to the engaged charge). The observed anodic C.E. was roughly theoretical (110 %) for Al but much higher for Mg (170 %) ; this means that there was also simultaneous chemical reduction of the dichlorosilane and of the monochlorodisilane at the anodically scoured metal, causing the loss of selectivity.

It is interesting to note that when mixtures of Ph_2SiCl_2 or Me_2SiCl_2 with Me_3SiCl or Ph_3SiCl were electrolyzed under different conditions, oxygen-free polysilanes were obtained in good yields; e.g. a 78% yield was obtained for a 1/1 mixture of Me_2SiCl_2 and Ph_3SiCl [58]. In this experiment, the monochlorosilanes were not in excess (molar ratios monochlorosilane/dichlorosilane 1/10 to 1/1), a divided cell fitted with Pt electrodes, a catholyte containing 0.4 M Bu_4NBF_4/DME with the passage of 2 F. mol^{-1} of charge were used.

Electrochemical synthesis of tetra and pentasilanes.

Larger polysilane oligomers could also be obtained using the aluminum sacrificial anode technique. Thus, the electrolysis of 1-chloro-1-phenyltetramethyldisilane, electrochemically prepared as described above, led to the corresponding tetrasilane in 85 % yield (Equation 14).

$$2\ Me_3Si\text{-}PhMeSiCl \xrightarrow[\substack{Al\ anode \\ THF + HMPA}]{1.2\ F.mol^{-1}} \underset{85\%}{Me_3Si\text{-}PhMeSi\text{-}SiMePh\text{-}SiMe_3} \qquad (14)$$

In the same manner chloropentamethyldisilane gave decamethyltetrasilane in 83 % yield (Equation 15).

$$2\ Me_3Si\text{-}Me_2SiCl \xrightarrow[\substack{Al\ anode \\ THF + HMPA}]{1.2\ F.mol^{-1}} \underset{83\%}{Me_3Si\text{-}Me_2Si\text{-}SiMe_2\text{-}SiMe_3} \qquad (15)$$

Finally, the electrolysis of a 1 : 6 mixture of dichloromethylphenylsilane and chloropentamethyldisilane gave 3-phenylundecamethylpentasilane in 75 % yield (Equation 16).

$$PhMeSiCl_2 \; + \; \underset{excess}{Me_3Si\text{-}Me_2SiCl} \; \xrightarrow[\substack{\text{Al anode} \\ \text{THF + HMPA}}]{4.2 \; F.mol^{-1}} \; \underset{75\%}{Me_3Si\text{-}Me_2Si\text{-}PhMeSi\text{-}Me_2Si\text{-}SiMe_3} \qquad (16)$$

Thus, the sacrificial anode method allows polysilane oligomers to be built up in a stepwise fashion.

Electrochemical Synthesis of Polydimethylsilanes (PDMS). The sacrificial anode technique was applied to the electroreductive polymerization of Me_2SiCl_2. Suitable sacrificial anodes are of interest in order to prevent the possible previously observed oxidative degradation of the starting Me_2SiCl_2 and/or polydimethylsilanes, formed either directly at an anode [59] or indirectly by reactive chlorine species anodically generated from Cl^- in an undivided cell [60]. Thus, an insoluble PDMS, similar to Yajima's, was obtained from the electrolysis of Me_2SiCl_2 (80 mmol, 10 mL, 1.1 mol.L^{-1}) with an Al anode and a stainless steel cathode in mixtures of either THF and HMPA or THF and TDA-1 (about 60 mL total volume). The amount of charge passed was 2.2 F. mol^{-1} (Equation 17, Table IV).

$$n \; Me_2SiCl_2 \; \xrightarrow[\substack{\text{Al anode} \\ \text{THF + HMPA or TDA-1} \\ \text{ii) Methanolysis}}]{\text{i) } 2.2 \; F.mol^{-1}} \; \underset{\substack{30\% \\ n>25}}{\text{-}(Me_2Si)\text{-}_n} \qquad (17)$$

Table IV. Electroreductive Polymerisation of Me_2SiCl_2 using Solvent/Co-solvent Mixtures, an Al Anode and a Stainless Steel Cathode[a]

Run	Solvent/cosolvent	(mL/mL) volume ratio	Supporting electrolyte	Yield %	Insoluble PDMS Current efficiency %	Ref
1	THF/HMPA	(30/30)	Et_4NBF_4	26	24	[33a]
2	THF/HMPA	(50/10)	Bu_4NBr	30	27	[33b,34]
3	THF/TDA-1	(30/30)	LiCl	17	16	[33a]
4	THF/TDA-1	(50/5)	Et_4NBF_4	21	19	[33b]
5	THF	(100/0)	LiCl	26	24	[33a]

[a] Initial and final cell voltages were respectively 15 and 36V.

The yield of insoluble PDMS, after methanolysis, was limited less than 30 % owing to the formation of oxygen-containing oligomers (from the ring opening of the THF by Me_2ClSi^- anions) which were recovered with PDMS oligomers (n < 9) in a soluble fraction. These oxygen-containing oligomers were clearly identified by ^{1}H NMR (which showed the presence of $-Me_2Si(CH_2)_4OSiMe_2$-patterns), IR (characteristic bands for the SiOC bonds) and GC-MS (many oligomers containing the $ClMe_2SiO(CH_2)_4^+$fragment). Moreover, it is interesting to note that the amount of these THF ring-opening products increased by 2- or 3- fold in the absence of a complexing agent (run 5). This indicates that $AlCl_3$, stoichiometrically formed during the electrolysis, plays a well known role as a polymerization initiator of THF.

Because of this, we optimized the reaction by performing the electrolysis "without solvent" but in the presence of a small amount of a complexing agent such as

HMPA or TDA-1, to obtain PDMS in almost quantitative current efficiency [61] (Equation 18, Table V).

$$n \ Me_2SiCl_2 \xrightarrow[\substack{\text{"without solvent"} \\ \text{ii) Methanolysis}}]{\text{i) } e^- \text{ (Al anode)}} \begin{array}{l} -(Me_2Si)-_n \\ CE = 90\% \\ n > 25 \end{array} \qquad (18)$$

Table V. Electroreductive Polymerization of Me_2SiCl_2 "Without Solvent"

Run	Me_2SiCl_2 (complexing agent)	Vol /mL	Charge passed /F	PDMS Current efficiency %	Al[a] %
1	Me_2SiCl_2 alone		0.09	72 (grey)	>5
2	Me_2SiCl_2 (HMPA)	70 (13)	0.07	76	2.9
3	Me_2SiCl_2 (TDA-1, HMPA)	70(15,10)	0.18	84 (white)	0.1
4	Me_2SiCl_2 (TDA-1, HMPA)	70(15,10)	0.24	90 (white)	0.05

[a] From elemental analysis

A solution of Bu_4NBr (0.03 M, supporting electrolyte) in 70 mL of Me_2SiCl_2 could maintain a current intensity of 0.1 A during one day (run 1), but, in the absence of the complexing agent, the PDMS obtained in 72 % current efficiency was grey, due to the presence of metallic aluminum resulting from reduction of $AlCl_3$. This was avoided by adding a complexing agent(s) in order to shift the reduction potential of $AlCl_3$ cathodically (runs 2 and 3). Moreover, sonication (35 kHz) resulted in a further increase in the yield of PDMS (90 %) (run 4).

Characterization of insoluble PDMS. Both samples (obtained either with or without a solvent) were similar, on the basis of their IR spectra, to a PDMS standard prepared by the Na-reductive polymerisation. They were also characterized by their melting point (260°C), UV (broad band, λ_{max} = 310-320 nm measured in liquid paraffin suspension) and MS. The highest-weight fragment observed in the MS (direct insertion, m/z = 961) corresponded to a minimum polymerization degree n = 16. Knowing that the melting point [62, 63] and λ_{max} [7, 12] values of polydimethylsilanes increase with the chain length, we can estimate, on the basis of the known standard curves, that n > 25.

As described above, PDMS is used as a precursor of polycarbosilanes in the industrial production of silicon carbide ceramics and fibers. We confirmed that our PDMS samples could be pyrolytically converted into the corresponding polycarbosilanes [33a] (Equation 19).

$$-(Me_2Si)-_n \xrightarrow{470°C} -(HMeSi-CH2)-_n \qquad (19)$$

Thus, the electrosynthesis can constitute an alternative route for the preparation of PDMS.

Electroreductive polymerization of $PhMeSiCl_2$. Initial, unoptimized results, show that soluble polyphenylmethylsilanes could be obtained from the electroreductive polymerization of $PhMeSiCl_2$ in a solution of Bu_4NBr in THF and

HMPA (or TDA-1) with an aluminum anode (Equation 20, Table VI). The polymer was identified by its IR (SiMe and SiPh absorptions) and ^{29}Si NMR spectra (= -38 to -43 ppm for -(PhMeSi)$_n$- and 14.6 ppm for terminal ClPhMeSi-) according to the literature data [64].

$$\text{nPhMeSiCl}_2 \xrightarrow[\text{THF+HMPA or TDA-1}]{\text{Al anode, e}^-} \text{Cl-(PhMeSi)-}_n\text{Cl} \qquad (20)$$

Table VI.   Electrochemical Polymerization of PhMeSiCl$_2$

PhMeSiCl$_2$ Concentration (M)	Q (F.mol^{-1})	Cl-(PhMeSi)$_n$-Cl Yield %	MeO-(PhMeSi)$_n$-OMe M_n	M_w / M_n
0.5	2.2	75	930	3.6
0.5	4.1	50	5000	1.8

These results showed that the yields of the chlorinated polysilanes and the molecular weights depended on the total current. Molecular weights were determined by GPC after methanolysis and removal of volatiles at 230°C at 0.5 mmHg (eluent : THF, reference : polystyrene standard).

Electroreductive Copolymerization of Me$_2$SiCl$_2$ and Ph$_2$SiCl$_2$.

Soluble copolysilanes could be obtained by electrolysis of a mixture of Me$_2$SiCl$_2$ and PhMeSiCl$_2$ in different ratios after passing 4 F/mol of the limiting chlorosilane (Equation 21, Table VII).

$$\text{Me}_2\text{SiCl}_2 + \text{x PhMeSiCl}_2 \xrightarrow[\substack{\text{Al anode} \\ \text{THF+HMPA}}]{4 \text{ F.mol}^{-1}} \text{Cl-(Me}_2\text{Si)-(PhMeSi)}_y\text{-Cl} \qquad (21)$$

Table VII   Electrochemical Copolymerization of Me$_2$SiCl$_2$ and PhMeSiCl$_2$

x = $\dfrac{\text{PhMeSiCl}_2}{\text{Me}_2\text{SiCl}_2}$	y	Cl-(Me$_2$Si)-(PhMeSi)$_y$-Cl Yield %	M_n	M_w / M_n
1	0.9	91	988	1.6
1.9	1.8	71	769	1.7
3	2.3	86	645	1.5

The conclusions were : i) when Me$_2$SiCl$_2$ was in excess (x < 1), the copolysilane obtained was insoluble, certainly due to the presence of too many Me$_2$Si groups, responsible for crystallinity ; ii) When PhMeSiCl$_2$ was equimolecular with Me$_2$SiCl$_2$ or in excess (x ≥ 1), the copolysilane obtained was soluble and the composition ratio y, determined by ^{1}H NMR spectroscopy, was almost equal to x, corresponding to the proportion of the starting materials. The yields, based on y = x were good but the molecular weights, determined by GPC, were relatively low.

The copolymers were also characterized by comparing their ^{29}Si NMR spectra with those reported in the literature [64, 65]. The spectra obtained reveal complex resonances between -36.6 and -42 ppm (internal Me$_2$Si and PhMeSi units) and two distinct resonances at +16.5 and +25.7 ppm (terminal PhMeSiCl and Me_2SiCl groups respectively).

Other Electrochemical Synthesis of Si-Si bond-containing compounds and polymers. The interest of the topic has involved, in recent years, a lively competition in the electrochemical formation of Si-Si bonds, especially from Japanese researchers and industrial companies. Thus, since 1990, papers and patents from Shono et al [66-70] have reported the electrochemical synthesis in THF of disilanes from monochlorosilanes, of polysilanes from $PhMeSiCl_2$, $ClMe_2Si-SiMe_2Cl$ and p-$ClMe_2Si-C_6H_4-SiMe_2Cl$ and of a germane-silane copolymer using an undivided cell with Al, Mg electrodes ; no complexing agent was used but sonication and polarity-alternated electrolysis. Polysilanes from $PhMeSiCl_2$, Ph_2SiCl_2 and Me_2SiCl_2 were also obtained in DME, with a constant current, by Umezawa et al (Nippon Carbon Co) and Nonaka : i) using a divided cell with Pt electrodes [37, 58, 71], ii) using and undivided cell with Al, Mg electrodes, and polarity-alternated electrolysis [72, 73]. Finally, Kunai, M. Ishikawa et al. have synthesized, using a constant current sacrificial anode technique, an undivided cell and DME : i) disilanes and trisilanes with Hg, Ag anodes and a Pt cathode [74a], ii) polysilane oligomers and poly(disilanylene)ethylenes with Cu-Pt and Cu-Cu electrodes [74b, 75].

Concluding Remarks

The sacrificial anode electrochemical synthesis under controlled current seems to be a simple alternative to the use of akali metals, especially for the synthesis of di-, tri-, polysilane oligomers and polymers in high current efficiency and yield. This method is easy and inexpensive; the reaction only needs a bar of a common metal such as Al, Mg, Cu ; it runs at room temperature ; a stabilized regulated supply is convenient not only because the reaction can be stopped instantly, but because, when needed, the dependency of the selectivity upon the quantity of electricity and upon the nature of the electrodes makes the reaction easy to control. Moreover, extrapolation to an industrial scale can be anticipated.

Acknowledgments.

The financial support for this work by "Electricité de France", Rhône-Poulenc Co. and Région Aquitaine (France) is gratefully acknowledged.

Literature Cited

1. Kipping, F. S. J. Chem. Soc. **1924**, 125, 2291.
2. Burkhard, C. A. J. Am. Chem. Soc. **1949**, 71, 963.
3. Yajima, S.; Hayashi, J.; Omori, M. Chem. Lett. **1975**, 931.
4. Yajima, S.; Okamura, K.; Hayashi, J. Chem. Lett. **1975**, 720.
5. Hasegawa, Y.; Iimura, M.; Yajima, S. J. Mater. Sci. **1980**, 15, 1209.
6. West, R. In Ultrastructure Processing of Ceramics, Glasses and Composites Hench, L.; and Ulrich, D. C., Eds.; John Wiley and Sons, Inc. : New York, **1984**.
7. a) West, R. L'actualité Chimique, March **1986**, 64 ;b) West, R. J. Organomet. Chem. **1986**, 300, 327, and references cited therein.
8. West, R.; Maxka, J. In Inorganic and OrganometallicPolymers; Zeldin, M.; Wynne, K. J.; Allcock, H. R.; Eds.; ACS Symposium Series 360 ; American Chemical Society ; Washington, DC, **1988** ; Chapter 2.
9. Miller, R. D.; Rabolt, J. F.; Sooriyakumaran, R.; Fleming, W.; Fickes, G. N.; Farmer, B. L.; Kuzmany, H. Ref 8, Chapter 4.

10. Michl, J.; Downing, J. W.; Karatsu, T.; Klingensmith, K. A.; Wallraff, G. M.; Miller, R. D. Ref 8, Chapter 5.
11. Matyjaszewski, K.; Chen, Y. L.; Kim, H. K. Ref 8, Chapter 6.
12. Miller, R. D.; Michl, J. Chem. Rev. 1989, 89, 1359, and references cited therein.
13. Harrod, J. F. Ref 8, Chapter 7.
14. a) Chang, L. S.; Corey, J. Y. Organometallics 1989, 8, 1885 ; b) Sakamoto, K.; Obata, K.; Hirata, H.; Nakajima, M.; Sakurai, H. J. Am. Chem. Soc. 1989, 111, 7641 ; c) Kim, H. K.; Matyjaszewski, K. J. Am. Chem. Soc. 1988, 110, 3321 ; Miller, R. D.; Thompson, D.; Sooriyakumaran, R.; Fickes, G. N. J. Polym. Sci. A : Polym. Chem. 1991, 29, 813 ; Miller, R. D.; Ginsburg,E. J.; Thompson, D. Polym.J. 1993, 25, 807 ; Cragg,R. H.; Jones, R. G.; Swain, A. C.; Webb, S. J., J. Chem. Soc., Chem. Commun. 1990, 1147 ; Zeigler, J. M. J. Inorg. Organomet. Polym. 1991, 1, 531.
15. Hengge, E.; Litscher, G. Angew. Chem. Int. Ed. Engl. 1976, 15, 370.
16. Hengge, E.; Litscher, G. Monatsh. Chem. 1978, 109, 1217.
17. Hengge, E.; Firgo H. J. Organomet. Chem. 1981, 212, 155.
18. Corriu, R. J. P.; Dabosi, G.; Martineau, M. J.Chem. Soc. Chem. Commun. 1979, 457.
19. Corriu, R. J. P.; Dabosi, G.; Martineau, M. J. Organomet. Chem. 1980, 188, 63.
20. Corriu, R. J. P.; Dabosi, G.; Martineau, M. ibid. 1981, 222, 195.
21. Boudjouk, P. Report 1983, AFOSR-TR 88-0178, AD-A190 042.
22. Tedoradze, G. A. J. Organomet. Chem. 1975, 88, 1.
23. Tuck, D. G. Pure and Appl. Chem. 1979, 51, 2005.
24. Hages, P. G.; Osmann, A.; Seudeal, N.; Tuck, D. G. J. Organomet. Chem. 1985, 291, 1.
25. Gambino, S.; Silvestri, G.; Filardo, G. J. Appl. Electrochem. 1982, 12, 549.
26. Silvestri, G.; Gambino, S.; Filardo, G.; Gulotta, A. Angew. Chem. Int. Ed. Engl. 1984, 23, 979.
27. Matschiner, H.; Rudorf, W. D.; Ruettinger, H. H. East Germany Patent 1983, 203537.
28. Sock, O.; Troupel, M.; Périchon, J. Tetrahedron Lett.1985, 26, 1509.
29. Chaussard, J.; Folest, J.-C.; Nédélec, J.-Y.; Périchon, J.; Sibille, S.; Troupel, M. Synthesis, 1990, 5, 369.
30. a) Pons, P.; Biran, C.; Bordeau, M.; Dunoguès, J.; Sibille, S.; Périchon, J. J. Organomet. Chem. 1987, 321, C27; b) Pons, P.;Biran, C.; Bordeau, M.; Dunoguès, J. ibid. 1988, 358, 31.
31. Bordeau, M.; Biran, C.; Pons, P.; Léger, M.-P., Dunoguès, J. ibid. 1990, 382, C21.
32. a) Bordeau, M.; Biran, C.; Pons, P.; Léger-Lambert, M.-P.; Dunoguès, J. J. Org. Chem. 1992, 57, 4705 ; b) Bordeau, M.; Deffieux, D.; Léger-Lambert, M.-P.; Biran, C.; Dunoguès, J. French Pat. 1993, FR 2,681,866, Appl. 1991, 91/11,922.
33. a) Pons, P. Thesis of the University of Bordeaux I, 1988, n° 203; b) Léger-Lambert, M.-P. ibid. 1991, n° 646.
34. Biran, C.; Bordeau, M.; Pons, P.; Léger, M.-P.; Dunoguès, J. J. Organomet. Chem. 1990, 382, C17.
35. Soula, G. J. Org. Chem. 1985, 50, 3717.
36. Corriu, R. J. P.; Dabosi, G.; Martineau, M. J. Organomet. Chem. 1980, 186, 19
37. Umezawa, M.; Takeda, M.; Ichikawa, H.; Ischikawa, T.; Koizumi, T.; Fuchigami, T.; Nonaka, T. Electrochim. Acta 1990, 35, 1867.
38. Boberski, W. G.; Allred, A. L. J. Organomet. Chem., 1975, 88, 73.

39. Duchek, P.; Ponec, R.; Chvalovsky, V. J. Organomet. Chem. **1984**, 271, 101 ; Collect. Czech. Chem. Commun.**1986**, 51, 967.
40. Dessy, R.E.; Kitching, W.; Chivers, T. J. Am. Chem. Soc. **1966**, 88, 453.
41. Shono, T.; Matsumura, Y.; Katoh, S.; Kise, N. Chem. Letters **1985**, 463.
42. Gilman, H.; Shiina, K.; Aoki, D.; Gaj, B. J.; Wittenberg, D.; Brennan, T. J. Organomet. Chem. **1968**, 13, 323.
43. Gilman, H.; Wu, T. C. J. Am. Chem. Soc. **1951**, 78, 4031.
44. Steudel, O. W.; Gilman, H. J. Am. Chem. Soc. **1960**, 82, 6129.
45. Fritz, G.; Grunert, B. Z. Anorg. Allgem. Chem. **1981**, 473, 59.
46. Wilson, G. R., Smith, A. G., J. Org. Chem. **1961**, 26, 557.
47. Sundmeyer, W. Z. Anorg. Allgem. Chem. **1961**, 310, 50.
48. Gerval, P., Thèse d'Etat, University of Bordeaux I, **1973**.
49. Ronald, A.; Maniscolo, J. (Nalco Chemical Co), Am. Pat.**1982**, n 4 309,556.
50. Boudjouk, P.; Han, B. H. Tetrahedron Lett. **1981**, 22, 3813.
51. Kumada, M.; Ishikawa, M.; Maeda, S. J. Organomet. Chem. **1964**, 2, 478.
52. Gilman, H.; Schwebke, G. L. J. Am. Chem. Soc. **1964**, 86, 2693.
53. Kumada, M.; Ishikawa, M. J. Organomet. Chem. **1963**, 1, 153.
54. Gilman, H.; Harrel, R. L. J. Organomet. Chem. **1966**, 5, 201.
55. Kumada, M.; Yamaguchi, M.; Yamamoto, Y.; Nakajima, J.; Shuma, K. J. Org. Chem. **1956**, 21, 1264.
56. Calas, R.; Frainnet, E. ; Dentone, Y. C. R. Acad. Sci.Paris, ser. C **1964**, 269, 3777.
57. Sakurai, H.; Tominaga, K.; Watanabe, T.; Kumada, M., Tetrahedron Lett. **1966**, 5493.
58. Umezawa, M.; Takeda, M.; Ichikawa, H.; Ishikawa, T. ; Koizumi, T.; Nonaka, T. Electrochim. Acta **1991**, 36, 621.
59. Diaz, A.; Miller, R. D. J. Electrochem. Soc. **1985**, 132, 834.
60. Hengge, E. Topics Curr. Chem. **1976**, 51, 24.
61. Bordeau, M.; Biran, C.; Léger-Lambert, M.-P.; Dunoguès, J. J. Chem. Soc. Chem. Commun. **1991**, 1476.
62. Kumada, M.; Ishikawa, M.; Maeda, S. J. Organomet. Chem. **1966**, 5, 120.
63. Wesson, J. P.; Williams, T. C. J. Polym. Sci. 1979, 17, 2833.
64. Wolff, A. R.; Nozue, I.; Maxka, J.; West, R. J. Polym.Sci., Part A, **1988**, 26, 701;
65. Wolff, A. R.; Maxka, J.; West, R. J. Polym. Sci. **1988**, 26, 713.
66. Shono, T.; Kashimura, S.; Ishifune, M.; Nishida, R. J.Chem. Soc., Chem. Commun. **1990**, 1160.
67. Shono, T.; Kashimura, S.; Nishida, R.; Kawasaki, S. Eur. Pat. Appl. **1991**, EP 446,578 A2.
68. Shono, T.; Kashiwamura, S.; Nishida, R. Jpn. Kokai Tokkyo Koho, **1991**, JP 03,264,683 A2 Heisi.
69. Shono, T. Kino Zairyo **1992**, 12, 15 ; C.A. 117: 7964 k.
70. Shono, T.; Kashimura, S.; Murase, H. J. Chem. Soc., Chem. Commun. **1992**, (12), 896.
71. Ishikawa, T.; Nonaka, T.; Tokutomi, K.; Ichikawa, H.; Takeda, M. Jpn. Kokai Tokkyo Koho, **1990**, JP 02,105,825 A2 Heisei; C.A. 113 (20): 180296 j.
72. Ishikawa,T.; Nonaka, T.; Ichikawa, H.; Umezawa, M. Jpn. Kokai Tokkyo Koho, **1991**, JP 03,104,893 A2 Hesei ; C.A. 115 (16): 169054 z.
73. Umezawa, M.; Ichikawa, H.; Ishikawa, T. ; Nonaka, T. Denki Kagaku Oyobi Kogyo Butsuri Kagaku, **1991**, 59, 421; ibid. **1992**, 60, 743; CA 118 : 7469 y
74. a) Kunai, A.; Kawakami, T.; Toyoda, E.; Ishikawa, M. Organometallics **1991**, 10, 893; b) ibid. **1991**, 10, 2001.
75. Kunai, A.; Toyoda, E.; Kawakami, T.; Ishikawa, M. Organometallics **1992**, 11, 2899.

RECEIVED April 14, 1994

Chapter 4

Stereostructure of Polysilanes by Ring-Opening Polymerization

Eric Fossum, Jerzy Chrusciel, and Krzysztof Matyjaszewski[1]

Department of Chemistry, Carnegie Mellon University, 4400 Fifth Avenue, Pittsburgh, PA 15213

The stereoisomers of 1,2,3,4-tetramethyl-1,2,3,4-tetraphenyl-cyclotetrasilane were identified using a combination of spin labeling and chemical derivatization. The dominating isomer obtained in the synthesis possesses an all-trans configuration of the methyl and phenyl substituents. Polymerization of this isomer using silyl cuprates proceeds with two inversions of configuration and results in poly(methylphenylsilylene) with 25% isotactic, no syndiotactic, and 75% heterotactic triads as determined by ^{29}Si NMR studies.

Polysilanes (polysilylenes) consist of a linear chain of silicon atoms carrying two substituents, generally, either alkyl or aryl.(1,2,3) Interest in these materials stems from their unique properties, such as sigma-catenation and thermochromic behavior. They have potential applications as photoresists, electro-optical devices, non-linear optical materials, and also as precursors to silicon carbide.

Polysilanes have been prepared by several methods including: a) the reductive coupling of dichlorosilanes,(1,2) b) dehydrogenative coupling of hydridosilanes,(4) c) anionic polymerization of masked disilenes,(5) and d) ring opening polymerization of cyclotetrasilanes.(6) Each of these methods possesses its own advantages and disadvantages. The ring opening route allows for the preparation of well-defined polysilanes with controlled molecular weights, relatively low polydispersities, and defect free structures. In addition, it allows for potential control over the resulting microstructure if the monomers, with known configurations of substituents, can be opened in a controlled manner.

Because the electronic properties of polysilanes are dependent on the conformation of the backbone, which may depend on the configuration of the substituents, it is necessary to prepare polymers with well-defined microstructures. In ring opening polymerization, the resulting polymer microstructure depends on two factors. The first one is the configuration of the substituents in the cyclic monomers. The second factor is the mechanism of the ring opening polymerization. This paper is mainly focused on the discussion of stereochemistry in the cyclotetrasilanes.

[1]Corresponding author

0097−6156/94/0572−0032$08.00/0

Results and Discussion

Monomer Synthesis.   The   synthesis   of   1,2,3,4-tetramethyl-1,2,3,4-
tetraphenylcyclotetrasilane   is   shown   in   Scheme   I.(6)   A   slurry   of
octaphenylcyclotetrasilane in methylene chloride is treated with four equivalents of
trifluoromethanesulfonic acid resulting in 1,2,3,4-tetrakis(trifluoromethanesulfonoxy)-
1,2,3,4-tetraphenylcyclotetrasilane. The solvent is then removed under vacuum and the
triflate derivative dissolved in benzene/toluene. This mixture is then reacted, at -30 °C,
with four equivalents of methyl magnesium bromide resulting in a mixture of
stereoisomers of 1,2,3,4-tetramethyl-1,2,3,4-tetraphenylcyclotetrasilane. Three of the
four possible isomers shown in Scheme I are obtained. The fourth isomer, having all
four methyl groups on one side of the ring plane, has not been detected.

The isomer possessing a 1,2,3-up-4-down, **1c**, configuration of the methyl
groups can be identified easily by the 1:2:1 pattern of peaks it gives rise to in ^{1}H, ^{13}C,

Scheme I.

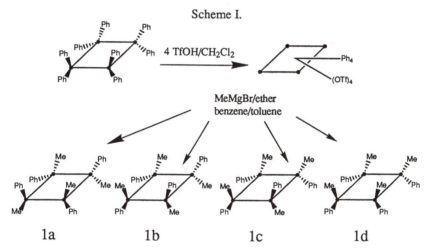

and ^{29}Si NMR spectra. The ^{29}Si NMR spectrum is shown in Figure 1(^{1}H and ^{13}C
show similar patterns). In this spectrum isomer **1c** is identified, along with tentative
assignments for isomers **1a** and **1b**, both of which would give rise to only one signal in
all NMR spectra. If this mixture is allowed to stand in hexane at sub-ambient
temperatures one of the isomers, either **1a** or **1b**, precipitates out of the solution in ≈
95% purity. This compound gives rise to the single peak shown in Figure 1b.

Preliminary Assignments

The compound, which can be purified to 95%, has previously been assigned to isomer
1a, based on intuitive arguments. Because isomer **1a** has an all-trans configuration of
substituents it can adopt a preferred conformation where all four phenyl rings occupy
pseudo-equatorial positions and the four methyl groups are in pseudo-axial positions.
Isomer **1b** cannot adopt such a conformation because it would always have two phenyl
groups in pseudo-axial positions and two in pseudo-equatorial postions. Therefore, **1a**
should be sterically less crowded, and present in a higher proportion. This
argumentation is based on thermodynamics and not kinetics because once the
cyclotetrasilanes are formed the configurations are permanent.

The line widths of the peaks for **1a** and **1b** also give some insight into the system. In the ^1H and ^{29}Si NMR spectra, the peak tentatively assigned to isomer **1a** has line narrower widths than those assigned to **1b** (1.5:1). This may be a result of the preferred conformation that can be adopted by isomer **1a**, whereas isomer **1b** can be rapidly equilibrating between two states of equal energy resulting in somewhat broader lines.

However, these intuitive arguments are not sufficient to unambiguously assign the configuration of the dominating isomer. Because the stereochemistry of the rings is highly important for a detailed study of the microstructure of the resulting polymer, it was necessary to take a closer look at the problem. Several attempts at obtaining crystals pure enough for an X-ray study were unsuccessful and thus, the methods of spin labeling and chemical derivatization were employed.

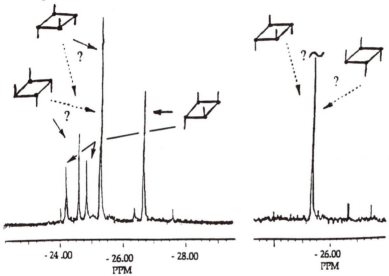

Figure 1. ^{29}Si NMR specta of a) the mixture of stereoisomers and b) purified to either isomer **1a** or **1b**.

Spin-Labeling Study

Because the triflate derivatives are so versatile, it is quite simple to introduce many different functionalities onto the ring. Thus, introducing methyl groups which have 99% ^{13}C labels was achieved simply by substituting ^{13}C labeled methyl magnesium iodide in the second step of monomer preparation. This afforded compound **1***, now with ^{13}C labeled methyl groups.(7) By examining the coupling patterns of the resulting NMR spectra some information about the configuration of the substituents around the ring of the starting compound could be obtained.

The ^1H NMR spectrum of **1*** in C_6D_6 is shown in Figure 2. The methyl region is a duplicate of the spectrum obtained from the non-labeled compound, **1**, but now split into doublets with a $^1J_{C-H}$ of 123 Hz. This spectrum served as confirmation that the desired product had indeed been prepared. However, no useful information about the configuration of substituents was obtained. The ^{13}C spectrum was an exact copy of that observed for **1**; no coupling from adjacent methyl groups was detected. This implies one of two possibilities: 1) the coupling constant $^3J_{C-C}$ is below the line width resolution (≈1 Hz), or 2) the structures are rapidly interconverting between conformations and therefore, averaging each methyl group into equivalent positions wherein no C-C coupling would be present.

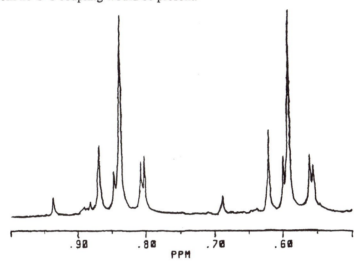

Figure 2. 300 MHz ^{1}H NMR spectrum of the methyl region of compound **1***.

The ^{29}Si NMR spectrum of **1*** is shown in Figure 3, in contrast to the spectrum of **1**, the spectrum for **1*** contains fine structure which provides deeper insight into the stereochemistry of each isomer. A simple splitting scheme with Lorentzian line shapes and spectral superposition was employed to derive the coupling constants shown in Table I.

In order to simplify the discussion of the coupling constants in the stereoisomers, it should be emphasized that the each of the four silicon atoms in isomer **1c** is in an environment resembling that of the silicon atoms in **1a**, **1b**, or **1d**. The two equivalent silicon atoms (**1c*-Si$_b$**) in isomer **1c** resemble the environment of the silicon atoms in **1b**, and therefore, should be present with twice the intensity of those resembling **1a** and **1d**.

Analysis of the observed coupling constants from Table I indicates large values of ^{1}J for all of the systems. Apparently, all of the values of ^{1}J for **1c** are higher than those for **1a** and **1b**, regardless of the stereochemical environment. These values can be strongly dominated by the hybridization in the corresponding Si-C bonds. The values of ^{2}J are quite similar in all cases, but smaller than ^{3}J values; the latter provide the most useful information.

The ^{3}J value for **1c*-Si$_b$** is the smallest (≈1.8 Hz). Values of ^{3}J for the **1c*-Si$_a$** and **1c*-Si$_d$** are 4.6 and 6.8 Hz. The ^{3}J values for the two synthetically available isomers **1a** and **1b** are 1.9 and 4.6 Hz, respectively. Similar values of the ^{3}J coupling constants for **1c*-Si$_b$**, present with twice the intensity of both **1c*-Si$_a$** and **1c*-Si$_d$**, and isomer **1b** enable the assignment to **1b**. The equivalent coupling constants for **1c*-Si$_a$** and **1a** confirm the assignment to **1a**. It is anticipated that ^{3}J value for the unavailable isomer **1d** could be similar to that for **1c*-Si$_d$** and equal to ^{3}J ≈ 6.8 Hz.

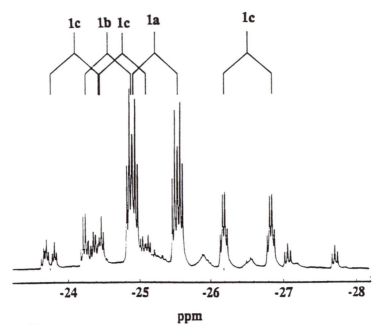

Figure 3. ^{29}Si NMR spectrum of the mixture of stereoisomers of compound **1c***.

Table I. Carbon-Silicon Coupling Constants for the Stereoisomers of **1***

Stereoisomer	^1J	^2J	^3J
1b*	37.8 ± 0.2 Hz	2.1 ± 0.2 Hz	1.9 ± 0.2 Hz
1a*	37.6 ± 0.2 Hz	2.1 ± 0.2 Hz	4.6 ± 0.2 Hz
1c*-Si$_a$	38.5 ± 0.2 Hz	1.8 ± 0.2 Hz	4.5 ± 0.2 Hz
1c*-Si$_b$ (x 2)	38.6 ± 0.2 Hz	2.1 ± 0.2 Hz	1.8 ± 0.2 Hz
1c*-Si$_d$	39.8 ± 0.2 Hz	1.7 ± 0.2 Hz	6.8 ± 0.2 Hz

$^1J_{C-Si}$ is the coupling constant between the observed silicon atom, and the methyl carbon directly attached. $^2J_{C-Si}$ is the coupling to the methyl carbons on two neighboring a silicon atoms, and $^3J_{C-Si}$ is the coupling constant to the methyl carbon on the b silicon atom.

Chemical Derivatization

Because the above study left some ambiguity as to the configuration of the dominating isomer, further evidence was needed. The relatively high symmetry of both **1a** and **1b** makes differentiating between them quite difficult. Therefore, if the symmetry of the molecules could be lowered or destroyed completely, some insight into the configurations could be found. One possibility to change the symmetry was to remove one more aromatic ring and replace it with a methyl group, thereby reducing the symmetry. This procedure is outlined in Scheme II.

If the dominating isomer is indeed **1a**, treating this compound with one equivalent of triflic acid, followed by methylation, should yield compound **2a** (Me5Ph3Si4). This new compound should give rise to NMR spectra which could be used to determine the structure of the starting material. A ^{1}H NMR spectrum with four peaks in the ratio 2:1:1:1 is expected, and also a ^{29}Si NMR spectrum with three peaks in the ratio 2:1:1 should be observed.

If the dominating isomer is **1b**, treatment with one equivalent of triflic acid, followed by methylation should lead to compound **2b**. This compound should give rise to a completely different pattern of peaks in the NMR spectra. The ^{1}H NMR

Scheme II.

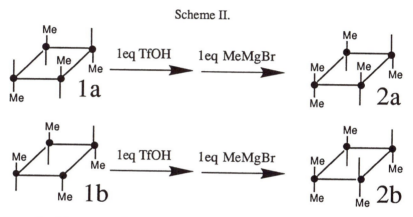

spectrum should appear as five resonances in the ratio 1:1:1:1:1 for the methyl region and the ^{29}Si NMR spectrum should appear as four individual peaks in the ratio 1:1:1:1.

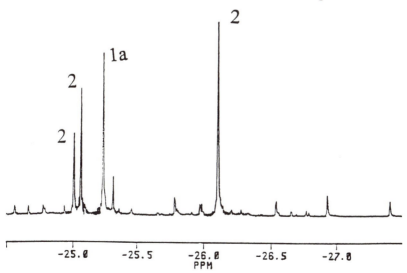

Figure 4. ^{29}Si NMR spectrum of the products of the reaction of **1a** with one equivalent of triflic acid, followed be methylation.

The resulting ^{29}Si NMR spectrum is shown in Figure 4. As is apparent, the dominating isomer possesses the expected all-trans configuration of substituents. In addition to unreacted Me$_4$Ph$_4$Si$_4$, there are present, as much lower intensity peaks, which can be attributed to Me$_6$Ph$_2$Si$_4$. The triflation reaction is known to occur with limited chemoselectively and this leads to the mixture of products.(8) However, the major peaks in both spectra give sufficient evidence for assignment of the dominating isomer to **1a**.

Microstructure of PMPS

With the configuration of the monomer known, it was then possible to obtain a clearer picture of the microstructures obtained from ring opening polymerization. The microstructure of PMPS presents an interesting problem.

The simplest description of the microstructure of PMPS consists of triad data where groups of three silicon atoms in the chain are considered. There are three different possibilities for the stereochemistry at each group. Schematic representations of the three possibilities are shown in Scheme III. Looking at the central silicon atom of the group of three, the configurations of the neighboring atoms are considered. If both adjacent atoms have the same configuration as the central atom, two meso junctions are present. This results in a mm triad, which corresponds to isotactic polymer.

Scheme III.

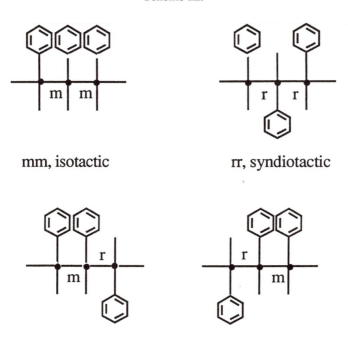

mm, isotactic   rr, syndiotactic

mr or rm, heterotactic

If the neighboring silicon atoms have different configurations from that of the central atom, two racemic junctions are then present, and a rr triad results, which corresponds to syndiotactic polymer. If the configurations of the two adjacent silicon atoms are not the same, then both meso and racemic junctures are present and this gives rise to either mr or rm triads result, corresponding to heterotactic polymer.

Assignments of Configuration. In the ^{29}Si NMR spectrum of PMPS prepared by the reductive coupling of dichloromethylphenysilane there are three broad resonances present at -38.5, -39.0, and -41.0 ppm. West et al. have assigned these resonances to isotactic, syndiotactic, and heterotactic PMPS respectively.(9) The assignments were based on the corresponding six membered rings where the molecules are forced into configurations which may be quite different from those in the linear chain; cis and gauche for the cyclics vs. trans and gauche for the linear chain. Therefore, the shielding effects of the aromatic rings may be quite different from those present in the polymer.

Stereoselective Ring Opening of 1a

Knowing the configuration of the monomer allows a detailed look at the microstructure of PMPS from ring opening polymerization. The all-trans monomer, **1a**, can be opened via three different pathways. Scheme IV depicts the possibilities. If ring opening occurs with two retentions of configuration at the attacked silicon atom and the newly formed reactive center, then a rrr sequence is built up with the possibility for defects to occur at the connections between monomer units because the monomer unit can be attacked on either of two faces of each of the prochiral silicon atoms in the ring. This results in a sequence of [(rrr)m/r], where the ratio m/r is equal to 1, leading to triad data which would consist of 75% syndiotactic, 25% heterotactic, and 0% isotactic triads, respectively.

Scheme IV.

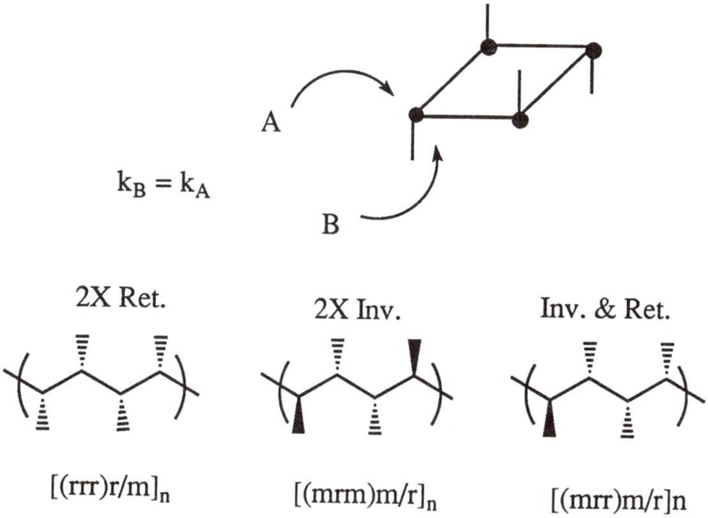

$$k_B = k_A$$

2X Ret. $[(rrr)r/m]_n$

2X Inv. $[(mrm)m/r]_n$

Inv. & Ret. $[(mrr)m/r]n$

If ring opening occurs via two inversions of configuration the sequence of [(mrm)m/r] would result. This corresponds to 75% heterotactic, 25% isotactic, and 0% syndiotactic triads, respectively.

The third potential pathway involves both a retention and an inversion of configuration. Regardless of whether the retention or inversion occurs first, the configuration sequence of [(mrr)m/r] is observed. This would result in a polymer with 37.5% syndiotactic, 12.5% isotactic, and 50% heterotactic triads, respectively.

Ring Opening Polymerization

The ^{29}Si NMR spectrum of the polymer prepared by ring opening polymerization of the mixture of stereoisomers, **1**, using BuLi/cryptand in benzene, was very similar to that observed for the polymer prepared via the reductive coupling route.

One possible explanation for this observation may involve scrambling (transsilylation) reactions. If this process occurs via an intramolecular pathway, the formation of macrocycles should be observed. However, the polymerizations were terminated before the scrambling could occur.

Apparently, with Li$^+$ as the counterion, very little, if any, control over the pathway of ring opening can be achieved. The result is then a random distribution of inversions and retentions leading to stereorandom polymer. Therefore, to achieve some degree of control over whether the polymerization occurs with retentions or inversion of configuration at the attacked silicon atom and the newly formed reactive center, a modified initiator must be employed.

Silyl Cuprates. The use of organocuprates in organic synthesis has been studied intensively.(10) More recently, silyl cuprates have received some attention for use in organic chemistry and also for polymer synthesis. Silyl cuprates are prepared by the reaction of varying numbers of equivalents of R$_3$SiLi with CuX. Oehlschlager and Sharma have studied the reaction of PhMe$_2$SiLi with CuCN in tetrahydrofuran.(11) They have found that there are many equilibria present in solutions of cuprates prepared with various ratios of PhMe$_2$SiLi to CuCN. The reaction of one equivalent of silyl lithium results in the monosilyl cuprate and no free silyl lithium present. At two equivalents, the majority of the cuprates exists as the disilyl compounds with only a small amount of both the mono- and trisilyl compounds present in solution. A third equivalent displaces the cyano group as LiCN giving the trisilyl cuprate. These equilibria are shown in Scheme V.

Scheme V.

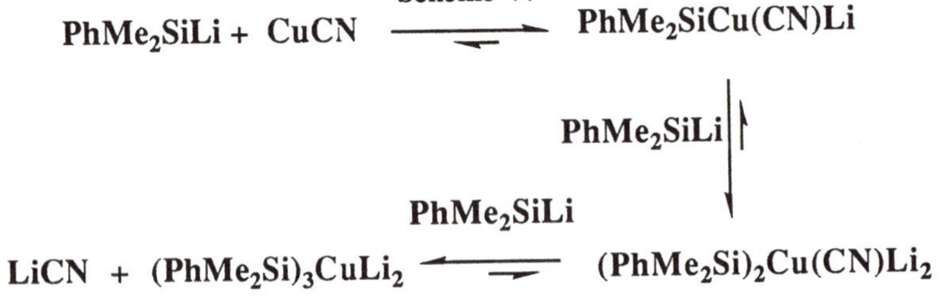

$$PhMe_2SiLi + CuCN \;\rightleftharpoons\; PhMe_2SiCu(CN)Li$$

$$\bigg\downarrow PhMe_2SiLi$$

$$PhMe_2SiLi$$

$$LiCN + (PhMe_2Si)_3CuLi_2 \;\rightleftharpoons\; (PhMe_2Si)_2Cu(CN)Li_2$$

Ring Opening Using Silyl Cuprates

Ring opening polymerization of **1a** was attempted with both the monosilyl and disilyl cuprates. The monosilyl cuprate was used unsuccessfully to initiate polymerization, however, use of the disilyl cuprate resulted in nearly quantitative conversion to monomodal polymer with polydispersities in the range of $M_w/M_n = 1.5$. The polymerizations were carried out in THF at room temperature, which is interesting in itself, because if the polymerizations using BuLi are carried out in THF there is rapid decomposition of the polymer to five and six-membered cyclic polysilanes, however, with the silyl cuprates, the formation of cyclics is not observed, even after 24 hours.

The most interesting result of the polymerizations employing silyl cuprates is the remarkably different microstructures of the formed polysilanes. The ^{1}H and ^{13}C NMR spectra have much sharper peaks than those observed using Li$^+$ as the counterion ion and also from polymers prepared by the reductive coupling route. However, the most dramatic differences can be seen in the ^{29}Si NMR spectrum shown in Figure 5.

The spectrum shows only two signals present at -38.5 and -41.0 ppm, in the ratio 3:1, no signal at -39.0 ppm is detected From the previously discussed possibilities of ring opening, the two microstructures which can be obtained are 3:1 syndiotactic:heterotactic and 3:1 heterotactic:isotactic, corresponding to two retentions and two inversions of configuration, respectively. On the other hand, polymerization of the mixture of all three steroisomers in the ratio 28% (**1a**): 14% (**1b**):58% (**1c**) provides three signals in the ratio 58% (-38.5 ppm): 15% (-39ppm): 27% (-41 ppm). Combination of the results obtained by the polymerization of the mixture and all trans isomer suggests the following assignment: -38.5 ppm (heterotactic), -39 ppm (syndiotactic), and -41 ppm (isotactic). This also indicates that ring-opening polymerization with silyl cuprates occurs with two inversions of configuration at the attacked silicon atom and at the newly developed growing center.

a)   b)

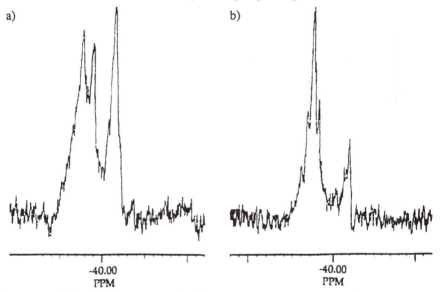

-40.00
PPM

-40.00
PPM

Figure 5. ^{29}Si NMR spectra of a) PMPS prepared by reductive coupling and b) PMPS prepared by the ring opening polymerization of **1a** with silyl cuprates.

Polymer Properties. The absorption and emission spectra of the polymer prepared by the silyl cuprate polymerization of **1a** are very similar to those of PMPS prepared by reductive coupling. The room temperature absorption maximum, $\lambda_{max} = 336$ nm, is 2 nm lower than for the polymer prepared by the reductive coupling process. No thermal transitions were observed during differential scanning calorimetry measurements, indicating a low degree of crystallinity in the polymer, which is consistent with a highly heterotactic polymer.

Conclusions

Ring opening polymerization of cyclotetrasilanes provides a unique and quite promising route to polysilanes. 1,2,3,4-Tetramethyl-1,2,3,4-tetraphenylcyclotetrasilane can be prepared and the dominating isomer possessing an all-trans configuration of substituents can be isolated in up to 95% purity. Ring opening polymerization of the monomer using silyl cuprates results in poly (methylphenylsilylene) with 75% heterotactic triads and 25% isotactic triads. The polymerization occurs via two inversions of configuration at the attacked silicon atom and also at the newly formed reactive center.

Acknowledgements. Partial support from the National Science Foundation and the Office of Naval Research is kindly acknowledged.

Literature Cited.

1. Miller, R.D.; Michl, J. *Chem. Rev.* **1989**, *89*, 1359.
2. West, R. *J. Organomet. Chem.* **1986**, *300*, 327.
3. Matyjaszewski, K.; Cypryk, M.; Frey, H.; Hrkach, J.; Kim, H.K.; Moeller, M.; Ruehl, K.; White, M. *J. Macromol. Sci.-Chem.* **1991** *A28*, 1151.
4. Tilley, T.D. *Acc. Chem. Res.* **1993**, *26*, 22.
5. Sakamoto, K.; Obata, K.; Hirata, H.; Nakajima, M.; Sakurai, H. *J. Am. Chem. Soc.* **1989**, *111*, 7641 .
6. Matyjaszewski, K.; Gupta, Y.; Cypryk, M. *J. Am. Chem. Soc.* **1991**, *113*, 1046 .
7. Fossum, E.; Gordon-Wylie, S. W.; Matyjaszewski, K. *Organometallics* , in press.
8. Chrusciel, J.; Cypryk, M.; Fossum, E.; Matyjaszewski, K.; *Organometallics* **1992**, *11*, 3257.
9. Maxka, J.; Mitter, F.K.; Powell, D.R.; West, R. *Organometallics* **1991**, *10*, 660.
10. Lipshutz, B. H. *Synthesis* **1987**, *4*, 325 .
11. Sharma, S.; Oehlschlager, A. *Tetrahedron* **1989**, *45*, 557.

RECEIVED April 8, 1994

Chapter 5

Polysilazane Thermosets as Precursors for Silicon Carbide and Silicon Nitride

J. M. Schwark[1] and A. Lukacs, III[2]

[1]Wacker Silicones Corporation, 3301 Sutton Road, Adrian, MI 49221
[2]Lanxide Corporation, 1300 Marrows Road, Newark, DE 19714

This chapter discusses the development of thermosetting preceramic polymers, emphasizing peroxide-curable polysilazanes. These polymers are excellent precursors for both silicon nitride and silicon carbide. Vinyl-substituted polysilazanes may be readily thermoset with peroxide initiators. In addition, a new class of polysilazanes which contain peroxide substituents directly bound to the polymer has been developed. The utility of peroxide-cured polysilazane precursors for the formation of silicon nitride articles has been demonstrated.

Polysilazanes and polycarbosilanes have been extensively used as precursors to silicon nitride and silicon carbide (1-7). Preceramic polymers offer an avenue to prepare ceramic articles, such as fibers or coatings, which cannot be made via traditional ceramic processing routes (2). Another potential advantage of this technology is improvement in current ceramic processing methods. For example, precursors may replace the organic polymer binders typically used in the injection molding of ceramic parts (8,9). A universal prerequisite for the utilization of preceramic polymers for the formation of a ceramic article is that they undergo a transformation from a processible state (generally soluble or thermoplastic) to an infusible state. Once rendered infusible, the fabricated ceramic "green" (unfired) body will not deform during the subsequent firing required to generate the finished ceramic article.

When the preceramic polymer is a liquid, or solid with a well-defined glass transition temperature, further crosslinking must occur before or during the pyrolysis process to produce a viable ceramic precursor. Numerous crosslinking methods, based on polymer structure, have been devised including: 1) exposure to a reactive gas such as a chlorosilane (10) or air (11,12), 2) irradiation with ultraviolet, gamma, or electron beam sources (4,13), 3) thermal addition reactions, and 4) catalytically-induced crosslinking via transition metal or peroxide compounds. Reactive gas exposure, while useful for high aspect ratio fibers, is not readily translated to bulk objects where the surface area is much lower. In addition, an exposure of this type

0097-6156/94/0572-0043$08.00/0

may introduce unwanted contamination, such as oxygen, into the final ceramic article. Irradiation techniques are also difficult to incorporate into traditional ceramic processing. While it might be feasible to irradiate a spun fiber, any bulk, molded article, such as an injection molded silicon nitride turbocharger rotor, could not be readily cured with such an exposure. Thermal addition and catalytic crosslinking have proven especially effective for rendering vinyl-substituted precursors infusible and this technology can be translated to the fabrication of ceramic parts.

This chapter will highlight several examples of these two methods, with an emphasis on free radical crosslinking. Vinyl crosslinking has been used by the authors to develop a family of peroxide-curable, polysilazane-based precursors for both Si_3N_4 and SiC (*14-19*). A further extension of this concept has been the preparation of new polysilazanes with peroxide groups bound directly to the polymer (*20*). The utility of these precursors for the manufacture of ceramic articles has been demonstrated.

Crosslinking Via Thermal Addition

Thermal crosslinking has been used as an effective technique to produce crosslinked polymers with good ceramic yield. For example, the liquid polysilane $Me_3Si(SiMeH)_x(CH_2=CHSiMe)_ySiMe_3$ showed a substantial exotherm when heated at 200°C, and both hydrosilylation and vinyl polymerization occurred (*21*). In a more detailed study of this system, Schmidt et al. reported that the primary thermal reaction mode was vinyl crosslinking (equation 1), which eventually yielded a glassy solid after heating the silane from ambient to 196°C in 57 minutes under nitrogen (*22*). In this case, the radical R•, responsible for vinyl polymerization, was presumably produced by thermal decomposition of an Si-Si, Si-H, or Si-CH₃ bond.

$$R\bullet \ + \ \underset{\underset{\overset{|}{CH=CH_2}}{1}}{-\overset{|}{\underset{|}{Si}}-\overset{|}{\underset{|}{Si}}-} \quad \longrightarrow \quad \underset{\underset{\bullet}{CH-CH_2R}}{-\overset{|}{\underset{|}{Si}}-\overset{|}{\underset{|}{Si}}-} \quad \overset{1}{\longrightarrow} \quad -\overset{|}{\underset{|}{Si}}-\overset{|}{\underset{|}{Si}}- \qquad (1)$$

Pyrolysis of the thermoset solid gave a mixture of SiC and C (*22*). This polymer may also be crosslinked more rapidly (10 minutes) by heating at 250°C (*13*).

Likewise, vinyl-substituted polysilazanes may be thermally crosslinked at elevated temperatures. A liquid oligovinylsilazane, $(CH_2=CHSiHNH)_x$, prepared by the ammonolysis of vinyldichlorosilane, thermoset to an infusible, insoluble solid after heating for 2 hours at 110°C (*23,24*). This solid had a char yield of over 85 wt% when pyrolyzed to 1200°C. For this system, infrared analysis showed that hydrosilylation, and not vinyl polymerization, was the predominant crosslinking mechanism.

For both the polysilane and oligovinylsilazane, low molecular weight liquids were converted to infusible solids with good ceramic yield by utilizing thermal activation of vinyl substituents. The major disadvantage of this type of crosslinking method is the slow kinetics.

Platinum-Catalyzed Crosslinking

An enhancement in the rate of precursor thermal crosslinking is obtained when transition metal catalysts are included in the thermal processing. The liquid oligovinylsilazane, $(CH_2=CHSiHNH)_x$, containing both vinyl and Si-H substituents, crosslinks to an infusible solid upon heating to 110°C for one hour in the presence of chloroplatinic acid (*23*). As mentioned in the previous section, the same solid was formed in two hours when no catalyst was present. Careful monitoring of the molecular weight and polymer structure by NMR spectroscopy showed that for this polysilazane, crosslinked oligomers were formed more rapidly in the presence of the catalyst. As in the thermal treatment, the predominant crosslinking mode was hydrosilylation.

Porte and Lebrun found that the ceramic yield of the polysilazane $[(MeSiNH_{1.5})_{0.45}(Me_2SiNH)_{0.25}(MeSiHNH)_{0.20}(CH_2=CHSiMeNH)_{0.10}]_x$ increased dramatically when a platinum complex of 1,3-divinyltetramethyldisiloxane was used during thermolysis (*25*). When the polymer alone was heated at 170°C for 20 hours, a resin with a char yield of 53% at 1400°C was produced. In comparison, when 100 ppm of the Pt catalyst were added, the same thermal treatment produced a solid with a thermogravimetric analysis (TGA) yield of 80%.

These examples illustrate that augmenting the thermal crosslinking reaction of vinyl-substituted precursors with platinum catalysts provides a means to obtain crosslinked precursors more rapidly or with enhanced ceramic yield. Still, the processing times described for both the thermal and the platinum-catalyzed cure methods are too long for rapid ceramic processing. In addition, hydrosilylation of nitrogen-containing silicon compounds is difficult at best.

Peroxide Crosslinking of Preceramic Polymers

Peroxide initiators are equally effective, but often kinetically faster, than platinum-based crosslinking systems. The crosslinking of silicon-based polymers with organic peroxides is well-known, and in this section several examples will illustrate the translation of this technology to preceramic polymer systems. For many vinyl-substituted precursors, and polysilazanes in particular, free radical-based crosslinking is the cure chemistry of choice.

Polysiloxanes. Siloxane rubbers are readily cured by heating in the presence of an appropriate peroxide and the technology is well-established (*26*). Indeed, this crosslinking chemistry has recently been extended into ceramic applications (*27*). Vinyl-substituted siloxanes were found to be very effective binders for SiC powder. With only about 1 wt% 2,5-dimethyl-2,5-bis(tert-butylperoxy)hexane (Lupersol 101), the $[(MeSiO_{1.5})_{0.50}(PhSiO_{1.5})_{0.25}(CH_2=CHMe_2SiO_{0.5})_{0.25}]$ polysiloxane cured to an infusible solid after one hour at 180°C (*27*). Subsequent pyrolysis gave β-SiC, and dense ceramic monoliths could be prepared using these polymer systems as SiC binders.

Polysilanes. Vinyl-substituted polysilanes have also been thermoset using peroxide-based free radical initiators (13). $Me_3Si(SiMeH)_x(CH_2=CHSiMe)_ySiMe_3$, a liquid polysilane, crosslinked in 20 minutes when heated at 200°C with 2-15 wt% dicumyl peroxide. A mechanism similar to that in equation 1 was postulated, with R• being the methyl radical formed in the decomposition of dicumyl peroxide. An interesting finding of this study was that benzoyl peroxide and azoisobutyronitrile were not effective initiators, even under reaction conditions in which they were certainly decomposed. It was speculated that an active alkyl radical was required for polymerization of the $Si-CH=CH_2$ substituents. Contrast this cure chemistry with the thermally activated vinyl polymerization of the same polymer described in the work of Schmidt in the previous section. In this case, the presence of the free radical generator allows the same vinyl crosslinking to take place more rapidly.

Polysilazanes with Vinyl Substituents. Little fundamental work has been described on the peroxide crosslinking of polysilazanes. In 1984, it was reported that simple organooligocyclosilazanes, prepared by the polycondensation of methylvinyl-cyclosilazanes, could be crosslinked by heating to 220°C with 0.5-2.5 wt% of the silylperoxide $(MeSiOO^tBuO)_x$ (28). Even for the highest levels of peroxide, only about 60% of the liquid silazane could be converted to a gel. Still, this work demonstrated the potential for crosslinking vinyl-substituted polysilazanes using a free radical approach.

Subsequently, several liquid polysilazane-based precursors have utilized peroxide-initiated crosslinking to obtain high char yield solids. Lebrun and Porte found that a vinyl-substituted liquid cyclic silazane, such as $(MeSiCH=CH_2NH)_x$, heated with 2.7 wt% dicumyl peroxide at 170°C for 24 hours produced a solid with a char yield of >70 wt% (29). Toreki et al. further studied this system and its utility for producing ceramic materials (30-33). In their work, $(MeSiCH=CH_2NH)_x$ was thermoset by heating with 0.1-1.0 wt% dicumyl peroxide for 12 hours at 135°C to a brittle, highly crosslinked polymer. Because this crosslinking occurred through vinyl groups, a network with an extended $Si-CH_2$ rather than Si-N backbone was obtained. Despite the polymerized vinyl network within the thermoset, and the relatively low nitrogen content of the precursor, α- and β-Si_3N_4 were the only observable crystalline ceramic products after pyrolysis to 1500°C under nitrogen. Elemental analysis showed, however, a high carbon content in the char. At higher temperatures, i.e., >1600°C, SiC formed at the expense of Si_3N_4 (33).

As reported previously for the vinyl-substituted polysilane, it was also observed that some free radical initiators were not effective for polymerizing the vinylsilazane. Benzoyl and lauroyl peroxide, and azoisobutyronitrile were ineffective polymerization initiators, even at elevated temperatures and extended heating times (32). Again, the efficacy of dicumyl peroxide was attributed to the formation of the reactive methyl radical.

Polysilazanes with Vinyl and SiH Substituents. In contrast to polysilazanes containing only $Si-CH=CH_2$ substituents, liquid polysilazanes containing both Si-H and $Si-CH=CH_2$ substituents thermoset very rapidly (14,16,18,19). For example, an isocyanate-modified polysilazane (2) was prepared by the reaction of 0.5 wt%

phenylisocyanate with the cyclic silazane [(MeSiHNH)$_{0.8}$ (CH$_2$=CHSiMeNH)$_{0.2}$]$_x$ at 70°C for 1 hour (*18*). The isocyanate addition provides controlled viscosity increase in the polysilazane system (*14,18*). When the isocyanate-modified polysilazane **2** was heated with 0.1 wt% dicumyl peroxide at 150°C, an exotherm occurred and within *30 seconds* the liquid cured to a glassy, insoluble solid with no discernible glass transition temperature.

Interestingly, it was also found in this work that certain peroxides were ineffective in the thermosetting reaction. For example, diacyl peroxides such as benzoyl, lauroyl, and decanoyl did not promote the thermoset reaction. Peroxyketals, peroxyesters, and dialkylperoxides, however, proved very effective. Examples from these classes include: 1,1-di(t-butylperoxy)-3,3,5-trimethylcyclohexane (Lupersol 231), 2,5-dimethyl-2,5-di(t-butylperoxy)hexane (Lupersol 101), t-butylisopropyl-monoperoxycarbonate (Lupersol TBIC-M75), t-butylperoxyperbenzoate, dicumyl peroxide, and α, α'-di(t-butylperoxy)diisopropylbenzene. Each of these peroxides gave excellent results at ≤0.5 wt% in the polysilazane. Because a range of peroxides is available, each having different decomposition kinetics, cure of the polymer may be effected over a range of temperatures.

Solid state ^{13}C NMR spectroscopy showed complete consumption of the vinyl moieties in the thermoset polymer **2**. Pyrolysis of the cured polysilazane under NH$_3$ from ambient to 1000°C, followed by heating under Ar to 1600°C gave a mixture of α- and β-Si$_3$N$_4$ (*19*). In contrast, pyrolysis under Ar from ambient to 1600°C gave β-SiC. Indeed, this thermoset polysilazane is an excellent precursor for *silicon carbide* as well as silicon nitride (*19*).

Contrast the rapid, peroxide-based thermoset cure of polysilazane **2** containing Si-H and Si-CH=CH$_2$ groups with that of (CH$_2$=CHSiMeNH)$_x$ (*29,30*) in which the cure was achieved only after heating for 12 hours or more. While the difference in this reactivity may be due to free radical induced hydrosilylation with the available Si-H groups, it is plausible that vinyl polymerization predominates with the rapid cure attributed to a high population of radicals sustained by Si-H scission. In either case, the polysilazane thermosetting reaction occurs rapidly enough to make formation of ceramic green bodies by injection molding techniques viable (*8,9*).

Peroxide-Substituted Polysilazanes

A further extension of the concept of thermosetting polysilazanes with peroxide initiators has been the preparation of new polysilazanes with peroxide groups bound directly to the polymer (*20*). Potential advantages of a peroxide-substituted polysilazane over systems in which the peroxide is simply admixed with the polymer include: 1) segregation of the peroxide upon storage cannot occur, 2) dissolution or dispersion of a peroxide in the polysilazane is not necessary, and 3) homogeneous distribution of the peroxide in solid, as well as liquid, polysilazanes is possible. We have prepared a new class of peroxide-substituted polysilazanes by the reaction of a hydroperoxide with a poly(methylvinyl)silazane. The liquid polymers may be thermoset, even with extremely low levels of peroxide substitution. This chemistry provides access to a class of polysilazanes previously unknown as ceramic precursors.

Experimental Section. All reactions were performed under nitrogen using standard inert atmosphere techniques and dried solvents. Reagents were obtained as follows: $(Me_2SiNH)_3$ (**3**) from Hüls and ^tBuOOH in a 3.0 M solution in isooctane from Aldrich. A cyclic silazane copolymer **4** was prepared by the coammonolysis of methyldichlorosilane and methylvinyldichlorosilane (4:1) mol ratio in hexane. The cyclic, vinyl-substituted polysilazane **4** has the formulation $[(MeSiHNH)_{0.8}$ $(CH_2=CHSiMeNH)_{0.2}]_x$ with $x \cong 8$. Thermogravimetric analyses (TGA) were performed at 20°C/min from 25-1000°C in nitrogen. Differential scanning calorimetry (DSC) was performed at 10°C/min under nitrogen from 40-320°C. ^{29}Si NMR spectra were obtained in C_6D_6 at 71.1 MHz with a gated decoupling pulse sequence and a 4 sec pulse delay. Cr(acetylacetonate)$_3$ was used as a Si atom relaxation agent and tetramethylsilane was used as an internal standard (0.00 ppm).

Cyclic Silazane/Hydroperoxide Reactions. Monomeric alkylsilylperoxides have been prepared by the reaction of hydroperoxides with linear (*34-36*) or cyclic (*36*) silazanes. Equation 2 shows the reaction of *tert*-butyl hydroperoxide with a silylamine to produce *tert*-butyltrimethylsilylperoxide and an amine (*34*).

$$Me_3SiNHR \ + \ Me_3COOH \ \longrightarrow \ Me_3SiOOCMe_3 \ + \ RNH_2 \qquad (2)$$

This idea was later extended to cyclic silazanes (equation 3) (*36*). The reactions proceeded readily at room temperature to give good yields of the desired silylperoxides. Reaction of one equivalent of hydroperoxide per Si-N bond in the silazane gave complete substitution of the cyclic silazane trimer.

$$(Me_2SiNH)_3 \ + \ 6\,ROOH \ \longrightarrow \ 3\,Me_2Si(OOR)_2 \ + \ 3\,NH_3 \qquad (3)$$

No examples, however, were reported in which a lower hydroperoxide/Si-N stoichiometry was used. Such a ratio would be expected to produce silylperoxides containing silylamine moieties. We adopted this strategy for the preparation of peroxide-substituted polysilazanes.

The reaction of the cyclic silazane trimer, $(Me_2SiNH)_3$ (**3**), with <u>one</u> equivalent of ^tBuOOH per mol trimer in hexane at room temperature proceeded with Si-N bond cleavage and ring-opening as shown in Scheme 1. The initial reaction is postulated to be a rapid ring-opening of $(Me_2SiNH)_3$ to give the linear intermediate **5** which was not observed. The Si-NH$_2$ end group is not stable in the presence of unreacted hydroperoxide, and further reaction of **5** with another equivalent of the hydroperoxide occurs to produce **6** and ammonia. By ^{29}Si NMR spectroscopy, the final mixture obtained had 50% unreacted $(Me_2SiNH)_3$ (δ -4.67, s, 3 Si) and 50% product **6** (δ 3.14, s, 2 Si; -5.71, s, 1 Si). The ratio of products indicates that preferential reaction of ^tBuOOH with the Si-NH$_2$ group in intermediate **5** occurs. Thus, instead of obtaining a 100% yield of **5** with one silylperoxide end group, the Si-NH$_2$ group of **5** further reacts with another equivalent of hydroperoxide to produce **6** (in a 50% yield based on ^tBuOOH) with two silylperoxide end groups.

Scheme 1

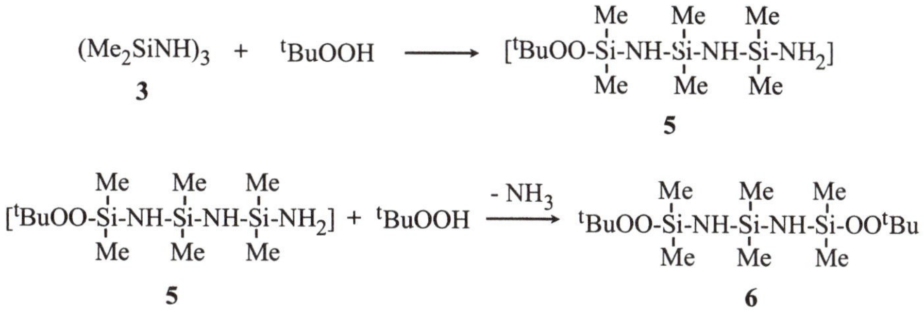

$$(Me_2SiNH)_3 + {}^tBuOOH \longrightarrow [{}^tBuOO\text{-}\overset{Me}{\underset{Me}{Si}}\text{-}NH\text{-}\overset{Me}{\underset{Me}{Si}}\text{-}NH\text{-}\overset{Me}{\underset{Me}{Si}}\text{-}NH_2]$$

3   **5**

$$[{}^tBuOO\text{-}\overset{Me}{\underset{Me}{Si}}\text{-}NH\text{-}\overset{Me}{\underset{Me}{Si}}\text{-}NH\text{-}\overset{Me}{\underset{Me}{Si}}\text{-}NH_2] + {}^tBuOOH \xrightarrow{-NH_3} {}^tBuOO\text{-}\overset{Me}{\underset{Me}{Si}}\text{-}NH\text{-}\overset{Me}{\underset{Me}{Si}}\text{-}NH\text{-}\overset{Me}{\underset{Me}{Si}}\text{-}OO{}^tBu$$

5   **6**

The reaction mixture which contained **6** had no other products by ^{29}Si NMR spectroscopy. This peroxide-substituted silylamine was successfully prepared by control of the reaction stoichiometry to supply the hydroperoxide as the limiting reagent, which prevented complete conversion to an alkyl silylperoxide as in equation 3. This approach was then used to prepare a peroxide-substituted polysilazane.

Synthesis and Characterization of a Peroxide-Substituted Polysilazane. Clearly, peroxides bound to silazane moieties could be readily prepared from cyclic silazanes. The polysilazane **4**, $[(MeSiHNH)_{0.8} (CH_2=CHSiMeNH)_{0.2}]_x$, of interest for thermosetting, is actually a mixture of cyclic oligomers with an average ring size of eight Si-N units. Thus, if the methodology shown in Scheme 1 is used with a very small quantity of hydroperoxide R"OOH, a small percentage of cyclics in the oligomeric polysilazane **4** will be ring-opened to provide linear polysilazanes end-capped with silylperoxide groups, e.g., R"OO-(R'RSiNH)$_x$-SiRR'OOR".

    To prepare a peroxide-substituted polymer, polysilazane **4** was reacted with 5 mol% (6.6 wt%) ^tBuOOH at room temperature in hexane to give silylperoxide-substituted polysilazane **7**. A high level of hydroperoxide was used so that the substituted polymer could be examined by NMR spectroscopy for formation of the ^tBuOOSi end groups. The substitution was confirmed by ^{29}Si NMR spectroscopy. To determine whether polymer **7** would thermoset, a DSC analysis was performed. The DSC showed an exotherm beginning at 118°C with a peak temperature of 155°C, coinciding with the initiation of cure. Because of the relatively high level of peroxide substitution in **7**, a heat of reaction of 105 cal/g was observed. When a bulk sample (10 g) was cured, an exotherm began at 124°C and reached 204°C within several minutes. The polymer thermoset to a hard, brittle glass with a char yield by TGA of 76 wt% at 1000°C under nitrogen.

    The level of silylperoxide substitution required to cure **4** and produce a solid with a good char yield was investigated. The data in Table I show the maximum exotherm reached during cure and TGA results obtained from the vinyl-substituted polysilazane **4** reacted in hexane at room temperature with the indicated weight percentage of ^tBuOOH. Accurate dispensing of the hydroperoxide was possible because it was supplied as a 3.0 M solution in isooctane. In each reaction, a 1-2°C exotherm and gas evolution (ammonia) accompanied the hydroperoxide addition. Following in vacuo removal of the hexane, each liquid polymer was thermoset in a preheated 160°C oil bath. In each case, the cure exotherm began when the sample reached about 130°C and the maximum exotherm temperature exceeded 200°C for a 10 g polymer sample (Table I).

Table I. Peroxide-Substituted Polysilazane Cure/TGA Results		
Wt. % ^{t}BuOOH Reactant	Thermoset Cure Exotherm T(°C)	TGA Yield (Wt.%)
0.03	217	71
0.16	207	75
0.32	210	73
1.60	241	75

A solid, crosslinked polysilazane was produced in about 5 minutes in each case. Thus, even very low levels of peroxide substituents on the polysilazane were effective in the thermoset cure. The TGA yields compare quite favorably with that obtained when $[(MeSiHNH)_{0.8}(CH_2=CHSiMeNH)_{0.2}]_x$ (4) is mixed with 0.5 wt% dicumyl peroxide and heated under the same conditions. In that case, a char yield of about 75 wt% is observed.

The effectiveness of this approach was somewhat unexpected in light of the general rule that silylperoxides are inefficient free radical initiators. Previous work with the crosslinking of polyethylene showed that silylperoxides require higher initiation temperatures, are less efficient and more thermally stable, than typical organic peroxides (37). Also, silylperoxides undergo oxygen extrusion reactions to form alkoxysilanes which are not active as free radical generators (equation 4) (34).

$$Me_3SiOOSiMe_3 \longrightarrow Me_2Si(OMe)OSiMe_3 \qquad (4)$$

Despite the expectation that the silylperoxide substituents might be inefficient initiators, the vinyl-substituted polysilazanes containing such moieties thermoset readily producing highly crosslinked solids which had good pyrolysis yields.

Ceramic Processing Utilizing Polysilazane Thermosets

Preceramic polymers have the potential to improve traditional ceramic processing methods (2). In particular, ceramic injection molding may be enhanced by the replacement of organic binders with thermosetting polysilazanes. Table II compares the attributes of the two binder systems.

Table II. Comparison of Ceramic Injection Molding Process with Organic Binder and Preceramic Polymer Binder	
Organic Binder	*Preceramic Polymer Binder*
Thermoplastic	Thermoset
Hot Mix/Cold Mold	Cold Mix/Hot Mold
Binder Must Be Burned Out	Binder "Burns In"
High Shrinkage During Firing	Lower Shrinkage During Firing

The complete molding process is further illustrated in Figure 1. In traditional ceramic injection molding, a ceramic powder and auxiliaries (e.g., sintering aids) are combined in a thermoplastic polymer (e.g., polyethylene) producing an extremely high viscosity mix. This mixture is heated until it flows and then injected into a cold mold to solidify the mixture and produce a ceramic green body. Once the green body is removed from the mold, it is subjected to a binder burnout step. During this phase, the green body is heated very slowly from ambient to a temperature at which the binder begins to volatilize. Binder removal must be very slow so that deformation of the molded object is avoided. Binder burnouts of up to one week are not uncommon. Once the binder is completely removed, the ceramic is fired to the temperature necessary to produce a fully sintered (densified) article.

By replacing the organic binder with a polysilazane thermoset, the binder burnout step may be essentially eliminated so that the cycle time from injection molded part to densified ceramic is greatly reduced. An injection molding scheme using the polysilazane binder system is depicted in Figure 1. The vinyl-substituted polysilazane, e.g. polymer **2**, is first combined with the ceramic powder and additives such as sintering aids and peroxide initiators in a Ross double planetary mixer. The peroxide-curable polysilazanes typically have very low viscosities (i.e. 25-150 cps), so mixtures with high loadings (50-55 vol%) of ceramic or metal particulate fillers are both fluid and non-dilatant at ambient temperature (*8,9*). The powder-filled mixture is injected at less than 500 ksi (ca. 3500 MPa) pressure into a heated (>150°C) mold. The polysilazane binder phase thermosets in less than 5 minutes to produce the molded green body which is quite strong. Silicon nitride powder-filled green parts having as little as 8 wt% binder, for example, have four point bend strengths which typically exceed 8 MPa (Matsumoto, R.L.K., Lanxide Corporation, presented at the 92nd American Ceramic Society meeting, Dallas, TX, April, 1990).

Because the thermoset polysilazane has no measurable glass transition, the green body will not deform upon heating. The molded part may be heated rapidly to the sintering temperature without going through a binder removal step. The high char yield (ca. 75 wt%) polysilazane decomposes during this heating to form a ceramic which "burns in" rather than burns out of the finished part. Since so little weight is lost during the firing, shrinkage (<10%) is lower than the 20+% obtained when organic binders are used. Flexural strengths and fracture toughnesses measured for sintered silicon nitride bodies fabricated using this processing method are comparable to those achieved using traditional, state-of-the art ceramic powder injection molding.

Conclusion

Several methods have been developed to crosslink vinyl-substituted preceramic polymers. While simple thermal treatment or heating in the presence of platinum catalysts does produce crosslinked systems, the cure kinetics are often slow. Peroxide initiators promote rapid vinyl crosslinking, especially for polysilazanes with both Si-CH=CH$_2$ and Si-H subtituents.

The peroxide may be added to the polysilazane or may be bound directly to the polymer through a simple synthetic procedure. In either case, a system is produced which thermosets very rapidly at less than 200°C. Because they cure so

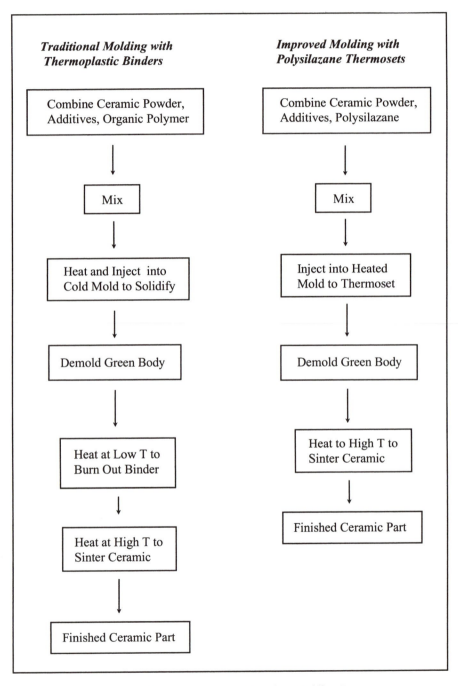

Figure 1. Comparison of Ceramic Injection Molding Processes

quickly, the peroxide-curable polysilazanes make ideal binders for ceramic powder injection molding. These preceramic polymers have fulfilled their potential to enable improvements in traditional ceramic processing technology.

Acknowledgments

The authors would like to thank J.A. Mahoney, D.A. Moroney, M.J. Centuolo, T.C. Bradley, and J. Brereton for their technical assistance. Discussions with Dr. J.A. Jensen and R.L.K. Matsumoto were most helpful.

Literature Cited

1. Peuckert, M.; Vaahs, T.; Brück, M. *Adv. Mater.* **1990**, *2*(9), 398-404.
2. Wynne, K. J.; Rice, R. W. *Ann. Rev. Mater. Sci.* **1984**, *14*, 297-334.
3. Laine, R. M.; Blum, Y. D.; Tse, D.; Glaser, R. In *Inorganic and Organometallic Polymers*; Zeldin, M., Wynne, K. J., and Allcock, H. R. Eds., ACS Symposium Series 360; American Chemical Society: Washington, DC, 1988, pp 124-142.
4. Seyferth, D.; Wiseman, G. H.; Schwark, J. M.; Yu, Y.-F.; Poutasse, C. A. In *Inorganic and Organometallic Polymers*; Zeldin, M., Wynne, K. J., and Allcock, H. R. Eds., ACS Symposium Series 360; American Chemical Society: Washington, DC, 1988, pp 143-155.
5. Pouskouleli, G. *Ceramics International* **1989**, *15*, 213-229.
6. Atwell, W. H. In *Silicon Based Polymer Science: Advances in Chemistry Series 224*; Zeigler, J. M., Fearon, F. W. G., Eds.; American Chemical Society: Washington, DC, 1990, pp 593-606.
7. Yokoyama, Y.; Nanba, T.; Yasui, I.; Kaya, H.; Maeshima, T.; Isoda, T. *J. Am. Ceram. Soc.* **1991**, *74*(3), 654-657.
8. Lukacs, A. US 5,190,709 (March 2, 1993).
9. Lukacs, A.; Matsumoto, R. L. K. US 5,240,658 (August 31, 1993).
10. Legrow, G. E.; Lim, T. F.; Lipowitz, J.; Reaoch, R. S. *Am. Ceram. Soc. Bull.* **1987**, *66*(2), 363-367.
11. Yajima, S. *Am. Ceram. Soc. Bull.* **1983**, *62*(8), 893-898.
12. Lipowitz, J.; Freeman, H. A.; Chen, R. T.; Prack, E. R. *Adv. Ceram. Mater.* **1987**, *2*(2), 121-128.
13. Lee, B. I.; Hench, L. L. In *Science of Ceramic Chemical Processing*; Hench, L. L., Ulrich, D. R., Eds.; John Wiley & Sons: New York, 1986, pp 345-353.
14. Schwark, J. M. US 4,929,704 (May 29, 1990).
15. Schwark, J. M. US 5,001,090 (March 19, 1991).
16. Schwark, J. M. US 5,021,533 (June 4, 1991).
17. Schwark, J. M. US 5,032,649 (July 16, 1991).
18. Schwark, J. M. *Polymer Preprints* **1991**, *32*(3), 567-568.
19. Schwark, J. M.; Sullivan, M. J. In *Better Ceramics Through Chemistry V*; Hampden-Smith, M. J., Klemperer, W. G., Brinker, C. J., Eds.; Materials Research Society: Pittsburgh, 1992, pp 807-812.
20. Schwark, J. M. *Polymer Preprints* **1993**, *34*(1), 294-295.
21. Schilling, C. L., Jr. *British Polymer Journal* **1986**, *18*(6), 355-358.

22. Schmidt, W. R.; Interrante, L. V.; Doremus, R. H.; Trout, T. K.; Marchetti, P. S.; Maciel, G. E. *Chem. Mater.* **1991**, *3*, 257-267.
23. Lavedrine, A.; Bahloul, D.; Goursat, P.; Yive, N. S. C. K.; Corriu, R.; Leclerq, D.; Mutin, P.; Vioux, A. *J. Eur. Ceram. Soc.* **1991**, *8*, 221-227.
24. Corriu, R.; Yive, N. S. C. K.; Leclercq, D.; Mutin, P. H.; Vioux, A. *New J. Chem.* **1991**, *15*, 85-92.
25. Porte, H.; Lebrun, J.-J. US 4,689,252 (August 25, 1987).
26. Noll, W. In *Chemistry and Technology of Silicones*; Academic Press, Inc.: New York, 1968, pp 392-395.
27. Burns, G. T.; Taylor, R. B.; Xu, Y.; Zangvil, A.; Zank, G. A. *Chem. Mater.* **1992**, *4*(6), 1313-1323.
28. Kazakova, V. V.; Avdeyeva, A. I.; Zhandov, A. A.; Akutin, M. S.; Osipchik, V. S.; Kotrelev, G. V. *Polymer Science USSR* **1984**, *26*(8), 1889-1897.
29. Porte, H.; Lebrun, J.-J. US 4,722,988 (February 2, 1988).
30. Toreki, W.; Batich, C. D. *Polymer Preprints* **1989**, *30*(2), 237-238.
31. Toreki, W.; Creed, N. A.; Batich, C. D. *Polymer Preprints* **1990**, *31*(2), 611-612.
32. Toreki, W.; Batich, C. D.; Sacks, M. D.; Morrone, A. A. *Ceram. Eng. Sci. Proc.* **1990**, *11*(9-10), 1371-1386.
33. Morrone, A. A.; Toreki, W.; Batich, C. D. *Materials Letters* **1991**, *11*(1,2),19-25.
34. Pike, R. A.; Shaffer, L. H. *Chemistry and Industry* **1957**, 1294.
35. Ostrozynski, R. L. US 3,700,712 (October 24, 1972).
36. Ostrozynski, R. L. US 3,843,703 (October 22, 1974).
37. Fan, Y. L.; Shaw, R. G. *Rubber World* **1971**, *164*(3), 56-62.

RECEIVED April 25, 1994

Chapter 6

Substituent Effects on UV Absorption of σ-Conjugated Polysilanes

Yu-Ling Hsiao, John P. Banovetz, and Robert M. Waymouth[1]

Chemistry Department, Stanford University, Stanford, CA 94305

Free radical chlorination of poly(phenylsilane) produces poly(chlorophenylsilane). These chlorinated polymers can be substituted with a variety of nucleophiles such as MeOH or MeMgBr with high selectivities. The spectroscopic properties of these materials are extremely sensitive to the nature of the substituent attached to the polymer backbone. The UV properties of a series of polysilanes containing Si-H, Si-Cl, Si-R and Si-OR functionalities are reported. The absorption maximum of poly(phenylsilane) appears at 294 nm (ε_{Si-Si} = 2489 cm^{-1}M^{-1}) whereas that of poly(methylphenylsilane) appears at 328 nm (ε_{Si-Si} = 4539 cm^{-1}M^{-1}). The absorption spectra of poly(methoxyphenylsilane) are red shifted considerably relative to the other polymers (λ = 348 nm, ε_{Si-Si} = 2710 cm^{-1}M^{-1}). These substituent effects are likely due both to conformational as well as electronic perturbations on the Si-Si backbone chromophore.

Soluble polysilane high polymers have recently attracted considerable attention because of their unusual electronic properties.(1, 2) Polysilanes absorb strongly in the UV spectral region, typically between 300-410 nm, and are strongly fluorescent. These transitions have been attributed to delocalized Si(σ–σ*) states. The spectroscopic properties of polysilanes are sensitive to the molecular weight, polymer conformations, and the substituents attached to the silicon backbone.(2-4)

The remarkable electronic behavior of these materials has stimulated considerable effort to develop selective syntheses of polysilanes with well-defined structure. Typical synthetic procedures include the Wurtz coupling of dichlorosilanes,(5-8) anionic polymerization of masked disilenes, (9-12) anionic ring-opening polymerization of strained cyclosilanes,(13-17) electrochemical reduction of chlorosilanes,(18-21) and transition metal catalyzed dehydrogenative coupling of phenylsilanes (eq. 1).(22-35)

The introduction of functional organic groups into polysilanes to develop polymers with specific electronic and photophysical properties is a challenging and important goal. The direct polymerization of functional monomers is somewhat limited in the Wurtz coupling process due to the the vigorous condensation conditions. Some examples have been reported utilizing suitably protected functional groups.(8, 36-39) Deprotection of the organic groups affords polysilanes with either hydrophilic or fluoro

---

[1]Corresponding author

0097–6156/94/0572–0055$08.00/0

substituents.*(40-45)* Poly(methylphenylsilane) can also be efficiently chloromethylated with chloromethylmethylether and stannic chloride.*(46)* Similarly, Stueger and West have reported that pendant phenyl groups of poly(methylphenylsilane-co-dimethylsilane) can be replaced by halogen through the reaction of a hydrogen halide and a Lewis acid.*(38, 41, 47)* Hydrogen halides can be added to the double bond of poly[(2-(3-cyclohexenyl)ethyl) methylsilane] in the presence of Lewis acids without chain degradation.*(38, 41)* Matyjaszewski et al. have shown that over 90% of the pendant phenyl groups of poly(methylphenylsilane) can be substituted by triflate groups by treatment of the polymer with trifluoromethanesulfonic acid.*(48, 49)*

The catalytic dehydrogenative coupling of hydrosilanes developed by Harrod and Tilley offers an appealing strategy for the synthesis of polysilanes with well-defined structure.*(22, 29, 35, 50, 51)* One of the major limitations of this procedure is the low molecular weights of the materials, typically less than 40 silicon atoms. The low molecular weights of these materials seems to be a serious synthetic limitation with regard to physical properties; however recent studies*(52-55)* would suggest that even low molecular weight polymers of 10-20 silicon atoms should exhibit electronic transitions similar to the high molecular weight materials. To date there have been few reports on both the functionalization and spectroscopic properties of polysilanes containing Si-H bonds.*(56-58)* We have recently reported a free-radical chlorination strategy to introduce chlorine into the polymer side chain through the reaction of CCl$_4$ with the Si-H moieties in the polymer backbone.*(59)* The resulted poly(chlorophenylsilane) are versatile precursors to a variety of functionalized polysilanes. Although there have been numerous studies of the effects of substituents the the conformational and electronic properties of polysilanes,*(39, 60-64)* most of the observed phenomena arise from the conformational transitions induced by steric interaction between the alkyl or aryl side chains with the polymer backbone.*(39, 60-64)* Only a handful of examples discuss the electronic effect of substituted aryl side groups.*(37, 41, 43)* In this paper, we focus on the functionalized polysilanes synthesized through the free-radical route developed in our group and report our studies on the spectroscopic properties of a series polysilanes containing Si-H, Si-Cl, Si-R and Si-OR moieties and the remarkable effect of substituents directly attached to the silicon backbone on the absorption spectra.

Results

Polymer synthesis. Polymerization of phenylsilane with a variety of zirconocene catalyst precursors leads to poly(phenylsilane) **1** with average degree of polymerization of 30-40 silicon atoms, as determined by vapor pressure osmometry. Treatment of poly(phenylsilane) with CCl$_4$ in room light selectively chlorinates the Si-H bonds to yield poly(chlorophenylsilane) **2** (eq. 2, Table I).*(59)* Substitution of the Si-Cl bonds with MeMgBr gives poly(methylphenylsilane) **3** containing 75% Si-Me / 15% Si-H / 10% Si-Cl, as determined by ^{1}H NMR and IR spectroscopy. Substitution of the Si-Cl bonds with MeOH in the presence of NEt$_3$ and dimethylaminopyridine at room temperature affords poly(methoxyphenylsilane) **4** of composition: 80% Si-OMe / 15% Si-H / 5% Si-Cl. On a separate sample, where we are able to substitute up to 90% of the Si-H bonds of the poly(phenylsilane). Subsequent substitution of Si-Cl bonds with nucleophiles are achieved with 100% selectivity (Table II).

The effect of substitution reactions on the number average molecular weight (M_n) of the polymer was investigated by vapor pressure osmometry. Chlorination of poly(phenylsilane) results in a decrease in the degree of polymerization (DP) from about 40 to 25 silicon atoms (Table I). Subsequent substitution of Si-Cl with Si-OMe further degrades the polymer to less than 15 monomer units. Methylation of poly(chlorophenylsilane) with MeMgBr affords, upon purification, a 41% yield of poly(methylphenylsilane) of approximately the same degree of polymerization as the original poly(phenylsilane).

$$\text{PhSiH}_3 \xrightarrow[\text{Cp}_2\text{Zr(H)Cl}]{\text{H}_2} \underset{\substack{| \\ \text{H}}}{\overset{\substack{\text{Ph} \\ |}}{\left(\!\!-\text{Si}-\!\!\right)_n}} \tag{1}$$

$$\underset{\substack{| \\ \text{H}}}{\overset{\substack{\text{Ph} \\ |}}{\left(\!\!-\text{Si}-\!\!\right)_n}} \xrightarrow[h\upsilon]{\text{CCl}_4} \underset{\substack{| \\ \text{Cl}}}{\overset{\substack{\text{Ph} \\ |}}{\left(\!\!-\text{Si}-\!\!\right)_n}} \begin{array}{l} \xrightarrow{\text{MeMgBr}} \underset{\substack{| \\ \text{Me} \quad \mathbf{3}}}{\overset{\substack{\text{Ph} \\ |}}{\left(\!\!-\text{Si}-\!\!\right)_n}} \\[2em] \xrightarrow[\text{DMAP}]{\text{MeOH}} \underset{\substack{| \\ \text{OMe} \quad \mathbf{4}}}{\overset{\substack{\text{Ph} \\ |}}{\left(\!\!-\text{Si}-\!\!\right)_n}} \end{array} \tag{2}$$

$$\mathbf{1} \qquad \mathbf{2}$$

Table I. Properties of Substituted Poly(phenylsilane)s

Sample	R	%[a]	M_n[b]	M_w/M_n[b]	M_n[c]	DP[d]	λ(nm)[e]	ε_{Si-Si}[f]
1	H		2590	1.70	4855	46	294	2489
2	Cl	85	2371	1.67	3329	25	330	1561
3	Me	75	2781	1.89	4890[g]	42	328	4539
4	MeO	80	2262	1.89	1563	12	348	2710

[a]Percentage of substitution on the polymers. [b]Values obtained from GPC using polystyrene as the standard. [c]Values obtained from Vapor Pressure Osmometry, in toluene, 60°C. [d]DP= degree of polymerization. [e]Si(σ-σ*) absorption maximum; measured in THF. [f]Values obtained from deconvoluted spectrum. Extinction coefficient in $cm^{-1}M^{-1}$. [g]Sample was washed with MeOH to remove magnesium salts.

Table II. Polysilanes with Same Percentage of Substitution

Sample	R	%[a]	M_n[b]	λ(nm)[c]
2	Cl	90	1828	324
3	Me	90	2310	328
4	MeO	90	1389	348

[a]Percentage of substituents on the polymers. [b]From GPC using polystyrene as the standard. [c]Si(σ-σ*) absorption maximum; measured in THF.

Spectroscopic Properties. The UV absorption spectrum of poly(phenylsilane) **1** contains a weak transition at 294 nm ($\varepsilon_{Si\text{-}Si}$ = 2489 cm^{-1}M^{-1}, Figure 1a). The absorption band assigned as the Si(σ–σ^*) transition occurs at 294 nm and likely contains considerable phenyl (π–π^*) character, as suggested by energy band calculations.(*65*) In contrast to other substituted polysilanes of similar molecular weight, we were unable to detect fluorescence for poly(phenylsilane) at room temperature in THF.

The absorption spectrum of the poly(chlorophenylsilane) **2** exhibits a long wavelength tail out to 380 nm with a transition centered at 330 nm ($\varepsilon_{Si\text{-}Si}$ = 1561 cm^{-1}M^{-1}, Figure 1b). Substitution of the Si-Cl bonds of the chlorinated polymers with methyl groups results in a red shift UV absorption relative to original poly(phenylsilane) coupled with an increase in the intensity of Si(σ–σ^*) transition (λ = 328 nm, $\varepsilon_{Si\text{-}Si}$ = 4539 cm^{-1}M^{-1}, Figure 1c). The absorption spectrum of poly(methylphenylsilane) **3** prepared by this two-step procedure is slightly blue shifted but comparable to the absorbance spectrum observed for high molecular weight poly(methylphenylsilane) prepared by Wurtz coupling (λ = 343 nm, $\varepsilon_{Si\text{-}Si}$ = 12,000 cm^{-1}M^{-1}).(*2*) The methylpolymers **3**, in contrast to the hydropolymers, fluoresce at room temperature in THF (λ_{em}= 360 nm).

Substitution of the Si-Cl bonds in the chlorinated polymers by MeOH leads to a red shift, relative to the other polysilanes, in the absorption maximum (λ = 348 nm, $\varepsilon_{Si\text{-}Si}$ = 1802 cm^{-1}M^{-1}, Figure 1d) along with a slight decrease in the extinction coefficient relative to the methylphenyl congener. The poly(methoxyphenylsilane) **4** fluoresces at room temperature (λ_{em}= 394 nm).

Discussion

The zirconocene catalyzed dehydrogenative coupling of phenylsilane provides a mild, room temperature method for preparing silicon catenates with degrees of polymerization up to 40 silicon atoms. Attractive features of catalytic methods include the ability to influence the stereochemistry of the polymerization reaction(*35, 66*) and the reactive functionality provided by the Si-H bonds.(*47, 56-59, 67, 68*)

Although the introduction of halogen-containing substituents into linear polysilanes have been reported,(*38, 41, 45*) further modification of structure and electronic properties of these polysilanes have not been extensively studied. We have recently reported that free-radical chlorination of poly(phenylsilane) with CCl$_4$ provides a facile synthesis of poly(chlorophenylsilane), a versatile synthon for the preparation of a variety of functionalized poly(phenylsilane).(*59*) This new synthetic capability has allowed us to investigate the influence of polymer substituent on the electronic properties of polysilanes.

The halogenation of phenylsilane polymers involves a free-radical chain reaction mediated by silicon and trichloromethane radicals. Our initial investigations of this reaction by gel permeation chromatography indicated that little degradation of the polymer occurred during the halogenation as the apparent number average molecular weight of the starting poly(phenylsilane) and the poly(chlorophenylsilane) product were similar.(*59*) Our reinvestigation of this reaction by vapor pressure osmometry reveals that some degradation of the polysilane backbone accompanies chlorination. The average degree of polymerization decreases from about 40 silicon atoms to 25 silicon atoms upon halogenation with CCl$_4$. Replacement of the Si-Cl bonds in poly(chlorophenylsilane) with nucleophiles also leads to polymer degradation. Although little change in molecular weight is indicated by GPC, addition of MeOH to poly(chlorophenylsilane) decreases the molecular weight substantially according to VPO measurements. These results illuminate the limitations of gel permeation chromatography for comparisons of polymers with different structures.

Although there has been numerous studies on the electronic properties of dialkyl, diaryl and alkylarylpolysilanes,(*39, 60-64*) few reports describe the spectroscopic

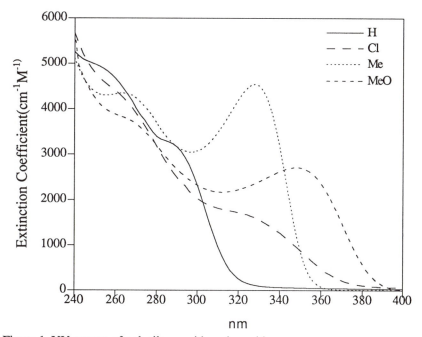

Figure 1. UV spectra of polysilanes with various side groups
(a) ——— : poly(phenylsilane) (b) — — — : poly(chlorophenylsilane)
(c) ·········· : poly(methylphenylsilane) (d) ‑ ‑ ‑ ‑: poly(methoxyphenylsilane)

properties of Si-H containing polysilanes.(*56, 58, 69-71*) The absorption spectra of poly(phenylsilane) **1** is unusual because it exhibits no strong absorption above 320 nm (Figure 1a). We had originally attributed this behavior to the low molecular weight of these materials; however, in light of results described by Michl and Hochstrasser,(*53-55, 62*) it is clear that dialkyloligosilanes with as few as 10 silicon atoms can show strong absorptions between 300-340 nm. Our current results suggest that the UV properties of hydropolymers are likely a consequence of the Si-H substituents since replacement of the backbone Si-H bond with alkyl or alkoxy subtituents results in significant red shift in the absorption maxima.

Substitution of the Si-H groups by Si-Cl results in a slight change in the absorption spectrum with the appearance of a weak transition centered at 330 nm ($\varepsilon_{Si-Si} = 1561$ cm^{-1}M^{-1}). A more dramatic change is observed following substitution of the chlorinated polymers **2** with MeMgBr. The absorption spectrum of the poly(methylphenylsilane) **3** prepared indirectly from poly(phenylsilane) exhibits a strong absorption at 328 nm, reminiscent of poly(methylphenylsilane) prepared via Wurtz coupling reactions.(*62*) Measurements by VPO show that the molecular weight of the poly(methylphenylsilane) **3** is similar to that of the poly(phenylsilane) starting material **1**. Also the small difference in the percentage of substitution (Table II) doesn't not seem to affect the absorption wavelength. This suggests the difference in the absorption spectra for the two materials is attributed primarily to the influences of the methyl and hydrogen substituent on the polysilane backbone. Substitution of the poly(chlorophenylsilane) by MeOH also results in a significant bathochromic shift in the absorption spectra. Poly(methoxyphenylsilane) **4** absorbs at 348 nm; a dramatic red shift considering the low M_n determined by VPO.

These results demonstrate that the electronic properties of polysilanes are extremely sensitive to polymer substituents. The origin of this behavior has still not been established, but is likely due to a combination of conformational effects as well as electronic perturbations on the σ-conjugated silicon backbone. The electronic spectra of high molecular weight polysilanes have been explained in terms of a segment model where the chromophores consist of segments of 10-20 silicon atoms in extended all-trans conformations interrupted by gauche or eclipsed twists in the polymer backbone.(*2*) At a first approximation, the position of the absorption maxima and absorptivity depend on the number of silicon atoms comprising the all-trans segment as well as the number of all-trans segments of a given length. In the context of this model, the difference in the absorption spectra of the hydro- and methyl-substituted polymer could be explained in terms of differences in overall conformation for the two polymers. Because the hydrogen substituent is small relative to methyl group, it might be expected that a larger number of gauche or eclipsed conformations would be energetically accessible, causing the polymer to exist as a tight random coil with few extended segments. In contrast, the methyl substituents might be expected to stiffen the polymer chain, resulting in a greater population of extended trans segments and thus a red-shift in the absorption spectrum. This purely conformational model is clearly an oversimplification; electronic effects are also likely involved. For example, Pitt has shown that a Si-H substituent in the 2-position of a trisilane results in a ipsochromic shift relative to the alkyl substituted congener.(*71*)

The red-shift observed upon incorporation of the methoxy groups is undoubtedly due to both conformational and electronic effects. Interaction of the oxygen lone pairs with the orbitals of the silicon backbone should increase the energy of the HOMO more than that of the LUMO, leading to a lower transition energy. This effect has also been observed for lower oligosilanes,(*71, 72*) but is particularly dramatic in light of the much lower molecular weight of the methoxy-substituted poly(phenylsilane) compared to the hydro- and methyl-substituted polymers.

Conclusion

The dehydrogenative coupling of phenylsilane yields low molecular weight silicon catenates containing reactive Si-H bonds that can be readily substituted by a wide range of functional groups. This synthetic strategy has allowed us to study the influence of substituents on the electronic properties of polysilanes. In this study we show that in spite of lower molecular weights, polysilanes produced by dehydrogenative coupling can exhibit electronic behavior similar to the high molecular weight materials produced via Wurtz coupling. The substituent effects on the absorption spectra are dramatic and are likely due to both electronic and conformational perturbations of the σ-conjugated silicon backbone. A clear understanding of these effects will ultimately allow us to prepare polysilanes with a wide variety of electronic and photophysical properties.

Acknowledgments

This work was supported by the NSF-MRL Program through the Center for Materials Research at Stanford University. We thank Marc Hillmyer and Professor R.H. Grubbs for assistance in carrying out the vapor pressure osmometry.

Literature Cited

(1)   Harrah, L. A.; Zeigler, J. M. In *Photophysics of Polymers*; C. E. Hoyle and J. M. Torkelson, Ed.; American Chemical Society: Washington, 1987; pp 482.

(2)   Miller, R. D.; Michl, J. *Chem. Rev.* **1989**, *89*, 1359-410.

(3)   Trefonas, P.; Damewood, J. R.; West, R.; Miller, R. D. *Organometallics* **1985**, *4*, 1318-1319.

(4)   Sun, Y.-P.; Miller, R. D.; Sooriyakumaran, R.; Michl, J. *J. Inorg. Organomet. Polym.* **1991**, *1*, 3-35.

(5)   Miller, R. D.; Thompson, D.; Sooriyakumaran, R.; Fickes, G. N. *J. Polym. Sci., Part A: Polym. Chem.* **1991**, *29*, 813-24.

(6)   Matyjaszewski, K.; Kim, H. K. *Polym. Bull.* **1989**, *22*, 253-9.

(7)   Zeigler, J. M.; McLaughlin, L. I.; Perry, R. J. *J. Inorg. Organomet. Polym.* **1991**, *1*, 531-43.

(8)   Zeigler, J. M. *Polym. Prepr., (Am. Chem. Soc. Div. Polym. Chem.)* **1986**, *27*, 109-110.

(9)   Sakamoto, K.; Obata, K.; Hirata, H.; Nakajima, M.; Sakurai, H. *J. Am. Chem. Soc.* **1989**, *111*, 7641-3.

(10) Sakamoto, K.; Yoshida, M.; Sakurai, H. *Macromolecules* **1990**, *23*, 4494-6.

(11) Sakurai, H. *Polym. Prepr., (Am. Chem. Soc. Div. Polym. Chem.)* **1990**, *31*, 230-231.

(12) Sakurai, H.; Sakamoto, K.; Funada, Y.; Yoshida, M. *Polym. Prepr. (Am. Chem. Soc. Div. Polym. Chem.)* **1993**, *34*, 218-219.

(13) Matyjaszewski, K.; Cypryk, M.; Frey, H.; Hrkach, J.; Kim, H. K.; Moeller, M.; Ruehl, K.; White, M. *J. Macromol. Sci., Chem.* **1991**, *A28*, 1151-76.

(14) Chrusciel, J.; Fossum, E.; Matyjaszewski, K. *Polym. Prepr., (Am. Chem. Soc. Div. Polym. Chem.)* **1993**, *34*, 221-222.

(15) Cypryk, M.; Gupta, Y.; Matyjaszewski, K. *J. Am. Chem. Soc.* **1991**, *113*, 1046-1047.

(16) Matyjaszewski, K. *Makromol. Chem., Macromol. Symp.* **1991**, *42*, 269-80.

(17) Cypryk, M.; Chursiel, J.; Flossum, E.; Matyjaszewski, K. *Makromol. Chem., Macromol. Symp.* **1993**, *73*, 167-176.

(18) Boberski, W. G.; Allred, A. L. *J. Organomet. Chem.* **1975**, *88*, 73-7.

(19) Biran, C.; Bordeau, M.; Pons, P.; Leger, M.-P.; Dunogues, J. *J. Organomet. Chem.* **1990**, *382*, C17-C20.

(20) Hengge, E.; Litscher, G. *Angew. Chem., Int. Ed. Eng.* **1976**, *15*, 370.

(21) Shono, T.; Kashimura, S.; Ishifune, M.; Nishida, R. *J. Chem. Soc., Chem. Commun.* **1990**, 1160-1161.
(22) Aitken, C.; Harrod, J. F.; Samuel, E. *J. Organomet. Chem.* **1985**, *279*, C11-C13.
(23) Li, H.; Gauvin, F.; Harrod, J. F. *Organometallics* **1993**, *12*, 575-577.
(24) Banovetz, J. P.; Suzuki, H.; Waymouth, R. M. *Organometallics* **1993**, *12*, 4700-4703.
(25) Corey, J. Y. *Adv. Silicon Chem.* **1991**, *1*, 327-387.
(26) Corey, J. Y.; Zhu, X. H.; Bedard, T. C.; Lange, L. D. *Organometallics* **1991**, *10*, 924-30.
(27) Tilley, T. D.; Woo, H. G. *Polym. Prepr., (Am. Chem. Soc. Div. Polym. Chem.)* **1990**, *31*, 228-229.
(28) Tilley, T. D. *Comments Inorg. Chem.* **1990**, *10*, 37-51.
(29) Woo, H.-G.; Tilley, T. D. *J. Am. Chem. Soc.* **1989**, *111*, 8043-8044.
(30) Woo, H. G.; Walzer, J. F.; Tilley, T. D. *J. Am. Chem. Soc.* **1992**, *114*, 7047-55.
(31) Hengge, E.; Weinberger, M. *J. Organomet. Chem.* **1992**, *441*, 397-410.
(32) Harrod, J. F. In *Acs Symp. Ser.* 1988; Vol. 360; pp 89-100.
(33) Harrod, J. F.; Mu, Y.; Samuel, E. *Polyhedron* **1991**, *10*, 1239-45.
(34) Hengge, E.; Weinberger, M. *J. Organomet. Chem.* **1993**, *443*, 167-173.
(35) Banovetz, J. P.; Stein, K. M.; Waymouth, R. M. *Organometallics* **1991**, *10*, 3430-3432.
(36) Frey, H.; Out, G. J. J.; Moller, M.; Greszta, D.; Matyjaszewski, K. *Macromolecules* **1993**, *26*, 6231-6236.
(37) Miller, R. D.; Sooriyakumaran, R. *Macromolecules* **1988**, *21*, 3120-2.
(38) Stueger, H.; West, R. *Macromolecules* **1985**, *18*, 2349-52.
(39) Yuan, C.-H.; West, R. *Macromolecules* **1993**, *26*, 2645-2646.
(40) Hayase, S.; Mikogami, Y.; Yoshizumi, A.; Nakano, Y. JP Pat. Appl. 05 39,358 **1993**.
(41) West, R. *J. Organomet. Chem.* **1986**, *300*, 327-46.
(42) Hayase, S.; Horiguchi, R.; Onishi, Y.; Ushirogouchi, T. In *Polymers in Microlithography-Materials and Processes*; E. Reichmanis, S. A. MacDonald and T. Iwayanagi, Ed.;; American Chemical Society: Dallas, Texas, 1989; pp 133-157.
(43) Ho-Yi, S.; Maeda, N.; Suzuki, T.; Sato, H. *Polym. J.* **1992**, *24*, 865-870.
(44) Horiguchi, R.; Onishi, Y.; Hayase, S. *Macromolecules* **1988**, *21*, 304-309.
(45) Kim, H. K.; Hove, C. R.; Ober, C. K. *J. Macromol. Sci., Pure Appl. Chem., A29* **1992**, *9*, 787-800.
(46) Ban, H.; Sukegawa, K.; Tagawa, S. *Macromolecules* **1987**, *20*, 1775.
(47) Hengge, E. *J. Organomet. Chem. Libr.* **1979**, *9*, 261-295.
(48) Hrkach, J. S.; Matyjaszewski, K. *Macromolecules* **1990**, *23*, 4042-6.
(49) Matyjaszewski, K.; Chen, Y. L.; Kim, H. K. In *Inorganic and Organometallic Polymers*; M. Zeldin, K. J. Wynne and H. R. Allcock, Ed.; American Chemical Society: Washington D.C., 1988; Vol. 360; pp Chapter 6.
(50) Gauvin, F.; Britten, J.; Samuel, E.; Harrod, J. F. *J. Am. Chem. Soc.* **1992**, *114*, 1489-91.
(51) Tilley, T. D. *Acc. Chem. Res.* **1993**, *26*, 22-29.
(52) Thorne, J. R. G.; Williams, S. A.; Hochstrasser, R. M.; Fagan, P. J. *Chem. Phys.* **1991**, *157*, 401-408.
(53) Sun, Y. P.; Michl, J. *J. Am. Chem. Soc.* **1992**, *114*, 8186-90.
(54) Sun, Y. P.; Wallraff, G. M.; Miller, R. D.; Michl, J. *J. Photochem. Photobiol., A* **1992**, *62*, 333-46.
(55) Sun, Y. P.; Hamada, Y.; Huang, L. M.; Maxka, J.; Hsiao, J. S.; West, R.; Michl, J. *J. Am. Chem. Soc.* **1992**, *114*, 6301-10.

(56) Hsu, T. M.; Sawan, S. P. *Polym. Prepr., (Am. Chem. Soc. Div. Polym. Chem.)* **1992**, *33(1)*, 1032-1033.
(57) Liu, H. Q.; Harrod, J. F. *Organometallics* **1992**, *11*, 822-7.
(58) Qiu, H.; Du, Z. *J. Polym. Sci., Part A: Polym. Chem.* **1989**, *27*, 2861-2869.
(59) Banovetz, J. P.; Hsiao, Y. L.; Waymouth, R. M. *J. Am. Chem. Soc.* **1993**, *115*, 2540-1.
(60) Trefonas, P.; West, R.; Miller, R. D.; Hofer, D. J. *J. Polym. Sci., Polym. Lett. Ed.* **1983**, *21*, 823.
(61) Harrah, L. A.; Zeigler, J. M. *J. Polym. Sci.: Polym. Lett. Ed.* **1985**, *23*, 209-211.
(62) Harrah, L. A.; Zeigler, J. M. *Macromolecules* **1987**, *20*, 2037-2039.
(63) Miller, R. D.; Hofer, D.; McKean, D. R.; Willson, C. G.; West, R.; Trefonas, P. T. In *Materials for Microlithography*; L. F. Thompson, C. G. Willson and J. M. J. Frechet, Ed.;; American Chemical Society: Washington, 1984; pp 293-311.
(64) Miller, R. D.; Hofer, D.; Fickes, G. N.; Willson, C. G.; Marinero, E.; Trefonas, P.; West, R. *Polym. Eng., Sci.* **1986**, *26*, 1129-1134.
(65) Takeda, K.; Teramae, H.; Matsumoto, N. *J. Am. Chem. Soc.* **1986**, *108*, 8186-8190.
(66) Gauvin, F.; Harrod, J. F. *Can. J. Chem.* **1990**, *68*, 1638-40.
(67) Gilman, H.; Atwell, W. H.; Sen, P. K.; Smith, C. L. *J. Organomet. Chem.* **1965**, *4*, 163-167.
(68) Sakurai, H.; Murakami, M.; Kumada, M. *J. Am. Chem. Soc.* **1969**, *91*, 519-520.
(69) Hsu, T. M.; Sawan, S. P. *Polym. Prepr., (Am. Chem. Soc. Div. Polym. Chem.)* **1992**, *33(1)*, 1040-1041.
(70) Bock, H.; Ensslin, W.; Feher, F.; Freund, R. *J. Am. Chem. Soc.* **1976**, *98*, 668-674.
(71) Pitt, C. G. *J. Am. Chem. Soc.* **1969**, *91*, 6613-6621.
(72) Pitt, C. G. In *Homoatomic Rings, Chains, and Macromolecules of Main-Group Elements*; A. L. Rheingold, Ed.;; Elsevier Scientific Publishing Company: Amsterdam, 1977; pp 203.

RECEIVED June 28, 1994

Chapter 7

Poly(dimethysiloxane)—Urea—Urethane Copolymers
Synthesis and Surface Properties

Kenneth J. Wynne[1], Tai Ho[2], Robin A. Nissan[3], Xin Chen[4,5], and Joseph A. Gardella, Jr.[4]

[1]Chemistry Division, Office of Naval Research, Arlington, VA 22217-5660 and Materials Chemistry Branch, Naval Research Laboratory, Washington, DC 20375-5320
[2]Department of Chemistry, George Mason University, Fairfax, VA 22030-4444
[3]Chemistry Division, Naval Air Weapons Center, China Lake, CA 93555
[4]Department of Chemistry, State University of New York at Buffalo, Buffalo, NY 14214

In connection with our interest in the development of minimally adhesive surfaces to discourage the settlement of marine organisms, we have investigated polyurethanes and polyureas containing polydimethylsiloxane (PDMS) segments. A two-step polymerization method was used to prepare dimethylsiloxane-urea-urethane copolymers with 1,4-benzenedimethanol as the chain extender. Thermal and mechanical properties of copolymers with chain extenders were found to be superior to those without chain extender, due to the additional hydrogen bonding interactions for the former. Surface composition was determined by angle-dependent electron spectroscopy for chemical analysis (ESCA). Effects of segmental length and annealing on the surface composition were investigated.

Our interest in the formation of minimally adhesive polymer surfaces has led us to an investigation of poly(urethane-ureas) containing unusual diols. Our goal is to discern the compositional and morphological features which create a surface minimally attractive to the settlement of marine organisms, an area pioneered by Griffith, et al.[1]

In addition to low surface energy, we postulate that a surface phase with a low T_g is desirable for minimizing mechanical locking of a prospective adherent to the surface. With the dual criteria of low surface energy and a low T_g surface phase in mind, we have prepared and characterized a series of polydimethylsiloxane containing poly(urethane-ureas) and have examined the effects of chain extenders on properties.

[5]Current address: Moore Research Center, Grand Island, NY 14072

0097–6156/94/0572–0064$08.00/0

The synthesis of siloxane urea urethane copolymers through step-growth polymerization has been explored previously,[2 - 5] and work performed before 1988 has been reviewed.[6] We have used a "hard segment first" two-step polymerization procedure related to that of Harrell[7] to achieve hard segment compositional control. The synthesis of these polymers[8] and initial studies of surface characterization[9] have been reported.

The pioneering work of Clark demonstrated the power of angle-dependent electron spectroscopy for chemical analysis (ESCA) in the characterization of polymer surfaces.[10] Subsequently, surface composition and morphology of block copolymers containing poly(dimethylsiloxane) (PDMS) segments have been studied using angle-dependent ESCA and other techniques. Polyurethane-PDMS,[11 - 13] Nylon-6-PDMS,[14] poly(α-methylstyrene)-PDMS,[15] polycarbonate-PDMS,[16,17] polysulfone-PDMS[18] and polystyrene-PDMS[19,20] of various block architectures and overall compositions have been investigated. Enrichment of PDMS segments in surface region was detected in each case due to the lower surface energy of PDMS component in these block copolymers. ESCA, a coinage by Kai Siegbahn,[21] is also known as X-ray photoelectron spectroscopy (XPS), which describes the physical process utilized in the instrument.

The present work builds on the observation that surface composition of multicomponent polymers depends not only on structure, but also sample history. Thus, surface composition of solution cast films of block copolymers can be controlled by using selective solvents[14,22] or by annealing.[14,15,17,23]

Experimental

Materials. Isophorone diisocyanate (IPDI, **1**, 5-isocyanato-1-(isocyanatomethyl)-1,3,3-trimethylcyclohexane, a mixture of isomers), and 1,4-benzenedimethanol (BDM), **2**, were purchased from Aldrich. 3-Aminopropyl endcapped dimethylsiloxane oligomers, $H_2N(CH_2)_3(Si(CH_3)_2)O)_nSi(CH_3)_2(CH_2)_3NH_2$, with different average molecular weights were kindly provided by Dr. I. Yilgör of Goldschmidt Chemical Corporation, Hopewell, VA. The nominal molecular weights of Tegomer A-Si 2120, **4a**, Tegomer A-Si 2320, **4b**, and Tegomer A-Si 2920, **4c**, were 1000 (n = 11), 2400 (n = 30) and 10,000 (n = 133), respectively. In addition, aminopropyl end-capped PDMS oligomer Hüls PS 513 (viscosity 2000 cs, MW $\cong$ 27,000, n = 363) was employed. All chemicals were used as received.

Polymerization. The two-step polymerization is shown in Scheme 1. Synthetic details have been described previously.[8] Polymers without chain extender were prepared in THF solution using only the second step of Scheme 1.

$$\begin{array}{ccccc} & & & 105 - 115°C & \\ (x+1)\ \text{IPDI} & + & (x)\ \text{BDM} & \text{------------} > & \text{IPDI-(BDM-IPDI)}_x \\ \textbf{1} & & \textbf{2} & & \textbf{3} \end{array} \qquad (1)$$

$$\text{THF}$$
$$\text{IPDI-(BDM-IPDI)}_x + \text{Tegomer A-Si} \text{--------} > \text{PDMS-urea-urethane copolymer} \qquad (2)$$

Scheme 1

Characterization. The new polymers were characterized by infrared and NMR spectroscopy, viscometry, and mechanical properties as previously described.[8] Molecular weights were determined by gel permeation chromatography, using a Hewlett-Packard Series 1050 pump, two Altex μ-Spherogel columns (size 10^3 and 10^4Å, respectively) connected in series, and THF as the solvent. Polymer in the effluent was detected with a Wyatt/Optilab 903 interferometric refractometer, and average molecular weights were determined relative to polystyrene standards. As previous workers have noted, polymers containing high weight fractions of poly(dimethyl siloxane) segments have low specific refractive increments (dn/dc) in THF solution.[24] High concentrations (about 2-3 wt%) were therefore used to enhance signals. Chromatograms were smoothed by the method of Fourier transform convolution.[25]

Film preparation for ESCA analysis involved solvent pretreatment to remove adventitious low molecular weight materials, exposure to high vacuum for solvent removal, and film casting from THF utilizing aluminum weighing pans.[9] Solvent was removed from the films (thickness ~50 μm) by a 2-day vacuum treatment at ambient temperature. Film annealing was accomplished by heating in a vacuum oven at 120°C for 15 minutes.

Angle-dependent ESCA experiments were performed on a Perkin-Elmer Physical Electronic Model 5300 ESCA with a hemispherical analyzer and a single channel detector.[9] For quantifying ESCA signals in carbon 1s, nitrogen 1s, silicon 2p, and oxygen 1s regions spectra were recorded at high resolution conditions. ESCA peak areas were measured by a Perkin-Elmer 7500 computer with the PHI ESCA version 2.0 software. An average of three independent runs was taken for all ESCA measurements. The ratio of N/C, instead of Si/C, is chosen to calculate the PDMS surface concentrations because of the larger variation in nitrogen concentration at different sampling depths. This results in higher sensitivity in PDMS wt % calculation. Concentrations were also calculated from Si/C ratios for a selected set of samples. The resultant concentration data fell within error limits as equivalent to that calculated from N/C ratios.

Photoelectron intensities detected by ESCA are convoluted signals, i.e. all atoms within the path of the probing X-ray contribute to the signal but the contribution of each decreases exponentially with the distance from the free surface.[10] The convoluted nature of the signal distorts depth profiles for samples with compositional gradients. To recover the depth profiles for such samples, a deconvolution procedure was applied to the ESCA data.

The deconvolution procedure is based on a well known correlation between photoelectron intensity and various factors, such as intensity of probing X-ray, atomic density distribution, material constants and instrument constants. One form of the correlation was given by Clark[10]

$$d\mathrm{I}(\theta) = F\alpha N(x) \ K \ e^{-x/(\lambda \ \sin\theta)} \ dx \tag{3}$$

where I is the detected intensity of photoelectron, θ the take off angle, F the X-ray flux, α the cross section of photoionization in a given shell of a given atom for a given X-ray energy, $N(x)$ the depth profile of the atomic density, x the vertical distance from the free surface, K a spectrometer factor, and λ the escape depth of the electrons. Using a suitable model for $N(x)$, one can calculate the photoelectron intensity as a function of the take off angle. By comparing the calculated values with the experimental data and modifying the model for N(x) with the methods of optimization, the atomic density profile can be deduced.[9,26] In this investigation, two models for N(x) were used. A discrete model treats the density profile as a summation of nine step functions, and a continuous model describes the profile with a four-parameter compound Gaussian distribution.[9]

Results and Discussion

Hard Segment Characterization. The preparation of copolymers of aminopropyl endcapped dimethylsiloxane oligomers and isophorone diisocyanate with 1,4-benzenedimethanol as the chain extender outlined in Scheme 1 is similar to the method used by Pascault and co-workers to prepare alkoxy-silane terminated macromers.[27] This "hard segment first" method allowed the separate isolation and characterization of the hard block. Thus, GPC analysis for the intermediate IPDI-BDM-IPDI-BDM-IPDI (desired "3:2" composition), also revealed peaks for IPDI-BDM-IPDI, and BDM-IPDI-BDM. The intensity ratio of the curves indicate that more than 85% of the molecules in the mixture were of the target "3:2" composition.

Tegomer Characterization. Assignment and determination of relative intensities of the NMR peaks due to the dimethylsilyl and the propyl amine end group protons was carried out to confirm GPC determined macromonomer molecular weight.[8] In addition to expected ^{1}H NMR signals, additional peaks were observed in 1, 2.4 and 10K Tegomer samples from Goldschmidt Chemical Co. A careful investigation of Tegomer A-Si 2120 was carried out to determine the identity of any additional materials present. In the ^{13}C gated spin echo (GASPE) spectrum for the 1K Tegomer (Figure 1) peaks expected for the n-propyl-terminated polydimethyl siloxane are seen at 15.22, 27.58 and 45.35 ppm, and are confirmed as CH_2's by down position. In addition, peaks at 11.51 (up), 26.31 (up), and 44.24 (down) ppm, were assigned as methyl, methine, and methylene carbons, respectively, based on GASPE attached proton test. Additional 2-d techniques allowed identification of all one-bond correlations and many two and three-bond correlations. All results, including chemical shifts and coupling patterns, indicate that the "impurity" is a 2-amino-1-methylethyl end group.

The 2-amino-1-methylethyl end group was likely introduced early in the synthesis stage in the preparation of the disiloxane "end-blocker". The disiloxane was presumably synthesized via hydrosilation (probably addition of allyl amine to hydrido dimethyl disiloxane, usually catalyzed by chloroplatinic acid). In this reaction, Si may add to the 3- or 2-carbon position. Addition to the 3-carbon gives the 3-aminopropyl dimethyl disiloxane, while addition to the 2-carbon gives 2-amino-1-methylethyl dimethyl disiloxane. The mixture of end groups would then be carried along into the Tegomers which are made by catalyzed equilibration reactions.[6] A review of the literature revealed surprisingly few quantitative studies of isomeric products from Pt-catalyzed allyl addition. Significant 2-addition was observed in the hydrosilation of allyl carbonates,[28,29] and the patent literature suggests that formation of straight chain and branched amino-propyl siloxanes has also been observed in the reaction of triethoxysilane with allylamine.[30]   Integrated ^{1}H spectra for $SiCH(C\underline{H}_3)CH_2NH_2$ vs.

SiC$\underline{H}_2$CH$_2$CH$_2$NH$_2$ revealed the relative amounts of these two end groups. In the respective Tegomers, the proportion of SiCH$_2$CH$_2$CH$_2$NH$_2$ is: 72%, Tegomer A-Si 2120; 73%, Tegomer A-Si 2320; 76%, Tegomer A-Si 2920. The presence of SiCH(CH$_3$)CH$_2$NH$_2$ end groups could not be detected in the aminopropyl end-capped PDMS oligomer Hüls PS 513. Molecular weights determined by NMR and GPC were found to be in excellent agreement.[8]

Polymer Synthesis and Characterization. Diamines or diols can be used as chain extenders in copolymerization reactions, but urea linkages formed from diamines often lead to polymer insolubility.[31] Poor solubility is avoided by using diols, but this requires a two stage preparation (Scheme 1) because of the disparity in reactivity of alcohols and amines with isocyanates. To form urea and urethane linkages separately, the reaction between isophorone diisocyanate and benzene dimethanol was carried out first in bulk at 115 °C without catalyst (Equation 1). Secondly, reaction between aminopropyl endcapped dimethylsiloxane oligomers and diisocyanate intermediates was effected in THF at room temperature (Equation 2). Completion of the polymerization reaction was established by NMR and infrared spectroscopy.[8]

Since the capability to form films is an important requirement for coating materials, we have focused on those compositions that form flexible films (Table I). Composition designations are as follows: first, the segmental average molecular weight of the siloxane oligomer is given; the two letters following the first hyphen identify IPDI as "IP"; the letter after the second hyphen represents BDM, the chain extender, "B"; and the next number shows the molar ratio of chain extender to siloxane oligomer. Thus, polymer PDMS2.4K-IP-B0 is made of one mole of Tegomer A-Si 2320 (MW≈2400) and one mole of IPDI with no BDM chain extender. The copolymers are also generically called PDMS-PU.

Mechanical Properties. Stress-strain tests were performed at room temperature at a strain rate of 1.67 min^{-1} on specimens cast from THF solutions (Figure 2, Table II). The average molecular weight of the siloxane segments has a dominant effect on the rigidity of the copolymers. Young's moduli for copolymers based on oligomer **4a** (MW≈1000) are above 40 MPa, those for copolymers based on **4b** (MW≈2400) are in the range of 1 to 12 MPa, and the moduli for copolymers based on **4c** (MW≈10,000) and Hüls PS 513 (MW≈27,000) are below 0.4 MPa.

The incorporation of BDM as the chain extender increases the Young's modulus significantly. The addition of two moles of BDM per mole of siloxane oligomer in copolymers based on **4b** and **4c** results in a tenfold increase in the modulus. On the other hand, the strain at break of the polymers decreases with the incorporation of BDM. For copolymers based on **4a**, the strain at break drops from 5.0 (500 %) to 2.7 with the incorporation of one half mole of BDM per mole of siloxane oligomer. In the series of copolymers based on **4b**, the strain at break decreases from over 21 to 4.2 with an increase in the content of the chain extender from none to two moles of BDM per mole of **4b**. Similarly, incorporation of BDM into copolymers based on **4c** at the ratio of two moles of BDM per mole of **4c** leads to a decrease in strain at break from 27 to 3. Toughness of a material is proportional to the area under the stress-strain curve. By this measurement, polymers with average

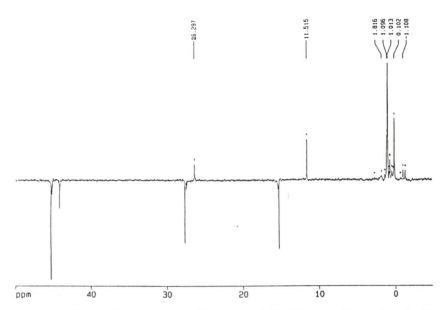

Figure 1. ^{13}C GASPE spectrum of Tegomer A-Si 2120, **4a**. (Reproduced with permission from ref. 8. Copyright 1993 ACS.)

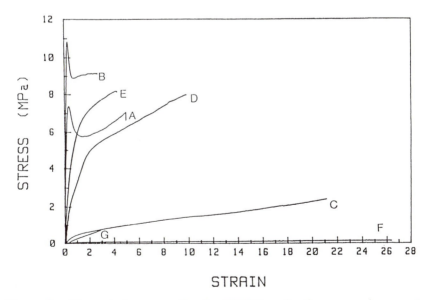

Figure 2. Stress-strain curves for the PDMS/urea/urethane co-polymers. A: PDMS1K-IP-B0; B: PDMS1K-IP-B0.5; C: PDMS2.4K-IP-B0; D: PDMS2.4K-IP-B1; E: PDMS2.4K-IP-B2; F: PDMS10K-IP-B0; and G: PDMS10K-IP-B2. (Reproduced with permission from ref. 8. Copyright 1993 ACS.)

Table I Molecular Weights of the Segmented Copolymers

No.	Sample ID	PDMS MW[a]	Reaction Stoichiometry PDMS: IPDI: BDM					Copolymer MW[b]		
								M_w	M_n	M_w/M_n
1	PDMS1K-IP-B0	1,000	1	:	1	:	0	16,700	12,000	1.4
2	PDMS1K-IP-B0.5	1,000	2	:	3	:	1	17,600	11,800	1.5
3	PDMS2.4K-IP-B0	2,400	1	:	1	:	0	65,000	30,700	2.1
4	PDMS2.4K-IP-B1	2,400	1	:	2	:	1	76,500	27,700	2.8
5	PDMS2.4K-IP-B2	2,400	1	:	3	:	2	42,700	19,100	2.2
6	PDMS10K-IP-B0	10,000	1	:	1	:	0	109,000	54,500	2.0
7	PDMS10K-IP-B2	10,000	1	:	3	:	2	81,200	29,200	2.8
8	PDMS27K-IP-B2	27,000	1	:	3	:	2	198,000	149,000	1.3

(a). Amino terminated siloxane oligomers with molecular weight provided by the supplier.

(b). Polymer molecular weight determined by GPC.

siloxane segmental molecular weights at 1K and 2.4K are tougher, by about one order of magnitude, than those with siloxane segmental molecular weights at 10K and 27K, regardless of the composition of the hard segment.

It has been recognized that many of the unique properties of the polyether or ester based polyurethanes are due to their phase-separated structure: the hard segment (urethane or urea) rich domains provide the "solid" attributes, while the soft segment (polyether or ester) rich domains account for the elastomeric behavior.[32] Previous work

Table II. Mechanical Properties of PDMS/Urea/Urethane Copolymers.

Designation	Young's Modulus (MPa)*	Stress at Break (MPa)*	Strain at Break*	Area under Curve (MPa)*
PDMS1K-IP-B0	43	7.4	5.0	31
PDMS1K-IP-B0.5	70	9.1	2.7	24
PDMS2.4K-IP-B0	1.0	2.4	21	31
PDMS2.4K-IP-B1	6.9	8.0	9.8	58
PDMS2.4K-IP-B2	12	8.1	4.2	28
PDMS10K-IP-B0	0.03	0.1	27	3
PDMS10K-IP-B2	0.4	0.7	3	1
PDMS27K-IP-B2	0.2	0.6	6	2

* Numbers shown are the average values of two to four measurements.

has shown that mechanical properties of PDMS urea (or urethane) copolymers also follow that pattern.[2,3,5] It has been shown, based on dynamic mechanical data, that phase separated morphologies exist in the current copolymers.[8] In such systems, an increase in the average molecular weight of the soft segments, which increases the average distance between two hard segment domains, will cause a decrease in the rigidity of the material, while an increase in chain extender content, which increases the proportion of the hard segment domain, will enhance the rigidity. The observed mechanical properties are in agreement with this reasoning. Based on this model the observed variation in toughness can also be explained. Materials based on **4a** are tough because of the proportionately greater contribution from the hard segment,

which also leads to high modulus. The toughness of materials based on **4b** is mainly due to the ductility of the materials facilitated by the longer soft segment. In contrast, the weakness of materials based on **4c** is probably due to failure of the long soft segment to transmit the external force to be born by the hard segment.

Surface Composition of PDMS-PU Films by ESCA.

Structural Effects. Previous studies have shown that surfaces of polymers containing PDMS segments display surface enrichment of the PDMS component.[11-20] PDMS surface phase separation is attributed to the lower surface energy of the PDMS segments at the polymer air interface, as compared with a surface presenting polar hard segments.[33] Our results for PDMS-PU segmented copolymer films cast from THF in air show a similar enrichment. ESCA results for the PDMS-PU segmented copolymers are angle dependent indicating a gradient in concentration of PDMS over the surface region. Through deconvolution of the ESCA data insights into effects of chain structure and processing effects have been obtained.

Some trends are apparent without extensive data analysis. Thus, Figure 3 shows surface compositions derived from direct measurement of N/C ESCA peaks as a function of take-off angle. The N/C ratio goes to zero when the surface is pure PDMS. The data in Figure 3 for as-cast films of PDMS2.4K-IP-B2, PDMS10K-IP-B2 and PDMS27K-IP-B2 shows that the copolymer with the highest average molecular weight for the soft segments has the highest PDMS surface concentration. This might be expected, as the PDMS weight fraction goes from 0.708 to 0.966 for this series of copolymers.

Figure 4 illustrates the change in PDMS surface concentration with increase in the length of hard segment. In this group of PDMS-PU samples, the average molecular weights for the PDMS segments are the same (2.4K), but the average lengths for the hard segment are different. The surface concentration of PDMS decreases as the length of the hard segment increases; the weight fraction of PDMS is in the 0.708 to 0.903 range for this series. Tezuka and co-workers have used ESCA and contact angle measurement to study the surface morphology of a series of polyetherurethane-PDMS segmented copolymers,[11] and their results are in good agreement with ours.

A deconvolution of the ESCA data has been carried out to recover depth profiles for the hard segment in each copolymer. The profiles reveal some features which are not obvious in the ESCA data derived from direct measurement of C/N peak intensities (Figures 3 and 4 discussed above). The deconvolution method takes into account the differing C atom densities in the PDMS (25.7M) and hard segment (ca. 54M) phases and utilizes an optimal fitting of the data.[9] The results of this analysis are seen qualitatively in Figure 5, which shows concentration depth profiles for hard segment in PDMS2.4K-IP-B2. The solid squares represent ESCA data derived from direct measurement of C/N peak intensities; curves A and B result from a continuous and a discrete model, respectively.

Both the discrete and continuous models indicate a *maximum* in the hard segment profile in contrast to the monotonically increasing profile suggested by N/C peak-ratio data. The hard segments of the materials are *chemically* bonded to the soft

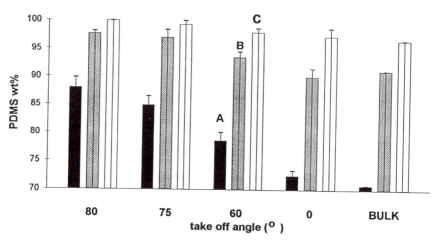

Figure 3. PDMS concentration in the surface region measured by ESCA for three copolymers. The distance between the horizontal bar and the top of the column represents one standard deviation. Data series: A, PDMS2.4K-IP-B2; B, PDMS10K-IP-B2; and C, PDMS27K-IP-B2.

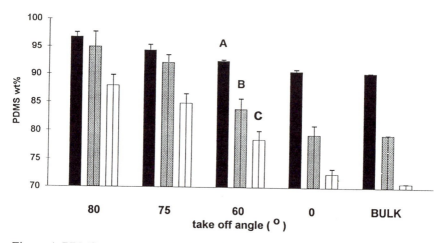

Figure 4. PDMS concentration in the surface region measured by ESCA for three copolymers. The distance between the horizontal bar and the top of the column represents one standard deviation. Data series: A, PDMS2.4K-IP-B0; B, PDMS2.4K-IP-B1; and C, PDMS2.4K-IP-B2.

segments. Since the soft segments are concentrated on the topmost layer of the surface, there should be an enrichment of hard segments right underneath that layer. The peak ratio ESCA data also give a lower hard segment content in the surface region than the models. These differences emphasize the fact that for polymers with partially or completely phase separated surface regions, the convoluted nature of the photoelectron intensity signals prevents a direct measurement of concentration depth profiles. Profiles produced by both models, when inserted into the convolution equations, lead to photoelectron intensity ratios in very good agreement with experimental data.

The usefulness of the continuous model may be seen in an analysis of surface profiles for samples with similar *bulk* PDMS concentrations. For such samples, the length ratios of PDMS segment to hard segment are similar. Copolymers PDMS2.4K-IP-B0 and PDMS10K-IP-B2 have similar bulk concentrations of PDMS (90.3% and 91.1%). Results based on ESCA peak-ratio data for these two copolymers are compared in Figure 6. Peak ratio data do not allow differentiation in surface composition within error limits although a slight trend is observed. Our continuous deconvolution model, on the other hand, resolves differences in these two concentration depth profiles, as illustrated in Figure 7.

In addition to entropically driven phase separation for longer soft segments, phase separation is driven by the enthalpic effect of stronger hydrogen bonding of larger hard segments. These two effects may act in a synergistic fashion, but we believe the enthalpic effect is of greater importance.

Annealing Effects. The surface compositions of PDMS-PU copolymer samples varied after a short annealing, the extent of the variation depends on the length of the soft segment. For samples with the shortest PDMS segments ($MW_{soft} = 1000$), annealing does not have any detectable effect on the surface concentration of PDMS. For samples with PDMS segments of moderate length ($MW_{soft} = 2.4K$), an increase in PDMS surface concentration was observed in the sample with the longest hard segments (Figure 8). If the PDMS segments are long ($MW_{soft} = 10K$), significant changes in the PDMS surface concentrations are observed regardless the composition of the hard segments. For example, the concentration of PDMS in the as-cast films of PDMS10K-IP-B0 and PDMS10K-IP-B2 measured at the lowest take off angle are 98.9 and 97.7 %, respectively, after annealing, the corresponding values are 99.9 and 99.4 %. Figure 8 reveals increases in the thickness of PDMS and hard-block enriched phases after annealing. The thickness of the topmost PDMS layer with less than 5% hard segment content increases from zero to 0.2 λ. (The escape depth λ is approximately 3.4 nm). By a comparison of peak widths, the thickness of the layer enriched with hard segments, which is responsible for the maximum in the profile, doubles from $\sim\lambda$ to $\sim2\lambda$. These observations again illustrate the positive effects of annealing in enhancing the surface concentration of PDMS.

The dominance of PDMS segments in the surface region is even more pronounced in copolymer PDMS27K-IP-B2. No hard segments were detected in measurements taken at the lowest take-off angle, and the topmost surface regions of both as-cast and annealed films contain only PDMS component.

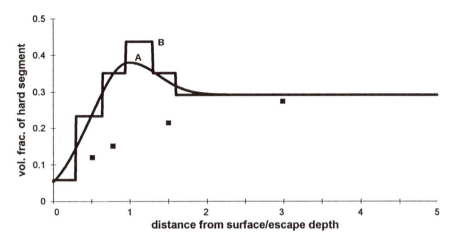

Figure 5. Concentration depth profiles for the hard segment in as-cast films of copolymer PDMS2.4K-IP-B2. Profiles: A, constructed based on a continuous model; B, constructed based on a discrete model; and (■) data measured by ESCA.

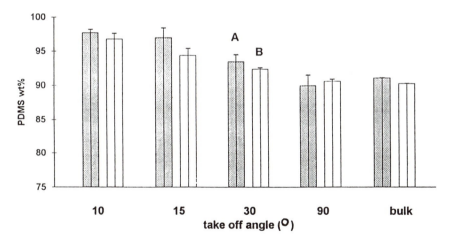

Figure 6. PDMS concentration in the surface region measured by ESCA for two copolymers of similar bulk PDMS concentration. The distance between the horizontal bar and the top of the column represents one standard deviation. Data series A, PDMS10K-IP-B2; and B, PDMS2.4K-IP-B0.

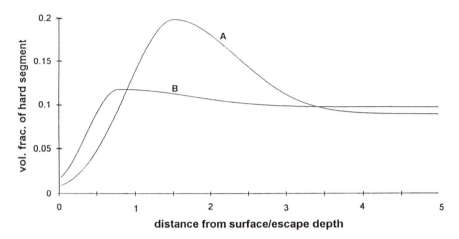

Figure 7. Hard segment concentration in the surface region for two copolymers of similar bulk PDMS concentration. The profiles are constructed based on a continuous model. Data series: A, PDMS10K-IP-B2; and B, PDMS2.4K-IP-B0.

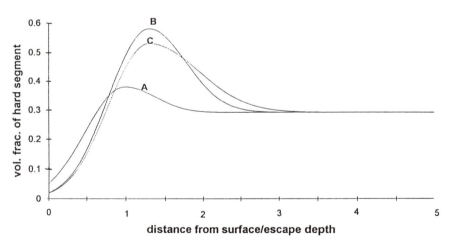

Figure 8. Concentration depth profiles of the hard segments in the surface region for the copolymer PDMS2.4K-IP-B2 under different thermal treatments. The profiles are constructed based on a continuous model. Curves: A, as-cast; B, annealed at 120°C for 15 minutes; and C, for two hours.

Effects of Structure and Molecular Weight on the Distribution of the Hard Segment. In addition to the difference in surface energies, which is the underlying driving force for the enrichment of PDMS segments at the surface, there are two additional factors that contribute to the observed surface phase separation. Long soft segments amplify entropically driven phase separation, and increased hydrogen bonding in hard segments augments the enthalpic driving force for phase separation. The interplay of these factors under different situations is discussed below. We define $v_N(x)$ as:

$$v_N(x) = \{v(x)/[1-v(x)]\} \, / \, \{v(\infty)/[1-v(\infty)]\} \tag{4}$$

Equation 4 represents the normalization of the volume fraction of the hard segment at the surface region, $v(x)$, with respect to the bulk value, $v(\infty)$. That is, $v_N(x)$ is the ratio of the volume fraction of hard/soft segments at the surface divided by the same ratio for the bulk. The quantity $v_N(x)$ can be visualized in the following way. Suppose a unit volume of soft segments with the entrapped hard segments moves from the bulk (location ∞) to the surface (location x), and during the movement the hard segments gradually drop out this volume due to the progress of phase separation, then $v_N(x)$ represents the fraction of initial hard segments remaining at the end of the movement. Using this measurement, materials can be compared based on their "effectiveness" in separating the soft and the hard segments.

In Figure 9, v_N depth profiles in the region less than one λ (ca. 3.4 nm) from the surface for copolymers PDMS10K-IP-B0 (A, C) and PDMS10K-IP-B2 (B, D) are shown. The two solid lines, representing concentration profiles in the as-cast films, indicate that the incorporation of chain extender promotes surface phase separation at ambient temperature presumably due to the increased interactions among the hard segments. This advantage is lost in annealing as both after-annealing profiles (dashed lines) are similar and shift to lower hard segment concentrations. Since the soft segments are the same and relatively large compared to the hard segments, the reason for the similar near-surface profiles after short annealing is driven by enthalpy. Thus, the rate of disadvantaged phase separation for PDMS10K-IP-B0 catches up to that of PDMS10K-IP-B2, as an increase in temperature enhances the higher-energy-barrier phase separation process for PDMS10K-IP-B0.

In Figure 10, v_N profiles for copolymers PDMS2.4K-IP-B2, PDMS10K-IP-B2, and PDMS27K-IP-B2 are plotted. Cast at ambient temperature, the hard segment concentration in the surface region decreases with increase in the average molecular weight of the soft segments. After annealing, the curves shift to lower hard segment concentrations while the relative positions to one another are maintained. The inverse correlation between surface hard segment concentration and bulk soft segment molecular weight as well as the insensitivity of the shape of the curve to temperature suggests strong entropic influence on the observed differences.

Conclusions

A hard-segment first two-step polymerization was used to prepare PDMS-urea-urethane copolymers with 1,4-benzenedimethanol as the chain extender. The chain

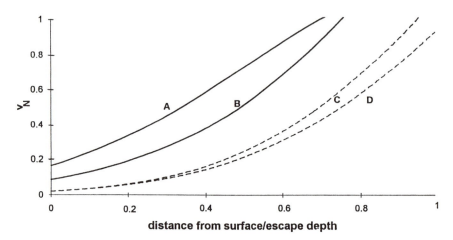

Figure 9. v_N as a function of depth in the topmost surface region for two copolymers. Curves: A, PDMS10K-IP-B0 as-cast; B, PDMS10K-IP-B2 as-cast; C, PDMS10K-IP-B0 annealed at 120°C for 15 minutes; and D, PDMS10K-IP-B2 annealed at 120°C for 15 minutes.

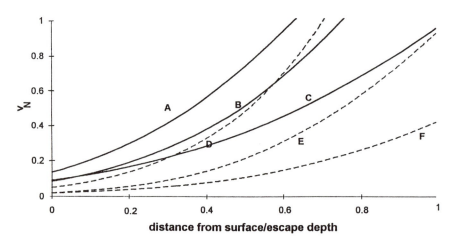

Figure 10. v_N as a function of depth in the topmost surface region for three copolymers. Curves: A, PDMS2.4K-IP-B2 as-cast; B, PDMS10K-IP-B2 as-cast; C, PDMS27K-IP-B2 as-cast; D, PDMS2.4K-IP-B2 annealed at 120°C for 15 minutes; E, PDMS10K-IP-B2 annealed at 120°C for 15 minutes; and F, PDMS27K-IP-B2 annealed at 120°C for 15 minutes.

extended copolymers were found to be superior to non-chain extended copolymers in thermal and mechanical properties due to the reinforcement effect of additional hydrogen bonding interactions between urea and urethane linkages. Details of thermal and mechanical behavior were consistent with those expected of a phase separated morphology, and were shown to result from the molecular structure of the materials.

ESCA analysis of films cast from THF surface show that surface concentration of PDMS increases with PDMS segment length. Annealing as-cast samples increases the PDMS surface concentration. ESCA data were deconvoluted to give depth profiles which typically revealed a hard segment maximum beneath the PDMS enriched surface layer. Annealing increases the thickness of both regions. Both the length of the soft segments and the interactions among the hard segments affect the phase separation in the surface region. Long soft segments amplify entropically driven phase separation, while increased hydrogen bonding in hard segments augments the enthalpic driving force for phase separation.

Acknowledgment: This research was supported in part by the Office of Naval Research. We are grateful to Dr. İ. Yilgör of Goldschmidt Chemical Corporation for a gift of the aminopropyl terminated polydimethylsiloxanes.

Literature Cited

1. Brady, R. F.; Griffith, J. R.; Love, K. S.; Field, D. E. *J. Coatings Tech.* **1987**,*59*, 113.
2. Tyagi, D.; Yilgör, İ.; McGrath, J. E.; Wilkes, G. L. *Polymer* **1984**, *25*, 1807.
3. Yu, X. H.; Nagarajan, M. R.; Grasel, T. G.; Gibson, P. E.; Cooper, S. L. *J. Polym. Sci., Polym. Phys. Ed.* **1985**, *23*, 2319.
4. Oishi, Y.; Kakimoto, M.; Imai, Y. *J. Polym. Sci., Part A: Polym. Chem.* **1987**, *25*, 2185.
5. Chen, L.; Yu, X. H. *Functional Polymer* **1991**, *4*, 19.
6. Yilgör, İ.; McGrath, J. E. *Adv. Polym. Sci.* **1988**, *86*, 1.
7. Harrell, L. L. Jr. *Macromolecules* **1969**, *2*, 607.
8. Ho, T.; Wynne, K. J.; Nissan, R. A. *Macromolecules,* **1993**, *26*, 7029.
9. Chen, X.; Gardella, J. A.; Ho, T.; Wynne, K. J. *in press.*
10. Clark, D. T. *Advances in Polymer Sciences,* **1977**, *24*, 126.
11. Tezkuka, Y.; Kazuma, H.; Imai, K. *J. Chem. Soc. Faraday Trans.,* **1991**, *87*, 147.
12. Shibayama, M.; Suetsugu, M.; Sakurai, S.; Yamamoto, T.; Nomura, S. *Macromolecules* **1991**, *24*, 6254.
13. Benrashid, R.; Nelson, G. L.; Linn, J. H.; Hanley, K. H.; Wade, W. R., *J. Appl. Poly. Sci.,* **1993**, *49*, 523.
14. Chen, X.; Gardella, J. A., Jr. *Polym. Prepr. Am Chem. Soc. Div. Polym. Chem.* **1992**, *33(2)*, 312.
15. Chen, X.; Gardella, J. A., Jr.; Kumler, P. L. *Macromolecules,* in press.
16. Mittlefehldt, E. R.; Gardella, J. A., Jr. *Appl. Spectrosc.* **1989**, *43*, 1172.
17. Chen, X.; Lee, H. F.; Gardella, J. A., Jr. *Macromolecules* **1993**, *26*, 4601.

18. Pertsin, A. J.; Gorelova, M. M.; Levin, V. Yu.; Makarova, L. I. *J. Appl. Poly. Sci.*, **1992**, *45*, 1195.
19. Chen, X.; Gardella, J. A., Jr.; Kumler, P. L. *Macromolecules*, **1992**, *25*, 6621.
20. Chen, X.; Gardella, J. A., Jr.; Kumler, P. L. *Macromolecules*, **1992**, *25*, 6631.
21. Siegbahn, K. *Science* **1982**, *217*, 111.
22. Thomas, H. R.; O'Malley, J. J. *Macromolecules*, **1979**, *12*, 323.
23. Coulon, G.; Russell, T. R.; Deline, V. R.; Green, P. F. *Macromolecules*, **1989**, *22*, 2581.
24. Veith, C. A.; Cohen, R. E. *Makromol. Chem., Macromol. Symp.* **1991**, *42/43*, 241-258.
25. We used programs included in *Mathematica* to perform this operation; Wolfram, S. *Mathematica: a system for doing mathematics by computer*, 2nd ed.; Addison-Wesley Publishing Company, Inc.: Redwood City, California, 1991, p 679.
26. Previous efforts in de-convoluting ESCA data include (a)Iwasaki, H.; Nishitani, R.; Nakamura, S. *Jpn. J. Appl. Phys.* **1978**, *17*, 1519; (b) Pijolat, M.; Hollinger, G. *Surf. Sci.* **1981**, *105*, 114; (c) Nefedov, V. I.; Baschenko, O. A. *J. Electron Spectrosc. Relat. Phenom.* **1988**, *47*, 1; (d) Tyler, B. J.; Castner, D.G.; Ratner, B. D. *Surf. Interface Anal.* **1989**, *14*, 443; (e) Jisl, R. *Surf. Interface Anal.* **1990**, *15*, 719; (f) Holloway, P. H.; Bussing, T. D. *Surf. Interface Anal.* **1992**, *18*, 251.
27. Surivet, F.; Lam, T. M.; Pascault, J. P. *J. Polym. Sci., Polym. Chem. Ed.* **1991**, *29*, 1977.
28. de Marignan, G.; Teyssie, D.; Boileau, S.; Malthete, J.; Noel, C. *Polymer* **1988**, *29*, 1318.
29. Yu, J. M.; Teyssie, D.; Boileau, S. *Polymer Bulletin* **1992**, *28*, 435.
30. Marciniec, B.; Gulinski, J.; Mirecki, J.; Nowicka, T. Polish Patent PL 251369 (841229) June (1988).
31. (a) Kajiyama, M.; Kakimoto, M.; Imai, Y. *Macromolecules* **1990**, *23*, 1244. (b) The authors, unpublished results.
32. Cooper, S. L.; Tobolsky, A. V. *J. Appl. Polym. Sci.* **1966**, *10*, 1837.
33. Wu, S. *Polymer Interface and Adhesion;* Marcel Dekker: New York, 1982; p 184.

RECEIVED June 27, 1994

Chapter 8

Silicon-Containing Resist Materials Based on Chemical Amplification

J. C. van de Grampel[1], R. Puyenbroek[1], B. A. C. Rousseeuw[2], and E. W. J. M. van der Drift[2]

[1]Department of Polymer Chemistry, University of Groningen, Nijenborgh 4, 9747 AG, Groningen, Netherlands
[2]Delft Institute of Microelectronics and Submicron Technology, Delft University of Technology, P.O. Box 5053, 2600 GB Delft, Netherlands

Two classes of siloxane polymers applicable as resist materials are being described. In the first series of materials acid-sensitive substituents as t-butoxycarbonyloxy, or t-butoxy are linked to a polysiloxane backbone. Preparation of these polymers occurs via hydrosilylation of polymethylhydrosiloxane with suitable olefins. The second class concerns graftcopolymers and/or tercopolymers in which acid-sensitive groups are attached to a carbon backbone in combination with grafted polysiloxane chains. These polymers are synthesized by free radical polymerization of olefins and silicon macromers. In microlithographic experiments two-layer resist systems have been irradiated with e-beam or DUV, the top layer being a silicon-containing polymer in combination with a catalytic amount of a photoacid generator. In general a high sensitivity was observed combined with a high oxygen RIE resistance.

The increasing demand for a smaller feature size in integrated-circuit devices has pushed the fabrication technology to dimensions and tolerances far into the nanometer-scale region. The key to success is the realization of details with maximal aspect ratio, both in pattern definition (1-2) and in pattern transfer (3). In pattern transfer anisotropy is obtained by plasma processing, by tuning properly the interplay of radicals and ions impinging on the substrate. However, a serious drawback of plasma processing is the limited selectivity with respect to the mask. Mask erosion inevitably leads to loss of dimension control but can be kept to a minimum when resist profiles are anisotropic as well. When feature sizes in resists come down to far into the submicron region or even to sub 100 nm details, a variety of problems are encountered, like resist swelling after development, proximity exposure effects, and substrate reflectivity. Moreover wafer topography can exert a negative influence on the resolution. Therefore new strategies in pattern definition technology have been developed in which the imaging and actual mask function are not longer combined in one resist layer. Two major approaches are being applied:

a) multilayer resist mask in which the image is defined in the top layer and the pattern is transferred anisotropically to the actual mask layer underneath.

b) single layer resist mask in which the pattern is defined by surface imaging followed by gas phase silylation of the exposed area and subsequently anisotropic pattern transfer to the bottom mask layer region (4).

In both methods anisotropic pattern transfer is achieved by plasma processing with a low pressure oxygen plasma process as the ultimate and decisive step. In this chapter attention will be focussed on the multilayer mask approach, in particular the development of new polymer materials for the imaging top layer. Key features in multilayer pattern definition processes are high lithographic resolution, high sensitivity and adequate dry etch resistance in oxygen plasma processing.

In a historical perspective multilayer mask technology started with a three-layer method. In this approach a conventional high resolution resist is used as the imaging top layer and irradiated by e-beam or X-rays. Depending on the chemical reactions upon irradiation and the developer used afterwards, the resist layer can exhibit a negative or positive tone. In the former case the irradiated area is less soluble than the neat material. A positive resist shows an enhanced solubility after exposure to radiation. The pattern obtained after irradiation and development is transferred to a very thin (about 10 nm) layer underneath the top layer by RIE, preferably based on fluorine chemistry (Figure 1). The composition of this intermediate layer is selected for superior dry etching resistance in oxygen plasma

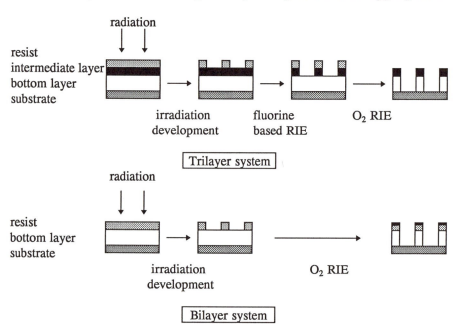

Figure 1. Schematic representation of lithographic techniques.

processing. In the next step the pattern is transferred to the bottom resist layer by etching very anisotropically in a low pressure oxygen plasma. Later developments led to the more efficient double layer technique (Figure 1). In this approach the resist material of the imaging layer is engineered such that it contains already the dry etching resistance characteristic in a precursor state. Silicon-containing resists are excellent candidates in this respect. After exposure and development the first few seconds of oxygen plasma processing leads to formation of a SiO_2 network which acts as a mask for the anisotropic etching of the resist bottom layer in the same oxygen plasma step.

Organosilicon polymers attract increasing attention as imaging layers in multilevel resist processes (5). Their solubility in common organic solvents ensures a facile spin coating, providing regular and planar surfaces. As stated before the presence of silicon offers an implicit resistance to oxygen RIE by the formation of silicon dioxide. To improve the sensitivity organic resists have been developed by Ito et al. (6-10) based on the concept of chemical amplification. In these systems the irradiation impact generates a cascade of reactions instead of a single event. The most common approach is the acid-catalyzed thermolysis of organic side groups (t-butoxy, t-butoxycarbonyloxy) or acid-catalyzed ring opening (epoxy).

In our studies siloxane-based polymers with acid-labile groups have been synthesized, thus offering the possibility for a high resolution combined with a high sensitivity and high dry etch resistance.

Polysiloxanes with acid-sensitive groups.

Synthesis of these polymers was carried out by means of hydrosilylation of poly(methylhydrosiloxane) with 35 monomer units (PMHS35) or 80 units (PMHS80). P-t-butoxycarbonyloxystyrene (TBCS), p-t-butoxystyrene (TBS) and t-butylmethacrylate (TMBA) were used as acid-sensitive components. On account of results obtained earlier for hydrosilylation platinum(divinyltetramethyl-disiloxane) (Karstedt compound) was used as a catalyst (11). Reactions were monitored by ^{1}H NMR and IR and were stopped when SiH groups were no longer detected.

Using both ^{1}H and ^{29}Si NMR it could be shown that TBCS and TBS gave α and β addition in about equal amounts (Figure 2) whereas TMBA only coupled via the β mode. Obviously steric considerations underlie the reaction pattern.

Table I. Glass Transition Temperatures (°C) of Some Polysiloxanes

	PMS35	PMS80
TBCS	37	32
TBS	4	1
TBMA	-50	-51

As may be seen from Table I the glass transition temperatures of the polysiloxanes are relatively low and almost independent of the chain length. Low T_g values are characteristic of siloxanes and only large and bulky groups can raise the T_g, providing the necessary stiffness. In principle all compounds can be used as negative or positive resists depending on the developer used (7). However the low glass transition temperatures of the TBS and TBMA addition products render them not suitable for microlithography.

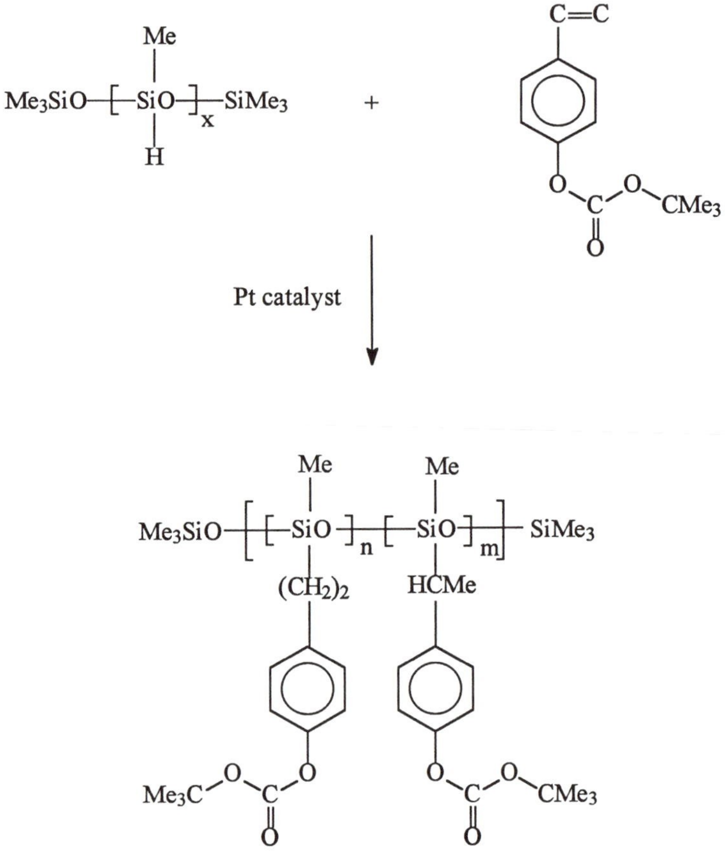

Figure 2. Hydrosilylation of PMHS with TBCS.

Graft copolymers with acid-sensitive groups

In analogy with well-known literature procedures (12-14) the syntheses of these polymers were carried out in two steps. First, methacrylate functional macromers were prepared by anionic ring opening polymerization of hexamethylcyclotrisiloxane by means of t-butyllithium followed by a termination reaction with 3-(methacryloxypropyl)dimethylchlorosilane. The molecular weight of the macromer can be easily controlled by the ratio cyclosiloxane versus butyllithium. Macromer functionality, as defined in ref. 15, appeared to be one in all cases. Table II gives two examples of macromers with about 30 and 220 repeating units, respectively. Figure 3 shows the general structure of macromers.

Table II. Characterization of a low and a high molecular weight macromer

Macromer	$\overline{M}_n$(calc)	$\overline{M}_n$(VPO) [a]	$\overline{M}_n$(UV) [b]	f [c]
m1	2.5×10^3	2.8×10^3	2.8×10^3	1.0
m2	17.0×10^3	15.4×10^3	14.0×10^3	1.1

[a] From vapor pressure osmometry; [b] From UV measurements;
[c] f (macromer functionality) = $\overline{M}_n$(VPO) / $\overline{M}_n$(UV).

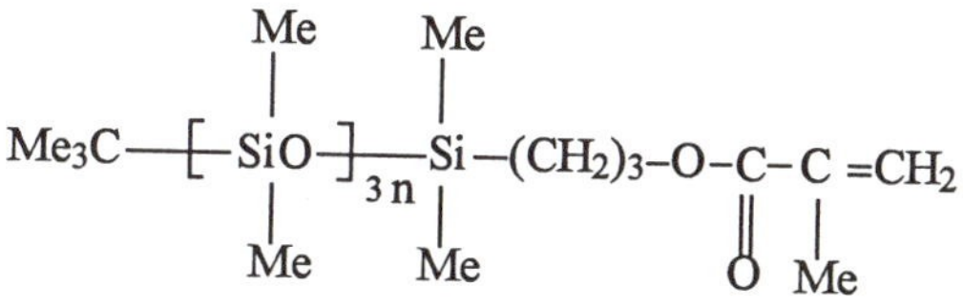

Figure 3. Structure of a macromer with a methacrylate end group.

The second step involved free radical copolymerizations of the macromers with TBCS, TBMA or glycidylmethacrylate (GMA) (16-17). The value of the functionality of the macromer points to a possible participation of each macromer chain in this reaction. A large variety of graft copolymers can be obtained for each of the monomers mentioned by changing the amount and molecular weight of the macromer used in the polymerization (Table III). In general the number of branches per polymer chain increases with decreasing molecular weight of the macromer, and increases with increasing feed ratio of macromer. The combination of low feed ratio combined with a high molecular weight of the macromer can even result in a number of branches lower than one which points to a mixture of graft copolymer and homopolymer (17). As can be seen from Table III the upper glass transition temperature of polymers having the same macromer content increases with increasing chain length of the macromer, pointing to increasing phase separation. Similar results were reported for poly(methylmethacrylate)-g-polydimethylsiloxane (14). Microphase mixing has been observed by TEM techniques for the copolymers derived from macromer m1 both in the cps and cpm

Table III. Synthesis and characterization of the graft copolymers

Graft co-polymer [a]	Yield in %	wt% of in feed	macromer found	$\overline{M}_n$ x 10^{-4}	D	Upper T_g in $^{\circ}$C	No.of bran-ches
cps A	60	33 (m2)	36	6.9	2.3	121	1.6
cps B	62	18 (m2)	19	7.0	2.1	121	0.9
cps C	67	33 (m1)	36	8.0	2.2	106	10.7
cps D	70	18 (m1)	22	7.1	2.1	113	5.5
cpm A	95	15 (m2)	16	4.9	2.5	114	0.5
cpm B	96	7 (m2)	7	4.9	2.3	112	0.2
cpm C	94	15 (m1)	15	4.7	2.2	106	2.5
cpm D	85	7 (m1)	7	5.3	2.3	110	1.2
cpa A	70	23 (m2)	19	9.6	2.3	78	1.1

[a] In the cps series the comonomer is TBCS; for the cpm series TBMA; for cpa GMA acts as comonomer.

[b] Molecular weight by membrane osmometry.

series (17), which ensures a homogeneous distribution of SiO$_2$ in the top layer Compared with polysiloxanes the silicon content of the graftcopolymers is lower, which may lead to a lower oxygen RIE resistance. However proper choice of macromer molecular weight and feed ratio offers the possibility to prepare copolymers with a high silicon content, combined with a homogeneous distribution of silicon grafts. Figure 4 shows the structure of a graft copolymer derived from TBCS.

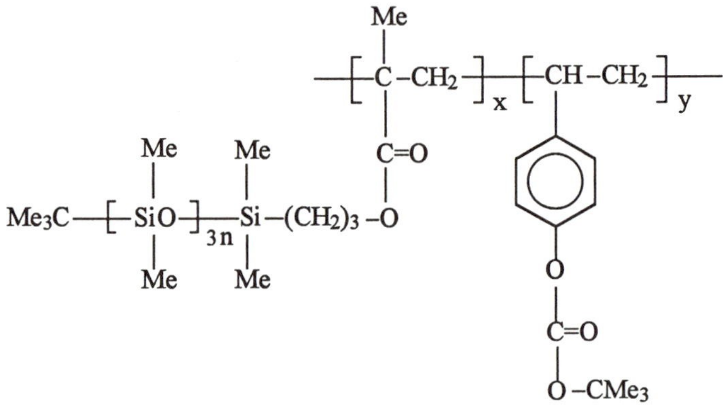

Figure 4. Graft copolymer formed by copolymerization of TBCS with a macromer.

Terpolymers

Using both GMA and bromostyrene (or chlorostyrene) in the procedure described above terpolymers were obtained in high yields (up to 90%). Incorporation of halogenated styrene units has two advantages. First, compared with the system with only GMA the glass transition temperature is raised by about 20 $^{\circ}$C by stiffening of the polymer chains. Secondly, the presence of halogen imparts better lithographic properties to the system (18). A novel compound was obtained by replacing bromostyrene by a chlorocyclophosphazene. Structures of two terpolymers are shown in Figure 5.

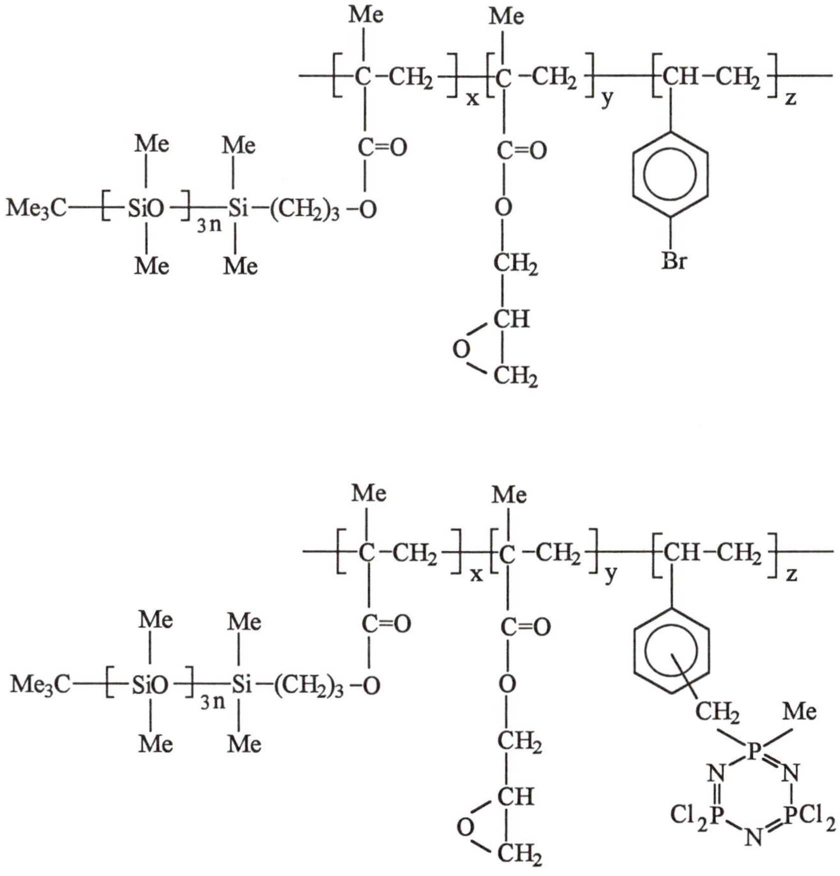

Figure 5. Two examples of terpolymers.

Microlithography

The resist systems currently under investigation consist of a bottom layer for the mask function with on top a thin layer of the silicon containing polymer for the imaging function. Triphenylsulfoniumhexafluoroantimonate was applied as the photoactive acid generator in the amplification process upon irradiation by e-beam or DUV. For the styrene derivatives PMS/TBCS or PMS/TBS the subsequent thermolysis leads to the formation of phenolic side groups which was confirmed by IR spectroscopy. For instance, after irradiation of PMS/TBCS and postbaking for 3 min at 90 °C the characteristic carbonyl band at 1750 cm^{-1} disappeared, accompanied by the appearance of the hydroxyl band at 3400 cm^{-1} (Figure 6).

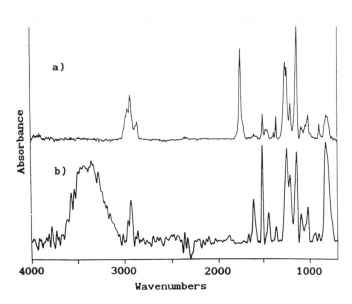

Figure 6. IR spectra of a 0.3 μm thick PMS/TBCS film with 5 wt% of triphenylsulfoniumhexafluoroantimonate. a) unexposed. b) exposed and postbaked.

All compounds described exhibit an oxygen RIE resistance ratio of about 10 compared to AZ-type resists. With respect to their sensitivity PMS/TBCS (in the negative mode) and the phosphazenic terpolymer appeared to be promising resist systems. For e-beam irradiation (50 keV) the dose for dissolution of 50% of the original film ($D^{0.5}$) equals to 10 μCcm^{-2} (mean value). Figure 7 gives a SEM photograph of negative images obtained from a phosphazene containing terpolymer (Figure 5) after irradiation, development and oxygen RIE.

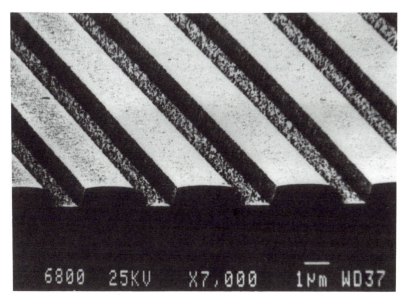

Figure 7. Bilayer imaging of a phosphazene containing terpolymer exposed to e-beam radiation of 20 μCcm^{-2} at 50 keV. Resist layer (300 nm) of terpolymer over bottom layer (750 nm) of a hard-baked AZ-type resist. O_2 RIE image transfer.

Experimental

Hydrosilylation. In a typical experiment a solution of PMHS35 (0.51 g, 0.23 mmol) and TBCS (3.6 g, 16.8 mmol) in 10 ml of toluene and 0.09 ml of Karstedt catalyst was stirred at 32 °C. After 24h the polymer was precipitated into methanol. The white powder obtained after decantation was washed three times with methanol. Subsequently the powder was dissolved in toluene (10 ml) and stirred for 3h with a small amount of activated charcoal. After filtration the precipitation procedure was repeated. The resulting white polymer was dried in vacuum at room temperature. Yield was 0.95 g (0.12 mmol, 53%). The same procedure was carried out for the reaction of PMHS80 and TBCS. The resulting polymer was obtained in a yield of 55%. Hydrosilylation reactions with TBS and TBMA were carried out in a similar way. In all cases the polymeric products were obtained as viscous oils in yields varying from 30% to 50 %.

Synthesis of graft copolymers with TBCS. Free radical copolymerizations of the macromers with TBCS were carried out in toluene during 3 days at 60 °C, using a feed of 33wt% or 18wt% of the macromer (with respect to TBCS) and AIBN as initiator. After addition of a small quantity of CH_2Cl_2 the copolymers were precipitated in 2-propanol and thoroughly washed and subsequently dried in vacuo.

Synthesis of graft copolymers with TBMA. The free radical polymerizations were carried out as described for the reaction with TBCS, however, using a feed of 15wt% or 7wt% of the macromers (with respect to TMBA). A reaction period of 4 days was used. Dialysis techniques were applied to remove any unreacted material. Standard procedures were used for purification.

Synthesis of a graft copolymer with GMA. The polymerization was carried out during 3h at 64 °C with a feed ratio of 23 wt% of macromer m2 in a 5:3 mixture of butanone and benzene, using benzoylperoxide (BPO) as radical initiator. After addition of CH_2Cl_2 stirring was continued for 15 min. Standard purification procedures were applied

Synthesis of a phosphazene containing resist. In a typical experiment 0.45 g of $(NPCl_2)_2 NPMe(CH_2C_6H_4CH=CH_2)$, 0.44 g of GMA, 0.45 g of m2 and 28 mg of BPO were dissolved in 3.0 ml of butanone and 2.0 ml of benzene. The polymerization reaction was carried out at 65 °C for 48 hours and terminated with 0.5 ml of CH_2Cl_2. Standard purification methods resulted in 0.96 g (72%) of a white, solid terpolymer. ($\overline{M}_n$ = 70 x 10^3, D=2.1; Tg = 90 °C. The amount of silicon in the terpolymer was 11% as calculated by ^1H NMR and elemental analysis.

Pattern fabrication in phosphazene containing resists. A solution of 9.5 wt% of the terpolymer in PGMEA with 3 wt% of $(Ph)_3S^+SbF_6^-$ was spun on a silicon wafer, casted with a 750 nm thick hard-baked AZ-resist layer, and prebaked at 100 °C for 3 min on a hot plate to form a 300 nm thick layer. Electron beam exposure was carried out using a vectorscan shaped beam writing machine, EBPG/03 (Philips) operating at 50 keV. After exposure the resist layers were postbaked at 90 °C for 3 min. The patterns were developed with PGMEA and subsequent etched in an oxygen plasma. Oxygen RIE was carried out using a low pressure oxygen plasma in a parallel plate plasma reactor (Leybold Heraeus, Z-401) operating at 13.56 MHz. in the RIE-mode. (pressure 0.3 Pa, rf power 20 Watt, oxygen flow 20 sccm).

Acknowledgments

The authors are much indebted to the Technology Foundation (STW) for funding this project. GE Plastics is acknowledged for providing the acid-generators.

Literature cited

1. Reichmanis, E.; Thomson, L.F. *Microelectronic Engineering* **1991**, 13, 3.
2. Wittekoek, S. *Microelectronic Engineering* **1994**, 23, 43.
3. Gottscho, R.A.; Jurgensen, C.W. *J. Vac. Sci. Technol.* **1992**, B10, 2133.
4. Roland, B, *Microelectronic Eng.* **1991**, 13, 11.
5. Reichmanis, E.; Novembre, F.A.E.; Tarascon, R.G.; Shugard, A.; Thompson, L.F. In *Silicon-Based Polymer Science*; Zeigler, J.M.; Gordon Fearon, F.M., Eds.; Advances in Chemistry. Ser. No. 224, ACS, Washington, DC, 1990, 265.
6. Ito, H.; Willson C.G; Frechet, J.M.J.; Farrall, M.J.; Eichler, E. *Macromolecules* **1983**, 16, 510.
7. Ito, H.; Willson C.G. In *Polymers in Electronics*. Davidson, T. Ed. ACS Symp. Ser. No. 242, ACS, Washington, DC, 1984, 11.
8. Willson, C.G.; Ito, H.; Frechet, J.M.J.; Tessier, T.G.; Houlihan, F.M. *J. Electrochem. Soc.* **1986**, 133, 181.
9. Houlihan, F.M.; Reichmanis, E.; Thompson, L.F.; Tarascon, R.G. In *Polymers in Microlithography*; Reichmanis, E.; MacDonald, S.A.; Iwayanagi, T., Eds., ACS Symp. Ser. No. 412, ACS Washington, DC, 1989, 39.
10. Reichmanis, E.; Houlihan, F.M.; Nalamasu, O.; Neenan, T.X. *Chem. Mater.* **1991**, 3, 394.
11. Puyenbroek, R.; Jekel, A.P.; van de Grampel, J.C. *J. Inorg. Organometal. Pol.* **1991**, 1, 105.
12. Smith, S.D.; McGrath, J.E. *Polym. Prepr. (Am. Chem. Soc., Div. Polym. Chem.)* **1986**, 27(2), 31.
13. Smith, S.D.; York, G.; Dwight, D.W.; McGrath, J.E.; Stejskal, J.; Kratochvil, P. *Polym. Prepr. (Am. Chem. Soc., Div. Polym. Chem.)* **1987**, 28(1), 458.
14. Smith, S.D.; DeSimone, J.M.; Huang, H.; York, G.; Dwight, D.W.; Wilkes, G.L.; McGrath, J.E. *Macromolecules* **1992**, 25, 2575.
15. Kawakami, Y.; Murthy, R.A.N.; Yamashita, Y. *Makromol. Chem.* **1984**, 185, 9.
16. Puyenbroek, R.; Werkman, P.; Jansema, J.J.; van de Grampel, J.C.; Rousseeuw, B.A.C.; van der Drift, E.W.J.M. *Polymer Prepr. (Am. Chem. Soc., Div. Polym. Chem.)* **1993**, 34(1), 238.
17. Puyenbroek, R.; Werkman, P.; Jansema, J.J.; Van de Grampel, J.C.; Rousseeuw, B.A.C.; van der Drift, E.W.J.M. *J. Inorg. Organometal. Pol.*, **1994**, in press.
18. Thompson, L.F.; Feith, E.D.; Heidenreich, R.D. *Polymer. Eng. Sci.* **1974**, 14, 529.

RECEIVED May 9, 1994

Chapter 9

Some Silicon Polymers of C_{60} (Buckminsterfullerene)

Robert West, Kunio Oka, Hideaki Takahashi, Michael Miller, and Takahiro Gunji

Department of Chemistry, University of Wisconsin, Madison, WI 53706

C_{60} undergoes hydrosilylation in the presence of platinum or cobalt catalysts. With $HSiMe(OEt)_2$, hydrosilylation is accompanied by some condensation of the silicon moiety, producing $Hn-C_{60}-[(SiMe(OEt)OR]_{n+m}$ where R=Et or SiMe(OEt)H. Hydrolysis of the product with $Si(OEt)_4$ gave a yellow-orange sol-gel glass. With $H(SiMe_2O)_3SiMe_2H$, hydrosilylation of C_{60} produced a black oil, with average composition $H_{12}C_{60}[(SiMe_2O)_3SiMe_2H]_{12}$. Hydrosilylation of C_{60} with a linear siloxane polymer containing Si-H bonds along the chain led to multiple substitution onto the C_{60} framework, with the siloxane effectively "wrapping up" the fullerene.

The chemistry of C_{60} and other fullerenes has developed rapidly since the first preparative-scale isolation of this molecule in 1990 (*1*). The new field of fullerene chemistry is leading to many novel derivatives some of which may have extraordinary optical and electronic properties. The many C_{60} derivatives reported so far include transition metal complexes, (*2-4*), complexes with aromatic rings (*4*), Diels-Alder products (*5*), addition compounds with carbenes (*6*), silylene (*7*) or oxygen (*8*) and other organic derivatives (*9*). Several polymers of C_{60} have also been reported (*10-15*).

The discovery of platinum complexes of C_{60} by Fagan and co-workers (*3*) raised the possibility that platinum or other transition metals might catalyze reactions of fullerenes. Here we report several transition metal catalyzed hydrosilylation reactions of C_{60} leading to polymeric structures of various kinds.

C_{60} with $HSiMe(OEt)_2$: A Sol-Gel Precursor

The sol-gel process is an important method for the preparation of glassy silicate materials, as bulk glass bodies or thin films (*16*). It would be of interest to make sol-

0097–6156/94/0572–0092$08.00/0

gel glasses containing fullerenes, but the low solubility of C$_{60}$ or C$_{70}$ in both organic and polar solvents provides a formidable obstacle. An attempt to prepare a silica sol-gel glass containing C$_{70}$ has been reported, leading to a silica gel in which the C$_{70}$ formed aggregates (*17*). Also, diffusion of C$_{60}$ into sol-gel glassed has recently been studied (*18*).

To make fullerene sol-gel glasses, it is therefore necessary to modify the structure of a fullerene in order to increase its compatability with solvents. The most suitable molecules for sol-gel formation are those containing alkoxysilyl groups (*16*). We therefore sought to introduce such groups onto C$_{60}$ by hydrosilylation, using methyldiethoxysilane, MeSiH(OEt)$_2$, as the silylating agent.

The reaction of C$_{60}$, with MeSiH(OEt)$_2$ was carried out in benzene with dicobalt octocarbonyl as catalyst. The methyldiethoxysilylated C$_{60}$, **1**, was isolated as a brown pasty solid soluble in benzene, hexane, diethyl ether, and chloroform. A GPC trace of the product (Figure 1) showed, besides a small peak due to unreacted C$_{60}$, five peaks with retention times between 5 and 14 minutes, indicating that **1** is a mixture of compounds with varying amounts of substitution. The silylated mixture **1** was mildly moisture sensitive; exposure to moist air led to gradual transformation into an insoluble solid.

The ^{1}H NMR spectrum of the product, shown in Figure 2, indicated the expected Si-CH$_3$, and Si-OC$_2$H$_5$ protons. The presence of Si-H can be explained by partial condensation of methyldiethoxysilane groups to give -SiMe(OEt)-O-SiMeH(OEt) moieties. The result from integration of the ^{1}H NMR spectrum in the presence of a standard was consistent with the average formula H$_n$C$_{60}$[SiMe(OEt)]$_{5.0}$[SiMe(OEt)OSiHMe(OEt)]$_{4.5}$. (The fullerene C-H hydrogens were obscured by the ethoxy protons and so could not be determined).

The UV-visible spectrum of **1** (Figure 3) showed an absorption maximum near 220 nm, tailing out to 550 nm. This featureless spectrum is characteristic for multiply-substituted C$_{60}$ derivatives and is quite different from that of C$_{60}$. The ^{13}C NMR spectrum of **1** showed signals due to the silylmethyl group (-0.6 ppm) and the ethoxyl carbons (18.4, 58.3 and 59.2 ppm). However, the signals due to the C$_{60}$ core were not observed clearly; The long relaxation time for the C$_{60}$ carbons makes them difficult to detect in ^{13}C NMR.

To obtain a C$_{60}$-silica hybrid gel, **1** (5%) was mixed with tetraethoxysilane (95%) and the mixture was hydrolyzed with aqueous HCl in THF. A solution of the product in THF was evaporated on a glass slide, which was allowed to stand at 25° for 3d and then heated to 100 °C for 4h, giving a yellow C$_{60}$-silica gel hybrid film with a thickness of about 50 μm. No aggregation was observed by SEM. The electronic spectrum was similar to that of **1**, indicating that the C$_{60}$ derivative was incorporated into the silica gel glass (Figure 4). By a similar process, monolithic orange-brown hybrid gels were obtained.

C$_{60}$ with a Hydrogen-Terminated Siloxane

This experiment was designed initially to make a polymer with fullerene units as part of the polymer chain. C$_{60}$ was mixed with the linear siloxane oligomer, H(SiMe$_2$O)$_3$SiMe$_2$H, in benzene and hydrosilylation was effected with a platinum catalyst. Because of the low solubility of C$_{60}$, it proved necessary to have the siloxane oligomer present in large excess. The reaction was followed by gel

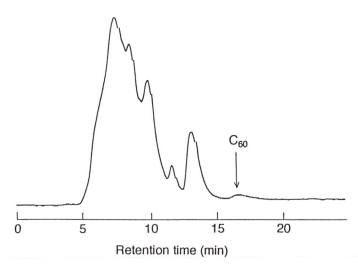

Figure 1. GPC trace for **1**.

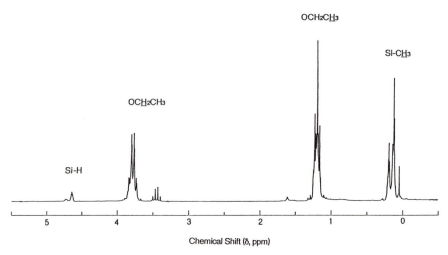

Figure 2. Proton NMR spectrum of **1**.

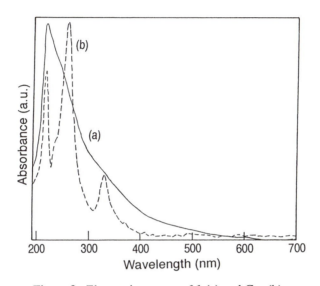

Figure 3. Electronic spectra of **1** (a) and C_{60} (b).

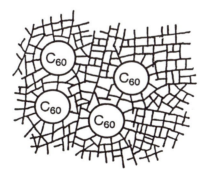

Figure 4. Schematic Representation of the hybrid C_{60}-sol-gel glass from **1** and $Si(OEt)_4$.

permeation chromatography (Figure 5). After 336h only a single peak was evident, indicating that the product, **2**, had a narrow molecular weight range.

The product **2** was isolated as a soluble black oil and purified by chromatography on silica gel and was analyzed by IR and ^{1}H and ^{29}Si NMR spectroscopy. The ^{1}H NMR spectrum (Figure 6) showed resonances assigned to Si-H, C_{60}-H and $SiMe_2$ protons, in relative abundance 1:1:26. The presence of Si-H bonds was confirmed by observation of an Si-H band in the infrared spectrum at ~ 2110 cm^{-1}.

Proton NMR in the presence of a standard was used to determine the amount of substitution, which was measured as 12 siloxane chains per C_{60} unit. The evidence is all consistent with formulation of **2** as $H_{12}C_{60}(SiMe_2OSiMe_2OSiMe_2OSiMe_2H)_{12}$. Some molecules may be present with substitution greater than or less than 12 and in addition several stereoisomers may be present. A drawing giving a possible structure is shown in Figure 7.

The electronic spectrum of **2** is similar to that of **1**: A featureless band from 220 nm tailing out to about 550 nm. Since **2** has Si-H functionality, it can be crosslinked to form an insoluble polymer. Crosslinking could be induced by addition of alcoholic alkali to a solution of **2**, or simply by adding a platinum catalyst to **2** in rather concentrated solution. Compound **2** is surface-active like other dimethylsiloxanes and has been used to make stable Langmuir-Blodgett films containing C_{60} moieties (*19*).

C_{60} with a Hydridosiloxane: Shrink-wrapped Fullerene

We also carried out the reaction of C_{60} with a siloxane oligomer containing Si-H units on the main polymer chain, having the average composition $Me_3SiO(SiMeHO)_3$ [SiMe(n-octyl)O]$_8SiMe_3$.

This oligomer was used for the hydrosilylation of C_{60} containing about 10% C_{70}. Reaction took place rapidly at 25° in the presence of a platinum catalyst. The product, **3**, was a gummy elastomer. A GPC trace of **3** showed a single, sharp peak (Figure 8), consistent with a narrow molecular weight distribution.

The Proton NMR spectrum of **3** is shown in Figure 9, along with that of the starting oligomer. The Si-H resonance at 5.19 ppm in the oligomer disappears upon hydrosilylation and is replaced by a C_{60}-H resonance at 4.0 ppm. The EI mass spectrum of **3** (Figure 10) shows a large peak centered near 3650, accompanied by a smaller peak near 3770 amu which we assign to the C_{70} polymer. All of the analytical data for **3** are consistent with the reaction of C_{60} with exactly two molecules of the oligomer. Since the Si-H hydrogens are lost, multiple attachment to the C_{60} core must take place. We propose that two molecules of the oligomer "wrap up" the C $_{60}$ (or C_{70}) units in a siloxane coat (Figure 11). The two oligomer molecules may be just sufficient to use up the easily available sites on the C_{60} core.

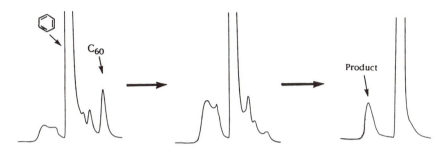

Figure 5. GPC trace for reaction of C$_{60}$ with H(SiMe$_2$O)$_3$SiMe$_2$H, after 48h (left); 168h (center); 336h (right).

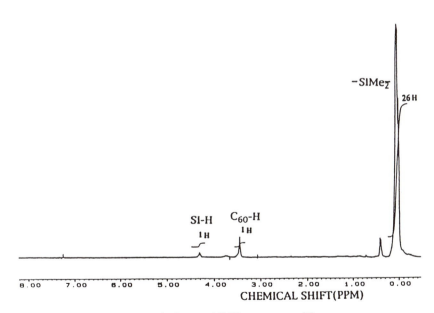

Figure 6. Proton NMR spectrum of **2**.

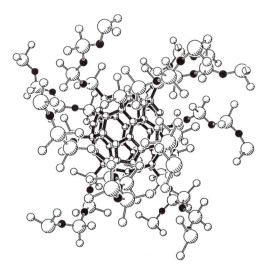

Figure 7. Schematic representation of a molecule of **2**.

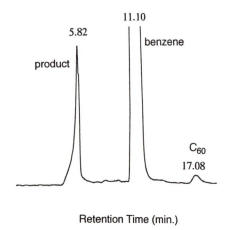

Figure 8. GPC trace for **3**, after 1 hour showing C_{60} still remaining.

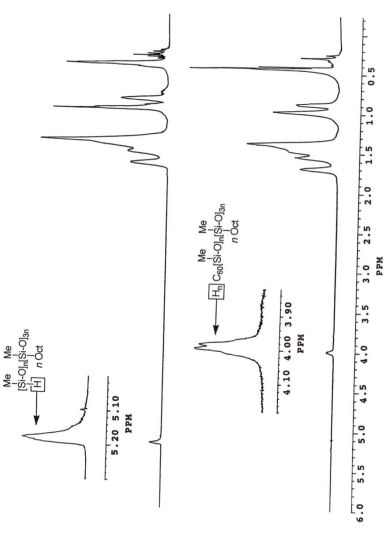

Figure 9. Proton NMR spectrum of siloxane oligomer (above) and of 3 (below).

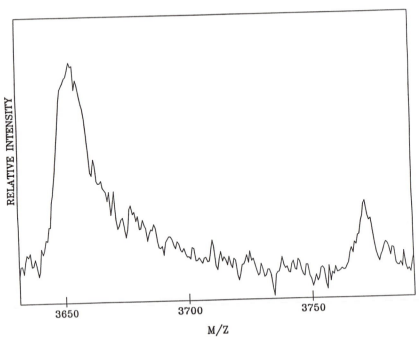

Figure 10.Mass spectrum for **3**.

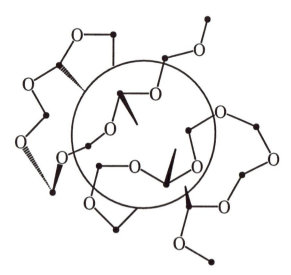

Figure 11. Schematic view of structure of **3**.

Literature Cited

1. Krätschmer, W., Lamb, L. D., Fostiropoulos, K.; Huffman, P.R. *Nature.* **1990**, *347*, 354.

2. Hawkins, J. M., Lewis, T. A., Loren, S. D., Meyer, A., Heath, J. R., Shibato, Y.; Saykally, R. I. *J. Org. Chem.* **1992**, *55*, 6250.

3. Fagan, P. J., Calabrese, J. C.; Malone, B. *J. Am. Chem. Soc.* **1991**, *113*, 9408.

4. Bach, A. L., Catalano, V. J., Lee, J. W.; Olmstead, M. M. *J. Am. Chem. Soc.* **1992**, *114*, 5455.

5. Hoke II, S. H., Molstad, J., Dilettato, D., Jay, M. J., Carlson, D.; Cooks, B. K. and R. G. *J. Org. Chem.* **1992**, *47*, 5069.

6. Suzuki, T., Li, Q., Khemani, K. C.; Wudl, F. *J. Am. Chem. Soc.* **1992**, *114*, 7300.

7. Akasaka, T., Ando, W., Kobayashi, K.; Nagase, S. *J. Am. Chem. Soc.* **1993**, *115*, 1605.

8. Elems, Y., Silverman, S. K., Sheu, C., Kao, M., Foote, C. S., Alvarez, M. M.; Whetten, R. L. *Angew. Chem. Int. Ed. Engl.* **1992**, *31*, 351.

9. Samulski, E. T., Desimone, J. M., Hunt, M. O., Menceloglu, Y. Z., Jarnagin, R. C., York, G. A., Lata, K. B.; Wang, H. *Chemistry of Materials.* **1992**, *4*, 1153.

10. Morton, J. R., Preston, K. F., Krusic, P. J., Hill, S. A.; Wasserman, E. *J. Am. Chem. Soc.* **1992**, *114*, 5454.

11. Stry, J. J., Coolbaugh, M. T., Turos, E.; Garvey, J. F. *J. Am. Chem. Soc.* **1992**, *114*, 7614.

12. Chiang, L. Y., Upasani, R. B.; Swirczewski, J. W. *J. Am. Chem. Soc.* **1992**, *114*, 10154.

13. Dennis, J. P., Hare, H. Kroto, W.; Walton, D. R. M. *J. Chem. Soc. Chem. Commun.*, **1992**, *665*.

14. Chen, K., Caldwell, W. B.; Mirkin, C. A. *J. Am. Chem. Soc.* **1993**, *115*, 1193.

15. Prato, M., Suzuki, T., Foroadian, H., Li, Q., Khemani, K.; Wudl, F. *J. Am. Chem. Soc.* **1993**, *115*, 1594.

16. *Sol-Gel Technology for Thin Films, Fibers, Preforms, Electronics, and Specialty Shapes;* Klein, L. C.; Park Ridge, Noyes Publications, **1988**.

17. Dai, S., Compton, R. N., Young, J. P.; Mamantòv, G. *J. Am. Ceram. Soc.* **1992**, *75*, 2865.

18. Zerda, T. W.; Brodka, A.; Coffer, J. *J. Non-Crystalline Solids* **1994**, *168*, 33.

19. Lee, W., Ph.D. Thesis, University of Wisconsin-Madison, 1994.

RECEIVED June 10, 1994

OXOPOLYMERS

Chapter 10

Structure–Property Relationships in Sol–Gel-Derived Thin Films

C. J. Brinker[1,2], N. K. Raman[2], M. N. Logan[1], R. Sehgal[2], R. A. Assink[3], D.-W. Hua[2], and T. L. Ward[2]

[1]Ceramic Synthesis and Inorganic Chemistry Department 1846, Sandia National Laboratories, Albuquerque, NM 87185
[2]University of New Mexico–National Science Foundation Center for Micro-Engineered Ceramics, University of New Mexico, Albuquerque, NM 87131
[3]Properties of Organic Materials Department 1812, Sandia National Laboratories, Albuquerque, NM 87185

The molecular-scale species distributions and intermediate-scale structure of silicate sols influence the microstructures of the corresponding thin films prepared by dip-coating. Using multi-step hydrolysis procedures, we find that, depending on the sequence and timing of the successive steps, the species distributions (determined by ^{29}Si NMR) and intermediate scale structure (determined by SAXS) can change remarkably for sols prepared with the same nominal composition. During film formation, these kinetic effects cause differences in the efficiency of packing of the silicate species, leading to thin film structures with different porosities.

The ability to control the microstructural features of the deposited film (e.g. pore volume, pore size, and surface area) is an outstanding advantage of sol-gel processing over conventional thin film processing techniques. Our previous research (*1*) has documented that the thin film microstructure is established by evaporation-induced aggregation of sol species within the thinning film to form a physical or chemical gel and the subsequent collapse of the gel due to capillary forces exerted during the final stage of drying (see Figure 1). The porosity of the film therefore depends on the porosity of the primary species - either polymers or particles - and that created or lost by the subsequent aggregation and drying processes associated with film deposition. In this paper we explore relationships between the size and structure of primary sol species as revealed by NMR and SAXS and the properties of the corresponding films as characterized by ellipsometry, TEM, and transport measurements. We limit our discussion to the deposition of silica sols where microstructural development can be qualitatively understood on the basis of fractal aggregation.

Aggregation of Fractals. Under many conditions of silicate sol synthesis, the primary species are characterized by a mass fractal dimension D (*2*) that relates the polymer mass M to its radius r_c as:

$$M \propto r_c{}^D. \tag{1}$$

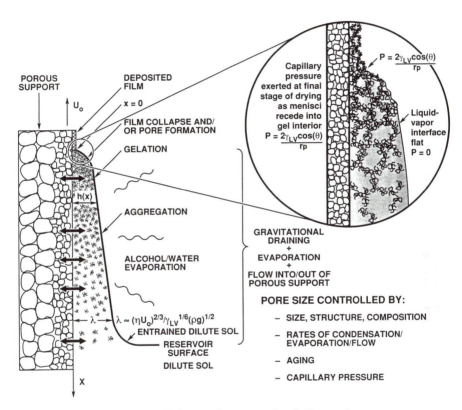

Figure 1. Schematic of the steady state sol-gel dip-coating process on a porous support, showing the sequential stages of structural development that result from draining accompanied by solvent evaporation and continued condensation reactions. Inset shows the collapse of the gel network during the final stage of drying.

The porosity Π of an object varies inversely with density, so in three dimensional space:

$$\Pi \propto r_c^3/r_c^D \text{ or } \Pi \propto r_c^{3-D}. \tag{2}$$

Thus, for mass fractals ($0 < D < 3$), the porosity of an individual object increases with its size, whereas for uniform objects ($D = 3$), porosity is invariant with size.

During film formation by dip-coating, primary sol species are rapidly aggregated by evaporation of solvent. The porosity of the secondary aggregate structure depends on the extent of interpenetration of the primary species. The extent of interpenetration is inversely related to the mean number of intersections $M_{1,2}$ of two polymers of radius r_c and mass fractal dimension D1 and D2 confined to the same region of three-dimensional space (2):

$$M_{1,2} \propto r_c^{D1+D2-3}. \tag{3}$$

From equation 3 we see that there are two classes of fractal polymeric sols. If the polymers exhibit fractal dimensions less than 1.5, the probability of polymer intersection decreases indefinitely with r_c. This situation should favor interpenetration of polymers within the thinning film during dip-coating. We refer to this class of sol species as *mutually transparent*, meaning that they move through one another freely. Alternatively, if the fractal dimension of each polymer exceeds 1.5, the probability of intersection increases algebraically with r_c. In this situation, polymer interpenetration is impeded (by points of intersection) as the polymers are concentrated during dip-coating. We refer to this class of sol species as *mutually opaque*.

Equation 3 assumes every point of intersection results in immediate and irreversible attachment, chemically equivalent to an infinite condensation rate. Finite condensation rates mitigate the criterion for mutual transparency, i.e. since every point of intersection does not result in immediate and irreversible "condensation", interpenetration may occur for structures characterized by $D > 1.5$. The value of equation 3 is that it provides a qualitative understanding of the effect of structure on the extent of polymer interpenetration and thus porosity: smaller polymer sizes and lower mass fractal dimensions favor denser films, and larger polymer sizes and greater mass fractal dimensions favor more porous films. In all cases increased condensation rates reduce the tendency toward interpenetration, favoring more porous films.

Multi-Step Hydrolysis Procedures. Two general multi-step procedures were used to prepare the precursor sols: a) a two-step acid-catalyzed hydrolysis of tetraethoxysilane (TEOS), using H_2O/Si molar ratio $r = 5$, or b) a two- or three-step acid/base-catalyzed hydrolysis of TEOS, using $r = 3.7$. Using multi-step hydrolysis procedures, it is possible to vary the sequence of the hydrolysis steps or the aging conditions between steps to obtain sols of the same nominal composition but with different species distributions and structures. Throughout this study we discovered examples of the role of such kinetic effects on the species distributions and structure of silicate sols. In many cases these effects are manifested in the properties of the corresponding films or membranes. We have confined our discussion to comparisons of sols (either acid- or acid/base-catalyzed) prepared with identical r values, pH and silica concentrations. The only differences are the sequence of the hydrolysis steps and the aging conditions. This comparison elucidates the role of kinetics in dictating the structures of silicate sols and the corresponding deposited films.

Experimental Procedure

Sol Preparation. A two-step acid-catalyzed process was used to prepare one series of silicate sols. In the first step tetraethoxysilane (TEOS), ethanol, H_2O and HCl were mixed in the molar ratio 1:3.8:1.1:0.0007, refluxed at 60°C for 90 min and cooled to room temperature (*3*). We refer to the silicate solution obtained after the first hydrolysis step as the *stock solution*. The stock solution was used immediately or stored in a freezer at 0°C. In the second hydrolysis step additional water and HCl were added at room temperature resulting in the final molar ratio of 1:3.8:5.1:0.056. The second hydrolysis step was performed on the stock solution after aging 0-7 days at 0°C or after aging 4-5 months at 0°C. These sols are identified as A2(fresh) and A2(aged), respectively. After stirring for 10 min, the A2 sols were allowed to age at 50°C for periods ranging from 2 to 35 hr and then diluted 2:1 with ethanol (vol EtOH:vol sol) prior to film deposition. Gelation times for the A2(fresh) and A2(aged) sols were determined at 50°C using undiluted A2 sols.

A two- or three-step acid/base-catalyzed process was used to prepare a second series of sols identified as B2 (*4*) or AAB (*5*), respectively. An aged (10 days to 8 months) stock solution was the starting point for both sols. For B2 the second hydrolysis step consisted of adding an aqueous solution of 0.05 M NH_4OH and diluting with ethanol, resulting in a final molar ratio of TEOS:EtOH:H_2O:HCl:NH_4OH of 1:48:3.7:0.0007:0.002, and a final sol pH of 5.5 as estimated using colorimetric pH indicator strips (EM Science). For AAB the second step consisted of adding 1M HCl diluted in ethanol, resulting in a H_2O/Si ratio $r = 2.5$, and refluxing at 60°C for 60 min. The third step consisted of adding an aqueous solution of 2M NH_4OH diluted in ethanol, resulting in a final molar ratio of 1:48:3.7:0.028:0.05. The AAB sols had the same final dilution, r value (3.7), and pH (5.5) as the B2 sols. The diluted sols were aged at room temperature or in a Class A (explosion-proof) oven at 50°C, and samples were removed at intervals up to the gel point or allowed to gel. The gelled samples were subjected to ultrasound to prepare sols of a consistency suitable for dip-coating.

^{29}Si NMR Investigations. ^{29}Si NMR was employed to determine the species distribution of the stock solution and A2, B2, and AAB sols after various periods of aging. ^{29}Si NMR spectra were recorded at 39.6 MHz, on a Chemagnetics console interfaced to a General Electric 1280 data station and pulse programmer. Chromium(III) acetylacetonate (Cr(acac)3, 15.7 mM) was added as a paramagnetic relaxation agent. Twenty millimeter diameter quartz sample tubes were used, which made negligible contributions to the baseline due to their very long spin relaxation times. Proton decoupling and a silicon free probe were employed during data acquisition (*6*).

SAXS Measurements. The effects of synthesis and aging conditions on the intermediate-scale sol structure (0.5 - 30 nm) were determined by small angle x-ray scattering (SAXS). SAXS data were collected on liquid samples, using a Rigaku-SAXS setup having a Kratky U-slit system. The incident beam was Cu-Kα radiation with a wavelength of 1.542Å. The data were corrected for slit collimation. The Guinier radius, R_G, and the mass fractal dimension, D, were calculated from the Guinier and Porod regions, respectively.

Film Formation. For all sols, films were deposited on polished single crystal <100> silicon substrates by dip-coating in a dry nitrogen atmosphere, using substrate withdrawal rates of 20 (A2 and AAB sols) or 25 cm/min (B2 sols). A2 films were dried at room temperature and annealed at 400°C for 10 min using a heating rate of 10°C/min. B2 and AAB films were dried at ~70°C for 20 min, using a heat lamp. Selected films were supported on edge in quartz trays and fired in air at 400°C for 10 min with an estimated 50°C/min ramp achieved by placing the quartz trays directly on ZrO_2 tiles maintained at the firing temperature. The A2 series of films were also prepared as supported membranes by sol deposition on commercial (U.S. Filter) Membralox alumina supports (7).

Film Characterization. Refractive indices and thicknesses of films deposited on silicon substrates were determined by ellipsometry using either a Rudolph AutoEL IV or Gaertner Model L116-C ellipsometer and assuming an absorption coefficient of 0 at $\lambda = 632.8$ nm. The thickness and morphology of films prepared as membranes were evaluated by cross-sectional TEM (7).

Transport Measurements. The permeabilities of the membranes were investigated using single gas (He, N_2, and C_3H_6) permeation measurements. A detailed description of the permeability equipment and technique used has been provided elsewhere (7).

Thermal Gravimetric Analysis. TGA experiments were performed on B2 and AAB xerogel powders dried at 50°C for 20.5 hours. Experiments were performed in flowing air (50 sccm) at a heating rate of 2°C/min to 550°C and 10°C/min to 1000°C.

Results

In this section we report the properties of the sols as determined by ^{29}Si NMR and SAXS and the properties of the corresponding films as deduced from ellipsometry and transport measurements. We use transport measurements in addition to ellipsometry to characterize the A2 series of films, because we observed little variation in the refractive index values of these samples but significant differences in permeance.

^{29}Si **NMR.** The Q-notation (8) was used to identify the silicic acid ester structures. The superscript on the Q-unit represents the number of siloxane bonds (-OSi) attached to the silicon under consideration. The subscript 4c identifies the silicon as being contained in a cyclotetrasiloxane ring composed of 4 SiO_4^{4-} tetrahedra. The assignments for the Q^n resonances were made by comparison with literature data (9, 10).

Stock Solution. Figures 2(a) and (b) show the ^{29}Si NMR spectra for the fresh and aged stock solutions along with the respective peak assignments. The most significant differences between the two spectra are the increased concentration of cyclic species, Q^2_{4c} (-95.5 ppm) and Q^3_{4c} (-103 ppm), and the decreased concentration of the hydrolyzed dimer species (-86.5 ppm) in the aged stock solution compared to the fresh stock solution. The aged stock solution also contains considerably more Q^3 species (-101 to -105 ppm) consistent with a greater overall extent of reaction. Both sols contain quite high concentrations of unhydrolyzed dimer (-89 ppm).

A2 Sols. The NMR spectra of the A2(fresh) and A2(aged) sols after 2 or 35 hours of aging at 50°C exhibit three broad peaks centered at -92, -101, and -110 ppm attributable to the highly varied environments of silicon contained in Q^2, Q^3, and Q^4 units, respectively. Figure 3 shows the distribution of the Q^n species as a function of

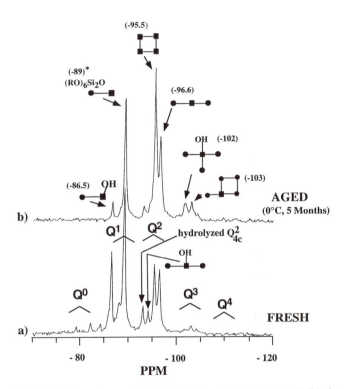

Figure 2. ^{29}Si NMR spectra of: (a) the fresh stock solution and (b) the aged stock solution. Peak assignments are from reference (9). The squares represent the Si atom under consideration and circles represent all other Si atoms in the molecule. Alkoxy groups are implied where not shown to make the Si 4-coordinate, and oxygen atoms are assumed between all Si atoms. *High resolution studies indicate that the -89 ppm peak is attributable to over 75% unhydrolyzed dimer.

aging time as determined by deconvolution of the corresponding ^{29}Si NMR spectra. Although the A2(fresh) sol is clearly less-highly condensed after 2 hours aging, the differences in overall extent of condensation largely vanish after 35 hours aging.

B2 and AAB Sols. The NMR spectra of the unaged and aged (39h at 50°C) B2 sols both show broad features due to Q^3 and Q^4 silicon species. However the spectra appear dominated by much sharper resonances associated mainly with monomers and weakly hydrolyzed Q^1 and Q^2 silicon species: in particular the unhydrolyzed monomer (-82.1 ppm), the unhydrolyzed dimer (-89.1 ppm), the unhydrolyzed cyclic tetramer Q^2_{4c} (-95.4 ppm), and an unhydrolyzed Q^2 species possibly associated with a trimer (-96.5 ppm). Figure 4 shows the evolution of the Q^n species distribution with aging time. We see that although the percent of Q^4 species increases at the expense of Q^2 and Q^3 species, consistent with further condensation, the percentage of Q^1 species remains virtually constant (30%), and the concentration of monomer increases from about 2 to 7%.

The ^{29}Si NMR spectrum of an AAB sol after the second hydrolysis step ($r = 2.5$, dilution = 2:1) exhibits two broad peaks attributable to Q^2 and Q^3 species. There is no evidence of monomer or end groups/dimers consistent with a more extensive state of hydrolysis and condensation promoted by the additional acid-catalyzed hydrolysis step. The spectrum of an AAB sol (diluted 5:1) three hours after the third (base-catalyzed) hydrolysis step shows Q^2, Q^3 and Q^4 species present in the proportions 11%, 46% and 43%, respectively. The solid state magic angle spinning spectrum of an AAB gel is composed of Q^3 and Q^4 resonances in the proportions 60% and 40%, respectively. It is clear from ^{29}Si NMR that, although AAB sols and gels have the same final pH and hydrolysis ratio r as the B2 sol, the Q^n species distribution is completely different: there is no evidence of monomer or end groups/dimers even after extensive aging of the AAB sol.

Small Angle X-ray Scattering (SAXS). SAXS was employed to investigate the growth and structure of the silicate species before gelation. The Porod plots (log scattered intensity vs log scattering wave vector, q) for the A2(fresh) sols show a limited range of fractal scaling with D increasing from about 1 to 1.7 over a period of 35 hours of aging. Above $q \approx 0.33$ Å^{-1}, all three curves exhibit Porod slopes of about -1.9. Guinier radii (R_G) evaluated at low q increase from 1.3 to 2.6 nm over the same time period. The A2(aged) sols exhibit power law scattering over a wider range of q than observed for the A2(fresh) sols. With aging, D increases from 1.18 to 1.4 and R_G increases from 2.0 to greater than 5.7 nm.

For the B2 sol, Porod slopes evaluated over the q range 0.01 to 0.1 Å^{-1} suggest that the sol is composed of fractal clusters with D increasing from about 2.3 to 2.4 over a 48 hour period of aging. Although the crossover to the Guinier region of scattering is not clearly established for the sol aged 48 hours, compared to the sol aged 0 hours, it is displaced to lower q consistent with further polymer growth accompanying the aging process. Compared to the B2 sols, after comparable periods of aging the AAB sols exhibit both power law scattering over a much greater q range and less negative values of the Porod slope. During 44 hours of aging, D increases from about 1.4 to 1.7. Compared to B2 sols, the crossover to Guinier scattering apparently occurs at lower q, consistent with larger average cluster size.

Ellipsometry and Thermal Gravimetric Analysis. Ellipsometry was performed on thin film samples deposited on silicon substrates to determine film thickness and refractive index. Table I lists values of thickness and refractive index for films prepared from A2(fresh) and A2(aged) sols after varying periods of aging (prior to

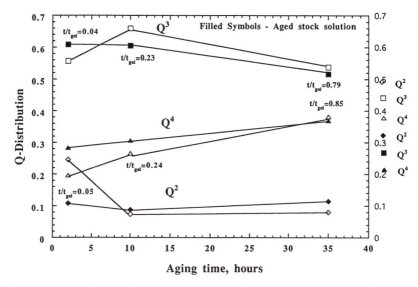

Figure 3. Q^n distribution of the A2(fresh) and A2(aged) sols with aging time at 50°C. The lines are drawn to guide the eye. The gelation time of the A2(fresh) and A2(aged) sols at 50°C prior to dilution was 41 and 44 hours, respectively. Different t/t_{gel} values for the same aging times are due to different t_{gel} times.

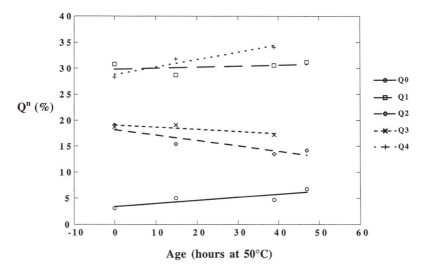

Figure 4. Q^n distributions determined by ^{29}Si NMR of the B2 sols (diluted 4:1) after various aging times at 50°C. The lines are linear regressions of the data. (Reproduced with permission from reference 5, Copyright 1992, SPIE)

film deposition) and annealing at 400°C. Volume fraction porosities were calculated from the refractive index using the Lorentz-Lorenz model assuming a skeletal refractive index of 1.45 for silica. We see that refractive index values are quite similar, corresponding to about 10% porosity, regardless of the pedigree of the sol or the aging times employed.

Table II lists refractive index and percent porosities for films prepared from B2 or AAB sols after various sol aging times (prior to film deposition). Whereas there is little effect of aging on the porosity of as-deposited (unannealed) B2 films, the as-deposited AAB films show a clear increase in porosity with aging time. For both B2 and AAB films, heating to 400°C causes a reduction in refractive index corresponding to an increase in porosity. After heating <u>both</u> AAB and B2 films exhibit an increase in porosity with aging time. Corresponding TGA results show about a factor of three greater weight loss for B2 specimens than for AAB specimens in the temperature range 150-300°C associated with pyrolysis of residual alkoxide groups.

Transport Characterization. In order to obtain additional information concerning the porosity of the A2 series of films, which showed no dependence of refractive index on sol aging time, we prepared A2 films in supported membrane configurations and measured the permeance of several gases. Figure 5 shows cross-sectional TEM images of the supported membranes prepared from A2(aged) sols after various periods of aging.

For both fresh and aged stock sols and all aging times, the deposition of a single coating led to a substantial reduction in permeance. The permeance of the support before and after membrane deposition was essentially independent of the mean pressure, indicating that the supports and membranes are free of large (viscous flow) defects. Considerable variability in the permeance of supports prior to coating was observed, however. This is believed to be due to variations in the thickness of the ~4.0 nm pore diameter γ-Al_2O_3 support layer. We have therefore evaluated the effect of membrane coating on permeance in terms of a normalized reduction in permeance, which we call the reduction ratio. The reduction ratio is the ratio of permeance of the support prior to coating to the permeance after depositing the membrane coating. The He permeance data along with the reduction ratios for supports coated with A2(fresh) and A2(aged) sols after different aging times are presented in Tables III(a) and (b). Nitrogen and C_3H_6 permeances were also measured for the membranes presented in Table III, and the reduction ratios exhibited trends similar to those found for He.

Discussion of Sols

Stock Solution. The stock solution is prepared by the acid-catalyzed hydrolysis of TEOS with sub-stoichiometric water ($r = 1$). ^{29}Si NMR studies of the fresh stock solution indicate that weakly-branched oligomeric species (Q^1 and Q^2) are the initial products of condensation. With aging, further condensation occurs, as is evident from the significant reduction in all hydrolyzed species and the appearance of branched (Q^3) species. Whereas the concentration of unhydrolyzed dimers remains virtually constant with aging, the concentration of the singly hydrolyzed dimer (-86.5 ppm) decreases dramatically. The concomitant increase in the unhydrolyzed cyclic tetramer suggests the reaction sequence:

$$2 \ (RO)_5(HO)Si_2O \rightarrow (RO)_9(HO)Si_4O_3 \rightarrow (RO)_8Si_4O_4.$$

Table I. Refractive index and thickness values determined by ellipsometry for films prepared from A2(fresh) and A2(aged) sols after various times of aging.

Aging at 50°C (t/tgel)	A2 sols from Fresh stock solution			A2 sols from Aged stock solution		
	Thickness (Å)	Refractive Index	Vol. Frac. Porosity	Thickness (Å)	Refractive Index	Vol. Frac. Porosity
0.05	1580	1.403	0.1	1376	1.398	0.1
0.24	1590	1.398	0.1	1359	1.415	0.08
0.82	2245	1.401	0.1	1369	1.409	0.09

Table II. Refractive indices and vol% porosities for films dip-coated from B2 sols and AAB sols (diluted 5:1) as a function of aging time (or aging time normalized by the gelation time). Mass fractal dimension values are for sols aged for comparable normalized aging times.

B2 Sol:			As-deposited:		After 400°C:	
Aging time (hr at 50°C)	Aging time (normalized)	D	Refractive Index	Porosity (vol %)	Refractive Index	Porosity (vol %)
0.0	0.00	2.27	1.425	4.9	1.369	16.0
24.0	0.32	2.32	1.424	5.0	1.346	20.8
48.0	0.63	2.40	1.421	5.6	1.325	25.1
72.0	0.95	--	1.418	6.2	1.292	32.1
94.0	1.24	--	1.417	6.4	1.240	43.5

AAB Sol:			As-deposited:		After 400°C:	
Aging time (hr at 23°C)	Aging time (normalized)	D	Refractive Index	Porosity (vol %)	Refractive Index	Porosity (vol %)
0.0	0.00	--	1.435	2.9	1.381	13.6
2.5	0.05	1.37	1.432	3.5	1.378	14.2
21.0	0.43	1.50	1.398	10.2	1.365	16.9
43.0	0.86	--	1.369	16.0	1.341	21.8
45.0	0.90	1.70	1.353	19.3	1.331	23.9

Table III(a). He permeance and reduction ratio for supports coated with A2(fresh) sols after different periods of aging followed by annealing to 400°C for 10 min.

Permeance $(cm^3/cm^2$-s-cm Hg) -Fresh stock solution			
A2 Sol Aging time (hr)	Support	After one coat	Reduction ratio
2	0.0906	0.0113	8.0
10	0.0617	0.0147	4.2
35	0.0646	0.0259	2.5

Table III(b). He permeance and reduction ratio for supports coated with A2(aged) sols after different periods of aging followed by annealing to 400°C for 10 min.

Permeance $(cm^3/cm^2$-s-cm Hg) -Aged stock solution			
A2 Sol Aging time (hr)	Support	After one coat	Reduction ratio
2	0.1017	0.0196	5.2
10	0.0791	0.0135	5.8
35	0.0557	0.0190	2.9

The preferential cyclization of the linear tetramer is a consequence of the steric bulk of the two pendant ethoxide groups per silicon, favoring cyclic over linear species, and the dearth of hydrolyzed sites, causing intramolecular condensation to be favored over intermolecular condensation.

A2 Sols. A2 sols were prepared by acid-catalyzed hydrolysis of the fresh or aged stock solutions with excess water ($r = 5$). Figure 3 indicates that, after two hours of aging, the A2(fresh) sol is less highly condensed than the A2(aged) sol. However, with additional aging time, both the Q-distributions and extents of reaction of A2(fresh) and A2(aged) sols calculated from the Q distributions become comparable. Despite these similarities in the average molecular-scale structure, SAXS results indicate that differences in the stock solution preparation significantly influence structure on intermediate length scales (0.2 - 10 nm). After comparable aging times, the A2(aged) sols are characterized by larger size, lower fractal dimension, and a wider range of fractal scaling behavior. The combined NMR and SAXS results are consistent with the greater incorporation of cyclic tetramers in siloxane polymers formed in the A2(aged) sols. Previous NMR studies have shown the cyclic tetramer and oligomers containing cyclic tetramer to be rather stable with respect to ring-opening hydrolysis or alcoholysis under both acid-catalyzed and neutral conditions (*12*). These stiffer structural units, when incorporated into the siloxane backbone, would favor more extended growth, so that after the same extent of reaction, polymers would exhibit larger size and lower mass fractal dimension, consistent with our experimental results. Unfortunately, the breadth of the Q^3 and Q^4 ^{29}Si NMR resonances of the A2 sols prevents quantification of the cyclic tetramer concentrations.

B2 and AAB Sols. B2 sols were prepared by base-catalyzed hydrolysis of aged stock solution with excess water ($r = 3.7$). Comparison of the ^{29}Si NMR spectra of the stock solution and the unaged B2 sol shows that the base-catalyzed hydrolysis step causes considerable condensation as is evident from the production of Q^4 species at the expense of Q^2 and Q^3 species. However, the concentrations of the unhydrolyzed dimer and unhydrolyzed Q^2 silicon species (associated possibly with trimer) remain largely unchanged, and the concentration of monomer has clearly increased. These results suggest that, as expected from both steric and inductive considerations, base-catalyzed hydrolysis promotes further hydrolysis and condensation of more highly hydrolyzed and condensed species, leaving unhydrolyzed and more weakly condensed species comparatively unreacted. The generation of monomer can be attributed to base-catalyzed siloxane bond hydrolysis or alcoholysis, leading to a redistribution of siloxane bonds according to reactions of the general type ($n \geq 2$) (*12*):

$$ROH + (RO)_{2n+2}Si_nO_{n-1} \leftrightarrow (RO)_4Si + HO(RO)_{2n-1}Si_{n-1}O_{n-2} \quad (4)$$

$$HO(RO)_{2n-1}Si_{n-1}O_{n-2} + (RO)_{2n+2}Si_nO_{n-1} \leftrightarrow$$

$$(RO)_{4n}Si_{2n-1}O_{2n-2} + ROH \qquad (5)$$

or

$$2\ Xmer \leftrightarrow monomer + (X+1)mer \qquad (6)$$

where R = alkyl (siloxane bond alcoholysis) or H (siloxane bond hydrolysis). Both continued polymerization and siloxane bond hydrolysis or alcoholysis could account for the observed reduction in concentration of the cyclic tetramer Q^2_{4c} species with aging. It is generally believed that siloxane bond hydrolysis is promoted under more basic conditions (*13*) consistent with our experimental results.

SAXS data for B2 sols show a general trend of increasing size and increasing mass fractal dimension with aging. These results are consistent with polymer growth accompanied by redistribution reactions. Continued cleavage of siloxane bonds to produce low molecular weight species followed by preferential re-condensation at more highly condensed sites is analogous to reaction-limited monomer cluster growth or ripening, mechanisms which in time should lead to more compact, uniform structures characterized by D→ 3.

AAB sols were prepared by a three-step process involving acid-catalyzed hydrolysis of aged stock solution (*r* = 2.5) followed by a base-catalyzed hydrolysis step (*r* = 3.7). The final hydrolysis ratio, pH, and silicon concentration were identical to those of B2 sols. Comparing the NMR spectra of the stock solution and the AAB sol, we observed that the effect of the second (acid-catalyzed) hydrolysis step was to promote extensive hydrolysis and condensation. Monomer and Q^1 species are consumed to produce primarily a variety of Q^2 and Q^3 species. Various cyclic tetramers are prominent sol species after the second hydrolysis step as was evident from a strong Q^2 resonance observed near -95 ppm.

The third (base-catalyzed) hydrolysis step promoted further condensation of Q^2 and Q^3 species to produce primarily Q^3 and Q^4 species. Aging caused practically complete elimination of Q^2 species including the cyclic tetramers. These results are consistent with the conclusion drawn above: base additions are quite effective in catalyzing the hydrolysis and condensation of more highly-hydrolyzed and condensed species (products of the second hydrolysis step). The absence of monomer and dimer/Q^1 species implied that monomer-producing redistribution reactions and siloxane bond cleavage reactions, in general, are inhibited for more highly condensed species. Overall, comparison of the NMR spectra of B2 and AAB sols proved that the hydrolysis and condensation pathways are strongly kinetically-limited for this composition.

The considerably lower mass fractal dimensions observed for AAB sols compared to B2 sols are consistent with the general expectation for reaction-limited cluster-cluster growth in the absence of monomers and oligomers (*13*). In addition, since cyclic tetramers are expected to promote the formation of extended polymers characterized by low values of the mass fractal dimension, the unusually low value of D (1.4) observed after 3 hours of aging may be attributable to the incorporation of cyclic tetramers, formed during the second hydrolysis step, into the siloxane backbone. Progressive hydrolysis or alcoholysis of cyclic tetramers after addition of base could partially account for the increase in D observed during aging. The absence of monomer and Q^1 species argues against ripening as being responsible for this behavior.

Discussion of Film Formation

This section examines the role of the molecular- and intermediate-scale structures of the sols in influencing the microstructure of the corresponding thin films formed by dip-coating. We consider the cases of mutually transparent and mutually opaque precursors.

Mutually Transparent Precursors. According to equation 3, we expect the conditions for mutual transparency to be fulfilled in general when D < 1.5. In addition, since equation 3 assumes an infinite condensation rate, mutual transparency may also be realized under conditions where D somewhat exceeds 1.5 if the condensation rate is low. The ellipsometry results obtained for both the A2(fresh) and A2(aged) series of sols show low volume percent porosities and essentially no dependence of porosity on sol age. These observations are completely consistent with the idea of mutual transparency: the ability of the polymers to freely interpenetrate as they are concentrated on the substrate surface should lead to dense packing and should effectively hide any differences due to size and fractal dimension of the individual precursor species. For example, the differences noted in the scattering behavior of the A2(fresh) and A2(aged) sols are not manifested as differences in porosity in the ellipsometry data (Table I).

The permeance data reported in Tables III(a) and (b) also show similar trends for both A2(fresh) and A2(aged) sols. Sol deposition causes a significant reduction in the permeance of the support, and the reduction ratios are greatest after two hours of aging and decrease with aging time. This behavior may be explained by consideration of factors governing flow through a porous medium. The molar flux due to Knudsen flow through the membrane plus support layer can be described by:

$$J_K = \frac{4\varepsilon r_p}{3\tau RT} \left[\frac{2RT}{\pi M}\right]^{1/2} \frac{\Delta P}{\Delta a}, \tag{7}$$

where ε is the porosity, τ is tortuosity, r_p is the pore radius, M is molecular weight, R is the gas constant, T is the temperature, ΔP is the partial pressure across the membrane or layer, and Δa is the membrane or layer thickness (14). Based on this expression the general reduction in permeance observed after membrane deposition is attributable to a decrease in porosity and/or pore size and an increase in tortuosity and/or thickness, consistent with the following observations. 1) The ellipsometry data indicates that the membrane layer should have a porosity of about 10 vol% compared to the 30-40 vol% of the support layer. 2) Cross-sectional TEM images of the support plus membrane (Figure 5) suggest that the pore radius of the membrane layer is considerably smaller than that of the support (~2.0 nm). 3) It is expected, and modeling has shown, that the permeation tortuosity of a material will become large and strongly dependent on porosity for volume fraction porosities near the percolation threshold (~16%) (15). The combined features of low porosity, small pore size, and large tortuosity attributable to all the membranes studied are responsible for the large reductions in permeance observed after membrane deposition.

The general trend of decreasing reduction ratios with increasing aging times can be explained by consideration of the relative sizes of the precursor species and the pores of the $\gamma-Al_2O_3$ support layer. Combined cross-sectional TEM and SAXS results indicate that when the characteristic polymer size R_G is appreciably less than the pore radius of the support, ~2.0 nm, deposition of silica occurs primarily within the ~4 µm thick support layer (as confirmed by EDS analysis (11)), whereas when $R_G \geq 2$ nm, an external membrane layer is formed on the support surface. For the same quantity of silica deposition, penetration into the support pores causes the effective thickness to be greater than when the membrane is formed as an external layer. (Since the support has a volume fraction porosity of about 33%, the depth of penetration into the support must be a factor of about 3 larger than the thickness of an external film for equivalent quantities of deposited silica). According to equation 7, greater film thickness results in reduced flux and correspondingly larger reduction

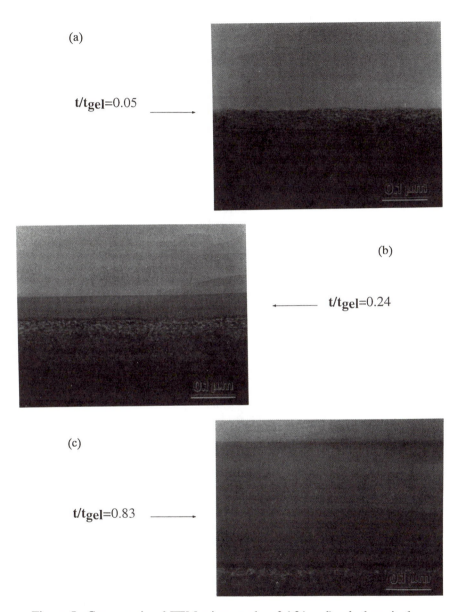

Figure 5. Cross-sectional TEM micrographs of A2(aged) sols deposited on γ-Al$_2$O$_3$ supports after various sol aging times: (a) t/t$_{gel}$ = 0.05, (b) t/t$_{gel}$ = 0.24, (c) t/t$_{gel}$ = 0.83. EDS analysis (11) indicated that for t/t$_{gel}$ = 0.05 silica deposition occurred exclusively within the pores of the γ-Al$_2$O$_3$ support.

ratios. The series of TEM micrographs shown in Figure 5 (a-c) indicate that with increased aging (increased R_G) silica is more efficiently filtered-out by the support to form an external membrane layer. This should cause a progressive decrease in overall membrane thickness (internal + external layer thicknesses) with aging time, explaining the observed trend. For the series of sols studied, there is no evidence that the porosity, tortuosity, or pore size of the deposited films or membranes depend on the pedigree of the stock solution or sol aging times.

Mutually Opaque Precursors. Based on equation 3, B2 sols fulfill the criterion for mutual opacity (D > 1.5), so we expect film porosity to increase with size of the primary sol species (equation 2). Although both NMR and SAXS data are consistent with polymer growth during aging, the as-deposited (unannealed) films showed essentially no dependence of refractive index on aging time (Table II). The expected reduction in refractive index with sol aging time was observed only after heating to 400°C. These results are explained by the presence of unreactive monomers and oligomers in the B2 sols. ^{29}Si NMR showed evidence for several unhydrolyzed or weakly hydrolyzed monomers and oligomers (especially dimer) in the B2 sols regardless of aging time. The presence of these small species should not influence the SAXS results performed under rather dilute conditions. However, as the sol is concentrated during film deposition, these small weakly reactive species will freely penetrate the larger fractal clusters, "filling-in" the cluster porosity. The heat treatment at 400°C pyrolyzes the residual alkoxide groups associated primarily with these small species revealing the trend of increasing porosity with increasing cluster size expected for mutually opaque precursors.

Compared to B2 sols, AAB sols exhibit larger polymer sizes, lower mass fractal dimensions, and no evidence for monomers or oligomers after aging for equivalent normalized times. Mass fractal dimensions increase from 1.4 to 1.7 with aging time, so we expect a transition from mutually transparent to mutually opaque behavior. Refractive index values of the as-deposited films (Table II) are consistent with this expectation. Films prepared from unaged AAB sols are quite dense. Porosity increases monotonically with aging time due to increased size and mass fractal dimension of the primary sol species. Increased size increases the porosity of a fractal object according to equation 2. Increased mass fractal dimension reduces the porosity of an individual object but reduces the probability of interpenetration (equation 3), which, in the absence of monomer and oligomers, allows the porosity of the primary species to be partially preserved in the as-deposited films. *In order to maximize film porosity, it is necessary to avoid small unreactive species and to chose an intermediate value of D that balances the effects of the primary cluster porosity and the secondary aggregate porosity.*

As shown in Table II, heating to 400°C causes an increase of about 6% in film porosity attributable to organic removal. The porosity created by pyrolysis of AAB films is considerably less than that created in B2 films consistent with TGA results, which show over a factor of three difference in weight loss in the temperature range 150-400°C. After heating, a similar trend of increasing porosity with increasing aging time is observed. It should be emphasized that this behavior and that of B2 sols is completely different from the behavior of the A2 series of sols where no effect of sol aging time on film porosity is observed after heating to 400°C. It is this distinction that provides a working definition of the concept of mutual transparency and mutual opacity in the context of film formation and that allows the rational design of thin film precursors.

Comparing the results of the A2 and AAB series of films, we observe what appears to be conflicting behavior, viz. A2 sols exhibiting D = 1.7 yield rather dense films consistent with mutual transparency, whereas AAB sols exhibiting D = 1.7 yield porous films consistent with mutual opacity. This apparent inconsistency is explained by the assumptions underlying Eq. 3 that condensation occurs immediately and irreversibly at each point of intersection and that the individual aggregating units are perfectly stiff. A2 sols exhibit lower condensation rates than AAB sols and are somewhat less-highly condensed. Thus although D exceeds 1.5 for the A2 sol, its behavior is consistent with mutual transparency. The AAB sol better fulfills the assumptions of the model, and the films exhibt behavior consistent with the model predictions.

Summary

This study provides several examples of how, using multi-step hydrolysis procedures, the sequence (acid-acid, acid-base, acid-acid-base) and timing of the successive steps influences the molecular and intermediate-scale structure of silica sols prepared from alkoxide precursors. These examples imply that the silica polymerization pathway is largely kinetically controlled at least for moderate values of r.

Under acid-catalyzed conditions with molar hydrolysis ratios $r = 1$ or 2.5, aging (at 0°C, room temperature, or under reflux) served to increase the concentration of cyclic tetrasiloxanes, which in turn influenced the intermediate-scale structure of polymers formed during subsequent acid- or base-catalyzed hydrolysis. Presumably incorporation of structurally rigid cyclic tetramers into the siloxane backbone leads to more extended structures characterized by larger size and lower mass fractal dimension.

It was shown that the species distribution resulting from base-catalyzed hydrolysis depends strongly on the extent of hydrolysis and condensation of the precursor sol to which the base is added. Base additions, when made to a pre-hydrolyzed sol, serve to catalyze condensation of more highly hydrolyzed and condensed species, leaving weakly hydrolyzed monomers and oligomers virtually unreacted. Even after long periods of aging, the species distribution identified by NMR differed profoundly for two sols prepared with the same value of r, pH, and concentration, but with a different sequence of steps.

With regard to thin film formation, we defined two classes of primary sol species referred to as mutually transparent or mutually opaque. Mutually transparent species are able to interpenetrate during dip-coating to produce rather dense films that exhibit no dependence of refractive index/porosity on polymer size or mass fractal dimension. Mutually opaque species do not freely interpenetrate during dip-coating. In the absence of monomer or oligomeric species that are able to fill-in small pores, we observe a systematic increase in film porosity with aging time attributable to increasing size and mass fractal dimension of the primary species. The ability to "tune" the refractive index by a simple aging procedure should allow the rational design of porous films.

Acknowledgments

Part of this work was performed at Sandia National Laboratories, supported by the U.S. Department of Energy Basic Energy Sciences Program under Contract # DE-AC04-94AL85000. Additional support has been received from the National Science Foundation (Contract # CTS9101659), the Electric Power Research Institute (Contract # RP8007-15), the Gas Research Institute (Contract # 5091-222-2306), and the Morgantown Energy Technology Center. The authors thank D.C. Goodnow for the TGA results.

Literature Cited

1. Brinker, C. J.; Hurd, A. J.; Schunk, P. R.; Frye, G. C.; Ashley, C. S. *J. Non-Cryst. Solids* **1992**, *147&148*, pp 424-436.
2. Mandelbrot, B. B. *The Fractal Geometry of Nature;* W. H. Freeman: New York, NY, 1983.
3. Brinker, C. J.; Keefer, K. D.; Schaefer, D. W.; Assink, R. A.; Kay, B. D.; Ashley, C. S. *J. Non-Cryst. Solids* **1984**, *63*, pp 45-59.
4. Logan, D. L.; Ashley, C. S.; Brinker, C. J. In *Better Ceramics Trough Chemistry V*; Hampden-Smith, M. J.; Klemperer, W. G.; Brinker, C. J., Eds.; Materials Research Society: Pittsburgh, PA **1992**, Vol. 271; pp 541-546.
5. Logan, D. L.; Ashley, C. S.; Assink, R. A.; Brinker, C. J. In *Sol-Gel Optics II,* Mackenzie, J. D., Ed.; SPIE - The International Society of Optical Engineering: Bellingham, WA, **1992**, Vol. 1758, pp 519-528.
6. Assink, R. A.; Kay, B. D. *Annu. Rev. Mater. Sci.* **1991**, *21,* pp 491-513.
7. Brinker, C. J.; Ward, T. L.; Sehgal, R.; Raman, N. K.; Hietala, S. L.; Smith, D. M.; Hua, D.-W.; Headley, T. J. *J. Membrane Science* **1993**, *77*, pp 165-179.
8. Engelhardt, V. G.; Altenburg, W.; Hoebbel, D.; Weiker, W. Z. *Anorg. Allg. Chem.* **1977**, *418*, p 43.
9. Kelts, L. W.; Effinger, N. J.; Melpolder, S. M. *J. Non-Cryst. Solids* **1986**, *83*, pp 353-374.
10. Balfe, C. A.; Martinez, S. L. In *Better Ceramics Through Chemistry II*, Brinker, C. J.; Clark, D. E.; Ulrich, D. R., Eds.; Materials Research Society: Pittsburgh, PA, 1986, Vol. 73; pp 27-33.
11. Brinker, C. J.; Sehgal, R.; Raman, N. K.; Schunk, P. R.; Headley, T. J. *J. Sol-Gel Sci. & Tech.,* in press.
12. Klemperer, W. G.; Mainz, V. V.; Millar, D. M. In *Better Ceramics Through Chemistry II;* Brinker, C. J.; Clark, D. E.; Ulrich, D. R., Eds.; Materials Research Society: Pittsburgh, PA, 1986, Vol. 73; pp 3-13.
13. Brinker, C. J.; Scherer, G. W.; *Sol-Gel Science*; Academic Press: San Diego, CA, 1990.
14. Keizer, K.; Uhlhorn, R. J. R.; Van Vuren, R. J.; Burggraaf, A. J.; *J. Membrane Science* **1988**, *39*, pp 285-300.
15. Zallen, R. In *The Physics of Amorphous Solids;* John Wiley and Sons: New York, NY, 1983, pp 183-191.

RECEIVED July 5, 1994

Chapter 11

Plasma Oxidation of Hydrocarbon Templates in Bridged Polysilsesquioxanes

Porous Materials by Design

Douglas A. Loy[1], Kenneth J. Shea[2], Richard J. Buss[1],
and Roger A. Assink[1]

[1]Sandia National Laboratories, Albuquerque, NM 87185–0367
[2]Department of Chemistry, University of California, Irvine, CA 92717

Hydrocarbon bridging groups in polysilsesquioxane network polymers were used as templates for porosity. Arylene- and alkylene-bridged polysilsesquioxanes were prepared by the hydrolysis and condensation of bis(triethoxysilyl)aryl monomers **1-4** and bis(triethoxysilyl)alkane monomers **5-9**. The bridged polysilsesquioxane xerogels (**X-1** through **X-9**) and monolithic gels (**M-1** through **M-3**) were treated with inductively coupled oxygen plasmas at low temperature ($\leq 100\ °C$). All of the bridged polysilsesquioxane gels save the microporous monolith **M-3** were quantitatively converted to silica gels. Non-porous alkylene-bridged polysilsesquioxanes were plasma oxidized to porous silica gels whose pore size and shape appeared to be related to the length of the bridging group.

Sol-gel processing of tetraalkoxysilanes leads to the formation of porous silica gels (xerogels and aerogels). Pore structure in these silica gels can be manipulated by changing the reaction and processing conditions used for their preparation (1). However, the pore size distributions of these materials are too broad for application as molecular sieves. Another strategy for manipulating porosity is to build a material with molecular-sized building blocks (2-7). According to this strategy the properties, including porosity, of the resulting synthetic materials may be dictated by the size, length, and shape of the monomeric building blocks. Most of these efforts have afforded non-porous materials or demonstrated an inability to control the materials' pore distributions. For example, arylene-bridged polysilsesquioxanes are mostly microporous materials, yet the length of the arylene spacer or building block has no significant effect on the mean pore diameter (7). Unlike porosity in microcrystalline molecular sieves, porosity in sol-gel processed amorphous materials is more dependent on macromolecular-scale architecture than molecular-based structure.

Bridging Groups in Polysilsesquioxanes as Pore Templates.

In this study, the organic components of bridged polysilsesquioxanes were used as disposable pore templates as well as structural spacers. Arylene- and alkylene-bridged polysilsesquioxane gels were prepared by sol-gel processing bridged triethoxysilyl monomers (Figure 1) (6-11). The resulting dried gels were then treated with an oxygen

0097–6156/94/0572–0122$08.00/0

plasma to create silica gels (Scheme 1). Loss of the hydrocarbon bridging groups generates new porosity within the material. In porous materials, loss of the templates would be expected to occur in the pore walls with a net effect of coarsening or enlarging the pores. In non-porous materials, the loss of templates would generate porosity where none existed previously. The size and shape of the pores would be dependent both on the identity of the template and how the template groups are associated with each other in the matrix. For example, formation of nano-phases of alkylene-bridging groups could generate a nano-sized pore template surrounded by silicon-rich regions.

Removing Pore Templates. In order to generate porosity through this approach, a method for removing the organic template without disturbing the inorganic framework must be used (Scheme 2). Organic phases in hybrid organic-inorganic materials have been used as templates for porosity before our effort. Chujo et. al. prepared non-porous nanocomposites of organic polymers and silica by a sol-gel method (12). These materials were then pyrolyzed at 600 °C in order to remove the templates and afford microporous silicas. Roger et. al. hydrolyzed non-porous tin-carboxylates to remove the organic pore templates leaving mesoporous tin oxides (13). We have focused on a low temperature oxygen plasma as a mild method for removing organic templates from bridged polysilsesquioxanes (14). In the bridged polysilsesquioxanes, the template is attached to two or more silicon atoms in the siloxane network through silicon-carbon bonds. Thermal gravimetric analysis of a phenylene-bridged polysilsesquioxane in air indicated that removal of the bridging group by thermal oxidation occurred only above 400 °C (10). We desired relatively mild conditions for template removal to minimize alteration of the existing siloxane framework.

Plasma Oxidation of Pore Templates. Low temperature, inductively coupled oxygen plasma treatment was selected as a "mild" method for removing the hydrocarbon templates in the bridged silsesquioxane network materials. However, it was not clear that an oxygen plasma would penetrate very far into the silsesquioxane network before forming a plasma-resistant silica overlayer (15). Therefore, the first bridged polysilsesquioxanes treated in this study were *porous* arylene-bridged polysilsesquioxanes that had been ground into powders. Once the oxygen plasma technique demonstrated removal of the bridging groups in these materials, non-porous alkylene-bridged polysilsesquioxanes were also treated to determine if the size or length of the bridging group could be used in a predictable way to affect porosity in the resulting silicas. Finally, monolithic aerogels and a xerogel were plasma processed to determine if the plasma would penetrate deep into these gels.

Experimental

Arylene Monomer Synthesis. Bis(triethoxysilyl)arylene monomers **1-4** were synthesized from the aryl dibromides and tetraethoxysilane (TEOS) or chlorotriethoxysilane using Barbier-Grignard or organolithium chemistry (Scheme 3). An example of the preparation of **1** from 1, 4-dibromobenzene and tetraethoxysilane is as follows:

1, 4-Bis(triethoxysilyl)benzene 1 (7). To a mixture of magnesium turnings (15.0 g, 0.62 moles) and TEOS (450 mL, 2 moles) in THF (300 mL) under nitrogen, a small crystal of iodine was added and the mixture brought to reflux. A solution of 1, 4-dibromobenzene (48 g, 20.4 mmol.) in THF (100 mL) was added dropwise over 2 hours [CAUTION: care must be exercised with the addition of dibromobenzene. Addition of more than 25% of 1, 4-dibromobenzene to the reaction mixture *before* the

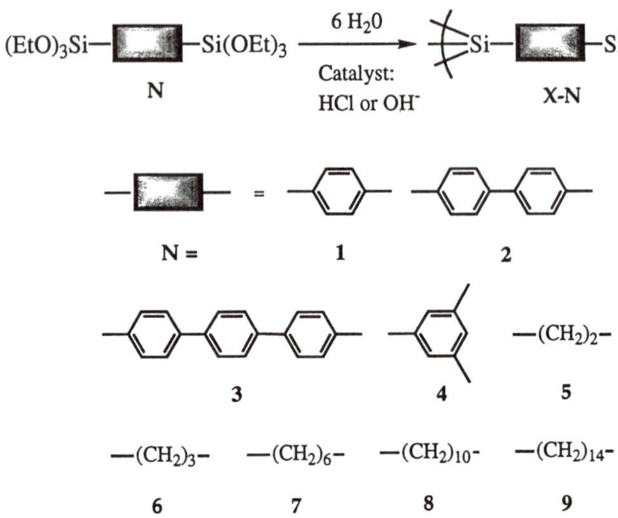

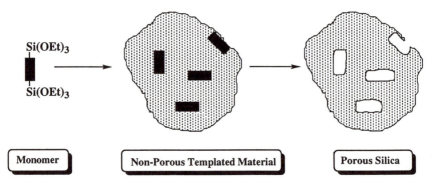

Figure 1. Preparation of bridged polysilsesquioxanes **X-1** through **X-9** from bis(triethoxysilyl) monomers **1-9**.

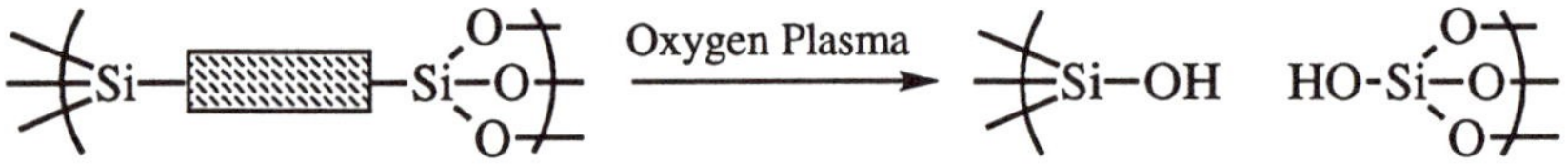

Scheme 1. Oxidative removal of hydrocarbon template from bridged polysilsesquioxanes.

Scheme 2. Organic template approach for creating porosity.

induction period of Grignard formation could result in a violent exotherm!] Within 30 minutes of initiating the addition, the reaction became mildly exothermic. The reaction mixture was kept at reflux for 1 hour after the completion of the addition of dibromobenzene. The grey-green mixture was cooled to room temperature before the THF was removed *in vacuo*. Hexane (200 mL, distilled from CaH_2) was added to precipitate any remaining magnesium salts in solution and the mixture was quickly filtered under nitrogen to afford a clear, light brown solution. Hexane was removed *in vacuo* and the remaining TEOS was distilled off leaving a brown oil. The residual oil was distilled (0.2 mm Hg, 130-5 °C) to give a clear colorless oil (**1**, 42 g, 55%). ^1H NMR (300 MHz, $CDCl_3$) δ 7.63 (s, 4H, ArH), 3.82 (q, 12H, J = 7.12 Hz, $SiOCH_2CH_3$), 1.19 (t, 18 H, J = 7.11 Hz, $SiOCH_2CH_3$); ^{13}C NMR (75.5 MHz, $CDCl_3$) d134.36, 133.45, 59.09, 18.56; ^{29}Si NMR (99.34 MHz, $CDCl_3$) d -57.72; IR (film on NaCl) 3058, 2976, 2886, 1391, 1168, 1147, 1102, 1079, 961, 778, 706 cm^{-1}; low resolution mass spectrum (EI, 70 eV) m/z 402 (M+), 357, 329, 297, 285, 147, 119, 107, 100, 91, 79; high resolution mass spectrum (EI, 30 eV) calcd. for $C_{18}H_{34}O_6Si_2$: 402.1890, found: 402.1909.

Alkylene Monomer Synthesis. Ethylene monomer **5** was prepared by the hydrosilation of vinyltriethoxysilane with triethoxysilane in the presence of chloroplatinic acid. Propylene monomer **6** was synthesized by hydrosilating allyltriethoxysilane. The remaining monomers **7-9** were prepared by hydrosilating the corresponding α, ω-dienes (11). An example of the preparation of 1, 6-bis(triethoxysilyl)hexane **7** by the hydrosilation of 1, 5-hexadiene (Scheme 4) is as follows:

1, 6-Bistriethoxysilyl)hexane 7 (11). To a mixture of 1, 5-hexadiene (100 g, 0.13 mol.) and chloroplatinic acid (20 mg) in benzene at room temperature (CAUTION: chloroplatinic acid is a cancer suspect agent and benzene is a carcinogen.) in a three-necked round-bottom flask equipped with stir bar, condenser with argon inlet, and rubber septa on the remaining two openings, triethoxysilane (100 mL, 0.27 mol.) was added by syringe (CAUTION: triethoxysilane is highly toxic!). Some bubbling was observed from the solid catalyst, which changed from orange to black. The solution was stirred at 50 °C and the reaction monitored by capillary gas chromatography. After four hours, the solution was cooled to room temperature; solvent and excess triethoxysilane were removed in vacuo, leaving a brown oil. Distillation afforded a clear colorless oil (33.9 g, 64 %). ^1H NMR (200 MHz, $CDCl_3$) δ 3.70 (q, 12 H, J = 7.02 Hz, $SiOCH_2CH_3$), 1.35 (m, 4 H, $CH_2CH_2CH_2Si$), 1.22 (m, 4 H, $CH_2CH_2CH_2Si$), 1.12 (t, 18 H, J = 6.98 Hz, $SiOCH_2CH_3$), 0.52 (t, 4 H, J = 7.81 Hz, $CH_2CH_2CH_2Si$); ^{13}C NMR (75.5 MHz, $CDCl_3$) δ 58.17, 32.71, 22.57, 18.20, 10.31; IR (film on NaCl) 2975, 2927, 2887, 2735, 1484, 1443, 1391, 1366, 1295, 1261, 1168, 1167, 1105, 1082, 958, 782 cm^{-1}; low resolution mass spectrum (EI, 70 eV) m/z 410 (M+), 364, 335, 321, 297, 291, 279, 253, 247, 163 (base peak), 135, 119, 107, 100, 91, 79.

Preparation of Gels. Bridged polysilsesquioxanes gels **X-1** through **X-9** were prepared by sol-gel processing monomers **1-9** (Scheme 5). Typically, the monomer was dissolved in about half the required solvent needed to make up the final concentration. To this solution of monomer in a volumetric flask, a mixture of aqueous catalyst in solvent was added; additional solvent was added to achieve the final volume of solution. The monomer and catalyst solutions were thoroughly mixed by shaking the flask. The contents of the volumetric were then transferred to a polyethylene bottle

Scheme 3. Barbier-Grignard preparation of 1, 4-bis(triethoxysilyl)benzene **1**.

Scheme 4. Hydrosilation of 1, 5-hexadiene to synthesize 1, 6-bis(triethoxysilyl)-hexane **7**.

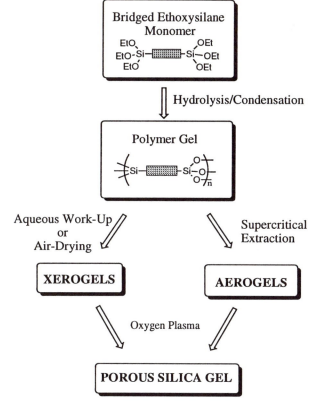

Scheme 5. Preparation of bridged polysilsesquioxanes by sol-gel techniques followed by oxygen plasma treatment to generate porous silicas.

which was tightly sealed to prohibit evaporation. The solutions were clear and colorless. With the onset of gelation a blue tint appeared. Gelation times were dependent on the monomer and type and concentration of catalyst, but were typically less than 1 day. The final gels were glass-like in appearance. The gels were aged in sealed containers as long as several weeks before either air drying, supercritical drying, or simply crushing the gel in water before drying under vacuum at 100 °C.

Arylene-bridged polysilsesquioxanes were prepared by reacting the arylene monomers (0.4 M) in tetrahydrofuran with an excess of aqueous ammonia (540 mol%) and water (>10 equivalents). The arylene-bridged gels were solvent-processed (7) and allowed to air dry. Alkylene-bridged polysilsesquioxane xerogels were prepared in tetrahydrofuran (0.4 M) with 10 mol% HCl as catalyst. The gels were air-dried, affording xerogels with approximately 5-10% of the original "wet" gel volume. The alkylene-bridged gels were aged two weeks before being crushed in water then washed with water (2X) and ethyl ether (11). The gels were ground into powders and dried under vacuum at 100 °C before treating with plasma. The propylene monomer **6** takes months to gel using the conditions described above. The plasma treatments in this study were carried out on a porous propylene-bridged polysilsesquioxane made with acid catalyst at 1.2 M monomer **6** concentration. Ethylene- and decylene-bridged monolithic aerogels (**M-1** and **M-2**) were prepared by extracting the wet gels with supercritical carbon dioxide (10). The phenylene-bridged polysilsesquioxane xerogel **M-3** was prepared by slowly air drying the wet gel over several months. The resulting monolith was a transparent, colorless glass.

Plasma Oxidation of Bridged Polysilsesquioxanes. Oxidations were carried out in an inductively-coupled rf plasma (Figure 2). In a typical experiment, a ground sample of decylene-bridged polysilsesquioxane **X-8** was placed in a plasma of pure oxygen (30 Watts, 138 mTorr, 2.4 standard cubic centimeters per minute (sccm) O_2, sample temperature ~50 °C). The plasma exhibited a strong carbon monoxide photoemission. After 16 hours of plasma exposure, no carbon monoxide emission was observed and the sample was removed. When enough sample was available, the plasma experiments were repeated three times.

Solid state ^{29}Si NMR spectra were obtained at 50.17 MHz with a Chemagnetics console interfaced to a Nicolet 1280 data station. Samples were spun at 4-5 kHz. Surface area analyses were carried out with a Quantachrome Autosorb-6 nitrogen porosimeter. Infrared spectra experiments were performed with a Perkin-Elmer 1750 FT-IR spectrometer.

Results

Bis(triethoxysilyl)arylene and alkylene monomers **1-9** (0.4 M in tetrahydrofuran) were hydrolyzed in the presence of six or more equivalents of H_2O (Scheme 5). In all cases, clear colorless solutions or "sols" were formed. As hydrolysis and condensation progressed, the sols became more viscous, finally forming gels. Acid catalyzed polymerizations always afforded transparent gels. The base catalyzed arylene-bridged gels were transluscent, while alkylene-bridged gels were always opaque white. The polysilsesquioxane gels were further processed to afford *xerogels* or *aerogels*. The former are dried gels which have shrunk considerably during processing, while aerogels undergo little shrinkage and are thought to retain more of the original wet gel's structure. In this study, xerogels are defined as gels that were air dried or processed by crushing in water. Gels that were extracted with supercritical carbon dioxide and generally exhibit pore distributions that extend well into the macropore regime are called aerogels. Solid state ^{29}Si and ^{13}C (CP MAS) NMR spectroscopy on the dried gels indicated the bridging groups were intact by the presence of the expected bridging

carbon resonances and the lack of ^{29}Si NMR peaks arising from silica (7-11, 14). Nitrogen sorption analyses revealed the arylene-bridged polysilsesquioxanes to be mostly microporous with surface areas ranging from 400-830 m^2/g (BET). The alkylene-bridged polysilsesquioxanes prepared with acid catalysts were non-porous (16). Monolithic aerogels of the ethylene-bridged (**M-1**) and decylene-bridged (**M-2**) polysilsesquioxanes prepared with base catalyst were porous with surface areas of 956 m^2/g and 547 m^2/g, respectively.

Plasma Treatment of Bridged Polysilsesquioxanes. Without any carbon source in the plasma chamber, the inductively coupled oxygen plasma appeared violet. However, when the oxygen plasma was generated in the presence of bridged polysilsesquioxane gels, an intense, light blue carbon monoxide emission was observed. The plasma color has proven to be a useful indicator for the completion of the oxidation process. Once the hydrocarbon removal was complete, the plasma reverted to a violet oxygen emission-dominated plasma. Powdered samples placed into oxygen plasmas did not undergo any apparent physical changes other than the appearance of a thin, light brown discoloration on the surface of the powders that persisted even after the carbon monoxide emission had disappeared. The powder underneath the brown surface was white. No evidence of residual hydrocarbon could be detected in the bulk samples and it is likely that the brown discoloration arose from "color centers" generated by ion or vacuum UV photon bombardment.

Weight losses from the plasma treatments were recorded (Table 1). Weight losses for the gels are difficult to compare with calculated values unless the degrees of hydrolysis and condensation in the precursor material and the degree of condensation in the resulting silica is determined. We are presently measuring these values by solid state ^{13}C and ^{29}Si NMR spectroscopy. As one would expect, weight losses for materials with large molecular weight contributions from the bridging groups or templates correspond more closely with the calculated values.

Plasma Oxidation of Monolithic Gels. In order to determine if structures more massive than micron-sized granules in the powdered gels could be plasma treated to remove the bridging groups or templates, we plasma treated a series of monolithic gels (**M-1**, **M-2**, **M-3**) possessing different mean pore sizes. After plasma treatment the porous ethylene-bridged aerogel **M-1** had lost 22 % of its volume and 16% of its mass. Surprisingly, the **M-1** suffered only from a single crack which ran through the middle of the gel. It retained the faint blue transluscence of the original ethylene-bridged aerogel. Scanning electron microscopy of the aerogel before and after the plasma treatment revealed little change in the gel's structure. Yet, no ethylene-bridging groups could be detected by infrared spectroscopy. The oxygen plasma had penetrated all the way through the monolith! Infrared spectroscopy also confirmed that only silica remained after the plasma oxidation of a mesoporous decylene-bridged polysilsesquioxane aerogel **M-2**. The opaque, white decylene-bridged monolith lost volume during plasma oxidation and, unlike the ethylene-bridged gel, suffered extensive shallow cracking over its exposed surface. The last gel to be plasma treated was a transparent, colorless phenylene-bridged xerogel **M-3**. Once the mostly microporous gel was exposed to the plasma, a white coating formed on the majority of the exposed surfaces. The remaining surfaces remained transparent, but turned dark brown. Scanning electron microscopy revealed that the silica coating extended slightly less than 1 micron into the bulk material. We can conclude from the results with the monoliths that an oxygen plasma can be used to convert monolithic polysilsesquioxane gels to silica as long as there is substantial meso- and macroporosity for the plasma to penetrate.

Infrared Analysis. Loss of the hydrocarbon template in powdered samples was also monitored by obtaining a series of infrared spectra on material withdrawn from the plasma chamber. In this experiment a porous phenylene-bridged polysilsesquioxane powder was placed in a watch glass in the plasma apparatus. At regular intervals, the plasma was stopped and samples of the powder removed for analysis. The aromatic C-H stretching band at 3060 cm^{-1} was used as the diagnostic peak. The plot of absorbance over time in the oxygen plasma indicates that nearly all of the phenylene spacer was gone within 15 minutes of exposure to the oxygen plasma (Figure 3). Plasma treatment was continued, however, until the carbon monoxide emission had disappeared, typically requiring an additional 4-12 hours.

Table 1. Silica Yields from Plasma Oxidation of Bridged Polysilsesquioxanes

Sample	Pore Template Bridging Group	% Weight Remaining after Plasma Theoretical %	Experimental%
X-5	ethylene	104.5	66.4[a]
X-6	propylene	94.5	76.0[c]
X-7	hexylene	73.4	59.8[c]
X-8	decylene	56.5	50.5[c]
X-9	tetradecylene	46.0	41.0[b]
M-1	ethylene	104.5	87.7[a]
M-2	decylene	56.5	51.2[a]
M-3	phenylene	76.6	80.0[a]

a) Sample oxidation was performed only once. b) Plasma experiment was performed twice. c) Experiment was performed three or more times. Theoretical % weight calculated based on fully condensed silsesquioxane oxidizing to Q^3 silica (Scheme 1).

Solid State NMR Analyses. Solid state ^{29}Si NMR confirmed conversion of the bridged polysilsesquioxanes to silica. The T resonances (7, 9, 17) in the solid state ^{29}Si CP MAS NMR spectrum of phenylene-bridged polysilsesquioxane **X-1** lie at -61, -70 and -78 ppm (Figure 4A). No Q resonances, representing Si-C bond cleavage to silica during the sol-gel preparation of the material, are evident. After plasma treatment (Figure 4B), the T resonances were greatly attenuated (< 5% remaining) and Q resonances at -90, -101 and -108 ppm indicate that silica is the major constituent in the product. Similar ^{29}Si NMR analyses of the arylene-bridged polysilsesquioxanes **X-2** through **X-5** before and after plasma treatment indicated near complete conversion to silica.

Nitrogen Sorption Porosimetry. Nitrogen sorption measurements of porous arylene- and alkylene-bridged polysilsesquioxanes and the post-plasma silica gels show coarsening of the existing porosity (Table 2). In general, mean pore diameters increased while surface areas decreased. In effect, the plasma treatments opened up the walls of existing pores to form mesoporous silicas. Interestingly, plasma treatment of non-porous alkylene-bridged polysilsesquioxanes (**X-7**, **X-8**, and **X-9**) generated pores that appear to increase in size with increasing spacer length. However, the

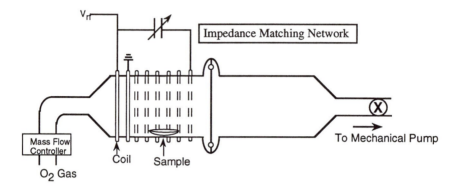

Figure 2. Glass Rf oxygen plasma apparatus for removing template groups.

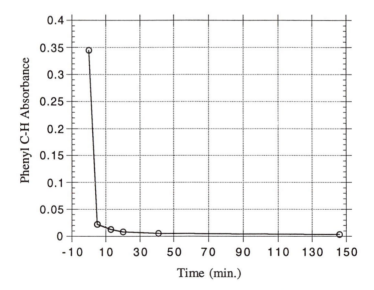

Figure 3. Decrease in phenyl C-H absorbance (3060 cm^{-1}) in infrared spectrum of **X-1** with time in oxygen plasma.

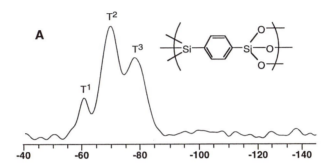

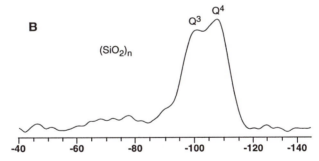

Figure 4. ^{29}Si MAS NMR spectra of phenylene-bridged polysilsesquioxane xerogel before oxygen plasma treatment, **A**, and the resulting silica after plasma treatment, **B**.

increase is small enough to warrant further characterization of the pore distributions. In addition, there was a significant increase in size of the hysteresis between the adsorption and desorption isotherms with the increase in length of the alkylene bridging groups. The silica gels prepared from the plasma treatment of the propylene-bridged gels exhibited almost no hysteresis. Thus, as the length of the bridging group increased, the size of the hysteresis increased. Hysteresis in a gas sorption isotherm is commonly attributed to bottlenecks in the pore structure. An increase in the hysteresis in these isotherms may be due to separation of a hydrocarbon nano-phase in the bridged polysilsesquioxanes with increasing template length and flexibility. After the plasma treatment, the hydrocarbon templates are replaced by pores whose dimensions mirror those of the original hydrocarbon nano-phase.

Table 2. Nitrogen Sorption Surface Areas and Mean Pore Diameters of Bridged Polysilsesquioxanes Before and After Treatment with an Oxygen Plasma *

Sample	Bridging Group	Before Plasma Treatment		After Plasma Treatment	
		Surface Area (m^2/g)	Mean Pore Diameter (Å)	Surface Area (m^2/g)	Mean Pore Diameter (Å)
X-1	phenylene	440	23	367	28
X-2	biphenylene	822	25	525	38
X-3	terphenylene	622	19	576	46
X-4	1, 3, 5- phenylene	756	34	450	53
X-5	ethylene	956	197	617	214
X-7	hexylene	8.7	NA	583	32
X-8	decylene	1.3	NA	582	34
X-9	tetradecylene	1.8	NA	647	38
M-2	decylene	547	34	254	72

*Surface areas are derived from the BET model (18).

Conclusions

Arylene and alkylene-bridged polysilsesquioxanes were treated with low temperature oxygen plasmas to remove the hydrocarbon components leaving porous silicas. Plasma oxidation was particularly efficient in removing the organic components from both porous and non-porous, powdered gels. Similarly, macro- and mesoporous monolithic aerogels were converted quantitatively into silica gels, but a microporous phenylene-bridged monolithic xerogel exhibited only surface etching on the order of 1 micron depth. Solid state ^{29}Si NMR and infrared spectroscopy were used to confirm the loss of the hydrocarbon bridging groups. We were successful in generating new porosity with the plasma treatment of the non-porous alkylene-bridged polysilsesquioxanes. There appears to be an effect on the dimensions of the resulting pores' size and shape from the length of the alkylene-bridging group used as evidenced by the increase in the isotherm hysteresis.

Acknowledgements

We would like to acknowledge Linda McLaughlin for the use of her porosimeter, Gary Zender for scanning electron micrographs, Duane Schneider for NMR spectroscopy, surface area analysis, and chemical synthesis, and Brigitta Baugher for surface area measurements and infrared spectroscopy. Edward Russick performed the supercritical CO_2 extractions. This research was supported by the United States Department of Energy under Contract No. DE-AC04-94AL85000.

Literature Cited

1. Brinker, C. J.; Scherer, G. W., Sol-Gel Science; Academic Press: London, 1990.
2. Agaskar, P. A. *J. Am. Chem. Soc.* **1989**, *111*, 6858.
3. Kaszynski, P.; Michl, J. *J. Am. Chem. Soc.* **1988**, *110*, 5225.
4. Day, V. W.; Klemperer, W. G.; Mainz, V. V.; Miller, D. M. *J. Am. Chem. Soc.* **1985**, *107*, 8262.
5. Dines, M. B.; DiaGiacomo, P. M. *Inorg. Chem.* **1981**, *20*, 92.
6. Shea, K. J.; Loy, D. A.; Webster, O. W. *Chem. Mater.* **1989**, 512.
7. Shea, K. J.; Loy, D. A.; Webster, O. W. *J. Am. Chem. Soc.*. **1992**, *114*, 6700.
8. Loy, D. A., Ph.D. Dissertation, University of California at Irvine, **1991**.
9. Small, J. H.; Shea, K. J.; Loy, D. A. *J. Non-Cryst. Solids* **1993**, *160*, 234.
10. Loy, D. A.; Russick, E.; Shea, K. J. "Better Ceramics Through Chemistry V," edited by M. J. Hampden-Smith, W. G. Klemperer, and C. J. Brinker (Mater. Res. Soc. Symp. 271 Proc. Pittsburgh, PA **1992**), p. 699.
11. Oviatt, H.; Shea, K. J.; Small, J. H. *Chem. Mater.* **1993**, 5, 943.
12. Chujo, Y.; Ihara, E.; Kure, S.; Susuki, K.; Saegusa, T. *Makromol. Chem., Macromol. Symp.*, 42/43 (Int. Symp. Ring Opening Cyclopolym., 6th, **1990**), 303-12.
13. Roger, C.; Hampden-Smith, M. J.; Brinker, C. J. "Better Ceramics Through Chemistry V," edited by M. J. Hampden-Smith, W. G. Klemperer, and C. J. Brinker (Mater. Res. Soc. Symp. 271 Proc. Pittsburgh, PA **1992**), p. 51.
14. Loy, D. A., Buss, R. J. , Assink, R. A., Shea, K. J., Oviatt, H. *Polymer Preprints* **1993**, *34*, 244.
15. Stability of polysiloxane-polyimides to oxygen plasmas was attributed to formation of a silica overlayer by Yilgor, I., Yilgor, E., Spinu, M. *Polymer Preprints***1987**, *23*, 84.
16. The absence of porosity in the alkylene-bridged polysilsesquioxanes may be related to nanophase separation and will be discussed in more detail, elsewhere.
17. T resonances in the ^{29}Si NMR spectra arise from silicon species with one substituent attached through a carbon-silicon bond and three substituents attached with silicon-oxygen bonds: $RSi(OH)_3 = T^0$, $[RSiO_{1.5}]_n = T^3$.
18. Brunauer, S.; Emmett, P.H.; Teller, E. *J. Am. Chem. Soc.* **1938**, *60*, 309.

RECEIVED April 14, 1994

Chapter 12

Sol–Gel Synthesis of Heterometallic Oxopolymers

F. Babonneau, S. Diré, L. Bonhomme-Coury, and J. Livage

Chimie de la Matière Condeseé, Université Peirre et Marie Curie, 4 place Jussieu, 75252 Paris, France

Heterometallic oxo-alkoxides provide convenient "building blocks" for the sol-gel synthesis of multicomponent ceramics. Heterometallic M-O-M' oxo bonds can be obtained when two alkoxides are mixed in solution. They are formed via Lewis acid-base reactions and are favored by a large electronegativity difference between both metal atoms and their tendency to increase their coordination number. Silicon has a rather high electronegativity and does not show any tendency toward coordination expansion. It is therefore difficult to prepare homogeneous multicomponent gels in silicate systems because of the large difference in hydrolysis and condensation rates between silicon alkoxide and other alkoxides. Several approaches are described to overcome this problem, including partial prehydrolysis of silicon alkoxides and matching of hydrolysis rates by chemical modification of other alkoxides with chelating ligands.

Sol-gel chemistry is based on the hydrolysis and condensation of metal alkoxides $M(OR)_z$. These reactions can be described as nucleophilic substitutions or additions. The reactivity of metal alkoxides toward hydrolysis therefore increases as the electronegativity of the metal atom decreases and its size increases. Silicon alkoxides are not very sensitive toward hydrolysis and condensation. Reaction rates have to be increased via acid or base catalysis. Most other alkoxides are highly sensitive to

0097–6156/94/0572–0134$08.00/0

moisture and their reactivity has to be tailored by chemical modification with complexing hydroxylated species XOH. These reagents decrease the effective functionality of alkoxide precursors leading to less condensed species. The formation of oxopolymers can then be chemically controlled with two parameters; the hydrolysis ratio h=H_2O/M and the amount of complexing ligand x=X/M (*1*).

For low hydrolysis ratio (h<z) condensation is mainly governed by the formation of μ-alkoxo and μ-oxo bridges. Solute molecular oxoalkoxides are first obtained. They increase in size as the hydrolysis ratio increases leading to the growth of oxopolymers. All alkoxide groups are hydrolyzed in the presence of a large excess of water (h>>z) giving rise to hydrous oxides $MO_{z/2}$,nH_2O. Less condensed species are obtained when nucleophilic ligands such as acetylacetone are added prior to hydrolysis. These strongly complexing ligands cannot be hydrolyzed easily. They remain bonded to the oxo-core of the molecule preventing further condensation. Colloidal particles can then be obtained in the presence of an excess of water and for small values of x. Their mean hydrodynamic diameter increases as h increases and x decreases (*2*).

The formation of oxide materials can be controlled via the chemical modification of metal alkoxides. However most advanced ceramics are multicomponent systems having two or more types of cations in the lattice. It is therefore necessary to prepare gels of high homogeneity in which cations of various kinds are uniformly distributed at an atomic scale through M-O-M' bridges. A major problem in forming homogeneous multicomponent gels is the unequal hydrolysis and condensation rates of the metal alkoxides. This may result in chemical inhomogeneities leading to higher crystallization temperatures, increased reaction times and even undesired crystalline phases. Several approaches have been attempted to overcome this problem, including partial prehydrolysis of less reactive precursors, matching of hydrolysis rates by chemical modification with chelating ligands and synthesis of double alkoxides. However, the organic ligands introduced for a better control of the solution chemistry should be removed during the heat treatment. This latter should be adapted especially with sufficient holding time at the temperature where the organic groups decompose. This paper mainly addresses the formation of heterometallic M-O-M' oxo bonds from molecular precursor solutions.

Non Silicate Oxopolymers

Heterometallic Alkoxides. In order to prepare crystalline materials at low temperature it is necessary to design metal-organic precursors such that metal ions are dispersed at the molecular level and the ligands undergo facile elimination during the transformation from molecular to bulk material. The development of precursor solutions which consist of polynuclear complexes with the metals in the stoichiometry of the desired oxide product will have a beneficial effect by lowering processing temperatures and times. Heterometallic alkoxides, which contain two or more different metal atoms linked by μ-OR bridges could therefore provide convenient precursors for the sol-gel synthesis of multicomponent materials.

Condensation typically follows the hydrolysis of metal alkoxides. However condensed species can also be observed, in the absence of hydrolysis, via the formation of oligomers (μ-OR bridges) or oxo-alkoxides (μ-oxo bridges). The driving force is then the coordination expansion of the metal atom. As a consequence most metal alkoxides, except $Si(OR)_4$, usually exhibit oligomeric structures. Their molecular complexity increases when the size of the metal atom increases and the steric hindrance of alkoxy ligands decreases (4).

The formation of heteroalkoxides is not only governed by coordination expansion, but also by acid-base properties (5). The nucleophilic addition of alkoxide groups between two different alkoxides can be described as a Lewis acid-base reaction. It is favored by a large difference in the electronegativity of metal atoms. Heteroalkoxides can therefore be easily formed by simple mixing of two alkoxides of low and high electronegativity. Heteroalkoxides incorporating an alkali metal such as $LiNb(OEt)_6$ represent, so far, the most important class of heteroalkoxides. However, since M-OR bonds are rather labile, the formation of heteroalkoxides can also be observed between alkoxides of metals as similar as Al and Ga or Nb and Ta. It is then favored by coordination expansion arising from large "N-z" values (5). N refers to the preferred coordination number of the metal. Heterometallic alkoxides are often used as "building blocks" for the sol-gel synthesis of multicomponent ceramics. The $LiNbO_3$ perovskite or the spinel $MgAl_2O_4$ for instance have been obtained via the hydrolysis of the bimetallic alkoxides $[LiNb(OEt)_6]$ and $[Mg(Al(OR)_4)_2]$ respectively (6)(7). Heterometallic alkoxides are not limited to two metals. A wide range of polymetallic alkoxides containing up to five different metals in the same molecular species have been reported (8). This explains the emphasis on heterometallic alkoxides. However polar M-OR-M' bonds could be broken upon hydrolysis and the question arises as to how the sol-gel synthesis really leads to the formation of a homogeneous oxopolymer or oxide network rather than nanophase materials (3).

Heterometallic Oxoalkoxides. Condensation between two alkoxide molecules can also proceed via ether elimination leading to the formation of μ–oxo bridges (5).

This often occurs when alkoxide solutions are heated under reflux. Oxo bridges are favored by the smaller size of oxo ligands and their ability to exhibit higher coordination numbers, up to six instead of three for OR groups. Oxoalkoxides are more stable than the corresponding alkoxides and are, of course, less reactive toward hydrolysis and condensation. They would therefore be better precursors for multicomponent ceramics or glasses. The tendency to form oxo bridges increases with the size and charge of metal ions. Large electropositive metals are known to give oxoalkoxides such as $Pb_4O(OEt)_6$, $Pb_6O_4(OEt)_4$, $Bi_4O_2(OEt)_8$, $Y_5O(OPr^i)_{13}$ or $Nb_8O_{10}(OEt)_{20}$ (9)(10). They are usually built of edge sharing $[MO_6]$ octahedra and their molecular structure is close to that of the corresponding polyoxoanions formed in aqueous solutions (1).

As for homometallic alkoxides, oxo-bridges can also be formed when two different metal alkoxides are refluxed together leading to heterometallic oxo-alkoxides. $Pb_4O(OEt)_6$ for instance undergoes complete dissolution in ethanol when $Nb(OEt)_5$ is added giving rise to $[Pb_6O_4(OEt)_4][Nb(OEt)_5]_4$ (9). Heterometallic alkoxides could provide molecular precursors with the correct M'/M stoichiometry in which some M-O-M' bonds are already formed. Whereas alkoxy bridges can be hydrolyzed during the sol-gel synthesis, oxo bridges are usually strong enough not to be broken. The crystallization of the perovskite $BaTiO_3$ phase synthesized via the sol-gel route occurs around 600°C when $Ti(OPr^i)_4$ and $Ba(OPr^i)_2$ are used as precursors. This crystallization temperature decreases down to 60°C when the mixture of alkoxides is refluxed prior to hydrolysis (11). This should be due to the in-situ formation of bimetallic oxo-alkoxides such as $[BaTiO(OPr^i)_4(Pr^iOH)]_4$. This bimetallic precursor was isolated as single crystals and characterized by X-ray diffraction. It exhibits the right Ba/Ti stoichiometry and Ba-O-Ti bonds are already formed in the solution (12).

Silica Based Oxopolymers.

Most ceramics and glasses are based on silica. The problem then is to form Si-O-M bonds in the molecular precursor and to keep them during the whole hydrolysis and condensation processes. However, silicon has a rather high electronegativity ($\chi_{Si}=1.74$) and does not show any tendency toward coordination expansion. Its chemical reactivity toward hydrolysis and condensation is therefore quite low. Silicon alkoxides $Si(OR)_4$ always exhibit a monomeric tetrahedral molecular structure and very few heterometallic alkoxides or oxo-alkoxides are formed with silicon (4). It is therefore difficult to prepare homogeneous multicomponent gels in silicate systems

because of the large difference in hydrolysis and condensation rates between silicon alkoxide and other alkoxides. Two examples will be discussed in order to illustrate this point, namely SiO_2-TiO_2 and SiO_2-Al_2O_3.

SiO_2-TiO_2 Based Oxopolymers. The key point, for the sol-gel synthesis of homogeneous SiO_2-TiO_2 materials, is to slow down the hydrolysis rate of the titanium precursors. It is possible to decrease their reactivity by complexation with acetic acid (AcOH) or acetylacetone (acacH). The condensation between hydrolyzed titanium alkoxides is then slowed considerably allowing reactions between silicon and titanium monomers. The reaction of $Ti(OPr^i)_2(acac)_2$ with silicic acid first provides solute poly-titanosiloxane species followed by further condensation to form cross-linked polymer gels. Heat treatment then leads to transparent monolithic SiO_2-TiO_2 glasses (*13*). In order to decrease the reactivity of titanium precursors without using chelating agents, alkoxides with bulky tertiary groups such as $Ti(OAm^t)_4$ can also be used. It has been shown that around pH$\approx$11, its hydrolysis rate is close to that of silicon methoxide (*14*)

By far the most used synthetic approach is to partially pre-hydrolyze silicon tetraethoxide (TEOS) with a small amount of acidic water in order to form $Si(OEt)_{4-x}(OH)_x$ species (x$\neq$0). The condensation between silicon species being very slow at low pH, hydrolyzed silicon precursors would then react with $Ti(OR)_4$ to form Si-O-Ti bonds as follows:

$$\equiv\text{Si-OH} + Ti(OR)_4 \rightarrow \equiv\text{Si-O-Ti}(OR)_3 + ROH$$

This reaction can be followed by ^{29}Si NMR (Figure 1). TEOS was pre-hydrolyzed with acidic water (pH=1) in ethanol (TEOS/EtOH/H_2O = 1/2/1). The ^{29}Si NMR spectrum reveals the presence of various Q_0, Q_1 and Q_2 species with Si-OH or Si-OEt groups. The purpose of this first hydrolysis step is to generate as many Si-OH groups as possible and to consume all the water. Thus the precipitation of titania particles can be avoided when the titanium alkoxide is added. The reaction occurs preferentially between Si-OH and Ti-OR groups. Indeed, after addition of titanium isopropoxide to the previously hydrolyzed TEOS sol (Ti/Si=1/1), all the peaks due to Si-OH species have disappeared and few peaks remain present in the ^{29}Si NMR spectrum corresponding to $Si(OEt)_4$ (δ=-82.1 ppm), -O-$Si(OEt)_3$ (δ=-88.9 ppm) and -O-$Si(OEt)_2$-O- (δ=-96.3 ppm). No clear evidence for new peaks was found, except a small one at δ=-89.9 ppm. Similar results have already been published with other titanium alkoxides such as $Ti(OBu)_4$ (*15*), (*16*) and $Ti(OEt)_4$ (*17*). The first paper suggests that no Si-O-Ti bonds are formed, but due to the high reactivity of Ti alkoxides towards hydrolysis, Si species with OH groups undergo re-esterification

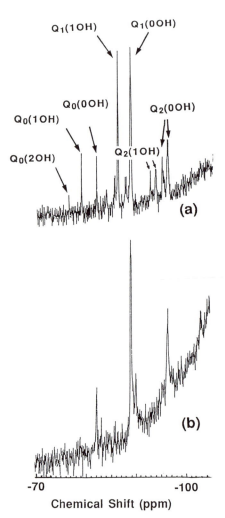

Figure 1. ^{29}Si solution NMR spectra.

a) - Pre-hydrolyzed TEOS solution [(TEOS/EtOH/H$_2$O 1/2/1; pH=1; t=6 hours].
The peaks are assigned and the number of Si-OH bonds is indicated in
parenthesis

b) - After addition of Ti(OPr^i)$_4$ [Ti/Si = 1/1; t=40 min.]

reactions to provide water (16). The second study points out that the overall intensity of the NMR spectrum and then the number of resonating species decreases when Ti is added. This suggests the formation of high molecular weight species that cannot be detected by solution NMR. In contrast with the SiO_2-B_2O_3 (18) and SiO_2-Al_2O_3 (19) systems in which the formation of Si-O-B and Si-O-Al bonds have been evidenced by ^{29}Si solution NMR, no Si-O-Ti bonds were clearly detected in the solutions.

Several techniques can be used to characterize the presence of Si-O-Ti bonds in the dried gels : Infrared or Raman spectroscopies (20), [29]Si Magic Angle Spinning (MAS) NMR ($21, 22$), X-ray absorption (23) and Photoelectron spectroscopy (24).

Si-O-Ti groups are usually characterized by a streching vibration around 945-950 cm^{-1}. (25) A silica gel prepared from the hydrolysis of TEOS is characterized by infrared absorption bands at 1190 and 1095 cm^{-1} (v_{as} Si-O), 945 cm^{-1} (v Si-OH), 800 cm^{-1} (v_s Si-O) (Figure 2a). The spectrum of a SiO_2-TiO_2 gel prepared by adding titanium isopropoxide to a pre-hydrolyzed solution of TEOS (Ti/Si=30/70) does not clearly show new bands (Figure 2b), but presents interesting differences: the relative intensity of the band at 945 cm^{-1} compared to the band around 1100 cm^{-1} strongly increases, presumably due to the presence of Si-O-Ti groups. All the bands due to the silica network become broader suggesting an increasing disorder in the oxide network due to the insertion of Ti^{IV} ions. Raman spectroscopy does not show the very strong band at 145 cm^{-1} which would arise from the formation of TiO_2 anatase domains (26).

^{29}Si MAS-NMR is a very powerful technique to detect Si-O-Al bonds in SiO_2-Al_2O_3 gels (27). In the SiO_2-B_2O_3 system, it is difficult to distinguish Si-O-B and Si-O-Si from the chemical shift values. The presence of Si-O-B bonds can be detected via the ^{11}B NMR peak position (28). In SiO_2-TiO_2 gels, the problem comes from the fact that Si-O-H and Si-O-Ti bonds lead to the same chemical shift of the Si resonance peak. Spectra of SiO_2-TiO_2 gels usually present peaks at $\approx$-90 ppm, $\approx$-100 ppm and $\approx$-110 ppm whatever the Ti/Si ratio is (Figure 3). However a peak at 100 ppm could be assigned to Q_3 units with one Si-OH group or Q_4 units with a Si-O-Ti bond. Indication for the formation of Si-O-Ti bonds comes from the increase in intensity of the peaks at -100 ppm and -90 ppm when the Ti content increases (Figure 4). However, the peaks are still sharp and do not reveal the existence of a large distribution of Si sites as expected if a large number of Si-O-Ti bonds were present.

SiO_2-Al_2O_3 Oxopolymers. The large difference in reactivity of aluminium and silicon alkoxides towards hydrolysis and condensation prevents the formation of many Si-O-Al bonds even when $Si(OR)_4$ is prehydrolyzed before mixing with

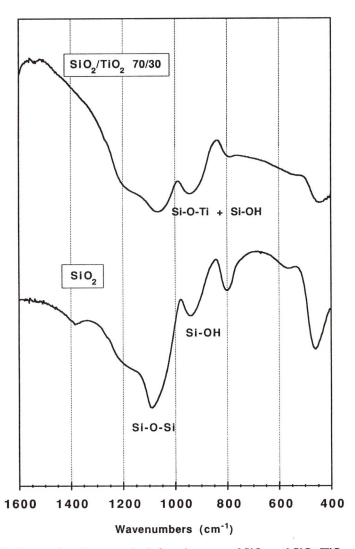

Figure 2. Comparison between the Infrared spectra of SiO_2 and SiO_2/TiO_2 gels.

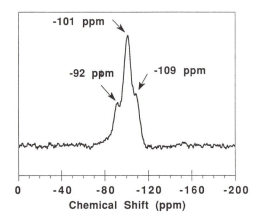

Figure 3. ^{29}Si MAS-NMR spectrum of a SiO$_2$/TiO$_2$ gel (Ti/Si = 20/80).

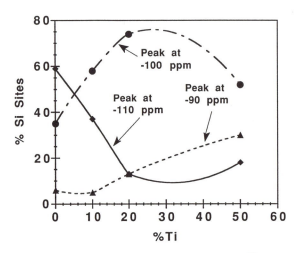

Figure 4. Evolution versus Ti content of the intensities of ^{29}Si NMR peaks in SiO$_2$/TiO$_2$ gels.

$Al(OR)_3$. Double alkoxides such as the commercially available $(Bu^sO)_2$-Al-O-$Si(OEt)_3$ have then been used for the sol-gel synthesis of SiO_2-Al_2O_3 systems (29). Si-O-Al bonds are already formed in the molecular precursor and it could be expected that they would remain in the gel. However, ^{29}Si and ^{27}Al NMR experiments show that Al-OBu^s groups are hydrolyzed much faster than Si-OEt groups. Condensation then occurs with Al-OBu^s rather than with Si-OEt leading to the formation of Al-O-Al bonds instead of Al-O-Si bonds (30).

Moreover Al^{III} exhibits several coordination numbers, from four to six, so that coordination expansion tends to occur as soon as water is added for hydrolysis. But in many crystalline structures, the aluminium atoms are tetracoordinated, and thus it is important to favor such coordination state in solution. This can be achieved either via complexation or reaction at high pH. Cordierite $Mg_2Al_4Si_5O_{18}$ is a typical example of the sol-gel synthesis of a silica based multicomponent ceramic in which four-fold coordinated Al^{III} behaves as a network former. Many sol-gel preparations have been proposed for cordierite (31). When aqueous solutions of aluminum nitrate $Al(NO_3)_3$ are used as precursors (31),(32) together with TEOS and magnesium nitrate, six-fold coordinated Al^{III} species characterized by a resonance peak at $\delta=0$ ppm are formed in the solution (Figure 5a). This coordination state remains in the gel as evidenced by the ^{27}Al MAS-NMR spectrum (Figure 6a). The ^{29}Si MAS-NMR spectrum presents features due to a silica network with Q_4 ($\delta\approx-110$ ppm) and Q_3 ($\delta\approx-90$ ppm) units (Figure 6c). This gel can thus be described as a silica gel in which Al^{III} and Mg^{II} ions are dispersed. No Si-O-Al bonds are formed at this stage. During a thermal treatment up to 800°C, Al^{III} reacts with Mg^{II} via solid state reactions leading to the spinel phase $MgAl_2O_4$. The μ-cordierite phase is formed at higher temperature, and still remains the main phase at 1000°C (31).

The formation of the spinel phase can be prevented when aluminium alkoxides are used as precursors (31-34). μ-cordierite is then obtained directly around 900°C and completely transforms into α-cordierite at 1000°C. However $Al(OBu^s)_3$ is very viscous and highly reactive toward moisture. Better results are obtained with aluminium alkoxides modified with acetylacetone (32) or ethylacetoacetate (etac) (31). The hydrolysis of $Al(OBu^s)_2(etac)$ was followed by ^{27}Al solution NMR (35). It leads to the formation of Al-OH groups, but their condensation with Al-OBu^s groups is partly prevented by the etac ligand. Hydroxylated Al-OH groups can then react with $Si(OR)_4$ to give Al-O-Si bonds.

$[Al(OBu^s)_2(etac)]_n$ was co-hydrolyzed with TEOS using an aqueous solution of magnesium acetate. The hydrolysis-condensation process was followed by ^{27}Al and ^{29}Si solution NMR (31). ^{27}Al NMR spectra show the formation of a transient species characterized by a sharp peak at $\delta \approx 51$ ppm due to tetracoordinated Al atoms in a highly symmetric environment (Figure 5b). Such a chemical shift is characteristic of the presence of Al-O-Si bonds (36) and the corresponding peak was attributed to $Al\{O[Si(OR)_3]\}_4$ species. These results are in agreement with the ^{29}Si NMR spectra showing a peak due to $Si(OAl)(OR)_3$ (19). Such species could behave as nuclei for the growth of a silico-aluminate network. Near the gelation point, this species has disappeared, but a broad component around $\delta \approx 55$ ppm increases in intensity. This peak is clearly visible in the ^{27}Al MAS-NMR spectrum of the dried gel (Figure 6b). Its chemical shift value is close to what is found for crystalline α-cordierite, in which all AlO_4 tetrahedra are surrounded by four SiO_4 tetrahedra (37). The ^{29}Si MAS-NMR spectrum confirms the formation of an alumino-silicate network (Figure 6d) with a broad signal centered at $\delta \approx -92$ ppm. It is characteristic of a distribution of Si sites and the chemical shift value suggests the presence of two or three Al-O-Si bonds per Si. When heated this gel does not lead to the formation of the spinel $MgAl_2O_4$, and the pure α-cordierite phase is obtained at 1000°C.

These sol-gel preparations of cordierite clearly demonstrate the importance of the formation of heterometallic Si-O-Al bonds in the solution. Starting from TEOS and two salts as precursors for Al^{III} and Mg^{II} ions leads to the formation of a silica network, with no evidence for the presence of heterometallic bonds in the final gel. Aluminum alkoxides provide reactive species containing Al-OH groups which can easily react with Si-OEt groups to build an alumino-silicate network.

A very interesting point is the change in the Al^{III} coordination state when bonded to Si atoms: the 4-fold coordination appears to be stabilized. The ability for Al^{III} to be 4-fold coordinated is certainly related to the large amount of Si-O-Al bonds formed in the gel. All SiO_4 tetrahedra seem to be bonded to at least one, but more probably two or three, AlO_4 tetrahedra.

Contrary to the aluminium case, in SiO_2-TiO_2 gels, prepared in two steps from a pre-hydrolyzed TEOS solution, Ti^{IV} ions are mainly hexacoordinated as shown by Ti K-edge XANES.(Figure 7). A triplet is present before the edge at 4964.6, 4966.1 and 4970 eV, characteristic of hexacoordinated Ti^{IV} ions (38). The main ^{29}Si resonance corresponds to SiO_4 tetrahedra bonded to at most one TiO_6 octahedron, about 15 to 30 % of the Si sites are present in a pure silica environment (^{29}Si resonance peak at -110 ppm). It seems thus difficult in this system to obtain very homogeneous oxopolymers, even when using experimental conditions to force the two alkoxides to react together.

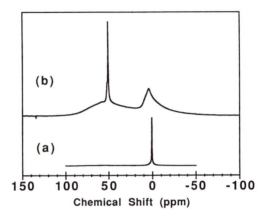

Figure 5. ^{27}Al Solution NMR spectra of two sol-gel preparations of cordierite
a) - preparation from TEOS, Aluminum and magnesium nitrates
b) - preparation from TEOS, Al(OBu[s])$_2$(etac) and magnesium acetate.
Adapted from reference 31.

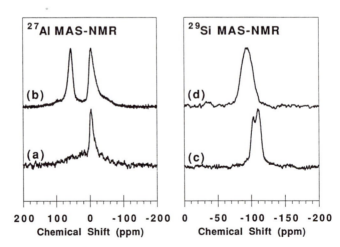

Figure 6. ^{27}Al and ^{29}Si MAS-NMR spectra of gels, precursors for cordierite
a) and c) - preparation from TEOS, Aluminum and magnesium nitrates
b) and d) - preparation from TEOS, Al(OBu[s])$_2$(etac) and magnesium acetate.
Adapted from reference 31.

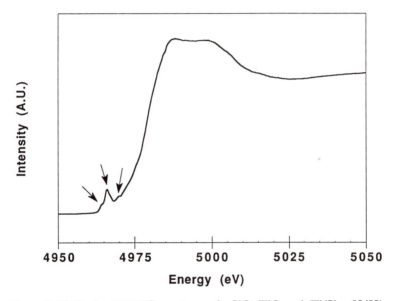

Figure 7. Ti K-edge XANES spectrum of a SiO_2/TiO_2 gel (Ti/Si = 50/50).

Literature Cited.

1 Livage, J.; Henry, M.; Sanchez, C.; *Prog. Solid State Chem.*, 1988, *18*, 259

2 Livage, J.; Sanchez, C.; *J. Non-Cryst. Solids*, 1992, *145*, 11

3 Chandler, C.D.; Roger, C.; Hampden-Smith, M.J.; *Chem. Rev.*, 1992, *93*, 1205

4 Bradley, D.C.; Mehrotra, R.C.; Gaur, D.P., in *Metal Alkoxides*, Academic Press, London (1978).

5 Caulton, K.C.; Hubert-Pfalzgraf, L.G., *Chem. Rev.*, 1990, *90*, 969

6 Eichorst, D.J.; Payne, D.A.; Wilson, S.R.; Howard, K.E., *Inorg. Chem.*, 1990, *29*, 1458.

7 Rai, J.; Mehrotra, R.C.; *J. Non-Cryst. Solids*, 1993, *152*, 118

8 Mehrotra, R.C.; *J. Non-Cryst. Solids*, 1992, *145*, 1

9 Papiernik, R.; Hubert-Pfalzgraf, L.; Daran, J.C.; Jeannin, Y., *J. Chem. Soc. Chem. Comm.*, 1990, 695

10 Hubert-Pfalzgraf, L.; Poncelet, O.; Daran, J.C., *Mat. Res. Soc. Symp. Proc.*, 1990, *180*, 73

11 Mazdiyasni, K.S.; Dollof, R.T.; Smith, J.S., *J. Am. Ceram. Soc.,* 1969, *52*, 523.

12 Yanovsky, A.; Yanovskaya, M.; Limar, V.; Kessler, V.; Turova, N.; Struchkov, Y., *J. Chem. Soc. Chem. Comm.*, 1991, 1605

13 Abe, Y.; Sugimoto, N.; Nagao, Y.; Misomo, T., *J. Non-Cryst. Solids*, 1988, *104*, 164

14 Yamane, M. Inoue, S.; Nakazawa, K., *J. Non-Cryst. Solids*, 1982, *48*, 153

15 Wies, C.; Meise-Gresch, K.; Müller-Warmuth, W.; Beier, W.; Göktas, A.A., Frischat, G.H., *Ber. Bunsenges. Phys. Chem.*, 1988, *92*, 689

16 Beier, W.; Götkas, A.A.; Frischat, G.H.; Wies, C.; Meise-Gresch, K.; Müller-Warmuth, W., *Phys. Chem. of Glasses*, 1989, *30*, 69

17 Lin, C-C.; Basil, J.D., in *Better Ceramics through Chemistry II*; Brinker, C.J.; Clark, D.E.; Ulrich, D.R., Eds. Mater. Res. Soc. Symp. Proc.; Pittsburgh, PA, 1986, Vol. 73, pp. 585-590

18 Irwin, A.D.; Holmgren, J.S.; Zerda, T.W.; Jonas, J., *J. Non-Cryst. Solids*, 1987, *89*, 191

19 Pouxviel, J.C.; Boilot, J.P., *J. Mater. Sci.*, 1989, *24*, 321

20 Schraml-Marth, M.; Walther, K.L.; Wokaum, A.; Handy, B.E.; Baiker, A., *J. Non-Cryst. Solids*, 1992, *143*, 93

21 Wies, C.; Meise-Greisch, K.; Müller-Warmuth, W.; Beier, W.; Göktas, A.A.; Frischat, G.H., *Phys. Chem. of Glasses*, 1990, *31*, 138

22 Walther, K/L/; Wokaum, A.; Handy, B.E.; Baiker, A. , *J. Non-Cryst. Solids*, 1991, *134*, 47

23 Morikawa, H.; Osuka, T.; Marumo, F.; Yasumori, A.; Yamane, M.; Momura, M., *J. Non-Cryst. Solids*, 1986, *82*, 97

24 Ingo, G.M.; Diré, S.; Babonneau, F., *Applied Surface Sci.*, 1993, *70*, 230

25 Andrianov, K.A. *J. Polym. Sci. ;* 1961, *52*, 257

26 Best, M.F.; Condrate Sr, R.A., *J. Mater. Sci.*, 1985, *4*, 994

27 Engelhardt, G.; Nofz, M.; Forkel, K.; Wishmzann, F.G.; Mägi, M.; Samoson, A.; Lippmaa, E., *Phys. Chem. of Glasses*, 1985, *26*, 157

28 Irwin, A.D.; Holmgren, J.S.; Jonas, J., *J. Non-Cryst. Solids*, 1988, *101*, 249

29 Pouxviel, J.C.; Boilot, J.P.; in *Ultrastructure Processing of Advanced Ceramics*, Mackenzie, J.D.; Ulrich, D.R., Eds.; J. Wiley & Sons, New York (NY), 1988, p.197

30 Pouxviel, J.C.; Boilot, J.P.; Dauger, A.; Hubert, L., in *Better Ceramics through Chemistry II*; Brinker, C.J.; Clark, D.E.; Ulrich, D.R., Eds. Mater. Res. Soc. Symp. Proc.; Pittsburgh, PA, 1986, Vol. 73, p.269

31 Bonhomme-Coury, L.; Babonneau, F.; Livage, J., *Chem. Mater.*, 1993, *5*, 323

32 Selvaraj, U.; Komarneni, S.; Roy, R., *J. Am. Ceram. Soc.*, 1990, *73*, 3663

33 Vesteghem, H.; Di Giampaolo, A.R.; Dauger, A., *J. Mater. Sci. Lett.*, 1987, *6*, 1187

34 Zelinski, B.J.J.; Fabes, B.D.; Uhlmann, D.R., *J. Non-Cryst. Solids*, 1986, *82*, 307

35 Bonhomme-Coury, L.; Babonneau, F.; Livage, J. *J. Sol-Gel Sci. Techn.*, (in press)

36 Mueller, D.; Gessner, W.; Behrens, H.J.; Scheler, G., *Chem. Phys. Lett.*, 1981, *79*, 59

37 Fyfe, C.A.; Gobbi, G.C.; Klinowski, J.; Putnis, A.; Thomas, J.M., *J. Chem. Soc., Chem. Comm.*, 1983, 556

38 Yarker, C.A.; Johnson, P.A.V.; Wright, A.C.; Wong, J.; Geegor, R.B.; Lytle, F.W.; Sinclair, R.N. *J. Non Cryst. Solids*, 1986, *79*, 117.

RECEIVED April 8, 1994

Chapter 13

Reaction of Boehmite with Carboxylic Acids

New Synthetic Route to Alumoxanes

Christopher C. Landry, Nina Pappé, Mark R. Mason, Allen W. Apblett,
and Andrew R. Barron

Department of Chemistry, Harvard University, Cambridge, MA 02138

Reaction of pseudo-boehmite, $[Al(O)(OH)]_n$, with carboxylic acids (RCO_2H) results in the formation of the carboxylate-alumoxanes, $[Al(O)_x(OH)_y(O_2CR)_z]_n$ where $2x + y + z = 3$ and $R = C_1 - C_{13}$. The physical properties of the alumoxanes are highly dependent on the identity of the alkyl substituents. The alumoxanes have been characterized by scanning electron microscopy, IR and NMR spectroscopy, and thermogravimetric analysis. A model structure of the alumoxanes is proposed, consisting of a boehmite-like core with the carboxylic acid substituents bound in a bridging mode. All of the alumoxanes decompose under mild thermolysis to yield γ-alumina.

The facile formation of ceramic materials from molecules has undoubtedly been one of the significant contributions made by chemistry to materials science (*1*). However, it is desirable not only to produce the ceramic *per se* but also to do so in a specific form, for example a fiber. Therefore, one of the key requirements for any ceramic precursor should be its processability. For this reason, there has been continued research effort aimed at the design of precursors with physical properties suitable for processing prior to pyrolysis. Two examples with significant commercial application are polyacrylonitrile and polyorganosilanes, both of which may be spun into fibers, and upon pyrolysis allow for the manufacture of carbon-graphite (*2*) and silicon carbide (*3*) fibers, respectively. Despite much effort, the extension of this polymer-type precursor strategy to other ceramic systems has only met with limited success.

In the case of alumina fibers a common synthetic route has involved the use of alumina gels, which are formed by the neutralization of a concentrated aluminum salt solution (*4*). However, the strong interactions of the freshly precipitated alumina gels with ions from the precursors' solutions makes it difficult to prepare the gels in pure form (*5*). Furthermore, the yield of alumina fibers from the gel is low because of the low processability of the precursor during spinning. To avoid these difficulties, alumina fibers have been prepared from alumoxanes (*6*). Alumoxane is the generic term given to aluminum oxide "polymers" (*7*) formed by the hydrolysis of aluminum compounds, AlX_3 (Eq. 1) where $X = $ alkyl, alkoxide, etc. (*8*).

0097–6156/94/0572–0149$08.00/0

$$AlX_3 \xrightarrow[- 2\ HX]{H_2O,\ \Delta} [Al(O)(X)]_n \qquad (1)$$

While early examples of alumoxane precursors for alumina fiber synthesis had processing characteristics superior to alumina gels, they were found to be unstable and decomposed during spinning. In addition their structures were a complete mystery, making further developments difficult.

The structure of alumoxanes have (despite contradicting spectroscopic data) traditionally been proposed to consist of linear or cyclic chains (**I**) (9).

$$\left(\begin{array}{c} X \\ | \\ -Al-O- \end{array}\right)_n$$

(**I**)

However, recent work from this laboratory (10) has redefined the structural view of alumoxanes and shown that, while the core structure is dependent on the identity of the organic substituent, alumoxanes are not chains or rings but three-dimensional cage compounds. Thus, alkyl-alumoxanes, $(RAlO)_n$, adopt cage structures analogous to those observed for gallium sulfides (11) and iminoalanes (12), while the structure of the hydrolytically stable siloxy-alumoxanes, $[Al(O)(OH)_x(OSiR_3)_{1-x}]_n$, consists of an aluminum-oxygen core structure (**II**) analogous to that found in the mineral boehmite, $[Al(O)(OH)]_n$, with a siloxide-substituted periphery (**III**) (10a).

(**II**)   (**III**)

We have reported that the physical properties of these siloxy-substituted alumoxanes are highly dependent on the relative organic content (10c). Low molecular weight clusters ($M \approx 2400$ g mol^{-1}) with high siloxide content (Si:Al ≈ 1.4) are soluble in hydrocarbon solvents. However, the alumoxanes formed from the equilibrium hydrolysis of aluminum compounds have a low siloxide content (Si:Al ≈ 0.14), are insoluble in all solvents, and infusible; a similar trend had previously been observed for carboxylate-alumoxanes (13). Without the advantage of hindsight, Kimura proposed that the instability of the alumoxanes used for fiber formation was due to the coordinative unsaturation at aluminum, and suggested that the use of carboxylate ligands in an appropriate ratio with aluminum would allow for the latter to be "properly coordinated". Kimura and co-workers subsequently demonstrated that carboxylate-alumoxanes were excellent precursors to alumina ceramic fibers with properties superior to those formed from alumina gels (14). These preceramic carboxylate-alumoxanes were prepared *via* a multi-step synthesis requiring accurate control over the reaction conditions (Eq. 2).

$$AlEt_3 \xrightarrow[- EtH]{+ HO_2CR} AlEt_2(O_2CR) \xrightarrow[- 2 EtH]{+ H_2O}$$

$$Al(O)(O_2CR) \xrightarrow{+ HO_2CR} Al(O)(O_2CR)(HO_2CR) \qquad (2)$$

Furthermore, while some of the carboxylate-alumoxanes formed gels in THF only those with long chain substituents (e.g., dodecanoic acid) and hence low ceramic yield were melt-processable. It would thus be desirable not only to prepare alumoxane preceramics in a one-pot bench-top synthesis from readily available starting materials, but also to determine if lower hydrocarbon substituents could yield better processability.

Synthetic Strategy

If we assume that all hydrolytically stable alumoxanes have the boehmite-like core structure (**II**) it would seem logical that instead of synthesizing alumoxanes from a low molecular weight precursor they could be prepared directly from the mineral boehmite. In the siloxy-alumoxanes we have shown the "organic" unit surrounding the boehmite core itself contains aluminum (**III**). Thus, in order to prepare the siloxy-alumoxane similar to those we have previously reported (*10a*) the anionic moiety the "ligand" $[Al(OH)_2(OSiR_3)_2]^-$ would be required as a bridging group; adding this unit would clearly present a significant synthetic challenge. However, the carboxylate-alumoxanes represent a more realistic synthetic target since the carboxylate anion, $[RCO_2]^-$, is an isoelectronic and structural analog of the organic periphery found in our siloxy-alumoxanes, *c.f.*, **III** and **IV**.

(**IV**)

In previous studies carboxylate-alumoxanes were prepared *via* a multi-step synthesis involving the reaction of a carboxylic acid with an alkoxy-alumoxane (*15*), Eq. 3.

$$[Al(O)(OR)]_n + RCO_2H \longrightarrow [Al(O)(O_2CR)]_n + HOR \qquad (3)$$

In order to simplify discussion, the alkoxy-alumoxanes were represented as having the general formula $[(RO)AlO]_n$. However, data from our laboratory has demonstrated that the alkoxide-alumoxanes have a high hydroxide content, i.e., $[Al(O)_x(OH)_y(OR)_z]_n$ (*16*). Based on spectroscopic characterization, the carboxylate-alumoxanes resulting from the synthesis have a lower hydroxide content; therefore, the carboxylic acid must also react with the hydroxide groups as shown in Eq. 4. It is entirely reasonable therefore that the reaction of boehmite, $[Al(O)(OH)]_n$, with a carboxylic acid, RCO_2H, should yield the appropriate carboxylate-alumoxane (Eq. 5).

$$\underset{\diagup\ \ \diagdown}{\overset{\overset{\displaystyle OH}{|}}{Al}} + RCO_2H \longrightarrow \underset{\diagup\ \ \diagdown}{\overset{\overset{\displaystyle O_2CR}{|}}{Al}} + H_2O \tag{4}$$

$$[Al(O)(OH)]_n + RCO_2H \longrightarrow [Al(O)(O_2CR)]_n + H_2O \tag{5}$$

Despite the fact that boehmite is a naturally occurring mineral, the majority of commercial samples are man-made, often by the hydrolysis/thermolysis of aluminum salts. In addition, they commonly contain significant quantities of gibbsite, $Al(OH)_3$, and are of variable porosity. Although we have shown the reactions discussed below to be applicable for samples from different commercial sources, to be self consistent we have chosen to discuss results obtained using a single source of low gibbsite (> 99 % boehmite) obtained from American Cyanamid.

Synthesis and Characterization of Carboxylate-alumoxanes

Refluxing powdered pseudo-boehmite, in air, with an excess of a carboxylic acid, RCO_2H, either neat (e.g., $R = CH_3$, acetic acid) or as a xylene solution (e.g., $R = C_5H_{11}$, hexanoic acid), results in the formation of the corresponding carboxylate-alumoxane (see Table 1). The alumoxanes are isolated by either filtration of the cooled reaction mixture or removal of all volatiles under vacuum followed by washing with Et_2O to remove traces of free acid. During the course of the reaction in xylene, gel formation is observed for most of the carboxylates; however, no gelation is observed if boehmite is refluxed in xylene in absence of an acid. Despite gel formation during synthesis only the hexanoate and octanoate alumoxanes form gels in organic solvents once isolated. A list of the alumoxanes synthesized and their appropriate synthetic routes is given in Table 1.

The as-received boehmite is a free flowing (fluid) white powder with no aggregation. The physical appearance and solubilities of the materials resulting from reaction of the boehmite with carboxylic acids are highly dependent on the identity of the carboxylate substituent. Thus, for $R = C_nH_{2n+1}$ (n = 1 - 3 and 13) and CH_2Cl the alumoxanes are white microcrystalline powders, insoluble in common organic solvents, whereas for $R = C_5H_{11}$ and C_7H_{15} the products are white solids, which readily form homogeneous gels in aromatic and other solvents. A summary of the physical appearance and solubility of each alumoxane is given in Table 1.

A sample of the boehmite examined by scanning electron microscopy (SEM), prior to reaction with the carboxylic acid, was found to consist of spherical particles varying in size from 10 - 100 μm in diameter (average ≈ 50 μm), see Figure 1. At higher magnifications the spheres may be seen to actually consist of small crystallites, packed together. From the SEI micrograph it is difficult to estimate a crystallite size; however, the largest distinct feature is *ca.* 0.1 μm in diameter, suggesting a crystallite size of < 0.1 μm. The average crystallite size as determined by XRD is in fact *ca.* 64 Å (020 plane) (*17*).

In contrast to the near-perfect spheres observed by SEM for the boehmite starting material, the carboxylate-alumoxanes exist as large "fluffy" conglomerates, 50 - 200 μm in size (e.g., Figure 2a), with a particle size estimated from SEM to be less than 0.1 μm in diameter. At higher magnification these constituent particles of the conglomerates can be more readily seen. Figure 2b shows some of the individual needle-like particles of the hexanoate alumoxane.

Table 1. Synthetic Routes and Selected Physical and Spectroscopic Data for Carboxylate Substituted Alumoxanes

Carboxylic Acid (RCO$_2$H), R	Synthesis[b]	Solubility	Ceramic yield[d] % (°C)	IR ν(O$_2$C), cm^{-1} antisym	sym	^{13}C NMR (ppm)	^{27}Al NMR (ppm)
acetic, CH$_3$	neat	none	30 (390)	1586	1466	179.3	1.0
chloroacetic, CH$_2$Cl	xylene	none	27 (410)	1619	1464	175.4	-4.7
propionic, C$_2$H$_5$	neat	none	28 (350)	1589	1473	180.0	e
butanoic, C$_3$H$_7$	neat	none	23 (347)	1586	1466	181.7	-0.5
pivalic, C(CH$_3$)$_3$	xylene	none	25 (351)	1586	1466	183.7	e
hexanoic, C$_5$H$_{11}$	xylene	THF, py aromatics[c]	30 (352)	1587	1465	180.0	1.5[f]
octanoic, C$_7$H$_{15}$[a]	xylene	aromatics[c]	30 (332)	1586	1466	180.8	2.4
dodecanoic, C$_{11}$H$_{23}$	xylene	pyridine, DMF	60 (465)	1587	1466	181.1	e

[a] waxy solid, [b] see Experimental Section, [c] gel formation in aromatic hydrocarbons, [d] from TGA, [e] not obtained, [f] solution spectrum.

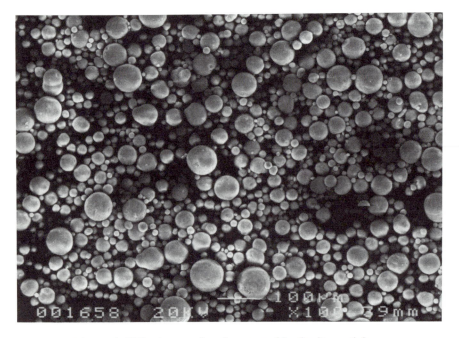

Figure 1. SEI micrographs of unreacted boehmite particles.

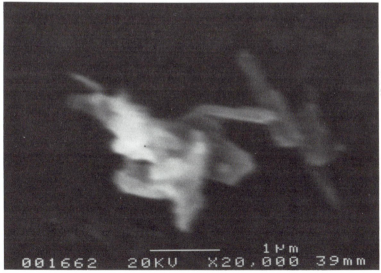

Figure 2. SEI micrographs of hexanoate-alumoxane particles.

Despite the particulate nature of the alumoxanes, homogenous continuous films and bodies may readily be prepared. Films of the hexanoate-alumoxane can be formed by dissolution of the alumoxane in either CH_2Cl_2 or THF followed by spin coating. For example, evaporation of a CH_2Cl_2 solution of the hexanoate-alumoxane on a glass slide yields a thin film, the SEM of which is shown in Figure 3. The homogenous nature of these films implies that they consist of an interpenetrating organic/inorganic matrix. While films of the alumoxanes are contiguous, many show significant shrinkage upon drying, resulting in cracking of the surface.

The surface area of the carboxylate-alumoxanes as determined by gas desorption was found to be significantly higher than the boehmite precursor (*ca.* 22 m^2). While the exact value is dependent on the acid and treatment conditions, all the alumoxanes were found to have a surface area of 120 - 150 m^2 (*18*).

Attempts to obtain self-consistent elemental analysis for the carboxylate-alumoxanes proved futile. For any specific alumoxanes the carbon and hydrogen analyses varied slightly between samples within a reaction run, and significantly between runs. Similarly, the determination of the hydroxide content by sodium napthalide titration/EDX analysis (*10a*) gave variable results. This means that while it is possible to define a general formula for the alumoxanes, i.e., $[Al(O)_x(OH)_y(O_2CR)_z]_n$, an exact representation of the product, i.e., the values for x, y and z, is not possible. Additional complications were found in determining the composition of the alumoxanes if the samples were not washed adequately with Et_2O to remove any excess carboxylic acid, which based on TGA data is adsorbed to the surface of the as-synthesized alumoxanes. However, all of the analytical techniques employed indicate that the value for z, the carboxylate content, was slightly greater than one (i.e., $1.0 \leq z \leq 1.3$) in all samples. This is in full agreement with the results of Kimura et al (*13*).

The variation in composition may be expected, since the reaction of the carboxylic acid with the pseudo-boehmite will be highly dependent on the reaction conditions; in particular the particle size, surface morphology, and the identity of surface groups (i.e., Al-O-Al *versus* Al-OH-Al). Thus, while it may be possible to obtain an average value for the number of acid groups per aluminum atom (z) one would still expect a large variation in this average from one particle to the next, even under identical reaction conditions. The variation of the acid:aluminum ratio between individual particles is best demonstrated by the EDX analysis of the alumoxane formed from the reaction of pseudo-boehmite with 6-bromohexanoic acid. The presence of the bromide allows the ready determination of the Br:Al ratio, and hence the number of carboxylate groups per aluminum atom, without disrupting the reactivity of the acid and/or the identity of the resulting alumoxane. EDX analysis shows a sizable variation in the Br:Al ratio from particle to particle.

Based upon their method of synthesis, the carboxylate-alumoxanes prepared by the route shown in equation 2 were proposed to be of the general formula of $[Al(O)(O_2CR)(HO_2CR)]_n$ in which one of the acid groups exists as the deprotonated form, while the other is protonated (*13a*). However, from ^{1}H NMR and IR spectroscopy the two carboxylate groups could not be differentiated (*13a*). In the present case the presence of greater than one equivalent of carboxylic acid per aluminum would possibly be consistent with this formulation. However, ^{13}C CPMAS NMR spectroscopy (see Table 1) of the carboxylate-alumoxanes indicates only a single carboxylate environment, consisting of the appropriate resonances due to the aliphatic carbons and one peak (*ca.* $\delta = 180$ ppm) due to the carboxylate α-carbon ($\underline{O_2C}R$).

If, as we have previously proposed, the core structure of alumoxanes is analogous to that of boehmite then one would expect the aluminum to retain exclusively six-fold coordination. This is clearly demonstrated to be the case by the

Figure 3. SEI micrograph of a hexanoate-alumoxane film spin coated on glass from methylene chloride solution.

^{27}Al MAS NMR spectra of the alumoxanes (Table 1), in which only a signal attributable to six-coordinate aluminum is detected (δ = 0 to -6 ppm, $W_{1/2}$ = 1800 - 2000 Hz), e.g., Figure 4. It is worth noting that the solution ^{27}Al NMR spectra of previously reported carboxylate-alumoxanes consists of a peak at *ca.* 0 ppm ($W_{1/2}$ = 80 - 100 Hz) (*15b*).

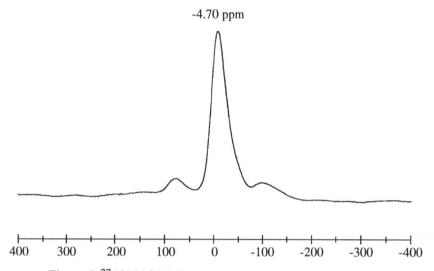

Figure 4. ^{27}Al MAS NMR spectrum of chloroacetate-alumoxane.

The IR spectra of all the carboxylate-alumoxanes contain bands at 1596 - 1586 and 1473 - 1466 cm^{-1} (Table 1), consistent with a bridging mode of coordination (**IV**). The carboxylate-alumoxanes previously reported have carboxylate bands at 1580 and 1470 cm^{-1} suggesting a similarity in the structures. Some samples with the highest carboxylate content contain additional bands at 1680 - 1640 and 1610 - 1570 cm^{-1}, indicative of unidentate coordination of a carboxylic acid group to aluminum (**V**) (*19*). In addition, all the IR spectra show a broad absorption bands between 3700 and 3400 cm^{-1}, consistent with our previous assignment for an aluminum-bound hydroxide group (*10a*).

(**V**)

A Structural Model for Carboxylate-Alumoxanes

While it is clear from the previous discussion that the carboxylate-alumoxanes are not a single species, we are able to define a structural model for the products formed from the reaction of boehmite with a carboxylic acid.

The structure of the boehmite precursor (**II**) may be considered as consisting of two parallel but staggered chains of six-coordinate aluminum atoms, linked by oxygens μ_3-capping alternate faces and μ_2-bridging the sides of the chain. The remaining coordination site on the aluminum centers are occupied by the μ_2-bridging oxygen atom of the neighboring chains. These chains pack in layers joined by hydrogen bonding, see Figure 5.

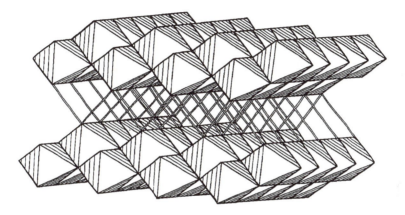

Figure 5. Structure of boehmite. Oxygen atoms are to be imagined at the vertices of each octahedron, which has an aluminum atom at the center. The double lines indicate hydrogen bonding.

Based on the ^{27}Al MAS NMR spectra of the carboxylate-alumoxanes we propose that the aluminum core retains the six-coordinate structure of the boehmite. Although direct crystallographic evidence to support this proposal is unavailable, it is entirely feasible based not only on the structure of the precursor but also on literature precedents for a wide range of main group and transition metal clusters; the best supporting evidence for our proposal are the recently reported clusters $Al_{10}(OH)_{16}(OSiEt_3)_{14}$ (*10a*), $Ni_8O_{14}[(OSiPh)_6]_2$ (*20*) and $Ca_9(OCH_2CH_2OMe)_{18}$ $(HOCH_2CH_2OMe)_2$ (*21*), all of which have boehmite-like metal-oxygen cores surrounded by a supporting organic framework.

It should be obvious, however, that whereas the mineral boehmite essentially has an infinite structure, carboxylate-alumoxanes must have a finite size. Thus there must be either end or edge groups encapsulating the boehmite-like core. We have shown previously that low molecular weight alkoxide- and siloxide-alumoxanes contain a significant proportion of four-coordinate aluminum centers, and that these four-coordinate aluminum atoms comprise the end groups. In the case of the carboxylate-alumoxanes no such four-coordinate environment is observed. Based upon ^{13}C NMR and IR spectral data we propose that the organic periphery consists of carboxylate groups occupying bridging positions across two adjacent aluminum atoms, such as shown in **VI**.

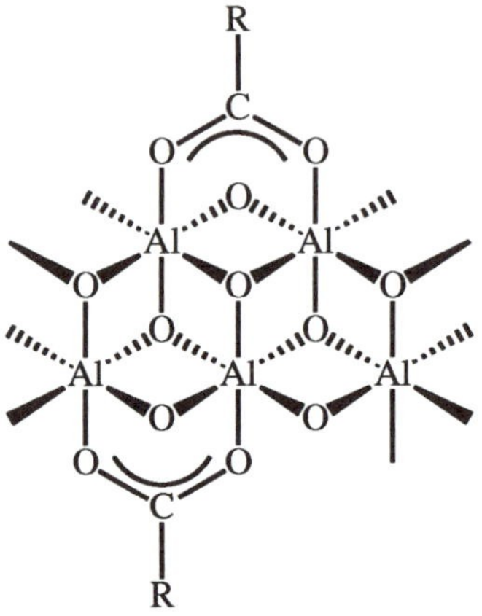

(**VI**)

Crystallographic precedent for such a structural unit exists for the aluminum dimer $[Al_2(OH)(O_2CMe)_2\{O=C(OMe)Et\}_6]^{3+}$ (22), in which two carboxylates bridge two six-coordinate aluminum atoms [**VII**, L = O=C(OMe)Et]; the carboxylate bands in the IR spectrum of the dialuminum cation (1590 and 1480 cm^{-1}) are within experimental error of the values for the acetate-alumoxane, 1587 and 1477 cm^{-1}.

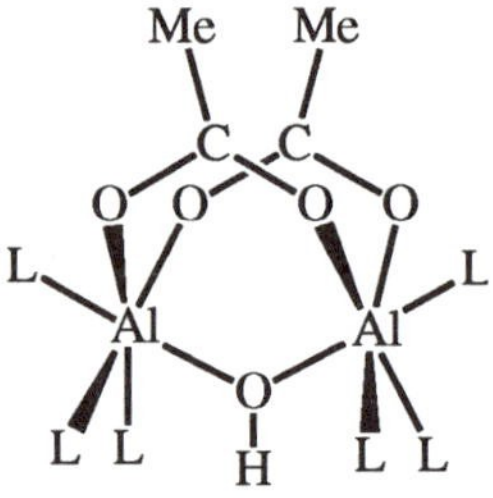

(**VII**)

   While we have no accurate data for the molecular weights of the carboxylate-alumoxanes, based upon the gas phase desorption measurements and the SEI micrographs, we can propose that the particle size of the alumoxanes is significantly smaller than the parent boehmite. Furthermore, the alumoxane particles are rod or sheet-like in shape, not linear polymers. This is due to the destruction of hydrogen-bonding within the mineral as hydroxide groups are removed and replaced with acid functionalities, as shown in Scheme 1.

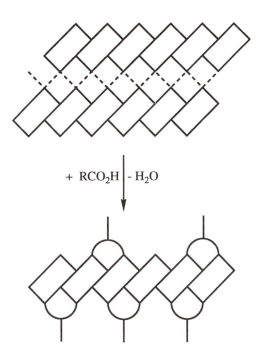

Scheme 1. Pictorial representation of the reaction of boehmite with carboxylic acids. The rectangles represent a side view of the aluminum-oxygen fused octahedra shown in Figure 5, while the carboxylate groups are represented by a semicircle and bar.

Thermal Decomposition of Carboxylate-Alumoxanes

All of the alumoxanes prepared from boehmite decompose between 180 and 385 °C to give Al_2O_3 in essentially quantitative yield. XRD spectra of the residues are consistent with their identity as γ-alumina. Based on X-ray photoelectron spectroscopic (XPS) analysis, carbon incorporation is found to be very low if the pyrolysis is carried out in an oxidizing atmosphere.

The thermogravimetric/differential thermal analysis (TG/DTA) of the boehmite starting material is shown in Figure 6a. Two distinct regions of mass loss are observed. First, between room temperature and 160 °C the mass decreases by 3 - 7 %, depending on the sample, its storage and age. Second, between 350 °C and 440 °C a loss of a further 14 - 16 % occurs. Based on mass spectroscopic analysis both of these are due to the loss of water. The first mass loss is due to water absorbed and/or hydrogen boded to the surface of boehmite, while the second is more likely due to the dehydration of boehmite giving γ-alumina with a theoretical mass loss of 15 %. The TG/DTA of the alumoxanes are distinct from that of boehmite. In general two regions are observed; the relative mass loss and temperatures at which these regions occur is dependent on the identity of the carboxylic acid. A representative example (the propanoate-alumoxane) is shown in Figure 6b.

The DTA shows a small endotherm at 140 °C consistent with the boiling point of propanoic acid (141 °C); however, the insignificant mass loss suggests that only traces of free acid remained in the sample. A second endotherm at 245 °C is accompanied by a sharp mass loss of 24 %. A third larger endothermic mass loss occurs at 330 °C. The carboxylate-alumoxanes prepared by Kimura et al (*13*) showed similar endothermic mass losses (220 - 260 and 300 - 500°C). However, in contrast to the carboxylate-alumoxanes prepared from boehmite the second endotherm occurred over a larger temperature range. The reasons for this difference are at present unclear.

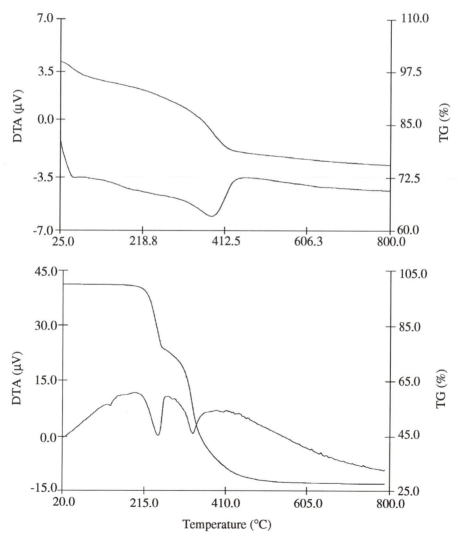

Figure 6. TG/DTA of (a) boehmite and (b) propanoate-alumoxane.

Conclusions

Alumoxanes of the general formula $[Al(O)_x(OH)_y(O_2CR)_z]_n$ have been prepared by refluxing boehmite with carboxylic acids. These carboxylate-alumoxanes have been shown by SEM to consist of conglomerates of tiny particles (less than 0.1 μm in diameter), while ^{27}Al NMR spectroscopy suggests that the boehmite core structure is retained. From ^{13}C NMR and IR spectroscopy, bridging carboxylate groups are proposed to encapsulate this boehmite-like core. The carboxylate-alumoxanes reported herein are spectroscopically similar to analogs prepared from small molecule precursors. In addition, several alumoxanes with moderate-length carbon substituents (e.g., hexanoate) form solutions or gels in various solvents, which may be readily spin-coated into films. Pyrolysis of the carboxylate-alumoxanes leads to the formation of γ-alumina with low carbon contamination. The boehmite-derived carboxylate-alumoxanes are spectroscopically similar, and have comparable ceramic yields, to analogs prepared from small molecule precursors.

Acknowledgments

This work was funded by a grant from the Office of Naval Research. We acknowledge NSERC (Canada) for a post-doctoral fellowship (A. W. A.), the University of Paris XI Orsay for a masters scholarship (N. P.), and NASA Lewis Research Center under the Graduate Student Researchers Program (C. C. L.). We are indebted to Dr. J. D. Carruthers at American Cyanamid for the generous gift of boehmite and Prof. Y. Kimura for useful discussions and hospitality.

References

1   See for example: (a) *Better Ceramics Through Chemistry IV*, Zelinski, B. J. J.; Brinker, C. J.; Clark, D. E.; Ulrich, D. R., eds., Mater. Res. Soc. Proc. *180*, Pittsburgh, PA 1990.
(b) *Chemical Perspectives of Microelectronic Materials II*, Interrante, L. V.; Jensen, K. F.; Dubois, L. H.; Gross, M. E., eds., Mater. Res. Soc. Proc. *204*, Pittsburgh, PA 1991.

2   Ezekiel, H. M.; Spain, R. G. *J. Polym. Sci., C.* **1967**, *19*, 249.

3   Yajima, S.; Hasegawa, Y.; Okamura, K.; Matsuzawa, T. *Nature* **1978**, *273*, 525.

4   See for example a) Serna, C. J.; White, J. L.; Hem, S. L. *Soil. Sci.* **1977**, *41*, 1009.
b) Hsu, P. H.; Bates, T. F. *Mineral Mag.* **1964**, *33*, 749.
c) Willstätter, R.; Kraut, H.; Erbacher, O. *Ber.* **1925**, *588*, 2448.

5   Green, R. H.; Hem, S. L. *J. Pharm. Sci.* **1974**, *63*, 635.

6   Ichiki, E. *Kagaku Kogyo* **1978**, *31*, 706.

7   While alumoxanes have often been classed as metalloxane polymers, this is misleading since they are not polymeric *per se* but exist as three dimensional cage structures, see Barron, A. R. *Comments Inorg. Chem.* **1993**, *14*, 123.

8   The first report of an alumoxane was that formed from the hydrolysis of $Al(OSiR_3)_3$. Andrianov, K. A.; Zhadanov, A. A. *J. Polym. Sci.* **1958**, *30*, 513.

9   (a) Storr, A.; Jones, K.; Laubengayer, A. W. *J. Am. Chem. Soc.* **1968**, *90*, 3173.
(b) Boleslawski, M.; Pasynkiewicz, S.; Kunicki, A.; Serwatoswki, J. *J. Organomet. Chem.* **1976**, *116*, 285.
(c) Siergiejczyk, L.; Boleslawski, M.; Synoradzki, L. *Polimery-Twoizywa Wielkoczasteczkowe* **1986**, 397.
(c) Pasynkiewicz, S. *Polyhedron* **1990**, *9*, 429.
(d) Bradley, D. C.; Lorimar, J. W.; Prevedorov-Demas, C. *Can. J. Chem.* **1971**, *49*, 2310.

10 (a) Apblett, A. W.; Warren, A. C.; Barron, A. R. *Chem. Mater.* **1992**, *4*, 167.
     (b) Mason, M. R.; Smith, J. M. ; Bott, S. G.; Barron, A. R. *J. Am. Chem. Soc.*
     1993, *116*, 4971.
     (c) Landry, C. C.; Davis, J. A.; Apblett, A. W.; Barron, A. R. *J. Mater. Chem.*
     1993, *3*, 597.
     (d) Apblett, A. W.; Barron, A. R. *Ceramic Transactions* **1991**, 35.
11 (a) Power, M. B.; Barron, A. R. *J. Chem. Soc., Chem. Commun.* **1991**, 1315.
     (b) Power, M. B.; Ziller, J. W.; Tyler, A. N.; Barron, A. R. *Organometallics* **1992**,
     11, 1055.
     (c) Power, M. B.; Ziller, J. W.; Barron, A. R. *Organometallics* **1992**, *11*, 2783.
     (d) Apblett, A. W.; Landry, C. C.; Mason, M. R.; Barron, A. R. *Mat. Res. Soc.,*
     Proc. Symp. **1992**, *249*, 75.
12 See for example, (a) Hitchcock, P. B.; Smith, J. D.; Thomas, K. M., *J. Chem. Soc.,*
     Dalton Trans. **1976**, 1433.
     (b) Cucinella, S.; Salvatori, T.; Busetto, C.; Perego, G.; Mazzei, A. *J. Organomet.*
     Chem. **1974**, *78*, 185.
     (c) Del Piero, G.; Perego, G.; Cucinella, S.; Cesari, M.; Mazzei, A. *J. Organomet.*
     Chem. **1977**, *136*, 13.
     (d) Amirkhalili, S.; Hitchcock, P. B.; Smith, J. D. *J. Chem. Soc., Dalton Trans.*
     1979, 1206, and references therein.
13 (a) Kimura, Y.; Sugaya, S.; Ichimura, T.; Taniguchi, I. *Macromolecules* **1987**, *20*,
     2329.
     (b) Kimura, Y.; Furukawa, M.; Yamane, H.; Kitao, T. *Macromolecules* **1989**, *22*,
     79.
14 Rees, Jr., W. S.; Hesse, W. *Mat. Res. Soc. Symp. Proc.* **1991**, *204*, 563.
15 (a) Kimura, Y.; Nishimura, A.; Shimooka, T.; Taniguchi, I. *Makromol. Chem.,*
     Rapid Commun. **1985**, *6*, 247.
     (b) Kimura, Y.; Tanimoto, S.; Yamane, H.; Kitao, T. *Polyhedron* **1990**, *9*, 371.
16 Apblett, A. W.; Barron, A. R. unpublished results.
17 Carruthers, J. D., personal communication.
18 Minnick, R. B.; Perrota, A. J., personal communication.
19 Gurian, P. L.; Cheatham, L. K.; Ziller, J. W.; Barron, A. R. *J. Chem. Soc., Dalton*
     Trans. **1991**, 1449.
20 Levitsky, M. M.; Schegolikhina, O. I.; Zhadanov, A. A.; Igonin, V. A.;
     Ouchinnikov, Yu. E.; Shklover, V. E.; Struchkov, Yu. T. *J. Organomet. Chem.*
     1991, *401*, 199.
21 Goel, S. C.; Matchett, M. A.; Chiang, M. Y.; Buhro, W. E. *J. Am. Chem. Soc.*
     1991, *113*, 1844.
22 Sobota, P.; Mustafa, M. O.; Utko, J.; Lis, T. *J. Chem. Soc., Dalton Trans.* **1990**,
     1809.

RECEIVED June 29, 1994

Chapter 14

Alucone Polymers and Their Pyrolytic Product Aluminum Oxides

William S. Rees, Jr.[1] and Werner Hesse

Department of Chemistry and Materials Research and Technology Center, Florida State University, Tallahassee, Florida 30306–3006

Alucone polymers of the general formula $[Al(OR) O\text{-linker-}O]_x$, where "linker" refers to an organic unit derived from an α, ω diol, have been prepared from the reaction of the diol with an alkoxy dialklylalane in a suitable medium. They have been characterized by elemental analyses, SEM, XRPD and TGA techniques. Following prolytic conversion into "alumina", the thermal evolution of phase has been monitored by XRPD. Both polymer and ceramic materials have been investigated by ss ^{27}Al NMR spectroscopy, and discussions of aluminum atom coordination environments are presented within the context of phase evolution for these materials.

The oxides of aluminum comprise a broad range of structural polymorphs, many of which have great technological utility (1). $\alpha\text{-}Al_2O_3$ finds numerous applications in ceramics, refractories, and abrasives, due to its high hardness, chemical inertness, high melting point, involatility, and oxidation and corrosion resistance. Due to its electrical insulating and high heat conducting properties, $\alpha\text{-}Al_2O_3$ is used widely in the microelectronics industry (2). The diverse $\gamma\text{-}Al_2O_3$ phases exhibit defect spinel-type structures and are termed "activated" aluminas (3). These open-structured materials are of value as catalysts, catalyst supports, ion exchangers, and chromatographic media. Compared to the relatively mature science of alumina refractories, a recent development is the production of alumina fibers. These fibers may be fabricated into a variety of textile forms such as paper, blankets and boards (4). Seemingly, no single precursor route allows the preparation of the whole spectrum of phases. The different applications demand specific macroscopic properties of each material, such as porosity, particle size and distribution and crystallinity. Therefore, the present investigation was initiated (5 - 8), wherein a new preparative procedure was sought, leading - by a suitable choice of reactants and conditions - to precursor compounds which pyrolytically convert to Al_2O_3 phases with the desired micro- and macro-structures. Such a preparative procedure should contain built-in adaptability for phase control over the resultant Al_2O_3.

[1]Correspondence address: School of Chemistry and Biochemistry and School of Materials Science and Engineering, Georgia Institute of Technology, Atlanta, GA 30332–0400

0097–6156/94/0572–0165$08.00/0

In the current approach to this problem, a choice was made to focus on polymers which were derived from diols. In part, this decision was based on their potential to form not only linear, possibly soluble polymers, but also their known reactivity towardalkyl alanes. For example, it is well-known that the reaction products of an alcohol and an alkyl alane are an alkoxy alane and a hydrocarbon (eq. 1). Another reaction of importance to this chemistry is the redistribution reaction which can occur between an aluminum alkoxide and an alcohol (eq. 2). In the presence of large excesses of HOR', this reaction will proceed to the right, as written.

$$\text{R-Al} + \text{HOR'} \longrightarrow \text{RH} + \text{R'O-Al} \tag{1}$$

$$\text{Al(OR)}_3 + \text{HOR'} \rightleftharpoons \text{Al(OR)}_2(\text{OR'}) + \text{HOR} \tag{2}$$

Utilizing the chemistry depicted in eq. 1 - 2, we designed a route to polymeric aluminum alkoxides with carbon-containing backbones. These materials have been termed "alucones", in parallel with the ubiquitous silicones, and were described briefly by Schlenker in 1958 (9). However, in the ensuing thirty years, they have been unexplored, to our knowledge. Mention in passing was given to materials of the general form $[\text{Al(OR)O-R'-O}]_x$ in technical reports from U.S. Borax in the early 1960's. However, no details of their preparation or properties were included. Therefore, due to the unknown potential of such alucone polymers either in fiber-forming reactions or as ceramic precursors, and a paucity of detail related to their composition and chemistry, a related series of organoaluminum macromolecules was chosen as a target to be examined as possible precursors useful in the preparation of ceramic materials.

Background

As briefly mentioned above, some prior work on alucones has been reported. One distinguishing feature of this class of materials, in contrast to their relatives the alumoxanes, is the presence of not only aluminum and oxygen atoms in the network backbone, but carbon atoms as well. Examples given in the U. S. Borax technical reports (**Figs. 1 - 3**) demonstrate this incumbent feature. Alumoxanes, and in particular methylalumoxane (MAO), lack this salient organic component of the polymer frame (10). Their basic repeat unit is $[(\text{R})\text{Al-O}]_x$, however, it is known now that such a simplistic formula is not representative of the magnitude of the components present in such compositions. Rather, clusters built up with Al-O-Al linkages are observed in MAO (11).

Results and Discussion
Syntheses
Polymers

The properties of the reaction products resulting from polymer syntheses differ strongly with changing preparative conditions. Presumably, these properties are dominated by the polymer structure, and, therefore, the structure is correlated closely with the synthetic procedure. For example, the key importance of the order of addition is demonstrated by comparison of the results of entries 1 - 4 in **Table I**. In experiment 1, ethylene glycol was added to an excess of diethylaluminumethoxide, resulting in the formation of $(\text{EtO})(\text{Et})\text{Al-OCH}_2\text{CH}_2\text{O-Al(Et)(OEt)}$. This molecule subsequently reacts with further ethylene glycol molecules to produce polymer 1. In experiments 2 - 4, diethylaluminumethoxide was added to an excess of ethylene glycol, leading to the formation of $\text{HOCH}_2\text{CH}_2\text{O-Al(OEt)-OCH}_2\text{CH}_2\text{OH}$. Some of the ethoxy groups also were substituted by glycol ligands in this initial step. Further reaction with diethylaluminumethoxide resulted, therefore, in partially cross-linked polymers.

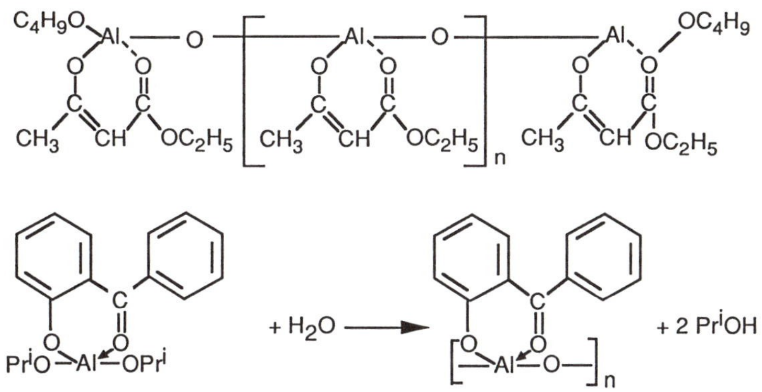

Figures 1 and 2: Examples of alumoxanes reported by U. S. Borax.

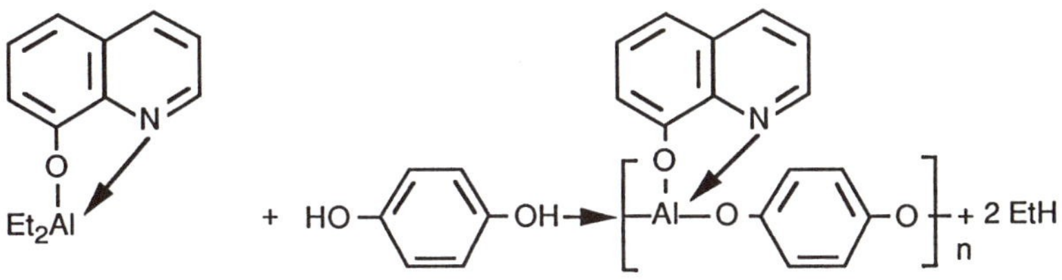

Figure 3: An example of an alucone, reported by U. S. Borax.

Table I

Observed and Calculated Mass Yields of Alucone Polymers,
Indicating Reactants and Conditions of Isolation

Experiment	Al-0.05mol	Diol x·0.05mol	Solvent	Order of addition (solvent volume)	Method of Isolation	Yield Observed (g)	Yield Calculated (g)
1	AlEt₂OEt	1 Glycol	Et₂O	G(0) Al(500)	Filt./Vt	6.2	6.4
2	AlEt₂OEt	1 Glycol	Et₂O	Al(500) G(1000)	Dist./Vt	8.9	9.1*
3	AlEt₂OEt	1 Glycol	Et₂O	Al(500) G(1000)	Dist. Th/Vt	7.3	7.6*
4	AlEt₂OEt	1 Glycol + 0.15 (SiOMe₂)ₓ	Et₂O	Al(300) G(1000)	Dist. Th/Vt	6.9	7.0
5	AlEt₂OEt	1 Glycol	THF	Al(300) G(600)	Filt./Vt	6.3	6.4
6	AlEt₂OEt	1.5 Glycol	Et₂O	G(0) Al(400)	Filt./Vt	6.9; (Vt/150°C) 5.9	5.9
7	AlEt₂OEt	1.5 Glycol	Et₂O	Al(500) G(1000)	Dist./Vt	7.3; (Vt/150°C) 6.1	5.9
8	AlEt₂OEt	1 Catechol	Et₂O	Al(200) C(300)	Filt./Vt	9.2	9.0
9	AlEt₂OEt	1 Catechol	(0°C) Et₂O	Al(250) C(1500)	a). Filt./Vt b). Dist./Vt	5.8 4.7	9.0
10	AlEt₂OEt	2 Catechol	Et₂O	Al(300) C(100)	1. Filt./Vt 2. Washed/Vt	14.4	14.5
11	AlEt₂OEt	Product from Experiment 10	THF	Al(500) 10(500)	Dist./Vt	ND	9.0
12	AlEt₃	1.5 Catechol	Et₂O	C(300) Al(500)	Filt./Vt	9.8	9.5
13	AlEt₃	1.5 Catechol	Et₂O	Al(200) C(300)	Filt./Vt	9.4	9.5

Key: G = Ethylene glycol; C = Catechol; Al = Aluminium reactant; Vt = Vacuum drying; THF = Tetrahydrofuran;
Th = with Teflon sleeves; * = Following the cross-linked formula from elemental analysis data; ND = not determined.

The accelerated ethylene glycol consumption at the beginning of this reaction sequence led to a polymeric product containing less than the nominal 1:1 stoichiometry found in polymer 1. In the products of reactions 2 - 4, the unreacted diethylaluminumethoxide is coordinated weakly. These polymers liberate a high boiling oily liquid when heated to 200°C. This oily liquid is a mixture of hydrocarbons with an even number of carbon atoms (as observed by MS data). The most probable formation mechanism is a radical-type chain-formation from ethyl ligands of unreacted diethylaluminumethoxide. Elemental analysis, weight loss upon thermolysis, and identification of trapped volatiles, each show that large amounts of ethanol were coordinated to the polymer. This observation of coordinated Lewis base is the result of incorporation of the reaction co-product from equation 2.

The choice of solvent also is important in determining the progress of reaction 1. For example, substitution of diethyl ether with the more strongly coordinating tetrahydrofuran led, under conditions identical to those of experiments 2 - 4, to a product very similar to the one from experiment 1. The results from experiment 2 indicate the influence of the method of product isolation; instead of isolating the polymer by filtration, the solvent was removed by distillation. The refluxing solvent dissolved a portion of the silicon stopcock grease used (Dow Corning high vacuum grease) which subsequently reacted with the diethylaluminumethoxide (*c.f.* (*12*)). This became evident for the first time when a mullite-type phase was detected in the pyrolysis product. The silicon content was confirmed by EDAX and elemental analysis. The substitution of silicon grease with Teflon sleeves (experiment 3) led to silicon-free polymers. The intentional incorporation of silicon was examined with stopcock grease as an added reagent in experiment 4.

The stoichiometric ratio of 1:1.5 of diethylaluminumethoxide and ethylene glycol led to three-dimensional cross-linked polymers with completely displaced ethoxy groups, independent of the order of addition (experiments 6 and 7). For confirmation of this degree of substitution, the composition of the filtrate of the reaction mixture (after isolation of the polymer) was investigated by GC/MS. In addition to the solvent (diethyl ether), ethanol was detected. As a consequence of both the coordinatively bound solvent and ethanol, these polymers must be dried at elevated temperatures (150°C) under vacuum.

Replacement of ethylene glycol with catechol led to air-sensitive polymers which turn from white to green when exposed to air. The product from experiment 8 is analogous to that from experiment 1. It was expected that an excess of catechol would lead to substitution of ethoxy ligands, due to the stronger acidity of catechol, compared to ethanol, whereas ethylene glycol is only slightly more acidic than ethanol (catechol pK_a = 9.25, ethylene glycol pK_a = 14.22, ethanol pK_a = 16.00). The result from experiment 10 indicates that a catechol excess of 5:1 does not yield a polymer, but, rather, a THF-soluble composition with all ethoxy ligands unsubstituted. The reaction of this product with an equimolar amount of Et_2AlOEt led, after distillation of the solvent, to a polymeric product (experiment 11), which should be similar to that obtained from experiment 8. However, the product from experiment 11 is soluble in THF and, therefore, is presumed to have a lower molecular weight than the product from experiment 8. Precipitation from THF solution with hexane yielded, besides the amorphous polymer, a small number of transparent crystals with rhombic habit and up to 0.8 mm in length. Unfortunately, these crystals were extremely air-sensitive and they quickly lost solvent under inert gas atmosphere in the absence of mother liquor. Therefore, it has not been possible, to this point, to determine the crystal structure. To obtain a highly cross-linked polymer analogous to the ethylene glycol polymers from reaction 2 (experiments 6 and 7), triethylaluminum, where all three of the ethyl groups easily are substituted by catechol, was used in place of $AlEt_2OEt$ (experiments 12 and 13).

Lastly, a general comment about our employment of the term "polymer" is in order. Unfortunately, we have, to this point, been unsuccessful in determining the molecular weights of these reaction products. From the observations of being infusible, poorly soluble, and amorphous (by XRPD) we concluded that these materials were of at least oligomeric character.

Ceramics

Preliminary investigation of the two groups, ethylene glycol-based polymers and catechol-based polymers, yielded reasons for the first group to be of greater interest for additional study. The decomposition temperatures from TGA experiments are about 100°C higher for the catechol-based polymers (300 - 500°C), compared with the ethylene glycol-based polymers (200 - 400°C). Therefore, the thermal decomposition of the ethylene glycol-based polymers was explored in considerably greater detail.

Pyrolysis of silicon-free polymers led to "activated" aluminum oxide phases. These "activated aluminas" often are simplified and collectively termed as γ-Al_2O_3. They crystallize in a distorted spinel-type of structure, where 21 1/3 of the 24 cation sites statistically are occupied. Stumpf $et\ al.$ described in 1950 seven different "γ-Al_2O_3" phases, all formed from pyrolysis of different aluminum hydroxides (13). He obtained the η-Al_2O_3 phase, which also is observed in the present exploration, from tempering Bayerite and transformation through Boehmite. At temperature above 800 - 900°C, this phase decomposed to θ-Al_2O_3, which transformed above 1200°C to corundum. θ-Al_2O_3 crystallizes in the β-Ga_2O_3 type of structure (14). In polymers 1 and 5, the η-Al_2O_3 phase persisted up to the highest temperature examined (1000°C), without observation of phase transformation. An explanation for this unexpected stability was found from the microtexture.

SEM-photographs (**Figs. 4-5**) show a lamellaer character of the polymer and the resulting oxide phase. The phase transformation from η– to θ-Al_2O_3, which previously had been observed only at temperatures above 600°C, demands a significant reorganization of the crystal lattice. The obtained lamellaer character hampers the aluminum atom diffusion perpendicular to the lamellae, and, therefore, the phase transformation. This explanation is supported by observations of Lippens and DeBoer, who determined an increased phase transformation temperature for lamellaer η-Al_2O_3 of 850°C (15). The authors compared the similar modification η- and γ-Al_2O_3, the last one transforming at high temperatures to δ-Al_2O_3, which shows a penultimate ordered super-spinel type of structure, before finally forming corundum. The spinel lattices of both modifications differ in the manner of distortion. While η-Al_2O_3 shows a strong one-dimensional distortion of the cubic closest packing, in γ-Al_2O_3 the tetrahedral aluminum atom sites are occupied statistically. A profile analysis of the powder diffractogram of both modifications was done by Ushakov and Moroz (16). A remarkable difference in the patterns is the clear splitting of the 400 and 440 reflections in the γ-form, while η-Al_2O_3 shows only a slight asymmetry of these reflections ($c.f.$ **Fig. 6**).

Polymers from experiments 2 and 4 contain silicon, as shown by EDAX examination. Thermolyses of these polymers yielded aluminum oxide powders, whose XRPD shows two distinct phases. These two phases were indexed as η– and m-Al_2O_3. m-Al_2O_3 first was reported by Foster in 1959 from quenching of cryolite/alumina melts (17). The name m-Al_2O_3 was chosen by Foster because of the great similarity of the powder pattern with the one of mullite. This phase also is named as ι-Al_2O_3 (iota) in the literature. Foster's samples were metastable and transformed at about 600°C into η–Al_2O_3, which changed at about 750°C into α-Al_2O_3. The silicon content was determined to be 1%. Other authors also described "silicon-free mullite phases", which contained sodium instead and were stable at least up to 1000°C (18 - 20). A silicon-free, as well as alkali metal-free, m-Al_2O_3

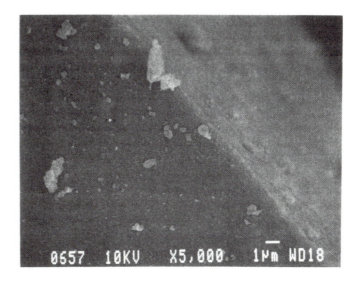

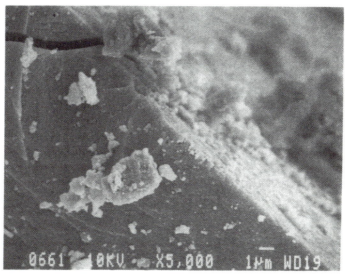

Figure 4: SEM photograph showing the morphology of the polymer (top) produced in experiment 2 (Table I) and its corresponding pyrolysis product (bottom).

Figure 5: SEM photograph showing the more "open" morphology of the polymer (top) produced by experiment 1 (Table I) and its corresponding ceramic (bottom).

was described by Cameron, but without experimental details (*21*). He constructed a (partially theoretical) phase diagram for the system $Al_2O_3 \cdot SiO_2 - Al_2O_3$ with a tetragonal mullite phase on the Al_2O_3-rich side. The lattice constants for the silicon- and alkali metal-free samples are in good agreement with the values measured for the pyrolysis product from polymer 2. A tetragonal mullite phase also was observed by Schneider and Rymon-Lipinski from the decomposition of tetraethylorthosilicate and aluminumbutylate at 950 - 1000°C (*22*). The samples contained about 26 weight percent SiO_2 and 74 weight percent Al_2O_3. Values between 10 and 15 mole percent silicon, compared to aluminum (this corresponds to >74 or 83 weight percent Al_2O_3, respectively) were detected for the silicon contents of the present investigation from EDAX measurements. In the phase diagram $SiO_2 - Al_2O_3$, pure mullite was only found up to an Al_2O_3 content of 63 mole percent (*23*). Mixtures of "mullite phase" and γ-Al_2O_3 were observed up to an Al_2O_3 content of 80 weight percent (66 mole percent), at higher Al_2O_3 concentrations only the pure γ phase was observed (*24*). McPherson describes a metastable gap of solubility in the system mullite-alumina at compositions between 70 and 100 weight percent Al_2O_3 (*25*). Therefore, the "mullite phase" of the current exploration is extremely silicon-poor and alkali metal-free, and would fit well in the phase diagram postulated by Cameron. In contrast to Foster's m-Al_2O_3, the phase in this work is stable up to the highest temperatures examined (1000°C).

Recently, substantial efforts were made to synthesize Al_2O_3 fibers with good physical properties, *e.g.*, at du Pont with the FP fiber (*26*). A common problem is the phase transformation from γ- to α-Al_2O_3 while sintering (*27*). The sudden change in volume, and onset of crystallite growth, deteriorate the mechanical properties of the fiber. Kimura, *et al.* were able to show that the incorporation of silicon by sintering agents leads to the formation of a mullite phase (*28*). Because of the thermal stability of this phase up to the highest sintering temperatures (1400°C), the phase transformation problem is inhibited, and remains unobserved. The precursor compounds of the present research allow the defined incorporation of silicon. The resulting polymers stay homogeneous and are single phase. Therefore, potential applications in this field may be possible. Inoue, *et al.* synthesized alumina phases with suitable textures for catalyst applications by "glycothermal" treatment of Gibbsite or aluminum *tris(iso*-propoxide) and subsequent calcination (*29*). The low pour density and high porosity of some of the oxides prepared in this study (see **Fig. 5**) could predestinate these phases for such catalytically-related applications.

Characterization
Polymers
In addition to bulk pyrolysis data (**Table II**), elemental analysis (**Table III**) and TGA (**Fig. 7**), the polymers were characterized further by derivatization with Hacac. The resulting completely soluble solution could be analyzed by ^{1}H NMR for a quantitative measure of each of the subunits present in the original polymer. Evaluation of peak area integrals, representative of the abundance of the species responsible for the signal, gave ratios of metal (as $Al(acac)_3$), difunctional linkers (HO~R~OH), and terminal groups (HO~R) (**Fig. 8**). From these ratios, as well as those found for free solvent incorporated into the polymeric structure (via Lewis base donation of the lone electron pairs on oxygen atoms to coordinatively unsaturated aluminum centers) a composite picture of the polymeric structures emerged (**Figs. 9 - 10**).

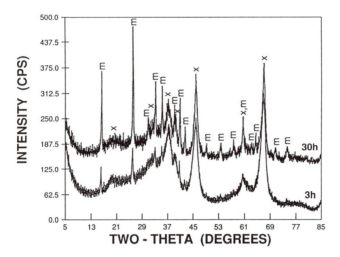

Figure 6: XRPD of the pyrolysis product from experiment 2 after 3 h and 30 h at 900°C. m indicates m-Al$_2$O$_3$ (JCPDS -No. 12-539) while x indicates η-Al$_2$O$_3$.

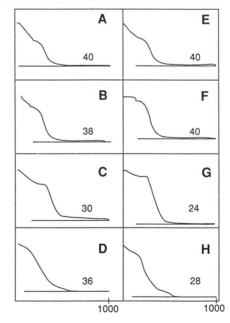

Figure 7: TGA of the polymers in flowing air, 5°C/min. Polymers are from experiments: A) 1, B) 7, C) 8, D) 3, E) 5, F) 7, with vacuum drying at 150°C, G) 13, H) 2.

Table II Observed and Calculated Weight Percent Residues Following Prolysis of Alcone Polymers in an Oxygen Atmosphere

Polymer from Experiment	Temperature (°C)	Obs. Residue (at %)	Calc. Residue (at %)
1	1000	39.9	38.64
2	1000	27.9	31.70
3	900	32.5	33.55
5	900	40.3	38.64
7	1000	38.1	43.59
7 dried @ 150°C	1000	42.3	43.59
9	1000	28.4	28.33
12	1000	24.3	27.00

Table III Observed and Calculated Elemental Analyses of Alucone Polymers, Indicating Proposed General Formula

Formula	Product from Experiment	Elemental Analyses				
		C	H	Al	Si	
OEt $\mid$ $(-Al-O-CH_2CH_2-O-)_n$	5	36.37 36.64	6.86 6.98	20.42 19.47		calc. found
$(OEt)_{0.57}$ $\mid$ $(-Al-O-CH_2CH_2-O-)_n$	1(after Vt/130°C)	33.45 33.07	6.08 6.06	23.97 22.87		calc. found
O-CH$_2$- $\mid$ $(-Al-O-CH_2CH_2-O-)_n$	6 7	30.77 30.82 31.38	5.13 5.63 5.49	23.08 21.93 20.49		calc. found found
O-CH$_2$- $\mid$ 2/3(-Al-O-CH$_2$CH$_2$-O)$_n$ · EtOH 1/3 · n AlEt$_2$OEt	3	42.11 42.76	8.55 7.10	17.76 16.61		calc. found
60%(Al(OCH$_2$)OC$_2$H$_4$O)$_n$ EtOH 30% Al(OEt)$_3$·0.5 Et$_2$O 10% (SiOMe$_2$)$_n$	2	40.59 41.24	8.33 7.39	14.84 12.62	1.71 1.71	calc. found
EtO-Al$<$ O-C$_6$H$_4$-OH O-C$_6$H$_4$-OH	9	57.93 58.87	5.21 4.36	9.30 5.03		calc. found
OEt $\mid$ $(-Al-O-C_6H_4-O-)_n$	8	53.34 52.76	5.03 5.46	14.98 13.00		calc. found
O-C$_3$H$_2$- $\mid$ $(-Al-O-C_6H_4-O-)_n$	13	57.14 57.24	3.17 4.60	14.29 12.59		calc. found

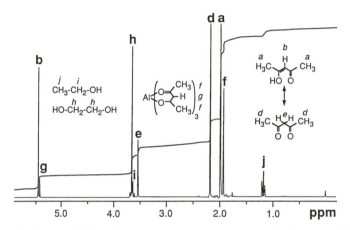

Figure 8: Proton NMR of an alucone polymer reacted with Hacac, in CDCl3.

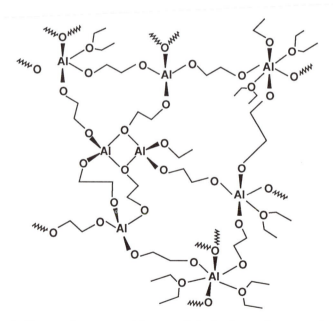

Figure 9: Proposed structure of the crosslinked polymer from experiment 2.

The differences in the apparent degree of crosslinking found between the two polymer forms can be explained on the basis of the mechanism of growth for each species (**Figs. 11 - 12**). In one instance, the transalcoholysis reaction (equation 2) is favored heavily due to the abundance of excess diol present in the reaction flask. In the reverse addition reaction, the preservation of telechelic growth dominates, as that the probability for introduction of a branching site is reduced greatly, with respect to the forward addition example.

Ceramics
In addition to the XRPD data (**Fig. 6**) available for the ceramic characterization, SEM (**Figs. 4 - 5**) data confirmed the distinction of "closed" and "open" structures for the highly crosslinked, and non-crosslinked materials, respectively. It was not possible to determine fully the structure of either the original polymeric compositions, or the resultant ceramics, by diffraction techniques or solution-based analyses alone, or in combination. Attention was turned, therefore, to solid state NMR.

Solid State NMR
Motivation
When the problem of determining a solid state structure arises, two traditional analytical techniques are turned to. X-ray powder diffraction (XRPD) and nuclear magnetic resonance (NMR) both are powerful tools for providing clear, primary data on the bulk composition and structure of solid state materials. Other analytical methods may provide parts of the puzzle, but in terms of the scope of the data provided, the ability to do bulk analysis, and the non-destructiveness of the techniques, NMR and XRPD are generally the methods of choice for structural determination. Each of these methods has its limitations, however, and as the field of materials science matures, these limitations increasingly are being felt. XRPD, a diffraction technique, gives precise information about a *crystalline* structure. For the technique to be of use, therefore, there must be some net, long-range order in the structure. In the absence of such order, *i. e.*, in an amorphous polymeric material, XRPD gives only limited data about the structural nature of the material.

Issues
One can envision a "worst case" substance of unknown structure which is amorphous, insoluble, and has as one of its principle components a quadrupolar nucleus. The alucones produced by the reaction of diethylaluminumethoxide and ethylene glycol fit into this category. Since the polymers are potential precursors for phase-specific aluminas, the primary structural concern is the relationship between the observed coordination environment present around the aluminum sites in the polymer and the corresponding ceramic (*5 - 8*). XRPD has shown that the structures are amorphous; there is no crystalinity within the microstructure which would diffract X-rays and divulge hints related to the structure (**Fig. 13**). ^{27}Al is a relatively sensitive NMR nucleus, but the polymers are insoluble, and the aluminum nucleus is quadrupolar; until recently NMR was all but useless for structural determination of these polymers (**Fig. 14**).

Quadrupolar Broadening: Second-order Effects
While magic angle spinning (MAS) has seen a number of applications in the last four decades, particularly in the field of organic polymers, the technique has been unable to provide high resolution results for all NMR-active nuclei. For nuclei possessing a non-zero quadrupolar moment, MAS fails to eliminate all anisotropies.

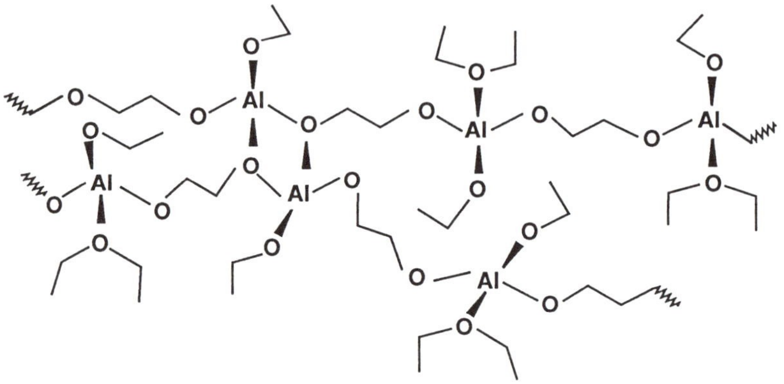

Figure 10: Proposed structure of the non-crosslinked polymer from experiment 1.

$$HO \sim OH + Al(OR)R_{2(xS)} \longrightarrow$$

$$+Al(OR) - O \sim O +_x \quad \overset{X}{\longrightarrow}$$

Figure 11: When the aluminum reagent is in excess no transalcoholysis occurs.

$$Al(OR)R_2 + HO \sim OH_{(xS)} \longrightarrow$$

$$+Al(OR) - O \sim O +_x \longrightarrow$$

$$+Al(O \sim O)_{0.5} - O \sim O +_x$$

Figure 12: In the presence excess alcohol, transalcoholysis is possible and cross-linking occurs.

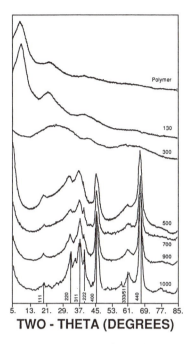

TWO - THETA (DEGREES)

Figure 13: XRPD of polymer and pyrolysis products heated from 25 - 1000°C.

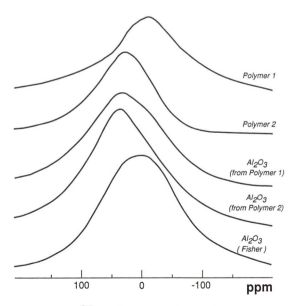

Figure 14: Solid state ^{27}Al NMR at 270 MHz, taken on static samples.

Many such nuclei -- ^{11}B, ^{27}Al, ^{23}Na, and ^{17}O, for instance -- are of considerable importance to the field of materials science. Further developments in high resolution solid state NMR of these nuclei, therefore, would provide a powerful new means of structural elucidation.

Quadrupolar interactions are considered "second-order" interactions, and the angular dependencies of quadrupolar interactions are described by both the second and the fourth Legendre polynomials. Unfortunately, the two polynomials do not have similar solutions. There is no single angle at which both the second and fourth Legendre polynomials can be averaged to zero. Some experiments have been performed using angles which solve the fourth Legendre polynomial, and these have been somewhat successful in improving resolution, depending on whether first- or second-order interactions dominate. Occasionally an intermediate angle has provided optimal resolution (30). These experiments have been described as VASS NMR, so-named for variable angle sample-spinning, but the improvement in resolution highly is sample dependent.

Other typically useful means of overcoming shortcomings in NMR also fail to simultaneously remove both quadrupolar and dipolar interactions. Since quadrupolar interactions are dependent inversely upon the magnitude of the Zeeman field, using more powerful magnets could provide better resolution (31). Such a solution technically is difficult as well as rather expensive. Radio frequency manipulations likewise fail to average out quadrupolar interactions (32).

Status

Pines, et al. realized that the averaging of first- and second-order broadening effects need not be truly simultaneous, only concurrent on the time scale of the NMR experiment (33, 34). Thus, the solution of the Legendre polynomials can be made time dependent, and, during the time of the experiment, both polynomials can be averaged to zero:

$$<P_l(\cos Q)> = \int P_l(\cos Q)\Omega(Q)dQ = \int P_l(\cos Q(t))dt = 0$$

where $\Omega(Q)$ is a weight function describing a distribution of angles. By solving for $l = 2$ and $l = 4$, and limiting the number of angles over which an experiment will take place to two, Pines has shown that for fractions k of time spent at one angle or the other, there are limitless solutions such that equations 3 and 4 are satisfied (32).

$$P_2(\cos Q_1) = -kP_2(\cos Q_2) \qquad (3)$$

$$P_4(\cos Q_1) = -kP_4(\cos Q_2) \qquad (4)$$

By allowing the magnetization to evolve for a certain time at one angle, then changing the angle and allowing further evolution, both spherical harmonics, and their associated broadening effects, can be brought to zero. Pines has called the technique dynamic angle spinning (DAS) and he has called the pairs of angles DAS complementary angles.

Thus, the DAS experiment consists of placing the sample in a specially-designed probe capable of rapidly reorienting the sample in the magnetic field. Doty Scientific has built such a probe, consisting of a 5 mm stator with transmit coil fixed on a pivot through the perpendicular axis, with a computer-controlled stepper motor capable of driving a connecting shaft up or down, thus changing the angle of the stator (and its spinning sample) through a possible 60° in ~30 ms. The complementary angles for the simplest DAS experiment, with two angles and equal evolution time at each angle, are 37.38 and 79.19° (35). These angles are calculated by the computer controlling the motor, after the magic angle is determined by standard calibration techniques. A 90° pulse is applied with the sample spinning at one angle, the magnetization is allowed to evolve for time t, a second 90° pulse is applied to store magnetization along the -z axis during the angle flip, and a final pulse returns the magnetization to the xy plane to allow it to evolve at the second angle for time t. The caveat is that T_1 for the sample must be sufficiently long enough to allow storage of the magnetization during the time it takes to flip the sample (*36*). Once this condition is met, a two-dimensional spectrum can be generated by varying t, with one projection showing the static powder pattern of the sample and the other projection showing a spectrum free from first- and second-order broadening effects, that is, the isotropic chemical shift spectrum.

Acknowledgments

We gratefully acknowledge the initial financial support provided this project by the Defense Advanced Research Projects Agency contract number MDA 972-88-J-1006. The DAS equipment was purchased under the National High Magnetic Field Laboratory. Professors Cross and Gullion and Dr. Lazo are acknowledged for stimulating discussions and ideas related to experimental design and execution. Werner Hesse was the recipient of a Deutsche Forschungsgemeinschaft postdoctoral fellow award. Mr. Tom Fellers obtained the SEM and EDAX data and Mr. Dick Rosanske, Dr. Tom Gedris and Mr. L. Chopin collaborated in obtaining the solid state ^{27}Al NMR data. Professor Ronald Clark assisted with the collections of the TGA data and Mr. Hyem Matthies participated with the XRPD data collection.

References

(1)   (a) Hoberg, H.; Gotor, V. *J. Organomet. Chem.* **1977**, *127*, C32.
(b) Hoberg, H.; Aznar, F. *J. Organomet. Chem.* **1979**, *164*, C13. (c) Hoberg, H.; Aznar, F. *J. Organomet. Chem.* **1980**, *193*, 161. (d) Hoberg, H.; Aznar, F. *Angew. Chem.* **1977**, *92*, 132.

(2)   (a) Hoberg, H.; Gotor, V.; Milchereit, A.; Krüger, C; Sekutowski, J. C. *Angew. Chem.* **1977**, *89*, 563. (b) Hoberg, H.; Aznar, F. *J. Organomet. Chem.* **1980**, *193*, 155.

(3)   Zura, Mac G.; Goodboy, K. P.; Koenig, J. J. "*Aluminum Oxide*" in "*Kirk-Othmer Encyclopedia of Chemical Technology*", 3rd ed. Vol. 2, Wiley, **1978**, New York, S. 218-244.

(4)   Greenwood, N. N.; Earnshaw, A. *Chemistry of the Elements*, Pergamon, **1984**, New York, S. 272-284.

(5)   Rees, W. S., Jr.; Hesse, W. *Chemical Perspectives of Microelectronic Materials II*, Interrante, L. V.; Jensen, K. F.; DuBois, L. H.; Gross, M. E., Eds.; *Material Research Society Symposium Proceedings*, Material Research Society: Pittsburgh, Pennsylvania, **1991**, Vol. 204, pp 563-570.

(6)   Rees, W. S., Jr.; Hesse, W. *Synthesis and Processing of Ceramics: Scientific Issues*, Rhine, W. F.; Shaw, T. S.; Gottschall, R. J.; Chen, Y., Eds., *Materials Research Society Symposium Proceedings*, Materials Research Society: Pittsburgh,Pennsylvania, **1991**, Vol. 249, pp 51-57.

(7)   Rees, W. S., Jr.; Hesse, W. *Polymer Preprints* **1991**, *32(3)*, pp. 573-574.

(8)   Rees, W. S., Jr.; Hesse, W. *Polymer Preprints* **1993**, *34*, 252-253.

(9)   Schlenker, F. *Farbe und Lack* **1958**, *64*, 174.

(10)   Andrianov, K. A.; Zhadanov, A. A. *J. Polym. Sci.* **1958**, *30*, 513.

(11)   Apblett, A. W.; Warren, A. C.; Barron, A. R. *J. Am. Chem. Soc.* **1993**, *115*, 4971.

(12) Apblett, A. W.; Barron, A. R. *Organometallics* **1990**, *9*, 2137.
(13) Stumpf, H. C.; Russell, A. S.; Newsome, J. W.; Tucker, C. M. *Ind. Eng. Chem.* **1950**, *42*, 1398.
(14) Kohn, J. A.; Katz, G.; Broder, J. D. *Amer. Mineral.* **1957**, *42*, 398.
(15) Lippens, B. C.; De Boer, J. H. *Acta Cryst.* **1964**, *17*, 1312.
(16) Ushakov, V. A.; Moroz, E. M. *Kinet. Katal.* **1985**, *26*, 968.
(17) Foster, P. A., Jr. *J. Electrochem. Soc.* **1959**, *106*, 971.
(18) Duvigneaud, P. H. *J. Am. Ceram. Soc.* **1974**, *57*, 224.
(19) Perrotta, A. J.; Young, J. E., Jr. *J. Am. Ceram. Soc.* **1974**, *57*, 405.
(20) Elliot, A. G.; Huggins, R. A. *J. Am. Ceram. Soc.* **1975**, *58*, 497.
(21) Cameron, W. E. *Amer. Mineral.* **1977**, *62*, 747.
(22) Schneider, H.; Rymon-Lipinski, T. *J. Am. Ceram. Soc.* **1988**, *71*, C162.
(23) Aramaki, S.; Roy, R. *J. Am. Ceram. Soc.* **1962**, *45*, 229.
(24) Ossaka, J. *Nature* **1961**, *191*, 1000.
(25) McPherson, R. *J. Am. Ceram. Soc.* **1980**, *63*, 110.
(26) Dhingra, A. K. *Phil. Trans. R. Soc. Lond.* **1980**, *A294*, 411 - 559. (b) Kimura, Y.; Nishimura, A.; Shimooka, T.; Taniguchi, I. *Makromol. Chem., Rapid Commun.* **1985**, *6*, 247. (c) Kimura, Y.; Sugaya, S.; Ichimura, T.; Taniguchi, I. *Macromolecules* **1987**, *20*, 2329. (d) Wallenberger, F. T. *Ceram. Bull.* **1990**, *69*, 1646.
(27) Kimura, Y.; Furukawa, M.; Yamane, H.; Kitao, T. *Macromolecules* **1989**, *22*, 79.
(28) Morita, H.; Yamane, H.; Kimura, Y.; Kitao, T. *J. Appl. Poly. Sci.* **1990**, *40*, 750.
(29) (a) Inoue, M.; Kondo, Y.; Inui, T. *Inorg. Chem.* **1988**, *27*, 215. (b) Inoue, M.; Tanino, H.; Kondo, Y.; Inui, T. *J. Am. Ceram. Soc.* **1989**, *72*, 352. (c) Inoue, M.; Kominami, T., Y.; Inui, T. *J. Am. Ceram. Soc.* **1990**, *73*, 1100.
(30) Ganapathy, S.; Schramm, S.; Oldfield, E. *J. Chem. Phys.* **1982**, *77(9)*, 4360.
(31) Chmelka, B. F.; Mueller, K. T.; Pines, A.; Stebbins, J.; Wu, Y.; Zwanziger, J. W. *Nature* **1989**, *339*, 42-43.
(32) Samoson, A.; Lippmaa, E.; Pines, A. *Molec. Phys.* **1988**, *65(4)*, 1013-1018.
(33) Llor, A.; Virlet, J. *Chem. Phys. Lett.* **1988**, *152*, 248.
(34) Mueller, K. T.; Sun, B. Q.; Chingas, G. C.; Zwanger, J. W.; Terao, T.; Pines, A. *J. Magn. Reson.* **1990**, *86*, 470-487.
(35) Mueller, K. T.; Chingas, G. C.; Pines, A. *Rev. Sci. Instrum.* **1991**, *62(6)*, 1445-1452.
(36) Jelinek, R.; Chmelka, B. F.; Wu, Y.; Grandinetti, P. J.; Pines, A.; Barrie, P. J.; Kilnowski, J. *J. Am. Chem. Soc.* **1991**, *113*, 4097-4101.

RECEIVED June 27, 1994

Chapter 15

Sol–Gel-Based Inorganic–Organic Composite Materials

H. Schmidt and H. Krug

Institut für Neue Materialien, Im Stadtwald, Gebäude 43,
66123 Saarbrücken, Germany

The sol-gel process allows the synthesis of inorganic non-metallic materials through a soft chemistry route and organic functions can be incorporated. While maintaining the phase size of the inorganic component on a molecular or nano range, inorganic molecular or nano composites can be fabricated. For this reason techniques that make use of surface interaction controlling ligands for colloidal sol-gel particles have been developed, allowing the inorganic part to be linked chemically to the organic one. Using these principles, new nanocomposites for optical application and for coatings with special properties (anti soiling, corrosion inhibition) have been developed.

Background

The sol-gel-process can be considerd as a chemical route for the synthesis of inorganic nonmetallic materials such as glasses or ceramics (*1-6*). The formation of the inorganic backbone is based on a condensation step including the formation of metal/oxygen/metal bonds. The condensation step can take place at low temperatures, e.g. at the boiling temperature of organic solvents like ethanol. For the formation of densified glasses or sintered ceramics, high temperatures are required (e.g. T_g of glasses) and at these temperatures, in general, organic groupings are oxidized or pyrolyzed. However, as shown elsewhere (*7-13*) the presence of non-volatile organic groupings, especially if they are linked to the inorganic backbone, leads to an interesting type of materials (inorganic-organic composite ormocers, organically modified ceramics, polycerams or ceramers). Organic groupings can be used for basically three types of structural effects. They can be used for the modification of the inorganic backbone reducing the network connectivity and leading to processing temperatures low enough to fabricate dense materials without distroying the organics, or they can additionally be used for building up a second type of network by polymerizing or polycondensing appropriate organic groupings, or they can be used for achieving special functions (acids, bases, electron donating or accepting functions).

Using this basic principle, a wide variety of possibilities exists for synthesizing new materials. The material properties, of course, should depend strongly on the

0097–6156/94/0572–0183$08.00/0

structures produced. This includes the type of interaction between the inorganic and the organic components (chemical bonds or physical bonds) and phase size and phase distribution of the inorganic and the organic part. For these reasons it is very important to control the mechanism of the formation of the inorganic part by the sol-gel-process as well as e.g. in case of polymerization reactions, the organic part for the formation of the organic chains or even networks. In this paper some principle possibilities will be discussed, and it will be shown how far these principles can be used for the synthesis of composites with interesting properties for applications.

General Principles

Sol-gel reactions can be characterized by several reaction steps. At first soluble precursors in form of monomers, oligomers or clusters, e.g. in several alkoxides (*14*) are reacted to active species mostly by addition of water and the subsequent formation of reactive ≡MeOH (Me≡ metal ion) groups. The reactivities depend on the thermodynamic parameters and on the electronegativity of the metal (*15-17*). The condensation rate of SiOH groups is slow (due to the electronegativity of the Si) compared to most other elements like aluminum, chromium, titanium or others. Spontaneous condensation reaction takes place in these cases started by a nucleation process and a subsequent growth reaction, which is, especially in not to highly concentrated solution, diffusion controlled. In general, the process leads to colloidal sols the particle size of which is dependant on the monomer concentration and the stabilization of the colloidal particles. Stabilization is necessary, otherwise the condensation process continues between the different particles leading to precipitates or gels.

In the case of silicon based compounds, e.g. organoalkoxysilanes the link between the inorganic and organic phase can be obtained by the SiC bond which is a very stable one. This principle, however, cannot be used for other elements since metal carbon bonds in general are not stable under hydrolytic conditions. As shown elsewhere, the use of double bond containing organoalkoxysilanes (methacryloxy or vinyl) leads to inorganic-organic composites on a molecular level, and coatings for various applications have been synthesized (*18-20*).

In case of titania or alumina the complexation of alkoxides with β-diketones containing polymerizable groupings has been proposed (*21*). The question arises how far this type of coordinative bond is stable against hydrolysis in synthesized sol-gel materials and how far within the resulting composites a molecular level distribution of inorganics and organics can be maintained. Another possibility for the complexation of alkoxides are carboxylic acids (*16*). As shown with methacrylic acid, complexes from zirconium alkoxides can be formed easily (*22*) and as shown further (*23*), after hydrolysis condensation and polymerization the resulting zirconia methacrylate composites are hydrolytically extremely stable. Using this principle nano scale particles can be formed from zirconia methacrylate systems by controlled hydrolysis and condensation and in combination with methacryloxy silanes they have been reacted to composites (*24*).

In Figure 1 the WAXS patterns are given for different compositions of this three component system. The scattering maxima correspond to particle sizes of about 2 nm showing that fairly narrow distribution can be established for various compositions. It has to be mentioned that the zirconia methacrylate colloids, having been formed after hydrolysis and condensation of the complexed alkoxide do not show any shift of the C=O frequency of the carboxy group towards the free acid pointing out that in the colloidal form the complex type of bond is still present (*24*).

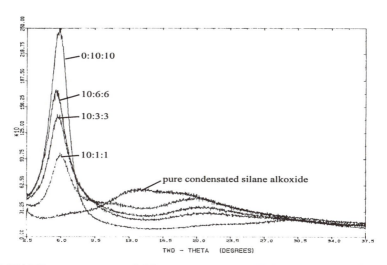

Fig. 1: WAXS-measurements of different composites (I/II/III) of a system alkoxisilane (I), Zr-alkoxide (II) complexed with methacrylic acid (III)

Similar results have been obtained from alumina with propionic acid (*25, 26*) even if commercially available boehmite with a particle size of 15 nm is used as starting material. Refluxing the boehmite for several hours in alcoholic solution with excess of propionic acid leads to a boehmite propionic acid surface compound which is hydrolytically stable and can be used as a stabilizer leading to sols with surprisingly high solid content (up to 40% by volume) without gelation. This type of surface modified boehmite has been incorporated into coatings for obtaining very high abrasion resistance (R. Kasemann, E. Wintrich, Institut für Neue Mat., 1993, priv. communication).

These examples show that colloidal systems produced by the sol-gel process can be surface modifed by organics and to obtain a new type precursors with organic functions for further processing.

If the modifying agents are bifunctional, the newly created colloidal particles surface as in the case of zirconia, can be used for obtaining specifically designed properties, e.g. polymerizable particles to be incorporated into organic polymers or ormocers. This principle has been used for the modification of a variety of different sol-gel-derived or precipitated particles, even for metal colloids (*27*) or semiconductor quantum dots (*28*). The new type of „colloidal" precursors are interesting candidates to synthesize new inorganic-organic nanocomposites where the ceramic, semiconductor or metal colloid provides special functions (e.g. mechanical, optical or electronical).

For material synthesis, hydrolysis and condensation as well as polymerization have to be controlled. For the control of hydrolysis and condensation, especially if silanols are used, ^{29}Si-NMR spectroscopy in combination with the determination of water content are interesting methods. In the composition shown in figure 1 this method has been used to monitor the state of hydrolysis and condensation of methacryloxy or epoxy group containing silanols for controlling the water and silanol content of the system to avoid precipitation of highly reactive alkoxides of other elements. If precipitation from alkoxides of zirconia titania or alumina has to be avoided it is necessary to incorporate the hydrolysed species as fast as possible into a sol-gel network before nucleation and particle growth can take plase.

For this reason, water required for hydrolysis has to be added in a very homogeneous or latent form. The problem can be solved in an elegant way by using silanols instead of molecular water. In this case, a cocondensation reaction between the silanol and the heteroalkoxide or a ligand exchange can take place, as shown in equation 1.

$$= AlOR + HOSi \equiv \ \rightarrow \ = Al - O - Si \equiv + \ HOR \qquad (1)$$
$$\equiv ZrOR + HOSi \equiv \ \rightarrow \ \equiv ZrOH + ROSi \equiv$$
$$\equiv ZrOR + HOSi \equiv \ \rightarrow \ \equiv Zr - O - Si \equiv$$

Whereas the formation of an =Al-O-Si≡ bond is very likely, and Si-Al hetero-esters (e.g. available from Dynamit Nobel chemical company) can be formed, in the case of zirconia, the formation of the corresponding bond could not be proved definitely. But it is well known that zirconia alkoxides or hydroxides are catalysts for the network condensation of silanes with a potential to immobilize the zirconia within the network to be formed on the level of very small particles or to maintain the colloidal distribution without further aggregation. This leads to a very specific strategy for the sol-gel reaction if alkoxysilanes are suitable for the formation of composites: at the point of the addition of the highly reactive alkoxide, the free H_2O content of the system should be as low as possible but should be present in the latent form as SiOH groups in concentrations as high as possible. In order to optimize the silanol group content, the hydrolysis and condensation reaction of methacryloxy group containing silanes (MTPS) has been monitored by Karl-Fischer titration and ^{29}Si-NMR spectroscopy. In Figure 2, the water balance of this system and the silanol concentration development are shown.

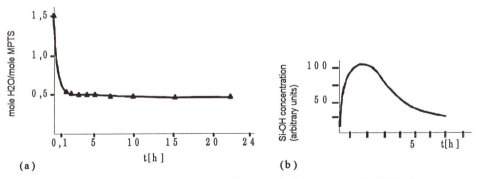

Fig. 2: H_2O and SiOH concentration in the hydrolysis reaction of MPTS-silane as a function of time in the hydrolysis reaction. (a): water content by Karl-Fischer; (b): SiOH by ^{29}Si-NMR.

As one can see, the water content decreases very rapidly at the beginning of the reaction and then stays almost constant, whereas the silanol concentration shows a maximum and then decreases slowly. The silanol concentration decrease even continues in the regime where the water concentration remains constant, which has to be attributed to the condensation reaction shown in equation 2.

$$\equiv SiOH + ROSi \equiv \;\rightarrow\; \equiv Si - O - Si \equiv + HOR \tag{2}$$

For the addition of fast-reacting alkoxides, the regime with the lowest possible water content and the highest possible silanol content is desirable. As shown elsewhere (*27*) by using this principle, rather high concentrations of zirconia (up to 50 mole %) can be added to the prereacted silane without precipitation, and a wide variety of compositions becomes possible (Wilhelm, R., Ph. D. Thesis, INM, Saarbrücken, 1994, in print).

ZrO_2 containing composites are of interest for tailoring the refractive index for optical applications, e.g. waveguides. Another interesting feature is mechanical properties resulting from colloidal zirconia material particles which leads to mechanical surface properties not obtainable by pure polymers.

Whereas the sol-gel reaction can be considered to be the first part of the inorganic-organic composite synthesis, the second part of the synthesis reactions consists of the polymerization of double bonds, if an additional organic backbone has to be built up. Since the formation of the inorganic backbone, for example in the case of zirconia, is supposed to exist in form of a colloidal network interpenetrated by a siloxane network. If methacrylic acid complexes are used, a rather complex structure of the unpolymerized system should be present, characterized by entangling methacryloxy groups linked to the silane and methacrylic acid groups linked to the zirconia. A model is shown in figure 3.

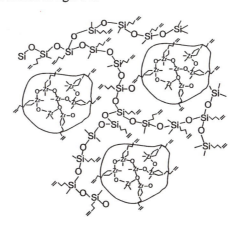

Fig. 3: Model for a colloidal organic-inorganic network interpenetrated by a siloxane network

The question arises how far it is possible to obtain conversion rates high enough to result in stable materials from these systems. For this reason, investigations have been carried out to define the degree of conversion as a function of reaction conditions and temperature.

In case of radical chain polymerization, the conversion behavior is characterized by three different stages of the polymerization process. In an initial stage, conversion is inhibited by oxygen absorbed in the material. A second period of constant reaction rate for the radical formation follows. C-C bonds are generated and polymerization takes place up to the so called gel point. At the gel point, the radicals lose their high mobility as the medium gets highly viscous and the diffusion controlled conversion process slows down. The conversion rate after the gel point is only controlled by the diffusion rate of the remaining small radicals and reactive groupings. The effect of the gel-point, which can also be compared to the glass transition in pure organic systems, can be reduced by increasing the temperature, and conversion and degree of polymerization can be shifted to high values during the polymerization process. This conversion of the C=C double bonds of up to 90 % can be obtained by a temperature treatment during photopolymerization in ZrO_2/methacrylic acid/methacryloxy silane systems, as shown in Figure. 4

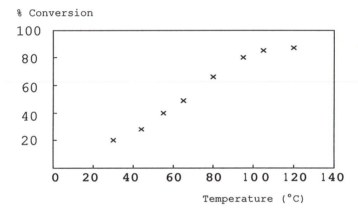

Fig. 4: Degree of conversion of C=C double bonds in dependance of temperature in the system ZrO_2/methacrylic acid/methacryloxy silane 1/1/10 mole.%.

The results given in Figure 4 show that, at higher temperatures, surprisingly high conversion rates are possible in a system which was expected to be extemely inflexible from the three-dimensional inorganic crosslinking. Furthermore, it means that the diffusion rates of the double bond-containing units are rather high and should be in the range of monomers in organic systems during photopolymerization. In order to investigate the diffusion coefficients, an optical method was developed (29). The method is based on the generation of local polymerization areas in order to decrease monomer concentration by polymerization within these areas. This should enhance diffusion of non-polymerized monomers into these spots. If the polymerized areas exhibit a diffraction index different from that of the unpolymerized areas (which should be very likely in case that the polymerizable groupings are linked to zirconia), the increase of the diffractive index with time can be used as a parameter for determining the diffusion rate. In the case of photopolymerization, a diffractive grating can be created by irradiating an area with a light beam the intensity of which is modulated by two wave mixing interference. The developing diffraction pattern can be visualized and the diffraction efficiency as a measure for the refractive index has been monitored as a function of time. The result of such a two-wave mixing experiment, which is also known as Forced Rayleigh scattering (FRS), is shown in Figure 5.

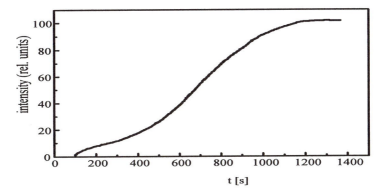

Fig. 5: Typical diffraction efficiency read in real time by FRS, system: ZrO2/methacrylic acid/methacryloxy silane 1/1/10 mole.% (Oliveira, P., private communication, INM, 1993).

From the time dependant changes of the diffracted intensity I(t) in the radical formation and conversion region (equation 4)

$$I(t) = \left(A + B\exp{-\frac{t}{\tau}} \right)^2 + C \tag{4}$$

where A and C are coherent and incoherent scattering backgrounds and B is a contrast factor, the characteristic relaxation time τ of the process can be calculated. This relaxation time is directly correlated to a Diffusion coefficient D (equation 5)

$$\tau^{-1} = 4\pi^2 D / \Lambda^2 \tag{5}$$

For a known grating period Λ of the diffraction grating, the diffusion coefficient D can be calculated and if the relaxing species has diffusive character, the diffusion coefficient will show an Arrhenius behavior of the form (equation 6)

$$\ln D = \ln D_0 - E_A / kT \tag{6}$$

and the activation energy E_A can be calculated. Figure 6 shows the experimental results of temperature dependant diffusion coefficients. For the described experiments, an activation energy of 29 kJ/mole is determined, which is typical for side chain relaxations of organic polymer systems (the diffusion coefficient calculated from equation 5 is in the order of magnitude of about 10^{-7} - 10^{-9} cm^2 sec^{-1} for the temperature range of 25-100 °C).

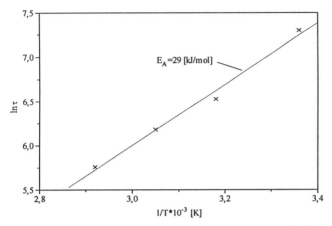

Fig. 6: Arrhenius plot of temperature dependant relaxation times in the system
ZrO$_2$/methacrylic acid/methacryloxy silane 1/1/10 mole.%.

These experiments show that it is possible to synthesize inorganic-organic
nanocomposites by appropriate control of the sol-gel process in combination with
photopolymerization. It has to be mentioned that this type of composites, due to the
control of phase separation, which is in the lower nanometer range, shows excellent
optical properties with respect to transparency and optical loss in the visible region.

Fabrication of Optical Waveguides. Direct laser writing (Figure 7) allows a locally
well defined polymerization by a focused laser beam. By using a x-y stage as a sample
holder, the experiment is very flexible and complex pattern can be prepared. After a
development step, in which the unpolymerized material is removed by alcohol, a three
dimensional structure is obtained. As can be seen from Figure 7, also rectangular
waveguide pattern can be obtained if laser intensity and moving velocity of the x-y
stage are correctly adjusted.

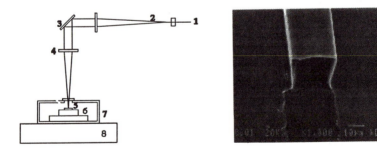

Fig. 7: Scheme of the direct laser writing experiment and SEM micorgraph a written
waveguide after development. (1: laser beam, 2: beam expander, 3: mirror, 4: lens, 5:
sample, 6: x-y stage, 7: closed chamber, 8: optical table).

Fabrication of Microlenses by Holographic Interference. The interference pattern of an object wave with a spheric and a planar wavefront results in concentric rings of maximum intensity. The sinusoidal form of this interference pattern leads to local changes of the degree of polyerisation of the same periodicity. After a development step, surface modulations are generated which act for inducing light as a Fresnel zone plate. The principal experimental setup and a patterned Fresnel lens is shown in Figure 8. The theoretical maximum diffraction efficiency of such Fresnel lens is about 34 %. The Fresnel lens by the described process shows a maximum diffraction efficiency of 29 % of the first order focus which is not far from the theoretical one.

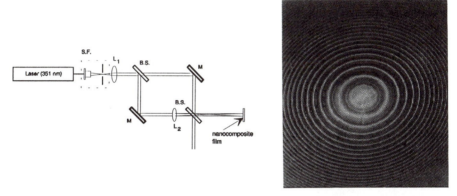

Fig. 8: Experimental setup for the fabrication of Fresnel lenses (L1,2:lens, M: mirror, BS: beam-splitter, SF: spatial filter) and developed Fresnel lens produced by this holographic interference technique (Oliveira, P., INM, 1993, private communication).

Protective Coatings. Corrosion protection of metals in most cases requires specially tailored interfaces. In the case of aluminum surface, the use of reactive silanes should lead to an interface characterized by the formation of chemical bonds. In the presence of wet atmosphere, Al-hydroxides are formed on the surface and SiOH and SiOR containing coating solutions should form a stable bond to the AlOH covered surfaces following equation 7

$$\equiv SiOH + HOAl \equiv \quad \rightarrow \quad \equiv Si\text{-}O\text{-}Al \equiv \quad + \quad H_2O$$
$$\equiv SiOR + HOAl \equiv \quad \rightarrow \quad \equiv Si\text{-}O\text{-}Al \equiv \quad + \quad HOR \quad \quad (7)$$

The cleavage of the Si-O-Al bonds as the reverse reaction should destroy the protective effect. For this reason, the immobilization of the silanes by organic crosslinking, as shown in (*30*), can prevent the interfacial degradation. In the present example, (a system based on an epoxy group containing silane, $Si(OR)_4$ and $Al(OR)_3$), the epoxy groupings have been crosslinked with bisphenoles, as shown in equation 8. The experimental details are shown in (*30*). The immobilized SiO-Al grouping should prevent the progress of the interfacial corrosion process and lead to stable interfaces.

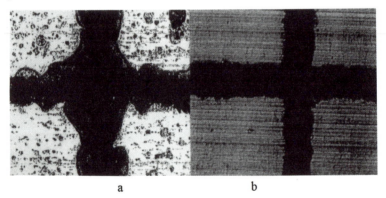

(8)

Applied to Al surfaces, dense coatings having a very efficient corrosion protection are obtained. Figure 9 shows two Al plates after a 14 days salt spraying test. Before starting the test, a cross cut was carried out on the coated specimen in order to test whether underpinning takes place. The results show that no underpinning takes place.

a   b

Fig. 9: Comparison of a conventional epoxy layer coated (a) and composite coated (b) Al plate after the salt spray test. Cross cuts show no corrosion along the interface and on the coated areas in the composite coated section.

The experiments demonstrate the high protection effect of the stable interface, especially since this type of coatings does not prevent the penetration of water molecules to the interface. If a corrosion reaction leads to the formation of aluminum hydroxides, as a stable solid phase, the reaction should, based on the rules of the mass balance law, continue in one direction. This is obviously not the case, so it has to be concluded that the reaction products, e.g aluminum hydroxide and silanols, cannot be removed from their "molecular" location, and the reaction stops at an equilibrium, representing an overall stable interface. The immobilization takes place by fixing the silane through the organic network. Colloidal Al_2O_3 particles have been introduced into the coating system using Al alkoxides to obtain high scratch-resistance additionally.

Low Surface-Free Energy Coatings. The state of an interface or surface always tends to achieve the lowest free enthalpy value. This has been used to tailor the interface and the surface of a coating from multicomponent systems with components with assigned special functions to combine good adhesion, high tranparency, high scratch resistance and low surface free energy as known from PTFE in one and the same system. For this reason liquid mixture containing components with silanol and SiOR groups of nano-scale particles from methacrylic acid complexed zirconia, silanes carrying perfluorinated carbon chains was used: if this sol is getting in contact with a polar surface as shows schematicaly in Figure 10 it can be expected that, thermodynamically driven, the polar groups (assumed that a sufficient diffusability is provided by the liquid phase) turn to a polar surface providing a good adhesion, and the perfluorinated groups turn to the atmosphere minimizing their surface-free energy and that the ZrO_2/methacrylic acid particles remain distributed in the system randomly to be imobilized by subsequent polymerization.

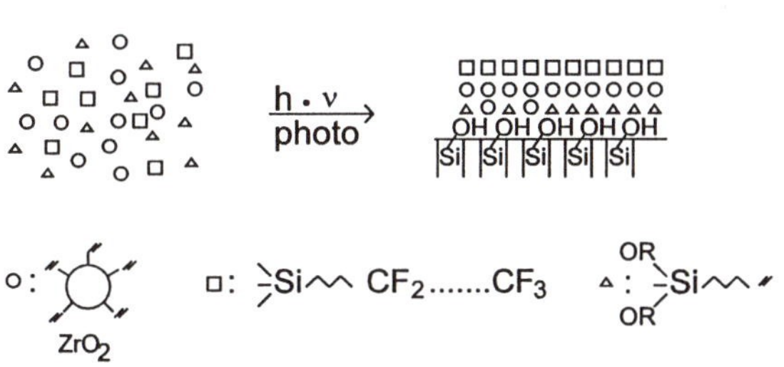

Fig. 10: Scheme of an organic-inorganic nanocomposite with polymerizable organic groupings

Experiments carried out with this type of systems (*31*) could prove that it is possible to obtain transparent coatings which are very scratch-resistant (zirconia colloids) with surface-free energy as low as 19 mJ/m^2, which is even below the surface free energy of polytetrafluoroethylene. The coatings can be prepared with high transparency so that they can even be used on transparent plastics or glasses, not disturbing the light transmission. As proven by ESCA measurements, the concentration of perfluorinated groupings in the surface layer is by the factor of 7 to 10 higher in the surface layer than in the bulk. The low surface-free energy (due to the surface concentration effect) can be obtained with concentrations of the fluorinated silane as low as one mole-%. The coatings can be photocured or even thermally cured at higher temperatures and are used as anti-soiling coatings on glasses and a variety of other substrates.

Conclusion

The investigations show that, using very specific features of sol-gel and colloidal chemistry in combination with organic polymerization chemistry, it is possible to develop inorganic-organic nanocomposites with interesting properties for application.

It also has been shown that the control of the reaction is a key to the material property tailoring. It is of special interest that even with colloids very stable links between inorganic and organic phases can be provided, thus preventing agglomeration of colloidal particles, which is an indispensible requirement for using these nanocomposites for optical applications.

Literature Cited

1.„Chemical Processing of Advanced Materials", Hench, L. L., West J.K. Ed.; Wiley John, New York (1992)

2. „The Physics and Chemistry of Sol-Gel Processing", Scherer G.W., Brinker C.J. Ed.; Academic Press, New York Boston 1990

3. „Sol-gel technology for thin film, fibers, preforms electronics and speciality shypes", Klein L. Ed., Noyes Publications, Park Ridge, New Jersey, 1988

4. „Sol-Gel Optics I", Mackenzie J.D., Ulrich D.R., Ed.; Proc. SPIE 1328, Washington (1990)

5. „Sol-Gel Optics II", Mackenzie J.D. Ed.; Proc. SPIE 1758, Washington (1992)

6. „Chemistry, Spectroscopy and Application of Sol-Gel Glasses", Reisfeld R., Jorgensen C.K., Ed.; Springer-Verlag, Berlin (1992)

7. Uhlmann D.R., Boulton J.M., Toewee G.: SPIE 1328 (1990), 270

8. Novak B.M., Davies C.: Macromolecules, 24, (1991), 5481

9. Huang H.H., Orler B., Wilker G.L.: Macromolecules, 20, (1987), 1322

10. Mackenzie J.D., Chung Y.J., Hu J.: J. Non-Cryst. Solids, 147, (1992), 271

11. Schmidt H.: MRS Proc., 171, (1990), 3

12. Schmidt H.in: „Ultrastructure Processing of Advanced Materials", Uhlmann D.R., Ulrich D.R., Eds.; John Wiley & Sons (1992), 409

13. Schmidt H., Wolter H.: J. Non-Cryst. Solids, 121, (1990), 428

14. Menrotra R.C., Bohra R., Gaur D.P.: „Metal β-diketonates and Allied Derivated" Academic Press, London, (1978)

15. Sanchez C., Ribot F., Doeuff S.: In: „Inorganic and Organometallic Polymers with Special Properties", Laine R.M., Ed.; Nato, ASI series, Kluwer Academic Pub., 206, (1992), 267

16. Bradley D.C., Mekrotra R.G., Gaur D.P., „Metal Alkoxides", Academic Press, London, (1978)

17. Zelinski B.J.J., Uhlmann D.R.: J. Phys. Chem. Solids, 45, (1984), 1069

18. Schmidt H., Seiferling B., Phillip G., Deichmann KI., in "Ultrastructure Processing of Advanced Ceramics", Mackenzie J. D., Ulrich D. R. Ulrich, Eds.; John Wiley & Sons, New-York, (1988), 651

19. Phillip G., Schmidt H., J. Non-Cryst. Solids 63, (1984), 283

20. Schmidt H., DVS Berichte 10, (1988), 54

21. Sanchez, C. in „Proceedings on „First European Workshop on Hybrid Organic-Inorganic Materials",Bierville (France), (1993), 9

22. Popall M., Meyer H., Schmidt H., Schulz J.: MRS Symp. Proc. 180, (1990), 995

23. Schmidt H., Krug H., Kasemann R., Tiefensee F.: SPIE Vol. 1590, (1991), 36

24. Krug H., Tiefensee F., Oliveira P.W., Schmidt H.: SPIE Vol. 1758 (1992), 448

25. Schmidt H., Nass R., Aslan M., Schmitt K.-P., Benthien T., Albayrak S., Journal de Physique IV, (1993), 1251

RECEIVED August 12, 1994

Chapter 16

Synthesis of Nanocomposite Materials via Inorganic Polymer Gels

Kenneth E. Gonsalves[1], Tongsan D. Xiao[2], and Gan-Moog Chow[3]

[1]Polymer Science Program, Institute of Materials Science and Department of Chemistry, and [2]Connecticut Advanced Technology Center for Precision Manufacturing, University of Connecticut, Storrs, CT 06269
[3]Center for Bio/Molecular Science and Engineering, Naval Research Laboratory, Washington, DC 20375

Materials with novel properties are obtained when the constituent phase morphology is reduced to the nanometer dimension. In this emerging area, investigations have involved the use of inorganic polymeric gels for the preparation of nanostructured materials containing Al-B-N or Fe-B-N. Inorganic polymer gels, or precomposite gels, were obtained by the ammonolysis of aqueous solutions of commercially available inorganic salts. The conversion of the inorganic polymer gels via thermochemical processing, resulted in the formation of nanostructured composite materials. The preparation of AlN/BN nanocomposite and FexN/BN magnetic nanocomposite materials via these chemical routes are outlined. The characterization and properties of these materials will also be discussed.

Recently, there has been considerable interest in the area of nanostructured materials. Nanostructured materials are usually referred as to materials having phases or grain structures less than 100 nm. Because of the large surface area to volume ratio of the nanostructures, a significant fraction of atoms (up to 50%) are found in the grain boundary regions. Compared to conventional materials with structures of micron size, nanostructured materials are anticipated to have superior mechanical, magnetic and physical properties [1-4]. For instance, paramagnetic and superparamagnetic nanocomposites may be produced for high density magnetic storage applications and for enhancing the efficiency of magnetic refrigeration cycles. The reduction of grain size to the nanometer regime also opens up novel processing possibilities for advanced materials. Examples are the brittle ceramics and intermetallics, which are current candidates for high temperature applications. The conventional micron size ceramics and intermetallics cannot be deformed plastically at ambient temperatures because of the small dislocation density and mobility, leading to undesirable brittle failure. However, low temperature deformability, superplasticity and ductility can be achieved for nanostructured materials that will allow an improvement of current processing methods for high temperature ceramic and intermetallic materials [4].

The starting point for the preparation of the nanostructured materials has been the synthesis of single phase metals and ceramic oxides using physical

0097–6156/94/0572–0195$08.00/0

evaporation methods [1]. However, detailed studies on the synthesis of silicon based nanostructured nonoxide ceramic composite materials have been particularly intriguing [5-7]. A synergistic approach was developed, by combining interdisciplinary concepts of organometallic polymer chemistry and materials processing for the preparation of nanocomposite materials. The nanophase Si-N-C containing ceramic particles were obtained by the ultrasonic injection of a silazane precursor into (i) the beam of an industrial CO_2 laser [5,6], and (ii) a hot-wall reactor [7]. The nanocomposites AlN/BN [8-9] and Fe_xN/BN [10-11] were prepared by the thermochemical conversion of a metal-organic polymeric gel. Using organometallic precursors, multi-component metallic alloy or steels have also been produced including nanostructured Fe-Cr-Mo-C M50 steel [12], Fe-Co alloy [13], metal carbides and metal silicides [14-17]. Methodologies for these syntheses involve the designing of suitable molecular precursors that facilitate the production of nanostructured materials or composites containing phases of desired compositions and microstructures. Organometallic and inorganic precursors are particularly attractive since they provide: (i) the synthesis of materials with the ultrafine structures by assembly of small atomic or molecular clusters, (ii) phases with selected stoichiometry, and (iii) mixing of constituent phases at the molecular level. Consequently, the possibilities now exist for the synthesis of nanocomposite materials in significant quantities. These may contain a broad range of phases, such as nitrides, carbides, borides, and intermetallic phases.

In this chapter, the preparation and charaterization of AlN/BN ceramic nanocomposite and Fe_xN/BN magnetic nanocomposite using inorganic polymeric gels are discussed. The latter precomposites, derived by the reaction of water soluble starting compounds, are converted to the final desired nanocomposite materials through thermochemical processing in a reactive gaseous environment. The methodology presented here has the following features: (i) using an aqueous solution mixture of aluminum- and boron-containing starting compounds (ii) ammonolysis of the aqueous solution, resulting in the formation of a pre-composite gel which converts into the AlN/BN composite on further heat treatment, and (iii) a potentially economically viable synthesis, and possibly also a process of low toxicity.

1. AlN/BN Composite System

The aluminum nitride/boron nitride (AlN/BN) composite material is of considerable interest in modern technological applications because of its excellent mechanical, chemical and physical properties. Aluminum nitride and boron nitride are essentially chemically inert materials, stable at elevated temperatures. Furthermore, they are electrically nonconductive, yet possess high thermal conductivity, together with optical transparency over a wide spectral range, and have excellent dielectric properties. These properties make the AlN/BN composites suitable as an electromagnetic window material, grinding media, heat sink, as well as electric and structural materials.

The synthesis procedure is outlined in a flowchart illustrated in Figure 1. Aluminum chloride hexahydrate $AlCl_3 \cdot 6H_2O$, urea $(NH_2)_2CO$, and boric acid H_3BO_3 were dissolved in water in molecular proportions with selected stoichiometry for Al, B, and N as shown in Table 1. After thorough mixing, ammonia gas was bubbled into the aqueous solution mixture, while it was vigorously stirred and heated to about 90°C. After the removal of water, a precomposite gel was obtained which was then transferred into a reaction chamber for post high temperature processing to obtain the final AlN/BN nanocomposite. The precomposite material was pyrolyzed under a constant flow of ammonia at a temperature range from 500°C to 1100°C.

Table 1: Materials Composition

NO.	AlN Concentration	BN Concentration
1	0%	100%
2	5%	95%
3	20%	80%
4	35%	65%
5	50%	50%
6	65%	35%
7	80%	20%
8	95%	5%
9	100%	0%

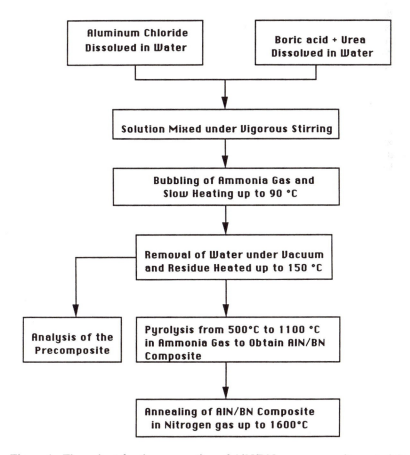

Figure 1. Flow chart for the preparation of AlN/BN nanocomposite materials.

1.1 Characteristics of the Precomposite Powders

The precomposite gel has a complex polymeric gel type structure: boron bonded with urea and aluminum bonded with N-H and chlorine atoms [8]. These structures have been tentatively derived by a combination of characterization techniques including NMR, FTIR, mass-spectrometry., and x-ray diffraction studies. Thermal analysis studies performed in argon using a simultaneous high temperature TGA/DTA technique (Figure 2) show that the precomposite gel undergoes a decomposition in the 100°C to 400°C temperature range, with a total weight loss of about 80%. This weight loss is mainly due to the formation and sublimation of NH$_4$Cl, and possibly other minor species such as CO. Only one peak was evident in the DTA curve, which corresponds to the major weight loss with respect to the TGA curve. The other two smaller peaks could probably be due to the loss of excess urea from the sample and the decomposition of carbonyl groups. A combination of the TGA/DTA measurement suggested that the thermal conversion of the precomposite materials could occur at a temperature as low as 400°C to obtain the final AlN/BN composite.

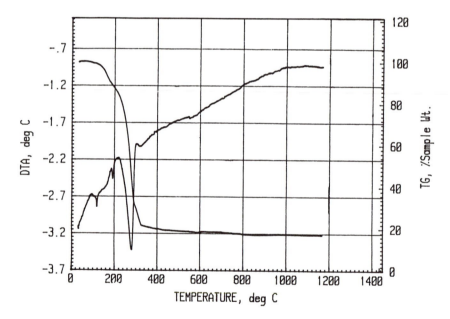

Figure 2. Simultaneous high temperature TGA/DTA measurement showing the conversion of the precomposite into the final AlN/BN nanocomposite.

1.2. Structure and morphologies of AlN/BN Composite Powders

Samples of the Al-B-N precomposite were heated in NH$_3$ from room temperature to 500°C, 800°C, and 1100°C to obtain the AlN/BN composite. As shown in Figure 3a, the precomposite material contains various chemical bonds. The relatively sharp peaks observed included one at 1620 cm.$^{-1}$ corresponding to the C=O bond, and one at 1400 cm.$^{-1}$ corresponding to the B-N bond [8, 18-21]. The peak at 1150 cm.$^{-1}$ may arise from a C-N bond [15], while the broad band from 800 to 400 cm.$^{-1}$ may arise from a combination of Al bonded with N, and C species [6]. The peaks located in the 3600 to 2800 cm.$^{-1}$ wavenumber range may be assigned to O-H and N-H bonds. The C=O group in the precomposite gel is eliminated upon heating to 500°C in ammonia, evidence of this is provided by the FTIR spectra in Figure 3b, where only three peaks are evident. The peak at 1400 cm.$^{-1}$ is due to the B-N bond [6], while the broad peak at 3600 to 2900 cm.$^{-1}$ is due to a combination of O-H and N-H. The broad peak centered at 700 cm.$^{-1}$ is due to Al-N bonded species. Following treatment at 800°C (Figure 3c), the small shoulder at ~1650 cm.$^{-1}$ on the BN peak disappeared, and the N-H peak at 3400 cm.$^{-1}$ became much narrower and substantially decreased in relative peak height.

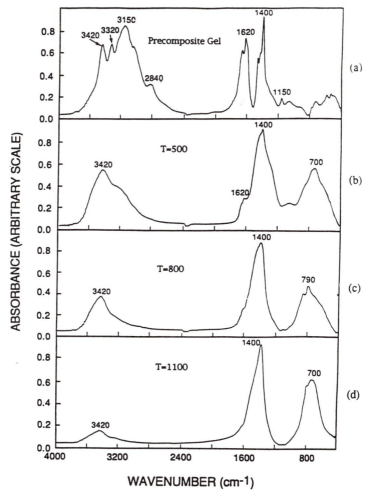

Figure 3. FTIR spectrum showing the chemical bonding information for the conversion of the precomposite into the final AlN/BN nanocomposite (a) precomposite, (b) T=500°C, (c) T=800°C, and (d) T=1100°C.

In the1100°C treated material (Figure 3d), two peaks at 1400 cm.$^{-1}$ and 700 cm.$^{-1}$ corresponding to the BN bond [8,18-21] and AlN bond [6] respectively, became evident. There is also a small peak at 3400 cm.$^{-1}$ due to the presence of a residual N-H bond in the sample. The N-H bond peak was eliminated when a sample initially heated to 1100 °C was heated further to 1400 °C in N_2. Finally, the AlN peak in the FTIR spectra appears to be rather broad, this is probably due to the presence of the B-N bond at a wavenumber of 790 cm.$^{-1}$ [18].

Figure 4 shows the XRD patterns for AlN/BN composites with varying compositions heat treated at 1100°C. When the BN concentration was below 30%, only a small amorphous BN signal was detected. For a BN content larger than 35%, broad peaks at two theta≈25° and 42° were seen. These peaks were assigned to turbostratic BN, which is a two dimensional variant of hexagonal BN and has a random stacking of layers [22]. An amorphous background ranging from 25° to 45° was also found in these samples, and its intensity increased with the BN concentration. Thus, for a BN concentration larger than 35%, BN existed as a mixture of crystalline and amorphous phases. The lattice constant values of BN were not determined, but the d-spacing (002) and (101) reflections agreed with those of turbostratic BN. The AlN phase was identified to be hexagonal for concentrations of 5% to 100% AlN. When AlN concentration was 5%, hexagonal AlN peaks were very small and broad. The intensity and sharpness of AlN peaks increased with its concentration. A calculation of the lattice constant from experimental data showed a good match for AlN with the known bulk value (i.e. a=3.11Å, c=4.98Å, c/a=1.601). Figure 5 shows an increase of average crystalline size of AlN as a function of its concentration, with a maximum at 52 nm at 100%AlN. The crystallite size of BN was estimated to be smaller than 10 nm from the very broad (002) peaks in the XRD patterns [9].

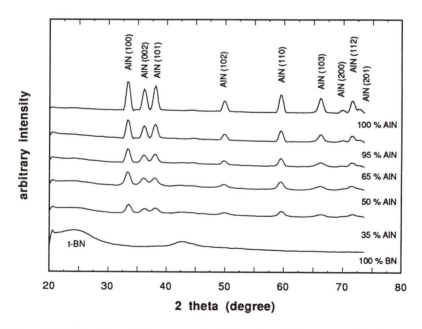

Figure 4. XRD pattern of the composite with various AlN concentration heat treated at 1100°C.

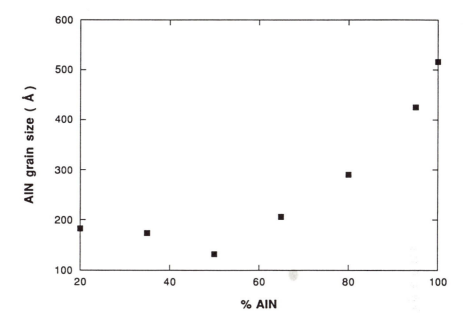

Figure 5. Average crystallite size of AlN as a function of its concentration in the nanocomposite.

The structure, morphology, and crystallite size of the composite powders were correlated with compositions and processing temperatures. Figure 6 shows the bright field transmission electron microscopy (TEM) studies of the composite powders of different compositions. There was a size distribution of AlN crystallite (dark contrast), but the average crystallite size increased with increasing AlN concentration as observed in XRD measurements. Selected area diffraction of powders showed a hexagonal AlN phase in all composite samples and a turbostratic BN phase in samples where the BN concentration was larger than 35%; below this concentration, the BN phase was amorphous. The structure of the turbostratic BN phase was also confirmed from studying the lattice fringes obtained by high resolution TEM (HRTEM) studies. Although both diffraction methods did not detect a crystalline BN phase at concentrations below 35%, the existence of BN was confirmed by FTIR and chemical analysis. Upon annealing, a more ordered hexagonal BN phase developed from the initial turbostratic and amorphous phases. Independent of compositions, grain growth of AlN and BN occurred, but both crystallites remained in the nanometer regime (e.g. crystallite sizes less than 60 nm for AlN, and less than 10 nm for BN) up to 1600°C, and the grain growth of BN was less sensitive to annealing temperature compared to AlN [9].

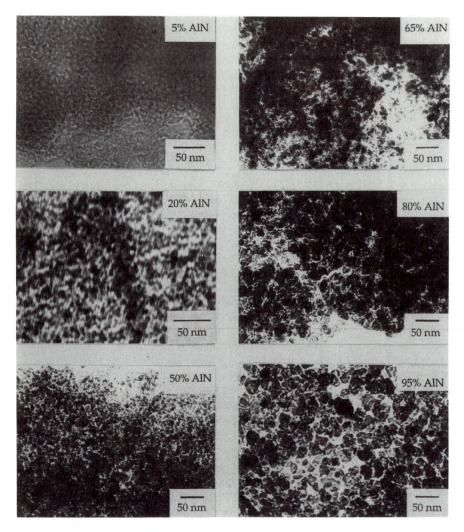

Figure 6. Bright field TEM micrographs showing the morphological changes with the materials composition.

2. Fe_xN/BN Magnetic Nanocomposite

For magnetic applications, nanocomposites are useful in high density information storage and magnetic refrigeration [2,23]. The magnetic properties of the material will change tremendously by reducing the particle size. Due to excellent magnetic properties combined with wear, oxidation and corrosion resistance over pure iron, iron nitride magnetic materials have recently received much attention [24]. The average magnetic moment per iron atom in Fe_4N and Fe_3N is 2.21 μ_B and 2.01 μ_B, respectively [24-26]. These magnetic moments of the nitrides are almost identical to the BCC α-Fe with a moment of 2.22 μ_B per Fe atom. These properties of iron nitrides are technologically important in potential high flux density application [24]. Magnetocaloric application requires a superparamagnetic nanocomposite whose magnetic particles are uniformly distributed in a nonmagnetic matrix phase. The superparamagnetic nanocomposite should have a large magnetic moment and a relatively small magnetocrystalline anisotropy. Iron nitride is very promising for this application. The non-magnetic matrix phase should be electrically nonconductive, but possess good thermal conductivity. Boron nitride is a good candidate because of its excellent dielectric and thermal properties [27].

The synthesis procedure for the preparation of Fe_xN/BN is similar to that of the AlN/BN nanocomposite. Experimentally, iron chloride hexahydrate $FeCl_3 \cdot 6H_2O$, urea $(NH_2)_2CO$, and boric acid H_3BO_3 were dissolved in water in the molecular proportion 1:2:1, respectively. After thorough mixing, ammonia was bubbled into the solution with vigorous stirring until the solution was strongly basic. The reaction mixture was then heated up to 150°C, and the water was removed under vacuum. The precomposite gel was then pyrolyzed under a flow of ammonia at a temperature of about 500°C. In a variation of the above experiment, thermal chemical conversion of the precomposite gel into the final nanostructured magnetic composite was also carried out in a fluidized bed reactor.

Similar to the AlN/BN precomposite material, the as-synthesized precomposite for Fe_xNBN magnetic nanocomposite also has a complex gel-type structure: boron bonded with urea, iron bonded with N-H and chlorine atoms. Also present is the NH_4Cl salt. Thermal analysis studies performed in argon using a simultaneous high temperature TGA/DTA technique show that the precomposite gel undergoes a decomposition in the 100 to 400 °C temperature range, with a total weight loss of about 74%. This weight loss was mainly due to the formation and sublimation of NH_4Cl, and possibly other minor species such as CO. Only one peak was evident in the DTA curve, which corresponds to the major weight loss with respect to the TGA measurement. The other small bumps could be probably caused by the loss of excess urea from the sample and the decomposition of carbonyl groups. Combination of the TGA/DTA measurement suggested that thermal conversion of the precomposite materials could occur at a temperature as low as 400°C.

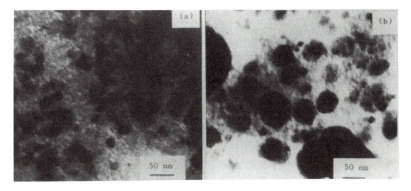

Figure 7. Bright field TEM micrographs for (a) Fe_4N/BN, and (b) Fe_3N/BN magnetic nanocomposite.

Figure 7 are typical TEM micrographs of the Fe$_4$N/BN and Fe$_3$N/BN nanocomposite. It can be seen that the iron nitride particles (dark contrast) range from 10 nm to 200 nm in size. The size distribution was confirmed by the observation of mixture of diffused rings and spots in the electron diffraction pattern. X-ray analysis shown in Figure 8 revealed that the precomposite had a crystalline NH$_4$Cl phase. No BN and Fe$_x$N phases were detected. When this material was heated in a furnace to 500°C in ammonia, crystalline Fe$_4$N and a-Fe x-ray peaks were detected. The weak and broad x-ray peak located at 2θ=26.5° was attributed to a hexagonal BN phase. The α-Fe phase was not detected when this precomposite material was heat treated in a fluidized bed in ammonia. The x-ray analysis revealed only the existence of Fe$_3$N, and Fe$_4$N was not detected in this case. Thermal conversion of the precomposite in a fluidized bed reactor would provide better uniformity for the particle-gas reaction , resulting in the full conversion of a-Fe into iron nitride. The reason for the formation of Fe$_3$N instead of Fe$_4$N however, is not fully understood at the present time.

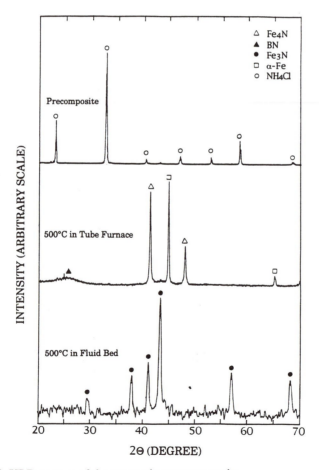

Figure 8. XRD pattern of the magnetic nanocomposite.

Figure 9 shows the magnetization curve of the nanocrystalline Fe4N/BN and Fe3N/BN composites were made at 4.5 K and room temperature, respectively. The saturation magnetization and coercivity measured at 4.5 K were 140 emu/g and 54 Oe for Fe3N/BN, and 110 emu/g and 150 Oe for Fe4N/BN, respectively.

Considering that the α-Fe, which existed in the sample as second magnetic phase, has a magnetic moment close to the average Fe moments of Fe4N and then comparing the measured value with theoretical saturation magnetization value 212 emu/g for bulk Fe4N, it was concluded that BN occupies approximately 50% of the total composition. A similar BN concentration was also obtained for Fe3N/BN composite. Although low temperature magnetic properties of these nanocrystalline Fe_xN/BN are similar to their bulk materials, a rather large reduction of saturation magnetization has been observed. This is typical for fine particle magnetic materials, due to the transition for some particles from ferromagnetic to superparamagnetic state [28].

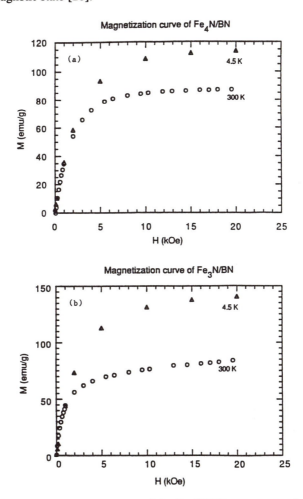

Figure 9. Magnetization measurement of the Fe_xN/BN nanocomposite.

ACKNOWLEDGMENT

We greatly acknowledge Professor Yide Zhang and Professor J.I. Budnick at the Physics Department of the University of Connecticut for the magnetic measurement. Partial funding from Connecticut Advanced Technology Center for Precision Manufacturing is gratefully acknowledged.

REFERENCES

1. H. Gleiter, Nanostructured Materials, 1992, 1, 1-19.
2. R.D. Shull, L.J. Swartzendruber, & L.H. Bennett, Proc. 6th Int. Cryocoolers Conf., eds. G. Green & M. Knox, David Taylor Res. Cntr. Publ. #DTRC-91/002, Annapolis, MD (1991) 231.
3. J.C. Parker& R.W. Seigel, Nanostructured Materials, 1992, 1, 53.
4. H. Hahn & R.S. Averback, Nanostructured Materials, 1992, 1, 95.
5. K.E. Gonsalves, P.R. Strutt, T.D. Xiao, & P.G. Klemens, J. Mater. Sci., 1992, 27 (12), 3231.
6. T.D. Xiao, Ph.D. Thesis Disertation, University of Connecticut, (1991).
7. T.D. Xiao, K.E. Gonsalves, P.R. Strutt, & P.G. Klemens, J. Mater. Sci., 1993, 28, 1334.
8. T.D. Xiao, K.E. Gonsalves, & P.R. Strutt, Am. Ceram. Soc., 1993, 76(4), 987.
9. G.M. Chow, T.D. Xiao, X. Chen, & K.E. Gonsalves, J. Mater. Res., 1994, 9, 168.
10. T.D. Xiao, Y.D. Zhang, P.R. Strutt, J.I. Budnick, K. Mohan, & K.E. Gonsalves, Nanostructured Materials, 1993, 2, 285.
11. K.E. Gonsalves, G.M. Chow, Y.D. Zhang, J.I. Budnick, & T.D. Xiao, Adv. Mater. (in press).
12. K.E. Gonsalves, T.D. Xiao, G.M. Chow, & C.C. Law, Nanostructured Materials 1994, 4 (2), 139.
13. K.E. Gonsalves, U.S. Patent, 4,842,641, June, 1989.
14. R.M. Laine, US Patent No. 4,789,534, Dec. 1988.
15. R.M. Laine, US Patent No. 4,826,666, May, 1989,. PCT Int. Appl. WO 88 01,603. Chem. Abstr. 109: 76196h.
16. R.M. Laine, US Patent No. 4,906,493, March, 1990.
17. R.M. Laine, US Patent No. 4,895,709, Jan., 1990.
18. D.Kwon, W.R. Schmidt, & L.V. Interrante, Inorganic & Organometallic Oligomers & Polymers, pp. 191-97, Edited by J.F. Harrod, & R.M. Laine, Kluwer Academic Publishers, Boston, MA, 1991.
19. K.E. Gonsalves & R. Agarwal, Appl. Organomet., 1988, 2, 245.
20. J. Kouvelakis, V.V. Patel, C.W. Miller, & D.B. Beach, J. Vac. Sci. Technol., 1980, A8(60), 3928.
21. R. Riedel & K.U. Gandl, J. Am. Ceram. Soc., 1991, 74(6), 1331.
22. J. Thomas Jr., N.E. Weston, & T.E. OConor, J. Am. Chem. Soc., 1963, 84, 4619.
23. R.D. Shull & L.E. Bennett, Nanostructured Materials, 1992, . 1, 119.
24. S.K. Chen, S. Jin, G.M. Kammlott, T.H. Tiefel, D.W. Johnson Jr., & E.M. Gyorgy, J. Magnetism & Magnetic Mater., 1958, 110, 751.
25. C. Guillaud & H. Creveaux, Compt. Rend., 1946, 222, 1170.
26. B.C. Frazer, Phys. Rev., 1958, 112, 751.
27. D.J. Twait, W.J. Lackey, A. W. Smith, W.Y. Lee, & J.A. Hanigofcky, J. Am. Ceram. Soc.,1990, 73(6), 1510.
28. C.P. Bean & J.D. Livington, J. Appl. Phys., 1959, 30, 120S.

RECEIVED August 1, 1994

Polyphosphazenes

Chapter 17

Macromolecular and Materials Design Using Polyphosphazenes

Harry R. Allcock

Department of Chemistry, Pennsylvania State University,
152 Davey Laboratory, University Park, PA 16802

Polyphosphazenes are among the most versatile of all polymers. This is a consequence of the unique properties of the phosphorus-nitrogen backbone and the ease with which a wide range of different side groups can be introduced mainly by macromolecular substitution, but also by the polymerization of different "monomers". This field has reached a stage where the fundamental chemistry now allows the design and synthesis of a broad range of new materials that are valuable for properties such as elasticity, high refractive index, liquid crystallinity, ferroelectric-, NLO-, and photochromic attributes, and for uses as solid ionic conductors, biomedical microencapsulation polymers, and materials with controlled surface properties.

Inorganic-Organic Polymers as Components of New Materials

Materials science covers a very broad range of substances within the four main classical categories of ceramics, metals, semiconductors/electro-optical solids, and petrochemical polymers. Most of the examples in the first three categories are inorganic in composition, while the last is mainly organic. All four have advantages and disadvanges. For example, the totally inorganic materials are heavy, often britttle, and difficult to fabricate. On the other hand, conventional polymers are tough and easy to fabricate but are unstable at elevated temperatures and lack many of the interesting electrical and optical properties of the inorganic materials.

The main concept behind the current interest in inorganic or inorganic-organic polymers is that they may provide a means for the preparation of new materials that combine the advantages and minimize the disadvantages of the classical materials. Thus, the long term interest lies in the design and synthesis of new polymers that have combinations of properties such as ease of fabrication coupled with ceramic-like thermal stability, toughness combined with electrical conductivity, elasticity plus biological compatibility, etc.

0097–6156/94/0572–0208$08.72/0

It is also believed that the incorporation of inorganic elements into macromolecules will generate property combinations that will be tunable over a wide range by alterations in the types of inorganic elements present and by changing the ways in which the inorganic components are incorporated into the structure. In order to be able to realize this potential it is necessary to understand the structural features of polymers that are responsible for their special properties and the properties to be expected from the presence of certain inorganic elements. The following paragraphs provide a summary of these factors.

Types of Macromolecular Structure.

Polymers can be found that are linear, macrocyclic, lightly branched, dendritic, lightly crosslinked, heavily crosslinked, ladder-like, or sheet-like in structure, and each geometry gives rise to a different set of properties. Furthermore, the properties are also influenced by the degree of polymerization of the molecules. Species that contain only a few monomer units (say 10 to 100 repeating units) lack the entanglement capabilities of higher polymers and therefore lack strength, toughness, and ease of fabrication. Such species are of only limited usefulness as materials, although they may serve as precursors to ceramics. Genuine polymeric properties become apparent only when 1000 or more repeating units are present

Perhaps the next most important determinant of properties is the degree to which the system is crosslinked. Uncrosslinked polymers are usually soluble in some solvent, although the existence of crystalline domains (brought about by molecular symmetry and efficient interchain packing) may make the process of dissolving difficult. Nevertheless, the existence of an average of only 1.5 crosslinks per chain is sufficient to totally prevent the separation of the polymer molecules in a liquid medium. The polymer may swell in a suitable liquid, but it will not dissolve. Increases in the number of crosslinks per chain will progressively reduce the degree of swelling, until eventually the system will be an unswellable, three-dimensionally crosslinked ceramic. Light crosslinking may favor elasticity by preventing individual chains fron sliding past each other when the bulk material is stretched. Heavy crosslinking not only confers resistance to solvents, but also provides rigidity and strength.

Macromolecular geometry, degree of crosslinking, and repeating unit symmetry affect other properties too. An important characteristic of any polymeric material is the way in which the physical properties change with temperature. An amorphous polymer will be a glass at temperatures below its glass transition temperature (Tg), but a rubbery material above that temperature, changing to a semi-fluid gum at still higher temperatures. The glass transition temperature is a measure of two factors - the ease of torsional flexibility of both the backbone and the side groups, and the degree of "free volume" generated by the shape of the side groups. Backbone segments that contain bonds with low barriers to torsion will often yield low Tg values. For example, some of the lowest Tg values known (-100°C to -130°C) are associated with polymers that possess silicon-oxygen or phosphorus-nitrogen backbone bonds, because these bonds have very low barriers to torsion. Thus, the incorporation of inorganic elements into the backbone structure provides opportunities to control the Tg over a very wide temperature range, and in turn to control whether the polymeric material will be an elastomer, a glass, or a gum throughout a particular temperature range.

Efficient chain packing, and the resultant presence of microcrystallites, will cause the elastomeric phase to be replaced by a so-called thermoplastic phase which is flexible, but has dimensional integrity under moderate stress. This phase gives way to a liquid or gum-like phase at temperatures above the melting point of the crystallites. The presence or absence of crystallinty depends on the shape and size of the side groups, on the regularity of their disposition along the chain, and on the absence of extensive chain branching or crosslinking.   Side groups that can form ordered structures in the liquid-like phase may give rise to liquid crystallinity.

The presence of inorganic elements in the main chain and in the side group structure can affect the properties through all of these influences and can impart other characteristics as well.   For example, properties such as flame retardance, electrical conductivity, biological compatibility, resistance to high energy radiation, or the appearance of ferroelectric behavior, may all be generated by the incorporation of inorganic elements.

The largest class of inorganic-organic polymers is found in the polyphosphazenes (**1**).   At least 300 diferent polyphosphazenes are known at the present time.   Thus, this system is an excellent example of the structural diversity and range of properties that can be generated by the incorporation of elements other than carbon into a polymer backbone.

$$\left[-N = P \!\!\begin{array}{c} R \\ | \\ | \\ R \end{array} \right]_n$$

1

Introduction to Polyphosphazenes

The most widely used synthesis route to stable poly(organophosphazenes), first developed in our program, is illustrated in Scheme I (*1-4*).   The ring-opening polymerization of hexachlorocyclotriphosphazene (**2**) [or its fluoro-analogue $(NPF_2)_3)$] leads to the formation of a high molecular weight poly(di-halogenophosphazene) such as **3**.   This polymerization takes place in the molten trimer or in solution. Solution state polymerizations are facilitated by the use of Lewis acid initiators (*5, 6*).

Polymer **3** is a reactive macromolecular intermediate for a wide range of halogen replacement reactions (Scheme I) (*4, 7, 8*). Many of these reactions are S_N2-type substitutions, although a few have S_N1 character. Replacement of the halogen atoms in **3** by most organic or organometallic side groups yields hydrolytically-stable derivative polymers. These reactions have led to the synthesis and study of a broad range of stable polyphosphazenes, the properties of which are frequently improvements over those of widely used totally-organic macromolecules (*4*).

Further structural diversity can be accomplished by the simultaneous or sequential replacement of the halogen atoms in $(NPCl_2)_n$ or $(NPF_2)_n$ (see structures **6-8**) by two or more different nucleophiles or by chemical reactions carried out on the organic, inorganic, or organometallic side groups without involvement of the polymer skeleton. Moreover, alkyl, aryl-, or organometallic side groups can be linked to the phosphorus atoms at the level of the cyclic trimeric "monomer". Subsequent ring-opening polymerization then provides a pathway to even more polymer structures, as shown in Scheme II.

Scheme I

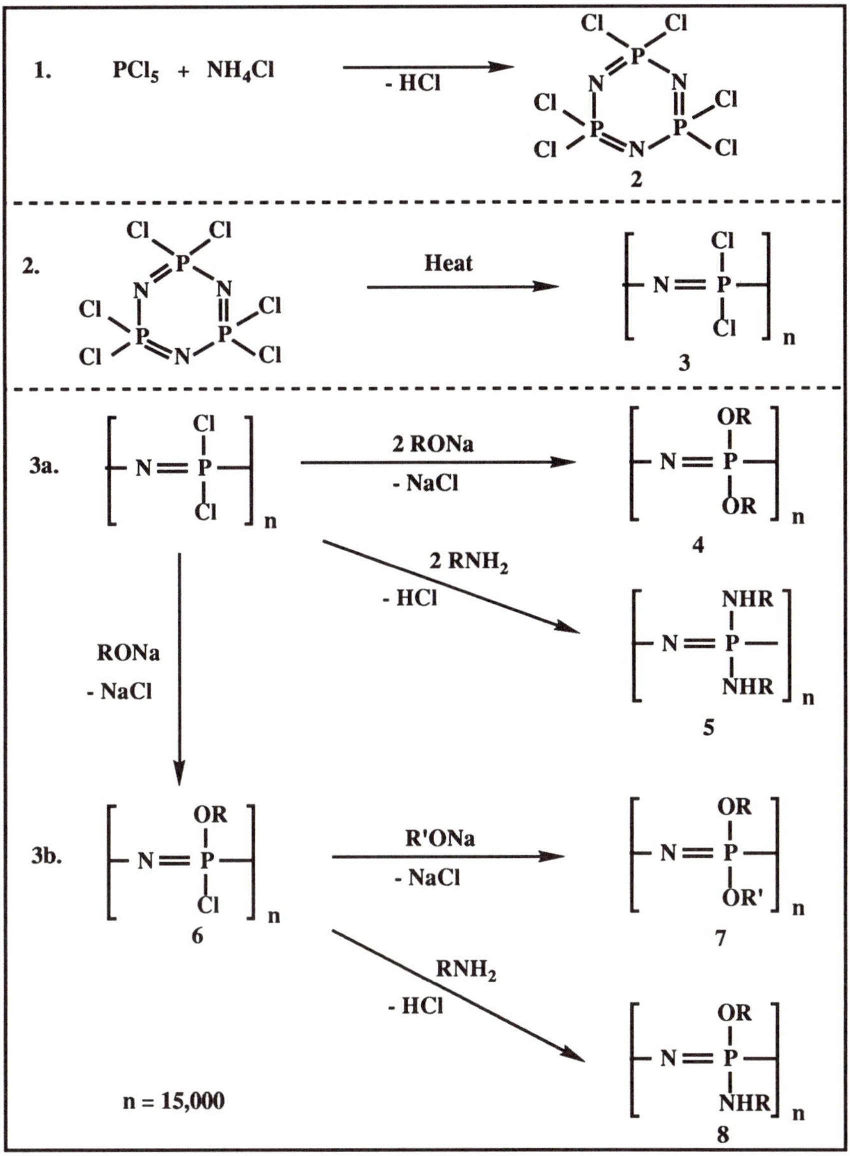

Scheme II

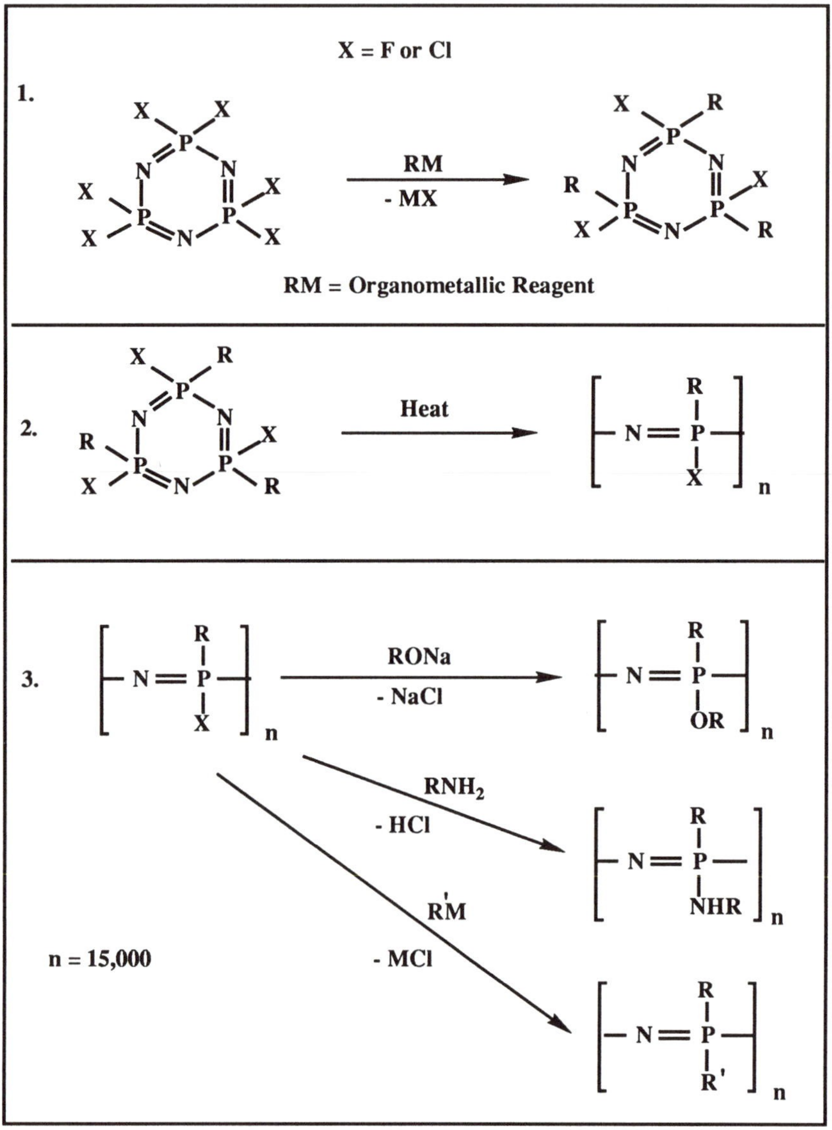
X = F or Cl
1.
RM
- MX
RM = Organometallic Reagent
2.
Heat
3.
RONa
- NaCl
RNH2
- HCl
R'M
- MCl
n = 15,000

A further variant of this preparative protocol involves the synthesis of polymers that contain phosphorus, nitrogen, and a third element in the main chain. For example, we have recently prepared the first poly(carbophosphazenes) and poly(thiophosphazenes) via the chemistry shown in Schemes III and IV (*9, 10*). Manners has reported the synthesis of poly(thionylphosphazenes) using the same approach (*11*).

Alternative methods of polyphosphazene synthesis also exist (*12-15*) and several of these approaches are discussed elsewhere in this volume. Finally, chemistry conducted at the surface of polyphosphazenes allows the tailoring of those surfaces to achieve the development of specific chemical, physical, or biological characteristics (*16-20*).

Inherent in all this work is the idea that these processes allow a rational control to be exercised over macromolecular and materials structure. This has led to the development of a wide range of structure-property relationships that form the basis for producing useful new materials. The following examples illustrate these principles.

Control of the Glass Transition Temperature (Tg) Through Skeletal and Side Group Changes

The polyphosphazene skeleton is one of the most flexible chains yet studied in polymer chemistry. T_g values as low as -105°C have been measured, and this is an indication of a very low barrier to torsion of the backbone bonds. This property is attributed to the absence of side groups on every other skeletal atom, to the wide P-N-P bond angle (which minimizes intramolecular interactions), and to the participation of phosphorus 3d orbitals in the d_π -p_π unsaturation of the skeleton. However, the introduction of carbon atoms into the skeleton raises the Tg by roughly 20°C when the same side groups are present, and this probably reflects the higher barrier to torsion of C=N compared to P=N double bonds (*9*).

Superimposed on the low Tg properties of the backbone is the influence of the side groups. Small or flexible side groups such as -F, -Cl, $-OCH_3$, $-OC_2H_5$, $-OC_3H_7$, $-OCH_2CH_2OCH_2CH_2OCH_3$, $-OCH_2CF_3$, etc. impose little or no additional barrier to torsion of the backbone, and the Tg's remain in the -100°C to -60°C region. However, phenoxy groups raise the Tg to -8°C, p-phenylphenoxy- to +93°C, and phenylamino to +91°C. Thus, steric bulk or the capacity for hydrogen bonding brings about a sharp rise in the Tg. The use of different side groups allows the Tg to be tuned over a 200°C temperature range, and this is a powerful design feature.

High Refractive Index, Liquid Crystalline, Ferroelectric, NLO, and Photochromic Side Groups.

Chart I illustrates five polyphosphazenes in which different side groups generate various optical or electro-optical properties.

First, the refractive index of a polyphosphazene can be altered by the introduction of side groups that contain different numbers of electrons. High numbers

Scheme III

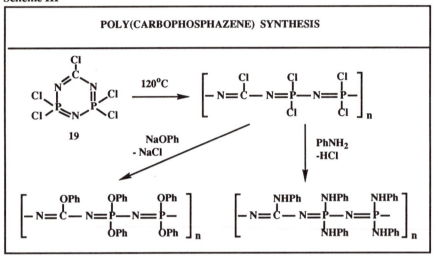

POLY(CARBOPHOSPHAZENE) SYNTHESIS

POLY(THIOPHOSPHAZENE) SYNTHESIS

Scheme IV

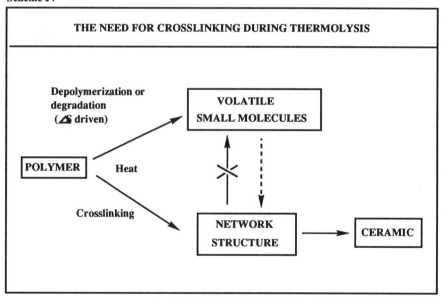

(Reproduced with permission from reference 60. Copyright 1992 John Wiley & Sons, Inc.)

Chart I

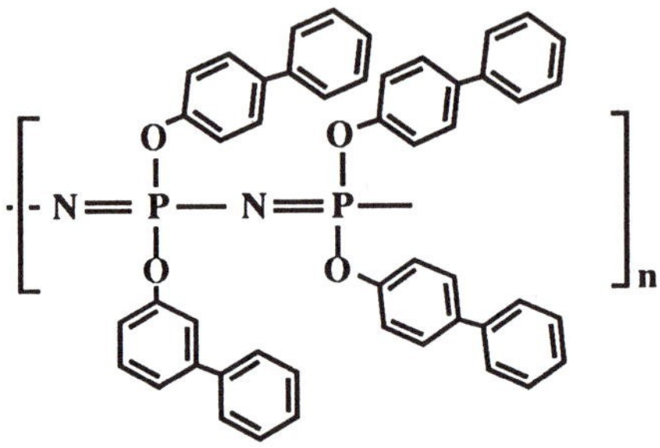

9

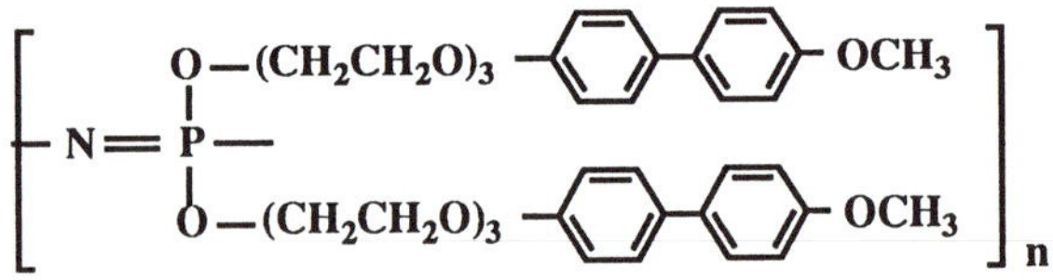

10

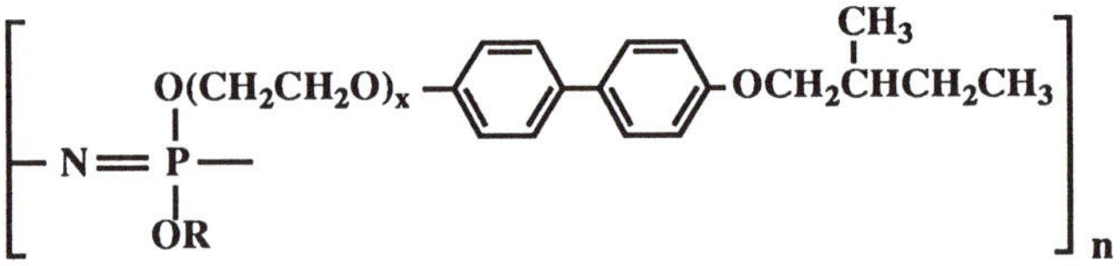

11

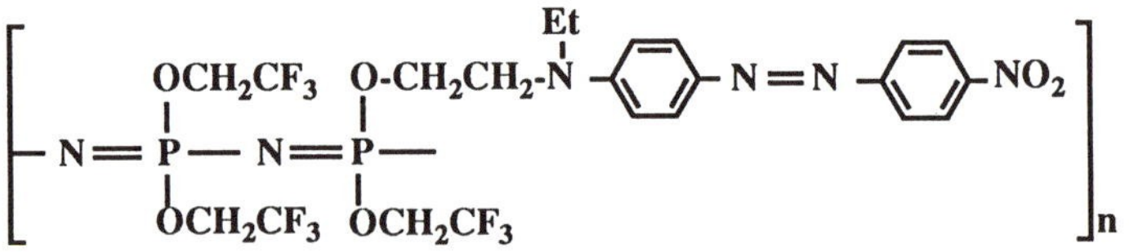

12

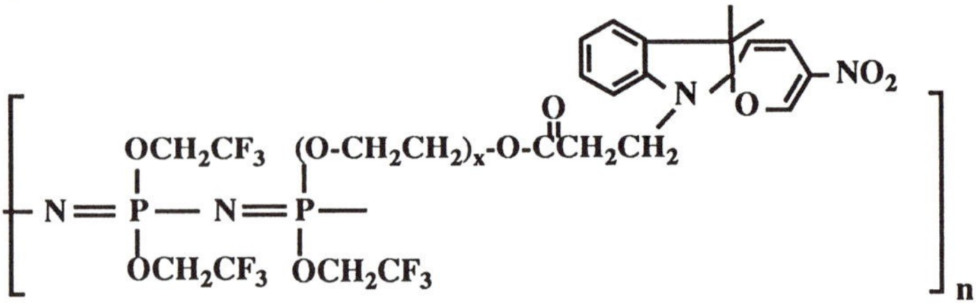

13

of electrons per side unit yield higher refractive indices. Hence, aromatic or polyaromatic side groups give some of the highest values. Polymer **9** has a refractive index of 1.686 (*21*) which can be compared, for example, to the value of 1.591 for polystyrene.

Second, the presence of rigid, mesogenic side groups, which can form ordered arrays above the crystallite melting temperature, can give rise to liquid crystallinity (*22-28*). Structure **10** represents a class of polymers that have this property.

Ferroelectric behavior has been detected for polymer **11**, as a consequence of the "chiral smectic-C" structure of the liquid crystalline phase (*29*). Polymer **12** is a second order nonlinear optical material, the polarizable side groups, in which can be poled in a 10-20 kV electric field to generate χ^2 values in the region on 34-45 picometers per volt (*30, 31*). Finally, species **13** is a photochromic polymer, which undergoes a spiropyran-merocyanine transformation when exposed to ultraviolet light (*32*).

Side Groups for Different Crosslinking Processes

The development of reactions for the crosslinking of polyphosphazenes is an important aspect of the materials science of this field. Crosslinking processes play a role in processes as diverse as elastomer formation, stabilization of NLO properties, photolithography, and the formation of hydrogels and organogels. Crosslinking processes fall into two categories--chemically-induced processes, and those brought about by radiation.

Commercial alkoxy- or aryloxy-phosphazene elastomers contain pendent allyl groups for free radical crosslinking in the final stages of fabrication (*33*). Methathetical side groups exchange reactions have also been used for crosslinking (*34*). Chemical crosslinking also plays a critical part in the conversion of several polyphosphazenes to ceramic-type materials. Unless crosslinking takes place at an early stage in the ceramicization process, the individual polymer molecules may depolymerize and the fragments volatilize before a useful material can be formed (Scheme IV).

For example, the n-propylaminophosphazene polymer shown in Scheme V undergoes thermal side group condensation and crosslinking between 250°C and 1000°C to yield phosphorus nitride and graphite (*35, 36*). The borazine-bearing polyphosphazene depicted in Scheme VI reacts in a similar fashion (*37*). Amino groups on borazine rings and others linked to the polymer backbone facilitate the crosslinking reactions that are required to prevent depolymerization and fragmentation at elevated temperatures. The initial pyrolysis leads to the formation of phosphorus-nitrogen-boron matrices, and above 1000°C to boron nitride.

Photo-crosslinking through organic side groups is an important process in polyphosphazene chemistry, especially because the backbone is insensitive to a wide spectrum of high energy radiation. Thus side group crosslinking can be effected without significant backbone cleavage.

Ultraviolet-induced crosslinking has been demonstrated for polyphosphazenes with -OCH$_2$CH$_2$OCH$_2$CH$_2$OCH$_3$ side groups (with and without photosensitizers) (*38*)

Scheme V

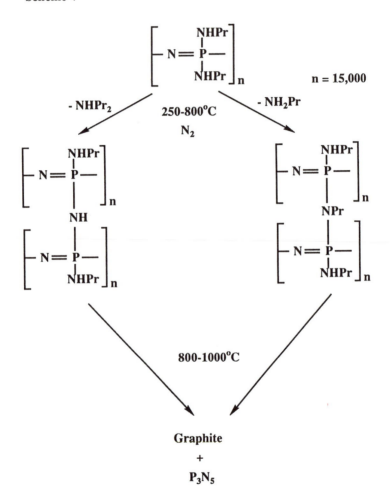

Also vapor deposition phosphorus nitrides

Scheme VI

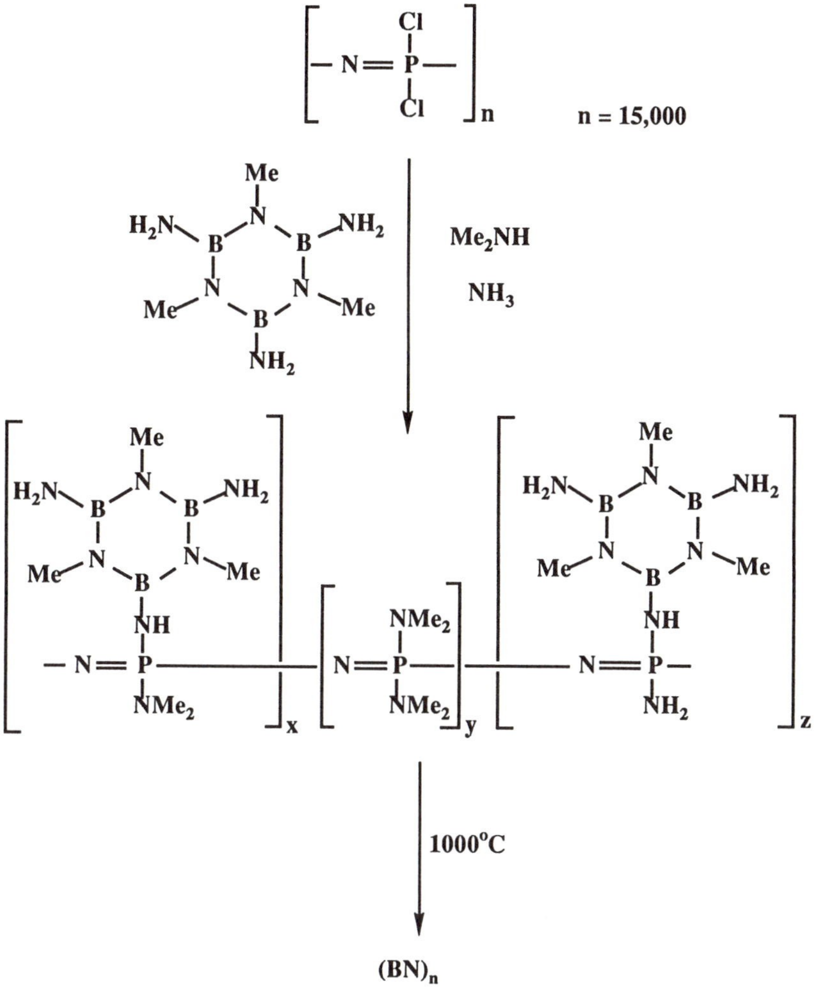

(Reproduced with permission from reference 60. Copyright 1992 John Wiley & Sons, Inc.)

and for the 2+2 cycloadditions of cinnamate and chalcone side groups (*39*). The UV-induced crosslinking of MEEP (**14**) is important both for the formation of hydrogels and for the dimensional stabilization of solid electrolytes (see later). Cinnamate and chalcone-based crosslinking reactions were developed for the stabilization of NLO activity, but are also potentially useful in photolithography.

Electron beam lithography is possible with allyl-containing arylaminophosphazene polymers (*40*). Work in the group of V. T. Stannett has demonstrated that these species have an e-beam sensitivity and sufficiently high Tg's that they are appropriate for microlithography.

Gamma rays are particularly effective for crosslinking alkyl ether phosphazene polymers such as $[NP(OCH_2CH_2OCH_2CH_2OCH_3)_2]_n$ or $[NP(NHCH_3)_2]_n$ (Scheme VII) (*41-44*).

The pattern is emerging that almost any polyphosphazene with a high loading of aliphatic C-H bonds in the side group system can be crosslinked by exposure to gamma-rays. This is a clean procedure that generates no impurities as side products and allows the degree of crosslinking to be controlled within fine limits by the radiation dose. Perhaps the most effective use of this technique so far is in the crosslinking of MEEP, to be discussed in the following section.

Solid Polymeric Electrolytes Based on Crosslinked MEEP-Type Materials

(a) General Features. The design and development of MEEP (14) as a solid polymeric electrolyte was initiated in 1983-84 through a collaborative program between our research group at The Pennsylvania State University and that of D. F. Shriver at Northwestern University (*45-49*). The possible use of this polymer in lightweight, rechargeable lithium batteries was a driving force for that work.

MEEP is synthesized by the process shown in Scheme VIII (*50*), in which poly(dichlorophosphazene) (**3**), is allowed to undergo chlorine replacement reactions in the presence of the sodium salt of methoxyethoxyethanol in a solvent such as tetrahydrofuran. MEEP is an excellent solid solvent for salts such as lithium triflate. Salts dissolved in this polymer undergo ion-pair separation as a result of coordination of the Li^+ cations by the oxygen atoms of the alkyl ether side groups. In this system the ions can migrate through moving polymer molecules under the pressure of an electric current. Use of this electrolyte in rechargeable lithium batteries is expected to provide an improvement in robustness, safety, and ease of fabrication compared to liquid electrolytes or polymers dissolved in liquid electrolyte systems.

It became clear at an early stage that MEEP has marked advantages as a solid electrolyte matrix over the classical standard in this field, which is poly(ethylene oxide). MEEP has a very low glass transition temperature of -84°C (the temperature below which molecular motion becomes restricted), it is non-crystalline, and has six oxygen atoms per repeat unit for coordination to cations in a way that facilitates ion-pair separation. By contrast, poly(ethylene oxide) contains crystalline domains that impede ion migration, so that electrolytic cells based on this polymer must be heated to 80-100°C before practical conductivities can be obtained. At room temperature, the

Scheme VII

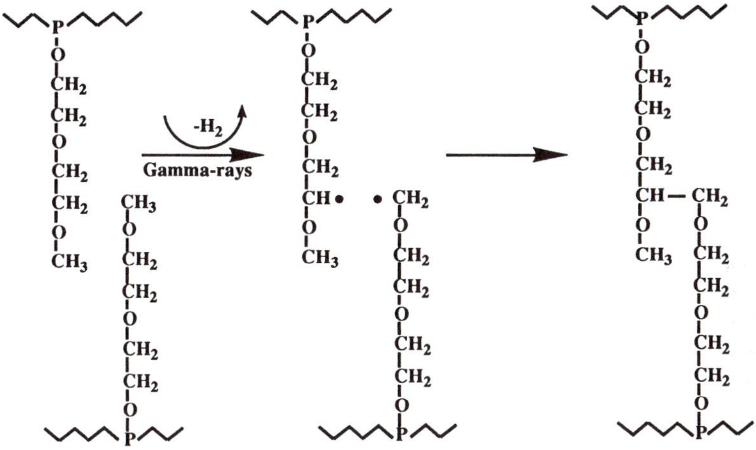

Scheme VIII

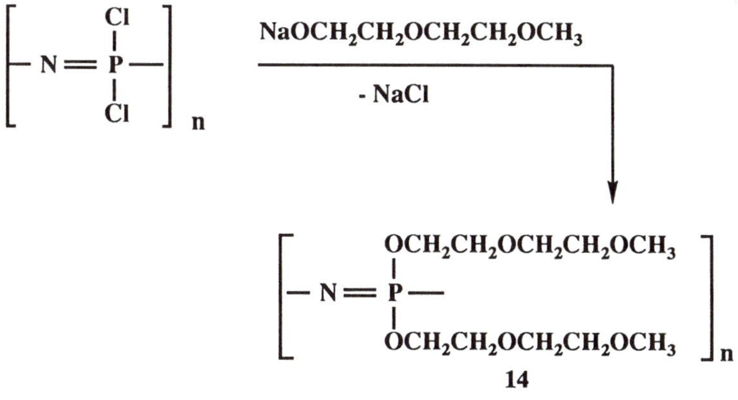

14

"MEEP"

conductivity of MEEP-based electrolytes is 2-3 orders of magnitude higher than comparable systems based on poly(ethylene oxide). This difference can be partly overcome by the use of random copolymers of ethylene oxide and propylene oxide, but such copolymers still contain fewer oxygen solvation sites per repeat unit than does MEEP.

The original studies of MEEP-based electrolytes revealed one problem - that the polymer molecules were so flexible that the bulk polymer had a tendency to undergo viscous flow under pressure, thus being prone to slow extrusion from electrolytic cells.

The work described here was designed to overcome this problem by the development of methods to crosslink the polymer chains while, at the same time, avoiding contamination by conductive impurities and maintaining the relatively high room temperature conductivity of MEEP/lithium triflate electrolytes. Crosslinking *after* the formation of the MEEP/lithium triflate films was considered to be an essential prerequisite to allow the fabrication of the electrolyte in the form of large surface area batteries or in intricately shaped devices.

(b) Gamma-Ray Crosslinking. A series of studies carried out in my group by Kwon, Gebura, Riding, Fitzpatrick, and Bennett *(41-44, 51)*. revealed that MEEP is sensitive to crosslinking when exposed to gamma rays. This behavior is a consequence of the high loading of aliphatic carbon-hydrogen bonds (22 per repeating unit), which are especially prone to radiation-induced homolytic cleavage, and this provides a mechanism for intermolecular carbon-carbon bond formation (Scheme VIII). An advantage of this technique is that the degree of crosslinking can be finely controlled by the radiation dose, and no impurities are introduced into the material.

Gamma-radiation crosslinked MEEP swells in water to form hydrogels *(41, 52)*. In the water-free state the polymer does not undergo viscous flow. Conduction experiments carried out by Bennett, Dembek, Allcock, Heyen, and Shriver *(51)* on gamma-ray crossliked MEEP/lithium triflate films indicated little if any diminution in conductivity following the introduction of an average of one crosslink per 100 repeating units (20 Megarads). Thus, the local segmental motion of the polymer is maintained under these conditions, and the ability of the system to facilitate ion migration is retained.

However, ^{60}Co gamma-ray crosslinking requires the use of specialized facilities, and this constitutes a possible impediment to expanded research on and development of this system.

(c) Ultraviolet Crosslinking Most totally organic polymers are affected detrimentally by exposure to ultraviolet light. Both main chain cleavage and side group free radical reactions are common, and these often lead to a deterioration in materials properties. However, the phosphazene backbone in MEEP is relatively stable to high energy radiation and it resists free radical cleavage processes.

The exposure of films of MEEP to 2200 - 4000 Å ultraviolet radiation results in crosslinking *(38)*. The reaction takes place readily in the presence of photosensitizers such as benzophenone, but also occurs in the absence of added photosensitizers.

Crosslinking can be monitored by changes in the ^{1}H and ^{13}C NMR spectra and by the degree to which the polymer imbibes solvents without dissolving.

The ultraviolet crosslinking method is inexpensive (compared to gamma-ray irradiation) and yields a dimensionally stable electrolyte uncontaminated with conductive chemical residues.   The characteristics are particularly useful in the design and development of rechargeable energy storage devices, several of which have been assembled and tested in our program.

Hydrogels Formed from MEEP

Water-soluble synthetic polymers are quite rare.   But they are extremely important in general technology and in biomedicine. One of their most significant properties is their use as precursors to hydrogels. A hydrogel is a water-soluble polymer that, when crosslinked, will imbibe water to a limit defined by the average number of crosslinks per chain. Hydrogels may be 90 or 95% water and are useful as biomedical soft tissue prostheses, soft contact lenses or intraocular lenses, membranes, and as platforms for the immobilization of biologically-active species.

MEEP forms excellent hydrogels when crosslinked by either gamma-ray or UV-irradiation techniques. One application--the immobilization of the enzyme urease--will be described here (*53*). Solutions of MEEP and urease in water were fabricated into films by water evaporation . These were crosslinked by gamma rays and the system allowed to absorb water to form a hydrogel (see Figure 1). The enzyme molecules could not be extracted from the hydrogels by water, confirming that each enzyme molecule was physically trapped within the polymer matrix and surrounded by water molecules. Exposure of the gels to aqueous solutions of urea indicated that the trapped enzymes retained their ability to convert urea to ammonia.   This is a prototype system for the preparation of a wide variety of enzyme "reactors."

A Polyphosphazene Polyelectrolyte for Biomedical Microencapsulation

Naturally-occurring polyelectrolytes, such as alginates, have been studied extensively for the microencapsulation of biologically-active species. The interest in these polymers arises from the fact that they can be crosslinked to hydrogels by ionic crosslinking in the presence of a divalent cation such as Ca^{++}. However, a severe need existed for a synthetic replacement for alginate, with reproducible chain length, low toxicity, and a higher loading of crosslink sites, such as carboxylic acid units.

Such a polymer has been designed and developed through a joint rersearch program between our group and the group of Robert Langer at M.I.T. The polymer is an aryloxyphosphazene with a carboxylic acid function at the para-position of every aryloxy side group (Structure **15**) (*54-59*).   Polymer **15** is soluble in water as its sodium salt, but undergoes ionic crosslinking when exposed to aqueous solutions of $CaCl_2$ (Scheme IX). Using appropriate equipment, this process can be used for the preparation of microspheres of the crosslinked polymer and for the microencapsulation of mammalian cells, proteins, and a range of other species (Figure 2) (*54-58*). A long-range objective is to use the microspheres in artificial organic research or for the controlled delivery of drugs or antigens.

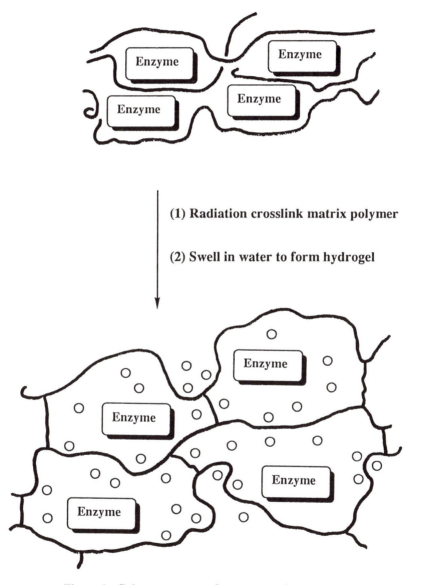

Figure 1. Gel-entrappment of enzymes such as urease.

Scheme IX

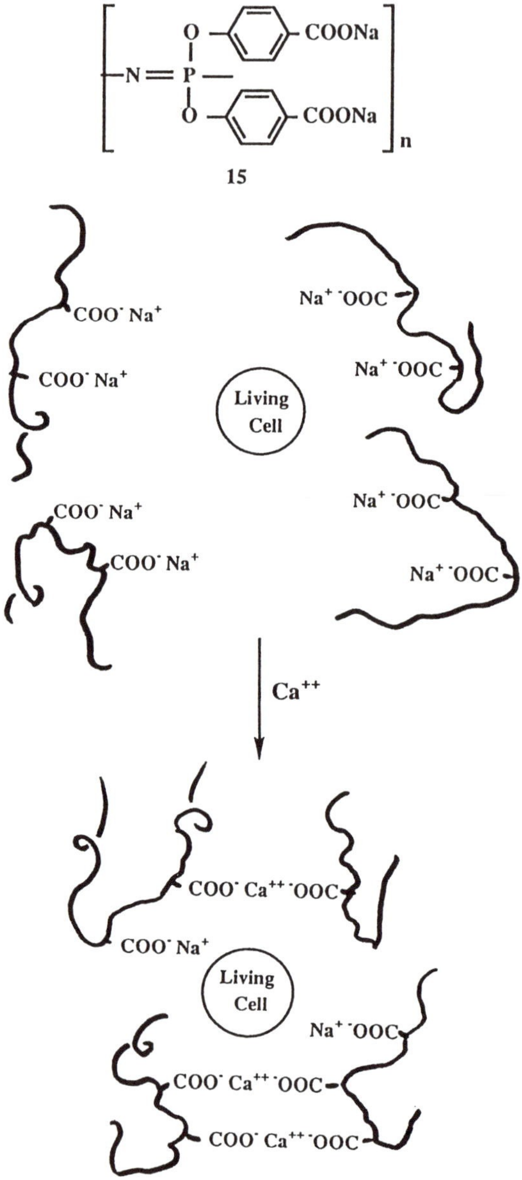
O
COONa
N
P
O
COONa
n
15
COO⁻ Na⁺
COO⁻ Na⁺
Na⁺ ⁻OOC
Na⁺ ⁻OOC
Living
Cell
COO⁻ Na⁺
COO⁻ Na⁺
Na⁺ ⁻OOC
Na⁺ ⁻OOC
Ca⁺⁺
COO⁻ Ca⁺⁺ ⁻OOC
COO⁻ Na⁺
Living
Cell
Na⁺ ⁻OOC
COO⁻ Ca⁺⁺ ⁻OOC
COO⁻ Ca⁺⁺ ⁻OOC

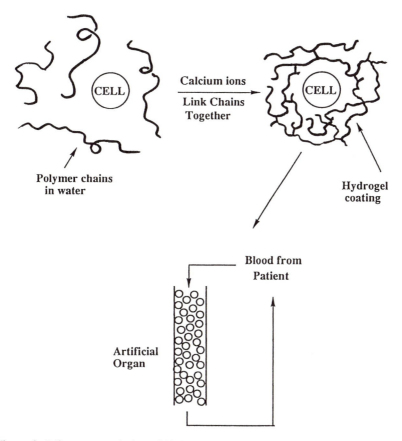

Figure 2. Microencapsulation of biologically active species, such as mammalian cells, using ionic crosslinking of a phosphazene polyelectrolyte.

Surface Reactions of Polyphosphazenes

The surface character of a polymer often controls its technological and biological properties. In the past, new surface characteristics could be generated mainly by the synthesis of entirely new polymers. The most modern approach is to choose a polymer for its "internal" properties (elasticity, rigidity, etc.) and then modify the surface by chemical reactions. We have developed a number of approaches to achieving this for polyphosphazenes (*16-20*).

First, the most important characteristic of a surface is its hydrophilicity, hydrophobicity, or amphiphilic behavior. The seven reactions shown in Scheme X illustrate approaches we have examined using hydrolysis, oxidation, sulfonation, and surface grafting to convert a hydrophobic polymer surface to a hydrophilic one. In a few of these cases, the new surface has an antibacterial character, which is important for the use of polymers in surgery.

A second type of polyphosphazene surface reaction is illustrated in Scheme XI (*59*). Here, poly[bis(phenoxy)phosphazene] is coated on to the surface of high surface area porous alumina particles, and phenoxy groups at the surface are first nitrated, and the nitro groups then reduced to $-OC_6H_4NH_2$ units. Enzymes, such as trypsin or glucose-6-phosphate dehydrogenase, were then linked to the surface amino groups using coupling reagents, such as glutaric dialdehyde, and the enzymes were shown to retain their activity and to be useable in continuous flow reactor devices. The wide variety of side groups available in polyphosphazenes is an asset for surface immobilizations of this type

Connection to Small-Molecule Chemistry

Classical inorganic Main Group chemistry has traditionally focussed on the synthesis, reactions, and structures of small-molecule heteroatomic and homoatomic ring systems, neglecting the macromolecular and materials aspects. As an end in itself, the study of small molecule rings has become a restrictive exercise with little relevance to the wider evolution of the chemical sciences. However, the small-molecule chemistry assumes a considerably greater importance if it is viewed as model chemistry for eventual translation to the more complex macromolecular and materials fields. This is an approach that has been used in our program for many years and which has provided a continuous stream of new reactions and structural interpretations for use at the polymer and materials levels. However, it requires a willingness on the part of the small-molecule scientist to make use of often unfamiliar concepts and techniques. This is a transition that has been made by the authors of all the articles in this volume, and it is hoped that the examples given here will encourage others to follow the same path.

Acknowledgment

The work described here was funded mainly through the generous support of the U.S. Army Research Office, Office of Naval Research, Johnson & Johnson, Corning Inc, Virus Research Institute, DOE, NSF, and EPRI. I would also like to thank my coworkers, many of whom are mentioned in the reference list.

Scheme X

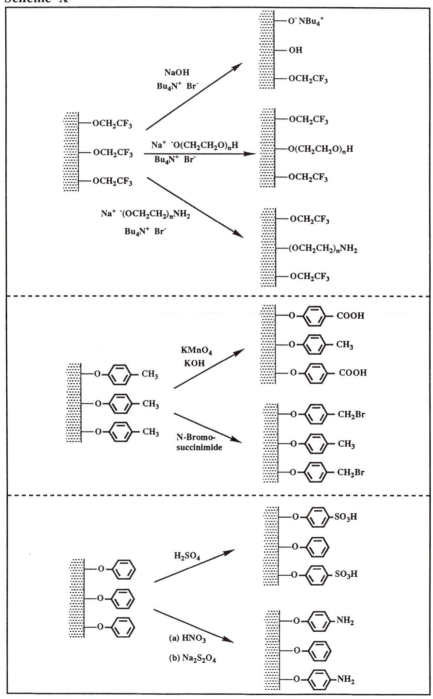

Scheme XI

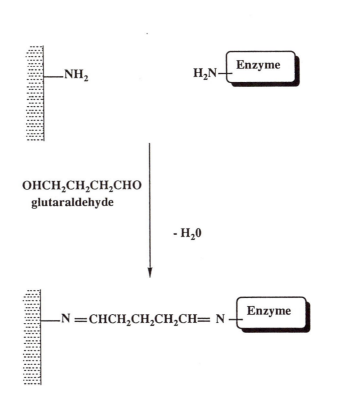

References

1. Allcock, H. R.; Kugel, R. L., *J. Am. Chem. Soc.* 1965, *87*, 4216-4217.
2. Allcock, H. R.; Kugel, R. L.; Valan, K. J., *Inorg. Chem.* 1966, *5*, 1709-1715.
3. Allcock, H. R.; Kugel, R. L., *Inorg. Chem.* 1966, *5*, 1716-1718.
4. Allcock, H. R. In *Inorganic Polymers*; Prentice Hall: Englewood Cliffs, New Jersey, 1992; pp 61-140.
5. Sennett, M. S.; Hagnauer, G. L.; Singler, R. E., *Polym. Mater. Sci. Eng.* 1983, *49*, 297-300.
6. Fieldhouse, J. W.; Graves, D. F., *U.S Pat.* 1980, *8*,
7. Singler, R. E.; Schneider, N. S.; Hagnauer, G. L., *Polym. Eng. Sci.* 1975, *15*, 321-38.
8. Tate, D. P., *J. Polym. Sci., Polym. Symp.* 1974, *48*, 33-45.
9. Allcock, H. R.; Coley, S. M.; Manners, I.; Nuyken, O.; Renner, G., *Macromolecules* 1991, *24*, 2024-8.
10. Allcock, H. R.; Dodge, J. A.; Manners, I., *Macromolecules* 1993, *26*, 11-16.
11. Liang, M.; Manners, I., *J. Am. Chem. Soc.* 1991, *113*, 4044-5.
12. Neilson, R. H.; Wisian-Neilson, P., *Chem. Rev.* 1988, *88*, 541-62.
13. Wisian-Neilson, P.; Neilson, R. H., *J. Am. Chem. Soc.* 1980, *102*, 2848-9.
14. Flindt, E. P.; Rose, H., *Z. Anorg. Allg. Chem.* 1977, *428*, 204-8.
15. Montague, R. A.; Matyjaszewski, K., *J. Am. Chem. Soc.* 1990, *112*, 6721-3.
16. Allcock, H. R.; Fitzpatrick, R. J.; Salvati, L., *Chem. Mater.* 1991, *3*, 1120-32.
17. Allcock, H. R.; Rutt, J. S.; Fitzpatrick, R. J., *Chem. Mater.* 1991, *3*, 442-9.
18. Allcock, H. R.; Fitzpatrick, R. J.; Salvati, L., *Chem. Mater.* 1991, *3*, 450-4.
19. Allcock, H. R.; Fitzpatrick, R. J.; Visscher, K.; Salvati, L., *Chem. Mater.* 1992, *4*, 75-80.
20. Allcock, H. R.; Fitzpatrick, R. J.; Salvati, L., *Chem. Mater.* 1992, *4*, 769-75.
21. Allcock, H. R.; Mang, M. N.; Dembek, A. A.; Wynne, K. J., *Macromolecules* 1989, *22*, 4179-90.
22. Allcock, H. R.; Kim, C., *Macromolecules* 1989, *22*, 2596-602.
23. Allcock, H. R.; Kim, C., *Macromolecules* 1990, *23*, 3881-7.
24. Kim, C.; Allcock, H. R., *Macromolecules* 1987, *20*, 1726-7.
25. Singler, R. E.; Willingham, R. A.; Lenz, R. W.; Furukawa, A.; Finkelmann, H., *Polym. Prepr.* 1987, *28*, 448-9.
26. Singler, R. E.; Willingham, R. A.; Lenz, R. W.; Furukawa, A.; Finkelmann, H., *Macromolecules* 1987, *20*, 1727-8.
27. Singler, R. E.; Willingham, R. A.; Noel, C.; Friedrich, C.; Bosio, L.; Atkins, E. D. T.; Lenz, R. W., *Acs Symp. Ser.* 1990, *435*, 185-96.
28. Singler, R. E.; Willingham, R. A.; Noel, C.; Friedrich, C.; Bosio, L.; Atkins, E., *Macromolecules* 1991, *24*, 510-16.
29. Allcock, H. H.; Kim, C., *Macromolecules* 1991, *24*, 2841-5.
30. Dembek, A. A.; Kim, C.; Allcock, H. R.; Devine, R. L. S.; Steier, W. H.; Spangler, C. W., *Chem. Mater.* 1990, *2*, 97-9.
31. Dembek, A. A.; Allcock, H. R.; Kim, C.; Devine, R. L. S.; Steier, W. H.; Shi, Y.; Spangler, C. W., *Acs Symp. Ser.* 1991, *455*, 258-66.
32. Allcock, H. R.; Kim, C., *Macromolecules* 1991, *24*, 2846-51.
33. Penton, H. R.; Pettigrew, F. A., *U.s.* 1986, 022.
34. Allcock, H. R.; Moore, G. Y., *Macromolecules* 1972, *5*, 231-5.
35. Allcock, H. R.; Kolich, C. H.; Kossa, W. C., *Inorg. Chem.* 1977, *16*, 3362-4.

36. Allcock, H. R.; McDonnell, G. S.; Riding, G. H.; Manners, I., *Chem. Mater.* 1990, *2*, 425-32.
37. Allcock, H. R.; Welker, M. F.; Parvez, M., *Chem. Mater.* 1992, *4*, 296-307.
38. Nelson, C. J.; Coggio, W. D.; Allcock, H. R., *Chem. Mater.* 1991, *3*, 786-7.
39. Allcock, H. R.; Cameron, C. G., *Unpublished work* 1994,
40. Welker, M. F.; Allcock, H. R.; Grune, G. L.; Chern, R. T.; Stannett, V. T. In *Polymers for Microelectronics*; L. F. Thomson, C. G. Wilson and s. Tagawa, Eds.; A. C. S. Symp. Series: 1994; Vol. 537; pp 293-303.
41. Allcock, H. R.; Fitzpatrick, R. J.; Gebura, M.; Kwon, S., *Polym. Prepr.* 1987, *28*, 321.
42. Allcock, H. R.; Kwon, S.; Riding, G. H.; Fitzpatrick, R. J.; Bennett, J. L., *Biomaterials* 1988, *9*, 509-13.
43. Allcock, H. R.; Gebura, M.; Kwon, S.; Neenan, T. X., *Biomaterials* 1988, *9*, 500-8.
44. Bennett, J. L.; Dembek, A. A.; Allcock, H. R.; Heyen, B. J.; Shriver, D. F., *Polym. Prepr.* 1989, *30*, 437-8.
45. Blonsky, P. M.; Shriver, D. F.; Austin, P.; Allcock, H. R., *Polym. Mater. Sci. Eng.* 1985, *53*, 118-22.
46. Blonsky, P. M.; Shriver, D. F.; Austin, P.; Allcock, H. R., *Solid State Ionics* 1986, *19*, 258-64.
47. Lerner, M. M.; Tipton, A. L.; Shriver, D. F.; Dembek, A. A.; Allcock, H. R., *Chem. Mater.* 1991, *3*, 1117-20.
48. Blonsky, P. M.; Shriver, D. F.; Austin, P.; Allcock, H. R., *J. Am. Chem. Soc.* 1984, *106*, 6854-5.
49. Shriver, D. F.; Tonge, J. S.; Barriola, A.; Blonsky, P. M.; Allcock, H. R.; Kwon, S.; Austin, P., *Polym. Prepr.* 1987, *28*, 438.
50. Allcock, H. R.; Austin, P. E.; Neenan, T. X.; Sisko, J. T.; Blonsky, P. M.; Shriver, D. F., *Macromolecules* 1986, *19*, 1508-1512.
51. Bennett, J. L.; Dembek, A. A.; Allcock, H. R.; Heyen, B. J.; Shriver, D. F., *Chem. Mater.* 1989, *1*, 14-16.
52. Allcock, H. R.; Pucher, S. R.; Turner, M. L.; Fitzpatrick, R. J., *Macromolecules* 1992, *25*, 5573-7.
53. Allcock, H. R.; Pucher, S.: unpublished work, 1993
54. Cohen, S.; Bano, M. C.; Visscher, K. B.; Chow, M.; Allcock, H. R.; Langer, R., *J. Am. Chem. Soc.* 1990, *112*, 7832-3.
55. Bano, M. C.; Cohen, S.; Visscher, K. B.; Allcock, H. R.; Langer, R., *Bio/Technology* 1991, *9*, 468-71.
56. Cohen, S.; Bano, C.; Visscher, K. B.; Chow, M. B.; Allcock, H. R.; Langer, R. S., *Pct Int. Appl.* 1992, *45*,
57. Cohen, S.; Bano, M. C.; Cima, L. G.; Allcock, H. R.; Vacanti, J.P.; Vacanti,C. A.; Langer, R., *Clinical Materials* 1993, *13*, 3-10.
58. Andrianov, A. K.; Payne, L. G.; Visscher, K. B.; Allcock, H. R.; Langer, R.,9 *Polym. Prepr., ACS Div. Polym. Chem.* 1993, *34*, 233-234.
59. Allcock, H. R.; Kwon, S., *Macromolecules* 1989, *22*, 75-9.
60. Allcock, H. R., *Chemical Processing of Advanced Materials* 1992, 704, 712, 714.

RECEIVED May 3, 1994

Chapter 18

Synthesis of Poly(alkyl/arylphosphazenes) and Their Precursors

Robert H. Neilson, David L. Jinkerson, William R. Kucera,
Jon J. Longlet, Remy C. Samuel, and Christopher E. Wood

Department of Chemistry, Texas Christian University,
Fort Worth, TX 76129

Poly(alkyl/arylphosphazenes), $[R_2P=N]_n$, are a class of phosphazene polymers in which all of the side groups are attached to the P-N backbone via direct phosphorus-carbon bonds. While other types of poly(phosphazenes) are usually prepared by the ring-opening polymerization of the cyclic trimer, $[Cl_2P=N]_3$, followed by nucleophilic substitution of the P-Cl groups along the polymer backbone, the poly(alkyl/arylphosphazenes) are directly obtained by the thermal *condensation polymerization* of N-silylphosphoranimines, $Me_3SiN=PR_2X$. This paper reviews (1) the overall scope and limitations of this condensation polymerization, (2) some general properties of the title polymers, (3) recent improvements in the synthetic methodology including catalysis of the polymerization process as well as simplification of the precursor synthesis, (4) observations relevant to the polymerization mechanism such as leaving group effects, and (5) the synthesis, structure, and reactivity of some related small molecule species including N-silylphosphoranimines with bulky silyl groups, acetylene-substituted Si-N-P systems, and new N-phosphoryl-phosphoranimines, e.g., $Cl_2P(O)-N=PR_2X$.

Since the mid-1960's, a large number of polyphosphazenes, $[R_2PN]_n$, with a great variety of substituents at phosphorus have been prepared and characterized. (*1*) Because of the diversity of possible side groups, phosphazene polymers exhibit a very broad spectrum of useful chemical, physical, electrical, optical, and biological properties. (*2-4*).

This diversity of side groups is a direct result of the fact that there are varied and complimentary methods for the synthesis of the phosphazene backbone and, more importantly, for the attachment of the desired side groups to the phosphorus centers along the P-N chain. The first, and most extensively studied, approach is the ring-opening/substitution method developed by Allcock and coworkers. (*1, 3*)   As illustrated in *Scheme 1*, this procedure involves the initial preparation of poly(dichlorophosphazene), $[Cl_2PN]_n$, by the ring-opening polymerization of the cyclic trimer, $[Cl_2PN]_3$, with subsequent nucleophilic displacement of the chlorine atoms along the polymer backbone.   Although a few significant variations involving certain substituted cyclics are possible, this method generally requires that the desired side groups be introduced after polymerization and that they be attached to phosphorus through oxygen or nitrogen linkages.

Scheme 1

Ring-Opening Polymerization

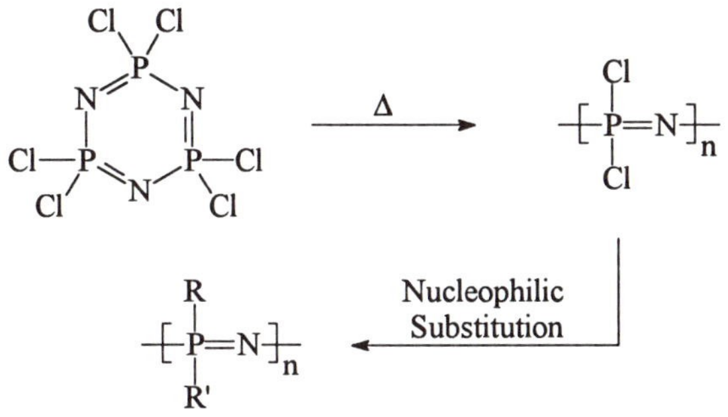

- Best suited for P-O and/or P-N bonded side groups
- Substituents attached *after* polymerization
- High molecular weights (ca. 10^6) but broad MW distribution

The second synthetic route to polyphosphazenes, which has been under investigation in our laboratories during the last 15 years, is based on the *condensation polymerization* (*Scheme 2*) of suitable Si-N-P precursors. (*4, 5*)   This method differs from and, therefore, compliments the ring-opening/substitution approach in two significant respects: (a) the desired substituents on phosphorus are introduced prior to polymerization, and   (b) the side groups are directly attached to phosphorus via thermally stable P-C bonds.   Furthermore, polymers such as $[Me(Ph)P=N]_n$ can be subsequently derivatized by deprotonation/substitution reactions of the pendent P-methyl groups to afford a much greater variety of functionalized materials. (*6*)

Scheme 2

Condensation Polymerization

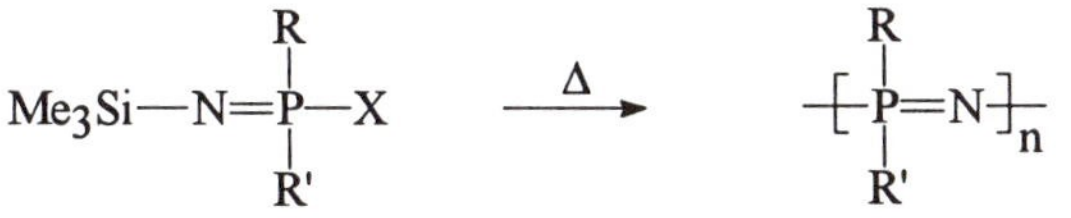

R, R' = alkyl, aryl;   X = OCH_2CF_3, Ph

- Best suited for P-C bonded side groups
- Substituents attached *before* polymerization
- Moderate molecular weights (ca. 10^5) but narrow MW distributions

Recently, two important variations of the condensation polymerization approach to poly(phosphazenes) have been reported. The first (7) involves an alternate synthetic route to poly(dichlorophosphazene) via the thermal elimination of phosphoryl chloride from a P-N=P precursor (equation 1). The second (8) involves the *catalyzed* condensation polymerization of certain N-silylphosphoranimines (equation 2) to afford various alkoxy- and/or aryloxy-substituted poly(phosphazenes).

$$
\underset{\substack{Cl\\|\\Cl}}{O=P}\!\!-\!\!N\!\!=\!\!\underset{\substack{|\\Cl}}{\overset{Cl}{P}}\!\!-\!\!Cl \quad \xrightarrow[\substack{-P(O)Cl_3}]{\Delta} \quad \underset{\substack{|\\Cl}}{\left[\!\!-N\!\!=\!\!\overset{Cl}{P}\!\!-\!\!\right]_n} \tag{1}
$$

$$
Me_3Si\!\!-\!\!N\!\!=\!\!\underset{\substack{|\\OR}}{\overset{OR}{P}}\!\!-\!\!OR \quad \xrightarrow[\substack{-Me_3SiOR}]{F^-\ catalyst} \quad \underset{\substack{|\\OR}}{\left[\!\!-N\!\!=\!\!\overset{OR}{P}\!\!-\!\!\right]_n} \tag{2}
$$

In regard to the poly(alkyl/arylphosphazenes), we are interested in investigating the overall scope and possible limitations of the condensation polymerization process as well as the physical and chemical properties of the resulting polymers. Some of our current efforts, which will be briefly reviewed in the following sections include: (1) study of the overall scope and limitations of this condensation polymerization, (2) recent improvements in the synthetic methodology including catalysis of the polymerization process as well as simplification of the precursor synthesis, (3) observations relevant to the polymerization mechanism such as leaving group effects, and (4) the synthesis, structure, and reactivity of some related small molecule species including N-silylphosphoranimines with bulky silyl groups, acetylene-substituted Si-N-P systems, and new N-phosphoryl-phosphoranimines, e.g., $Cl_2P(O)-N=PR_2X$.

Chemistry of Polymer Precursors

The successful synthesis of poly(alkyl/arylphosphazenes), $[R_2P=N]_n$, via the condensation polymerization process (*Scheme 2*) requires availability of the appropriate "monomers," the N-silylphosphoranimines, $Me_3SiN=PR_2X$. These compounds can be readily prepared from commercial starting materials such as PCl_3, $(Me_3Si)_2NH$, and Grignard reagents by way of the three-step process outlined in *Scheme 3*. This process affords good overall yields (60 - 70%) of the final phosphoranimine product, $Me_3SiN=PR(R')X$ [X = OCH_2CF_3, OPh], and is routinely carried out on a 1-2 mole scale in the laboratory. (9) As currently practiced, however, this synthetic methodology does have some practical limitations. For example, it seemed to be necessary to: (a) isolate and purify the (silylamino)phosphine product of *Step 1*, (b) use benzene as a solvent for the oxidative bromination (*Step 2*), (c) completely remove the Me_3SiBr byproduct of *Step 2* prior to introduction of the OCH_2CF_3 or OPh leaving group in *Step 3*.

Scheme 3

Synthesis of N-Silylphosphoranimines

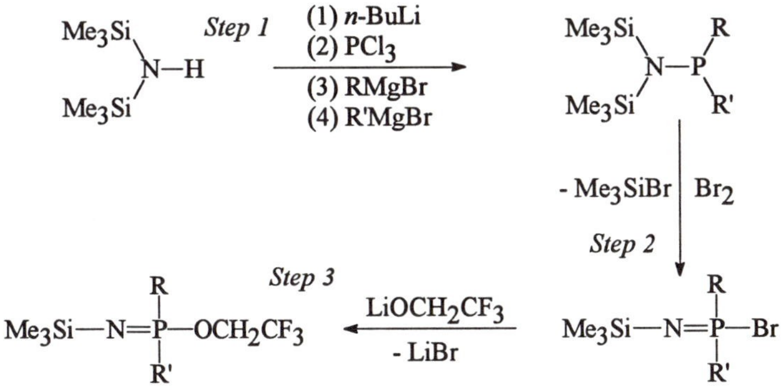

Variations:

Step 1: Use PhPCl$_2$ instead of PCl$_3$ to yield -P(Ph)R derivatives or 2 equivalents of the same Grignard to yield symmetrical -PR$_2$ compounds.

Step 2: Use C$_2$Cl$_6$ instead of Br$_2$ to yield P-Cl analogs.

Step 3: Use a wide variety of LiOR or LiOAr reagents.

Improved Synthesis of N-Silylphosphoranimines. Recent work in our laboratory demonstrates that all of the above problems can be circumvented by use of a modified "one-pot" reaction sequence. (*10*) A similar 3-step process is employed but with some significant variations. First, it is not necessary to isolate the (silylamino)phosphine (*Step 1*) if hexachloroethane, C$_2$Cl$_6$, is used as the oxidative halogenation agent in *Step 2*. (11) The reaction with C$_2$Cl$_6$ (equation 3) is essentially quantitative and works equally well in a variety of solvents, thus eliminating the hazards of using benzene in the bromination reaction. Second, we find that it is not necessary to isolate the P-chlorophosphoranimine product of *Step 2* (equation 3) and, in fact, removal of the Me$_3$SiCl byproduct is not even required. Addition of *one equivalent* of LiOPh or LiOCH$_2$CF$_3$ (*Step 3*) to the mixture containing both the P-chlorophosphoranimine and Me$_3$SiCl results in a surprisingly selective reaction with only the P-Cl electrophile.

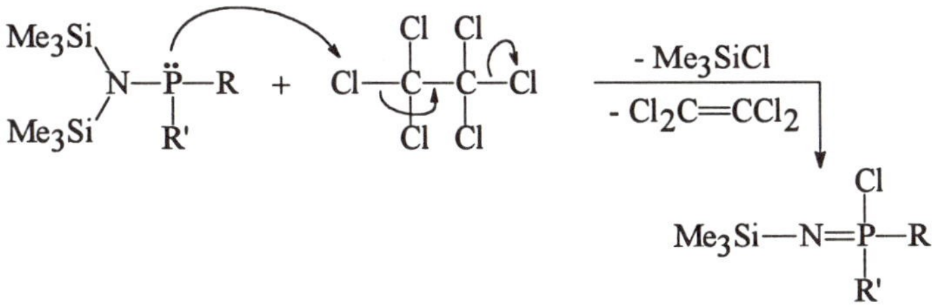

$$(3)$$

This greatly simplified "one-pot" reaction sequence, employing C$_2$Cl$_6$ instead of Br$_2$ in *Step 2*, and one equivalent of the nucleophile in *Step 3*, affords overall yields of ca. 75% and requires only a single fractional distillation to isolate the final phosphoranimine product. Complete experimental details of this process will be published elsewhere. (*10*)

Si-N-P Compounds with Bulky Silyl Groups. Some of the important features of the condensation polymerization process (*Scheme 2*) in which we are interested include the possibility of controlling the degree of polymerization and/or the chemistry of the polymer's end groups. In one aspect of this effort (*10*), we have recently prepared a series of N-silylphosphoranimines that contain sterically hindering N-silyl groups (equation 4). In this work, both the (silylamino)phosphines as well as the N-silylphosphoranimines were characterized by NMR spectroscopy and elemental analysis. The compounds which contain the very bulky *t*-butyl*diphenyl*silyl moiety are often crystalline solids and, to date, three phosphine derivatives and one P-phenoxyphosphoranimine have been characterized by single crystal X-ray diffraction, thus affording detailed and useful molecular structural information. (*10*) Another noteworthy point about this series of compounds is that the less bulky Me_3Si group is selectively cleaved from nitrogen during the oxidative bromination reaction. The polymerization behavior of these sterically hindered N-silylphosphoranimines generally parallels that of the Me_3Si analogs, although higher temperatures and longer reaction times are required for their complete thermolysis.

$$t\text{-}BuR_2Si \diagdown$$
$$\qquad\qquad N\text{—}H \quad \xrightarrow[\substack{(2)\ PCl_3\ \text{or}\ PhPCl_2 \\ (3)\ MeLi}]{(1)\ n\text{-}BuLi} \quad t\text{-}BuR_2Si \diagdown \qquad \diagup R$$
$$Me_3Si \diagup \qquad\qquad\qquad\qquad\qquad\qquad N\text{—}P$$
$$Me_3Si \diagup \qquad \diagdown Me$$

R = Me, Ph

$$t\text{-}Bu\!-\!\!\overset{\displaystyle R}{\underset{\displaystyle R}{Si}}\!-\!N\!=\!\!\overset{}{\underset{\displaystyle Me}{P}}\!-\!OR' \quad \xleftarrow[\substack{(2)\ LiOR' \\ R'\ =\ CH_2CF_3,\ Ph}]{(1)\ Br_2}$$

(4)

Synthesis of Acetylene Substituted Si-N-P Compounds. Some of our other related efforts in the area of small molecule Si-N-P chemistry include the synthesis, characterization, and possible polymerization of novel P-functionalized derivatives. These studies normally begin with the preparation of the appropriate (silylamino)phosphines, some novel examples of which are the P-acetylenic compounds (*Scheme 4*). Subsequent oxidative bromination, followed by derivatization reactions of the P-Br bond affords the $P\text{-}OCH_2CF_3$ or P-OPh substituted phosphoranimines as potential precursors to acetylene substituted poly(alkyl/arylphosphazenes). These new compounds, which are obtained in good yields as distillable liquids, have been fully characterized by multinuclear NMR spectroscopy and elemental analysis. (*12*) At the small molecule level, the study of these acetylene derivatives is part of our larger program of research on novel Si-N-P compounds bearing reactive functional groups. (*13*) It is especially noteworthy that the bromination reaction (*Scheme 4*) occurs exclusively at the phosphorus center, leaving the C-C triple bond unchanged. We had earlier made similar observations about the reactivity of P-*vinyl* (silylamino)phosphines. (*14*)

Scheme 4

Synthesis of Acetylene-Substituted Si-N-P Compounds

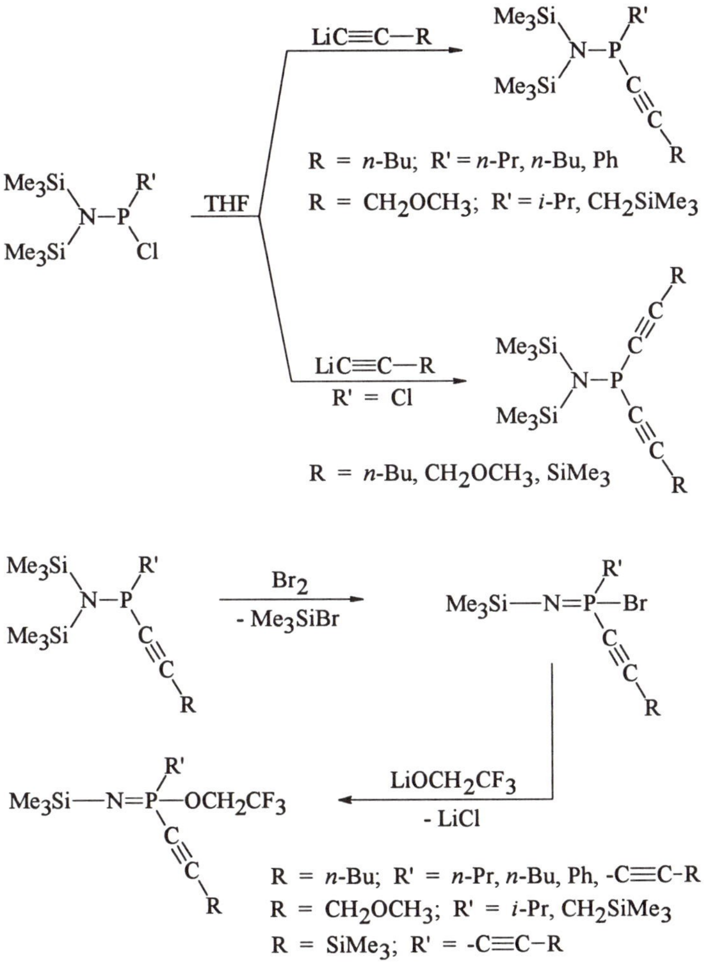
LiC≡C—R
THF
R = n-Bu; R' = n-Pr, n-Bu, Ph
R = CH2OCH3; R' = i-Pr, CH2SiMe3
LiC≡C—R
R' = Cl
R = n-Bu, CH2OCH3, SiMe3
Br2
- Me3SiBr
LiOCH2CF3
- LiCl
R = n-Bu; R' = n-Pr, n-Bu, Ph, -C≡C-R
R = CH2OCH3; R' = i-Pr, CH2SiMe3
R = SiMe3; R' = -C≡C-R

N-Phosphoryl Substituted Phosphoranimines. In addition to being useful as condensation "monomers" (*Scheme 2*), the N-silylphosphoranimines have a rich and varied chemistry of their own. Three distinct modes of reactivity (*Scheme 5*) are well documented. Cleavage of the Si-N bond is typified by a series of transsilylation reactions with substituted chlorosilanes (*15*) while the deprotonation-substitution chemistry has led to a wide variety of P-CH_2-E derivatives containing silyl, (*16*) phosphinyl, (*17*) and organic (*18*) functional groups. Some of these latter reactions serve as useful model systems for similar derivative chemistry of the preformed phosphazene polymers such as $[Me(Ph)P{=}N]_n$. (*6*)

Scheme 5

Reactivity of N-Silylphosphoranimines

silicon-nitrogen
bond cleavage
            nucleophilic substitution
            at phosphorus

$$Me_3Si\text{---}N\text{=}P\text{---}X$$

with substituents R (top) and CH$_3$ (bottom)

deprotonation / substitution
at P-methyl group

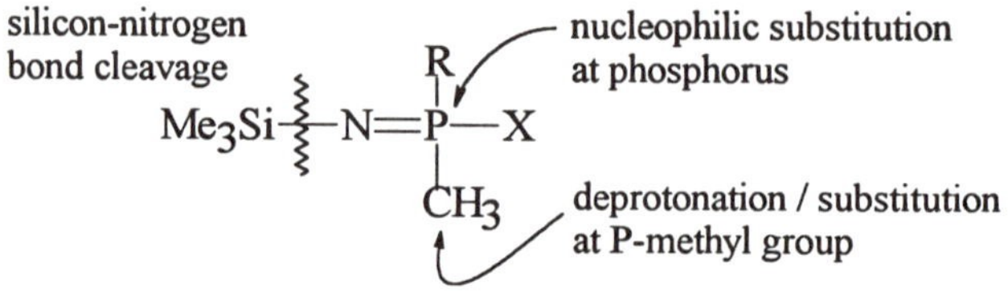

As part of our overall program on the development of new precursors to poly(alkyl/arylphosphazenes), we are currently studying a variety of possible Si-N bond cleavage reactions of the N-silylphosphoranimines with reactive halides of phosphorus, boron, and other main group elements. For example, we find that phosphoryl halides such as P(O)Cl$_3$ and MeP(O)Cl$_2$ react very smoothly with both the dimethyl- and the phenyl(methyl)phosphoranimines (*Scheme 6*) to afford high yields of new P-N=P systems that contain chlorophosphoryl, Cl-P(=O)R-, functional groups. (*19*) Although these P-Cl derivatives are routinely obtained as fully-characterized, distillable liquids, they are also easily converted to the P-dimethylamino derivatives via treatment with Me$_3$SiNMe$_2$.

Scheme 6

Synthesis of N-Phosphoryl-Substituted Phosphoranimines

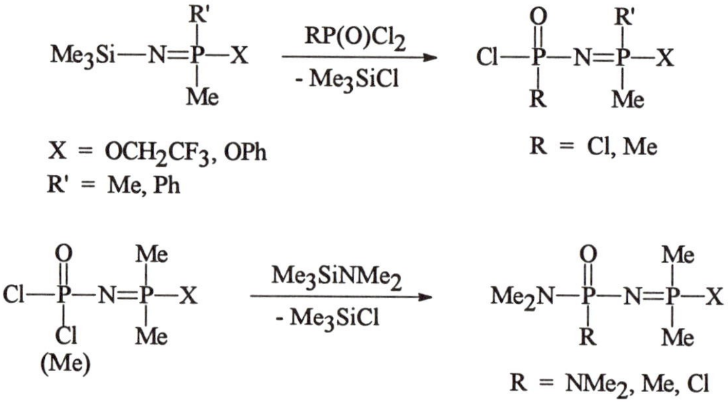

$$Me_3Si\text{---}N\text{=}P\text{---}X \xrightarrow[-Me_3SiCl]{RP(O)Cl_2} Cl\text{---}P\text{---}N\text{=}P\text{---}X$$

with R' and Me on first P; O, R, R' and Me on product

X = OCH$_2$CF$_3$, OPh
R' = Me, Ph

R = Cl, Me

$$Cl\text{---}P\text{---}N\text{=}P\text{---}X \xrightarrow[-Me_3SiCl]{Me_3SiNMe_2} Me_2N\text{---}P\text{---}N\text{=}P\text{---}X$$

(Me)

R = NMe$_2$, Me, Cl

Only a few other examples of N-phosphoryl-phosphoranimines have been previously reported. Most notable of these is the perchloro analog which has been shown to be an effective precursor to poly(dichlorophosphazene) via the thermal elimination of P(O)Cl$_3$ (equation 1). (*7*) In contrast, our phosphoryl derivatives, which contain various substituents at the P=N phosphorus center other than chlorine, do not decompose in the same manner. Instead, we find that the only significant volatile products are phosphine oxides (from the phenoxy compounds) or, more surprisingly, CF$_3$CH$_2$Cl (form the trifluoroethoxy analogs). These products probably result from the types of fragmentations indicated in *Scheme 7*. The non-volatile, possibly oligomeric, products of this decomposition process are not readily identifiable and must await further investigation.   In any event, it appears that additional studies of this kind will at least provide information relevant to the possible mechanism for the condensation polymerization of $Cl_2P(O)-N=PCl_3$ (equation 1).

Scheme 7

Thermolysis of N-Phosphorylphosphoranimines

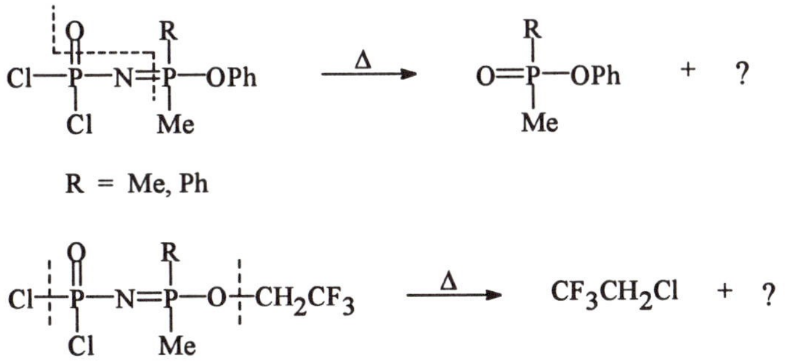

R = Me, Ph

Poly(alkyl/arylphosphazenes)

Having surveyed some of the relevant preparative chemistry of the N-silylphosphoranimines, we now turn attention to the products of their condensation polymerizations, the poly(alkyl/arylphosphazenes). (*Scheme 2*) We will focus here on the overall scope of this polymerization process, the types of polymers that it produces, and some of our recent efforts to modify and improve it.

Scope of the Condensation Polymerization Process. The thermolysis reactions of many different N-silyl-P,P-dialkyl-P-trifluoroethoxyphosphoranimines, and several of their P-phenoxy analogs, have been studied both individually and in a variety of copolymerizations with the dimethyl- or phenylmethyl-substituted precursors. These polymerizations (routinely conducted in sealed glass ampules at ca. 180-200°C for 7-12 days) (5, 20) provide a large series of new polymers that can be grouped into three general categories as summarized in Table 1.   Our studies of the physical, solubility, and thermal properties of these materials represent the first systematic analysis of the structure-property relationships among the poly(alkyl/arylphosphazenes).   Some of these properties are summarized briefly in the following sections.

Solubility.   In general, poly(alkyl/arylphosphazenes) are quite soluble in various organic solvents with the only exceptions found among the symmetrical dialkyl homopolymers (**A**).   Interestingly, while the parent member of this series, poly(dimethylphosphazene), $[Me_2PN]_n$, is soluble in CH_2Cl_2 EtOH, and THF/H_2O mixtures, all of the other symmetrical *dialkyl*phosphazenes, $[R_2PN]_n$, are insoluble in all common solvents.   This insolubility is attributed to the microcrystalline nature of these symmetrical polymers as confirmed earlier by X-ray diffraction studies (5) of the dimethyl and diethyl compounds and implied for all of them by intense melt transitions in their DSC traces.   In fact, these studies (5) indicate that the di*ethyl* polymer, $[Et_2PN]_n$, is more highly crystalline than the di*methyl* anaolg, consistent with its reduced solubility. It is possible, however, to dissolve all of these materials in organic solvents if a small amount (ca. 1%) of a weak acid such as acetic or benzoic acid is added to partially protonate the polymer backbone. (*21*) By using these acidified solvents, it is possible to obtain reliable NMR spectroscopic and solution viscosity data for such polymers. The unsymmetrical homopolymers (**B**) and the copolymers (**C**) are generally amorphous in nature and are very soluble in common organic solvents, especially CH_2Cl_2 and THF, thus facilitating their characterization by size exclusion chromatography (SEC). Notably, the homopolymers with relatively long alkyl chains (**B**: R = *n*-Bu, *n*-Hx) are even soluble in hexane and other nonpolar hydrocarbons.

Table 1. Classes of Poly(alkyl/arylphosphazenes)

A	*Symmetrical Homopolymers* R = Me, Et, *n*-Pr, *n*-Bu, *n*-Hx	$-[P(R)(R)=N]_n-$ (R above and below P)
B	*Unsymmetrical Homopolymers* R = Me, Et, *n*-Pr, *n*-Bu, *n*-Hx R' = Me, Ph	$-[P(R)(R')=N]_n-$
C	*Random Copolymers* R = Me, Et, *n*-Pr, *n*-Bu, *n*-Hx R' = Me, Ph	$-[N=P(Ph)(Me)]_x-[N=P(R)(R')]_y-$ (Me, Me / R, R')

Molecular Weights. Except for the higher homologs in the dialkyl series, $[R_2PN]_n$ (**A**: R ≠ Me), all of these polymers and copolymers are soluble in THF. Thus, by employing SEC conditions described elsewhere (*5*), we are able to obtain reliable molecular weight distributions for these polymers.   Typically, they have molecular weights (M_w) in the range of 50,000 - 150,000 and polydispersity indices (M_w/M_n ≈ 1.5-3.0) that are much smaller than those observed for the ring-opening polymerization (*Scheme 1*) but somewhat larger than those obtained in the fluoride-catalyzed condensation polymerizations. (*8*)   There are no obvious trends in molecular weight among these series of polymers and the factors which control the chain length in the condensation polymerization process are still not fully understood.

Thermal Transitions.   As noted earlier, the polymeric dimethyl- and diethylphosphazenes are microcrystalline powders with melting points of 143° and 217°C, respectively. Similar characteristics are observed for the other symmetrical di-*n*-alkyl derivatives although the di-*n*-hexyl polymer is a more wax-like solid that exhibits both a glass transition (-29°C) and a melt transition (129°C). On the other hand, the unsymmetrical homopolymers (**B**) have considerably different physical properties with only the Me/Et derivative being a true solid (T_m = 80°C). With their longer *n*-alkyl chains, the other members of this series are amorphous gums or waxes.

All of these Me/alkyl polymers (**B**: R' = Me) exhibit broad, poorly defined glass transitions around -50°C as compared to the very distinct T_g at -46°C observed for $[Me_2PN]_n$. (5) The copolymers (**C**) exhibit a very broad spectrum of physical properties as would be expected for such a large degree of structural variation of the side groups. These materials vary from microcrystalline solids (e.g., some of the Me_2/R_2 systems) to gums (e.g., the Me/Ph-containing copolymers) to viscous fluids such as the dimethyl/methyl-*n*-hexyl copolymers. Accordingly, a wide range of glass transition temperatures from ca. +10 to -70°C is observed, with the presence of phenyl or *n*-hexyl substituents favoring the higher or lower values, respectively. Thus, it is now possible to essentially "tailor make" a given alkyl/arylphosphazene polymer with whatever physical characteristics are desired for a particular application.

Leaving Group Effects on the Condensation Polymerization Process. In studies of partially polymerized reaction mixtures, we find that only relatively high molecular weight polymers ($M_w \approx 50,000$) and unreacted monomer are observed in the early stages of the reaction (4). Thus, *the condensation polymerization of N-silylphosphoranimines appears to involve a chain growth mechanism*, probably initiated by heterolytic cleavage of the polar P-X bond.

While the details of the polymerization mechanism are not completely understood, considerable information is available on how the nature of the leaving group (X) at phosphorus affects the thermal stability and the decomposition products of the N-silylphosphoranimines, $Me_3SiN=PR_2X$. (*Scheme 8*)

Scheme 8

Leaving Group Effects in Condensation Polymerization

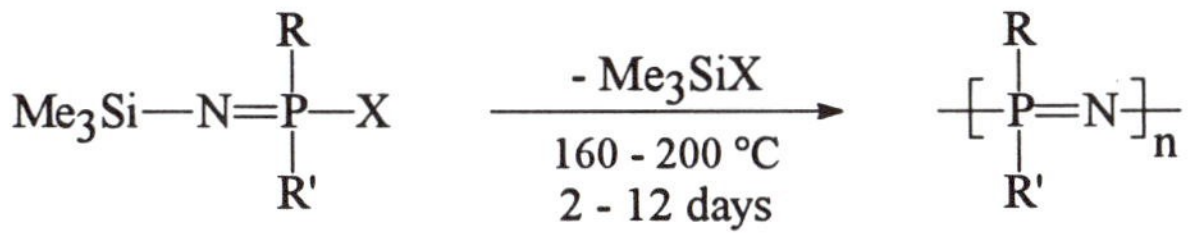

Reaction Products:

X = Br, Cl, alkoxy ⟶ cyclic phosphazenes

X = OCH_2CF_3, aryloxy ⟶ polymeric phosphazenes

Relative Rates:

X = Br, Cl > OCH_2CF_3, aryloxy > alkoxy, amino

The tendency of these precursors to thermally eliminate Me_3SiX follows the approximate leaving group (X) order: halogen > aryloxy $\approx$ fluoroalkoxy > alkoxy $\approx$ amino. (22) Greater polarity of the P-X bond and the corresponding leaving-group ability of X^- generally favors the elimination of Me_3SiX. As a result, the P-*halo*phosphoranimines often decompose during distillation while most of the other types of phosphoranimines can routinely be distilled without decomposition. The leaving group also has a pronounced influence on the type of phosphazene (cyclic or polymeric) that is formed in these condensation processes. For example, the N-silyl-

P-*halo*phosphoranimines (X = Br, Cl) decompose to yield *cyclic* phosphazenes (*20*, *22*), normally as mixtures of trimers and tetramers, rather than polymers as are obtained in the thermolyses of the trifluoroethoxy analogues.

At first glance, such a dramatic leaving group effect (i.e., X = Cl, Br vs. X = OCH_2CF_3, OPh) might be interpreted as an indication that two different mechanisms are involved. Other work, however, has shown that Me_3SiBr reacts with *polymeric* $[Me_2PN]_n$ and $[Me(Ph)PN]_n$, prepared from the P-OCH_2CF_3 derivatives, to cause chain degradation and formation of cyclic phosphazenes. (*22*) This observation illustrates the principle that condensation processes must produce an "inert" byproduct in order to be useful for polymerization reactions. (*4a*) Some recent work (*22*) indicates, in fact, that removal of the halosilane byproduct under a dynamic vacuum as the polymerization proceeds might eliminate this problem of chain degradation, although the process has not been optimized.

Other questions about the polymerization mechanism prompted us to investigate alkoxy substituents other than -OCH_2CF_3 as possible leaving groups. (*23*) When non-fluorinated *alkoxy* groups are attached to phosphorus, the precursors are much more thermally stable than the OCH_2CF_3 derivatives. Furthermore, when more drastic conditions are used, the P-*alkoxy* compounds decompose to yield *cyclic* phosphazenes along with other products. For example, in the thermolysis of the *t*-butoxy derivative, the formation of the (silylamino)phosphine oxide, $Me_3SiN(H)P(=O)(Ph)Me$, probably occurs via a β-elimination of *iso*butylene. (*4a*)

In contrast to the results obtained with the alkoxy analogues, the thermal decomposition of the *aryloxy* substituted monomers is an efficient, high yield, and relatively inexpensive preparative route to the poly(alkyl/arylphosphazenes). (*24*) The difference between these systems (i.e., alkoxy vs. aryloxy leaving groups) can be attributed to the inability of the aryloxy (or OCH_2CF_3) groups to undergo β-elimination and to the fact that OAr⁻ is a better leaving group than is OR⁻.

Catalysis of the Polymerization. In order to simplify, and improve the practicality and cost-effectiveness, of the condensation polymerization route to poly(alkyl/arylphosphazenes), we are currently striving to achieve (1) faster, lower-temperature polymerizations via the use of catalysts/initiators (*Scheme 9*), and (2) major improvements in the synthesis of the Si-N-P precursors. Some very promising preliminary results in the latter area have been obtained in terms of the "one-pot" synthesis as discussed earlier. We will now briefly summarize our results in the former area, i.e., catalysis of the polymerization. (*24*)

Several different nucleophilic reagents, including both anionic species (e.g., F⁻ or PhO⁻) as well as neutral molecules [e.g., $(Me_2N)_3P=O$, HMPA] have been studied in considerable detail as possible polymerization catalysts. (*24*) We find that many such reagents do, in fact, accelerate the condensation process. In most cases, the polymerizations are complete within 24 hours at temperature of ca. 150°C as opposed to several days at 180-200°C for the uncatalyzed process. Considerable variation, however, is found in regard to the molecular weight distribution (Figure 1) of the resulting polymeric products depending on which particular catalyst is employed. For example, with fluoride ion as the catalyst, the condensation reaction is generally complete in only a few hours, but the molecular weight distribution is invariably bimodal with major amounts of low molecular weight species. (see traces **B** and **C** in Figure 1) Interestingly, such bimodal distributions are not common for the fluoride-catalyzed polymerization of fully *alkoxy* substituted phosphoranimines, $Me_3SiN=P(OR)_3$. (8) Of the catalytic systems studied thus far in our laboratory, the best results have been obtained with sodium phenoxide. (Figure 1, **A**) While the reactions are not quite as fast as those catalyzed by F^-, the molecular weights of the poly(phosphazene) products are consistently high and the molecular weight distributions are narrow. Further studies of potential catalytic routes to poly(alkyl/arylphosphazenes), $[R_2PN]_n$, are in progress.

Scheme 9

Nucleophilic Catalysis of Condensation Polymerization

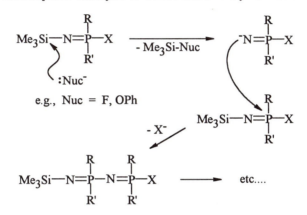

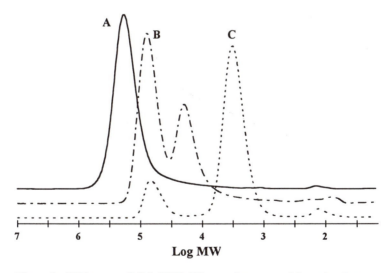

Figure 1. SEC traces of $[Me(Ph)P=N]_n$ samples prepared form the phenoxy precursor, $Me_3SiN=P(Ph)(Me)(OPh)$ under various conditions: **(A)** 1.0% NaOPh, 150°C, 24 hrs (peak MW ≈ 126,000). **(B)** 1.6% CsF, 150°C, 4 hrs (peak MW's ≈ 53,400 & 15,200). **(C)** 1.6% n-Bu$_4$NF, 150°C, 4 hrs (peak MW's ≈ 9,400 & 3,200)

Conclusion. The poly(alkyl/arylphosphazenes), $[R_2P=N]_n$, are a major, relatively new, class of phosphazene polymers that are readily prepared by the condensation polymerization of various N-silylphosphoranimines, $Me_3SiN=PR_2X$. The physical and chemical properties of these polymers can be varied, in a controlled manner, over a wide range depending on the nature and combination of the organic side groups. Recent developments in the preparative chemistry that were reviewed here include significant improvements in the polymerization process itself (i.e., catalysis) and in the synthesis of the Si-N-P precursors (e.g., the "one-pot" reaction). Future studies in this area will undoubtedly continue to enhance the prospects for development and application of this class of inorganic polymers.

Acknowledgment. The authors thank the U. S. Army Research Office for major financial support of the poly(phosphazene) work. Related aspects of some of the small molecule chemistry is supported by the National Science Foundation, the Robert A. Welch Foundation, and the Texas Christian University Research Fund.

Literature Cited

1. See the paper by H. R. Allcock and others in this *ACS Symposium* volume. See also: (a) Allcock, H. R. *Chem. Eng. News* **1985**, *63(11)*, 22. (b) Allcock, H. R. *Angew. Chem., Int. Ed. Engl.* **1977**, *16*, 147. (c) Allcock, H. R. *Science* **1992**, *255*, 1106.

2. (a) Singler, R. E.; Sennett, M. S.; Willingham, R. A. *ACS Symposium Series* **1988**, *360*, 268. (b) Penton, H. R. *ACS Symposium Series* **1988**, *360*, 277.

3. Allcock, H. R. *ACS Symposium Series* **1988**, *360*, 277. (b) Allcock, H. R. *ACS Symposium Series* **1983**, *232*, 49.

4. (a) Neilson, R. H.; Wisian-Neilson, P. *Chem. Rev.* **1988**, *88*, 541. (b) Neilson, R. H.; Ford, R. R.; Hani, R.; Roy, A. K.; Scheide, G. M.; Wettermark, U. G.; Wisian-Neilson, P. *ACS Symposium Series* **1988**, *360*, 283.

5. Neilson, R. H.; Wisian-Neilson, P.; Meister, J. J.; Roy, A. K.; Hagnauer, G. L.; *Macromolecules* **1987**, *20*, 910.

6. See the following paper by P. Wisian-Neilson in this *ACS Symposium* volume. See also: Wisian-Neilson, P.; Schaefer, M. A.; *Macromolecules* **1989**, *22*, 2003; and references cited therein.

7. D'Hallum G.; De Jaeger, R.; Chambrette, J. P.; Potin, Ph. *Macromolecules* **1992**, *25*, 1254; and references cited therein.

8. See the phosphazene paper by K. Matyjaszewski in this *ACS Symposium* volume and references cited therein. See also: (a) Matyjaszewski, K.; Dauth, J.; Montague, R. A.; Reddick, C.; White, M. *J. Am. Chem. Soc.* **1990**, *112*, 6721. (b) Matyjaszewski, K.; Moore, M. K.; White, M. L. *Macromolecules* **1993**, *26*, 674.

9. (a) Wisian-Neilson, P.; Neilson, R. H. *Inorg. Synth.* **1989**, *25*, 69. (b) Neilson, R. H.; Wisian-Neilson, P. *Inorg. Chem.* **1982**, *21*, 3568. (c) Wisian-Neilson, P.; Neilson, R. H. *Inorg. Chem.* **1980**, *19*, 1975.

10.   (a) Samuel, R.   Ph. D. Dissertation, Texas Christian University, 1993.   (b) Neilson, R. H.; R. Samuel, in preparation for submission to *Inorg. Chem.*

11.   For a review of the reactivity of phosphines toward chlorocarbons, including CCl_4 and C_2Cl_6, see: Appel, R. *Angew. Chem., Int. Ed. Engl.* **1975**, *14*, 801.

12.   Kucera, W. R.; Neilson, R. H., unpublished results.

13.   Neilson, R. H.; Angelov, C. M. *ACS Symposium Series* **1992**, *486*, 76.

14.   Waters, K. R.; Neilson, R. H. *Phosphorus and Sulfur*, **1988**, *39*, 189.

15.   Wettermark, U. G.; Neilson, R. H. *Inorg. Chem.* **1987**, *26*, 929.

16.   Wettermark, U. G.; Wisian-Neilson, P.; Scheide, G. M.; Neilson, R. H. *Organometallics* **1987**, *6*, 959.

17.   Roy, A. K.; Hani, R.; Neilson, R. H.; Wisian-Neilson, P. *Organometallics* **1987**, *6*, 376.

18.   Roy, A. K.; Wettermark, U. G.; Scheide, G. M.; Wisian-Neilson, P.; Neilson, R. H. *Phosphorus and Sulfur* **1987**, *33*, 147.

19.   Longlet, J. J.; Neilson, R. H., unpublished results.

20.   Jinkerson, D. L. Ph. D. Dissertation, Texas Christian University, 1992.

21.   For a recent study of the backbone coordination chemistry of $[Me(Ph)P=N]_n$, see: Wisian-Neilson, P.; Garcia-Alonso, F. J. *Macromolecules* **1993**, *26*, 7156.

22.   Hani, R.; Jinkerson, D. L.; Wood, C.; Neilson, R. H. *Phosphorus, Sulfur, and Silicon* **1989**, *41*, 159.

23.   Neilson, R. H., Hani, R., Samuel, R.; Wood, C. E., unpublished results.

24.   Wood, C. E. Ph. D. Dissertation, Texas Christian University, 1992.

RECEIVED April 18, 1994

Chapter 19

Derivatives of Poly(alkyl/arylphosphazenes)

Patty Wisian-Neilson

Department of Chemistry, Southern Methodist University,
Dallas, TX 75275

The synthesis of poly(phosphazenes) from preformed poly(alkyl/aryl-phosphazenes) such as [Me(Ph)PN]$_n$ is discussed. This approach affords polymers with a variety of side groups all of which are attached to the backbone by direct phosphorus-carbon bonds. The methods used include deprotonation-substitution reactions at methyl groups, electrophilic aromatic substitution at phenyl groups, and coordination of Lewis acids at the basic backbone nitrogen. The side groups incorporated by these approaches include silyl, alcohol, ferrocene, thiophene, carboxylic acids, carboxylate salts, fluorinated alkyl, ester, and simple transition metal moieties. Several graft copolymers are also discussed.

The formation of the poly(phosphazene) backbone can be accomplished by either ring-opening of selected cyclic phosphazenes, e.g., [Cl$_2$P=N]$_3$, or by condensation polymerization of certain X-P(Z$_2$)=N-Y compounds. As discussed in greater detail in other chapters of this book, there are numerous variations of these two general methods. Generally, ring-opening polymerizations involve cyclic phosphazenes in which at least a few of the phosphorus sites have halogen substituents. The halogens on the poly(phosphazenes) formed in this manner are then readily replaced with a variety of alkoxy, aryloxy, and/or amino groups using nucleophilic substitution reactions. (*1-3*) Condensation polymerization, on the other hand, is a newer, less well-studied approach that has yielded fully P-C-bonded alkyl/aryl substituted poly(phosphazenes) (*4, 5, 6*) as well as several simple dialkoxy, diaryloxy, (*7*) dichloro (*8*), and diaryl (*9*) substituted polymers.

A major factor in the diversity of the phosphazene polymer system, and indeed the reason poly(phosphazenes) are unique among polymers, is the broad range of chemistry that can be carried out on the poly(phosphazenes) once the backbone has been formed. As mentioned above, the best examples of this are the simple, yet very

0097–6156/94/0572–0246$08.00/0

well-developed, nucleophilic substitution reactions of P-halogen substituted phosphazenes. Such reactions have afforded hundreds of poly(phosphazenes), which, for the most part, have P-N and P-O bonded side-groups. Subsequent reactions on these polymers serve to further extend the diversity of substituents as well as the range of properties of these polymers. (*1, 2*) More recently, the derivative chemistry of poly(alkyl/arylphosphazenes) such as $[Me(Ph)PN]_n$, **1**, has also been under investigation as a means of expanding the range of properties of poly(phosphazenes) with P-C bonded substituents. The synthesis, characterization, and properties of polymers prepared by this approach are discussed below.

Poly(methylphenylphosphazene), **1**, which is prepared by the thermal condensation polymerization of N-silylphosphoranimines, $(Me)(Ph)P(OR)=NSiMe_3$ [R = OCH_2CF_3 (*4, 5*) or OPh (*6, 10*)], offers three sites for derivatization. These are (a) the methyl group which is suitable for deprotonation-substitution reactions, (b) the phenyl group which is susceptible to electrophilic aromatic substitution, and (c) the lone pair of electrons on the backbone nitrogen atom which is an excellent site for coordination of Lewis acids.

As discussed below, the reactivity of one site often interferes with desirable chemical reactions to be conducted at another site. It should also be noted that model reactions can sometimes be performed on the N-silylphosphoranimines precursors, (*6*) but, because these small molecules also contain reactive N-Si and P-O bonds, they are not always suitable prototypes. Furthermore, while derivatization of these precursors can be a suitable means of preparing new polymers with substituent diversity, some groups interfere with or inhibit the thermal polymerization. Thus, our work as described here focuses on the chemistry of the preformed polymer systems.

Deprotonation-Substitution Reactions

The most straightforward way to derivatize poly(methylphenylphosphazene) is through initial formation of the anionic intermediate **2** (eq 1). This reaction has generally been carried out by treatment of a 0.7 to 1.0 M THF solution of **1** with *n*-BuLi at -78 °C, followed by stirring at low temperature for 1 to 2 h. In most cases, the -78 °C solutions of **2** have been treated with electrophiles and then allowed to warm to room temperature. This affords maximum *x:y* ratios of 1:1 or substitution of 50% of the methyl groups. Using less than 0.5 equivalents of *n*-BuLi results in correspondingly lower degrees of substitution. (*11-13*)

$$\left[\begin{matrix} \overset{Ph}{\underset{Me}{P}}=N \end{matrix}\right]_n \xrightarrow[\text{THF}]{n\text{-BuLi}} \left[\begin{matrix} \overset{Ph}{\underset{Me}{P}}=N \end{matrix}\right]_x \left[\begin{matrix} \overset{Ph}{\underset{CH_2^-Li^+}{P}}=N \end{matrix}\right]_y \qquad (1)$$

1   **2**

Silyl Substituted Polymers from Chlorosilanes. One of the simplest deprotonation-substitution reactions studied in our labs was the reaction of **2** with Me_3SiCl (eq 2). Typically substitution of 10 and 50 % of the methyl groups, which was confirmed by NMR spectroscopy and by elemental analysis, was easily obtained by controlling stoichiometry. More importantly, gel permeation chromatography (GPC) and intrinsic viscosity measurements clearly indicated that no chain degradation occurred in the deprotonation-substitution process. (*11*)   NMR spectroscopy also showed the presence of small quantities of disubstituted methyl groups, $CH(SiMe_3)_2$. Sequential treatment with n-BuLi and Me_3SiCl or using more dilute THF solutions of **1** increased the degree of disubstitution. Differential scanning calorimetry (DSC) measurements of the glass transition temperature (T_g) showed only one transition (ca. 73 °C, when $x = y$) indicating that the substitution was random.

$$\left[\begin{matrix} \overset{Ph}{\underset{Me}{P}}=N \end{matrix}\right]_x \left[\begin{matrix} \overset{Ph}{\underset{CH_2^-Li^+}{P}}=N \end{matrix}\right]_y \xrightarrow[-\text{LiCl}]{+\ Me_3SiCl} \left[\begin{matrix} \overset{Ph}{\underset{Me}{P}}=N \end{matrix}\right]_x \left[\begin{matrix} \overset{Ph}{\underset{CH_2SiMe_3}{P}}=N \end{matrix}\right]_y \qquad (2)$$

      2   **3**

   Reactive Functional Groups. Other simple chlorosilanes provided access to poly(phosphazenes), **4**, [R = H, CH=CH$_2$, and (CH$_2$)$_3$CN] with potentially reactive SiH, vinyl, and cyanopropyl groups (eq 3). (*11*)   Currently, we are using hydrosilylation reactions of the vinyl substituted polymers for the attachment of $Si(Me_2)[CH_2CH_2O]_nC_6H_4C_6H_4R$ groups.   Several attempts to use the Si-H substituents in hydrosilylation reactions have had limited success. A number of simple ether groups have also been attached through the Me_2Si spacer group using $ClSiMe_2R$ where R = $Si(Me)_2C_3H_6OC_4H_9$ or $Si(Me)_2C_3H_6(OC_2H_4)_2OCH_3$. (*14*)

$$\left[\begin{matrix} \overset{Ph}{\underset{Me}{P}}=N \end{matrix}\right]_x \left[\begin{matrix} \overset{Ph}{\underset{CH_2^-Li^+}{P}}=N \end{matrix}\right]_y \xrightarrow[-\text{LiCl}]{+\ RMe_2SiCl} \left[\begin{matrix} \overset{Ph}{\underset{Me}{P}}=N \end{matrix}\right]_x \left[\begin{matrix} \overset{Ph}{\underset{CH_2SiMe_2R}{P}}=N \end{matrix}\right]_y \qquad (3)$$

      2   **4**

   Alkyl and Fluoroalkyl Groups.. A series of silylated poly(phosphazenes) (eq 3) with long chain alkyl [R = (CH$_2$)$_m$CH$_3$, m = 2, 3, 7, 9, 17] and fluoroalkyl [R = CH$_2$CH$_2$(CF$_2$)$_m$CF$_3$, m = 0, 5, 7] groups were prepared recently in our laboratory. (*15*) These polymers, where the degree of silyl substitution was between 40 and 70 %, were characterized by NMR spectroscopy, GPC, and elemental analysis. The higher degrees of substitution were achieved by using elevated reaction temperatures and higher proportions of n-BuLi to form the anion as discussed later in this paper. The T_g values in both series decreased with increasing chain lengths to a minimum of

-20 °C for the polymers with decyl and fluorodecyl substituents. With longer side chains, the T_g values were higher as has been observed with other systems. (*16, 17*) Contact angle measurements showed that the polymers became more hydrophobic as the length of the alkyl side group increased (ca. 90 to 96 °) . The polymers with fluoro alkyl groups were slightly less wettable (96 to 100 °). Both of these are significantly less wettable than the parent polymer **1** where the contact angle is 73 °.

Alcohol Substituted Polymers from Aldehydes and Ketones. As demonstrated by the chemistry in the previous section, the polymer anion **2** can be viewed as a polymeric version of an organolithium reagent. This is further illustrated by the successful reactions of **2** with aldehydes and ketones (eq 4). (*12, 13*) The alcohol derivatives **5** (R, R' = H, Me, Ph, thiophene, ferrocene) were prepared by initial reaction of **2** with the electrophiles at either 0 or -78 °C followed by quenching of the alkoxy anion with aqueous NH$_4$Cl. Typically the degree of substitution in these reactions was between 30 and 45% with best results with the aldehydes. In addition to attaching potentially electroactive groups such as ferrocene and thiophene, this approach has also been used to attach organofluoro [R= R' = CF$_3$; R = H, R'= C$_6$F$_5$] groups that may serve to modify the polymer surface. (*18*) The T_g values of the alcohol derivatives ranged from 49 to 102 °C which is significantly higher than the 37 °C of the parent polymer **1**. This can be attributed to both increased steric size of the substituents and to hydrogen bonding of the hydroxy groups. The T_g values increased with higher degrees of substitution and larger R and R' groups. (*13*) The fluorinated alcohols have T_g values near those of the non-fluorinated analogs and exhibit slightly lower onsets of decomposition in TGA analysis. (*18*)

$$ \text{(4)} $$

Ferrocene Substituents. Some of the most interesting poly(phosphazenes) that have been prepared from the reactions of the polymer anion **2** with carbonyl compounds are those with ferrocene substituents.

First, thermal gravimetric analysis showed up to 50 % retention of weight even on heating to 900 °C. (*13*) Second, the T_g values of a series of these polymers [$x = 0.94$, $y = 0.06$, R = H; $x = 0.80$, $y = 0.20$, R = H; $x = 0.64$, $y = 0.36$, R = Me; and $x = 0.56$, $y = 0.44$, R = H) increased systematically from 40 to 89 °C. Third, the electrochemistry of this series was examined both in solution and as films on electrode surfaces. Reversible electrochemistry was observed for each polymer and the charge-transfer efficiency increased as the degree of ferrocene substitution increased. (*19*)

Ester Derivatives. In addition to simply altering the properties of the poly(phosphazenes), attaching reactive functional groups extends the types of derivative chemistry that may be used to modify the polymers. An example of this is the preparation of a series of ester derivatives **6** from the alcohol obtained from acetone (eq 5). The T_g values of the esters were lower than that of the alcohol precursor (55 °C) presumably because hydrogen bonding was eliminated. These values also decreased systematically as the length of the alkyl ester group increased until a minimum value of 10 °C was reached for $m = 8$. Then, like poly(dialkoxyphosphazenes) (*17*), as the chain length became longer ($m = 10$ to 14), the T_g values became higher reaching 27 °C for $m = 14$. (*16*)

Carboxyl Substituted Polymers from Carbon Dioxide. Reactions of **2** with carbon dioxide afforded a series of carboxylate salt, **7**, carboxylic acid, **8**, and ester, **9**, substituted polymers [where $x = y$; $x = 3$, $y = 1$; and $x = 9$, $y = 1$] (Scheme 1). (*20*) The 50% substituted salt (where $x = y$, R = Li$^+$) is the first water soluble derivative of a poly(alkyl/arylphosphazene). The salts are readily converted to the carboxylic acids upon protonation or into esters when treated with an activated species such as 4-NO$_2$C$_6$H$_5$CH$_2$Br. Fluorescence studies of these simple acids indicate that they form a moderately hydrophobic environment in aqueous media. (*21*) The salts and acids are easily cross-linked by addition of divalent metal cations which suggests applications as hydrogels. (*20*) It should be noted that spectral data suggest that a better representation of the acid **8** is the zwitterionic form **8b** shown in **Scheme 1**. This is substantiated by studies of coordination of the backbone nitrogen as discussed later.

Scheme 1

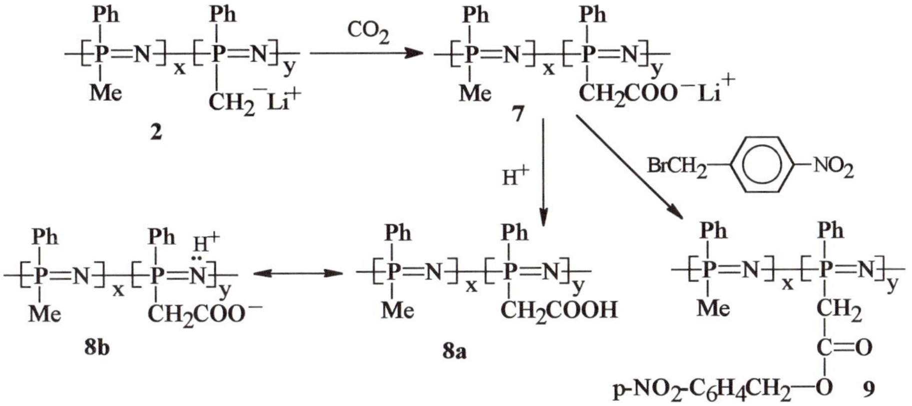

Graft Copolymers from Anionic Polymerization.

The similar reactivity of the polymer anion **2** and organolithium reagents suggested that the anion sites could be used to initiate anionic polymerization reactions. In this manner, both organic and inorganic graft copolymers of poly(phosphazenes) have been prepared.

Polystyrene Graft Copolymers. The anionic polymerization of styrene was initiated with **2** giving poly(phosphazene)-*graft*-polystyrene copolymers **10** (eq 6). Both the number of grafts ($y = 1$, $x = 4 - 9$) and the length ($z = 20 - 150$) of the graft were varied such that copolymers with 65 to 90 % styrene were prepared. Grafting was demonstrated by marked changes in GPC molecular weight data in terms of higher molecular weights and larger polydispersity values, and by increased absolute molecular weight as determined by membrane osmometry. Two distinct glass transition temperatures were observed for each graft at ca. 36 to 43 °C and 96 to 108 °C. These correspond to the T_g's of each of the homopolymer components and indicate that two well-separated phases form in the graft systems. (*22*) This is also evident in SEM's of these copolymers. (*23*)

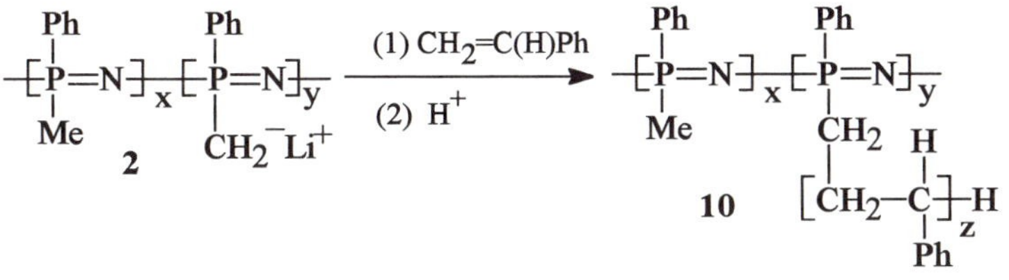

Poly(methyl Methacrylate) Graft Copolymers. A second type of inorganic-organic graft system was prepared using anionic polymerization of methyl methacrylate (MMA). The reaction of **2** with MMA resulted in insoluble materials due to reactions at both vinylic and carbonyl carbons. Thus the reactivity of the anion sites was modified by intial reaction with diphenylethylene (eq 7). The more sterically hindered anion reacted cleanly to give the poly(methyl methacrylate) (PMMA) graft copolymers. These were characterized as above. Unlike the polystyrene grafts, the PMMA graft copolymers exhibited some phase miscibility. Although two distinct glass transition temperatures were observed, the T_g of the PMMA segment (ca. 100 °C) was somewhat lower than that of PMMA homopolymers and the second T_g at 60 to 65 °C was significantly higher than that of the phosphazene segment. Similar merging of the T_g's of a blend of PMMA and $[Me(Ph)PN]_n$ were also observed. (*24*) The increased homogeneity of the components was also demonstrated by SEM. (*27*)

$$(7)$$

Poly(dimethylsiloxane) Graft Copolymers. The isoelectronic polysiloxane and poly(phosphazene) backbones were combined in graft copolymers prepared by ring opening polymerization of hexamethylcyclotrisiloxane initiated by the anion sites in **2** (eq 8). Grafting was demonstrated by significantly higher molecular weights and larger molecular weight distributions. Some chain scission of the polysiloxane chain was also observed but these low molecular siloxanes were readily removed by fractional precipitation. The poly(phosphazene) and poly(siloxane) components were phase separated as evidenced by the observation of two T_g's at -123 and 38 °C for the silicone and poly(phosphazene), respectively. (*25, 26*) A third, stronger transition attributable to T_m of poly(dimethylsiloxane) (-45 °C) was also observed in the DSC scan. By thermal gravimetric analysis (TGA), the onset of decomposition of these grafts was as high as 436 °C in air with 50% weight loss at 468 °C. This is considerably higher than the onset of decomposition of the parent polymer **1** at 350 to 360 °C.

$$(8)$$

Polymers with Sulfur-Containing Substituents. Sulfide groups have also been attached to the poly(phosphazenes) using either elemental sulfur (eq 9) or disulfides (eq 10). The reaction with sulfur proceeded through an intermediate, $[Me(Ph)PN]_x[(CH_2S^-Li^+)(Ph)PN]_y$ which was characterized by ^{1}H and ^{31}P NMR spectroscopy. In addition to simple reactions with electrophiles such as MeI (eq 9), this intermediate formed S-H terminated groups and $CH_2S\text{-}SCH_2$ crosslinks upon addition of acids. Addition of strong base removed crosslinking, but the reaction could successfully be repeated over only a few cycles. The sulfide groups are also of interest as potential sites for coordination of transition metals. (*27*)

$$\begin{array}{ccc} & (1)\ S_8 & \\ & (2)\ RI & \text{(9)} \\ \hline & & \\ & RSSR & \text{(10)} \end{array}$$

Other Polymer Derivatives with Reactive Functional Groups. The polymer anion **2** reacts with a number of other electrophiles such as Ph_2PCl, CH_2=$CHCH_2Br$, and Br_2 all of which contain reactive element-halogen bonds (eq 11). While both phosphine and allyl groups provide sites for coordination of transition metals, the unsaturated allyl group may also be a useful site for hydrosilylation or crosslinking reactions. The bromomethyl group has, however, proven to be quite unreactive. (*24*)

$$\text{(11)}$$

Variations in the Deprotonation-Substitution Reactions. As discussed in the previous paper in this volume, the preparation of copolymers with combinations of alkyl and aryl groups attached to the backbone by P-C bonds is readily achieved by the condensation polymerization of appropriate mixtures of N-silylphosphoranimines. The highly methylated copolymers **15** are of particuluar interest in terms of deprotonation-substitution reactions for several reasons. First, with even a low proportion of phenyl groups, these copolymers remain soluble in THF, a solvent suitable for deprotonation-substitution reactions. Second, the T_g's of the copolymers are significantly lower than in the homopolymer, $[Me(Ph)PN]_n$. Third, the copolymers show a strong affinity for water, and like the dimethyl homopolymer, $[Me_2PN]_n$, are often solubilized by only slightly acidified solutions. These differences in properties relative to poly(methyl-phenylphosphazene) are reflected in the derivatives of the copolymers. Furthermore, additional control of properties is enhanced by the ease of variation of the proportion of methyl to phenyl groups.

Using copolymers with nearly equal portions of $[Me(Ph)P$=$N]$ and $[Me_2PN]$ units, **15** ($m \cong n$), we have found that the conditions needed to deprotonate the methyl groups differ substantially from those needed for the homopolymer, $[Me(Ph)PN]_n$. Not surprisingly, since the copolymers are less soluble in THF, generation of the anion at -78 °C requires at least 3 hours to get the highest degrees of substitution. When this intermediate anion is treated with Me_3SiCl (eq 12), the resulting silylated polymers, **16**, have Me_3SiCH_2 groups on up to 60% of the backbone phosphorus atoms. Unlike the derivatives of $[Me(Ph)PN]_n$, the silylated copolymers are soluble in hexane and acetone as well as in THF and chlorinated hydrocarbons, and they are partially soluble in water. The glass transition temperatures of the silylated derivatives are also higher than the parent copolymer (e.g., 37 °C for the 40% substitution product versus 0 °C for **15**). (*28*)

The carboxylic acid and salt derivatives, **17**, of the copolymers also exhibit altered solubility (eq 13). Both the 50% substituted acid and the lithium salt of the 25% substituted acid are water soluble whereas the only the 50 % salt derivative (i.e., **7**) of $[Me(Ph)PN]_n$ was soluble in water (Scheme 1). (*28*)

A series of ferrocenyl derivatives, **18**, of the copolymers were also prepared and fully characterized (eq 14). Glass transition temperatures of these polymers range from 62 to 73 °C for polymers with 25 to 40 % substitution, which reflects the high degree of hydrogen bonding between hydroxy groups. Because these derivatives have significantly lower T_g's than observed for the homopolymer derivatives discussed above, their electrochemistry is of interest.

In most of the deprotonation-substitution reactions discussed thus far, the degree of substitution has usually not exceeded 50 %. Although this could be due to either electronic factors associated with the formation of charged sites along the chain or to simple steric effects, a recent ^{31}P NMR spectroscopic study of the anion indicates that steric size of the electrophile is the limiting factor in substitution. When one-half equivalent of *n*-BuLi was added to $[Me(Ph)PN]_n$ and the mixture was stirred at room temperature, two signals at ca. δ 4 and δ 33 were observed in the ^{31}P NMR spectrum. These correspond to simple P-Me and to PCH_2^- groups, respectively. However, when one equivalent of *n*-BuLi was added, only the signal at δ 33 was observed. When this solution was treated with the relatively small electrophile MeI, essentially all of the methyl groups were converted to ethyl groups giving $[Et(Ph)PN]_n$ (eq 15). (*27*)

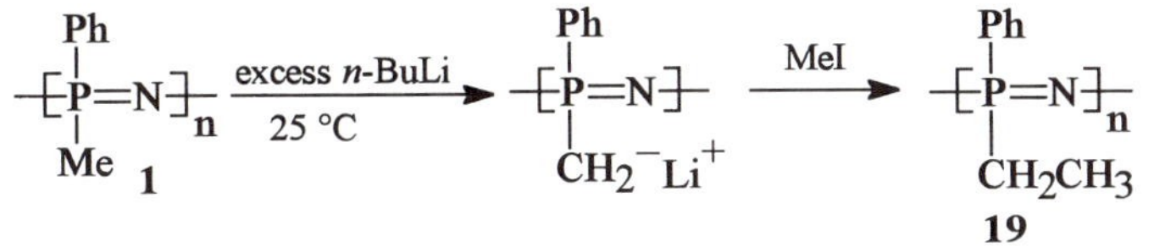

$$\tag{15}$$

Electrophilic Aromatic Substitution Reactions

The phenyl ring in $[Me(Ph)PN]_n$ appears to be well-suited to electrophilic aromatic substitution reactions. However, numerous attempts to acylate the ring were unsuccessful, presumably due to the coordination of catalysts such as $AlCl_3$ with the backbone nitrogen atom. Nonetheless, the phenyl ring can be nitrated at the meta position using a mixture of nitric and sulfuric acids (Scheme 2). (*29*) Between 15-55 % of the phenyl rings were derivatized by varying the reaction times from 30-60 min. Characterization of the nitrated polymers **20** was complicated by coordination of the backbone nitrogen atoms to protons. Even upon workup with Et_3N, inconsistent elemental analyses were common. This problem, which can be attributed to the presence of counter anions such as SO_4^{2-}, was later circumvented by workups involving NaOH. The nitrated polymers **20** were subsequently reduced to the amino substituted polymers **21** with Lalancette's reagent, $NaBH_2S_3$. Conversion to the amides **22** was accomplished with acid chlorides, which allowed for better characterization by ^1H NMR spectroscopy. The T_g's of the new polymers were as high as 80 °C (parent polymer **1**, 37 °C) as expected for the incorporation of more polar substituents.

Scheme 2

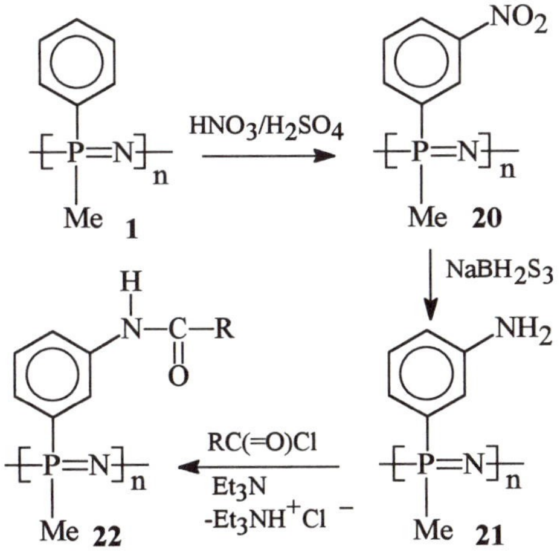

Coordination of Backbone Nitrogen Atoms

Throughout the studies of deprotonation-substitution and electrophilic aromatic substitution reactions, a recurring problem with coordination of the backbone nitrogen atoms to various reagents has been encountered. We have also made several attempts to attach transition metal substituents to the poly(alkyl/arylphosphazenes), but these were also complicated by coodination. Thus we recently conducted a more thorough investigation of this phenomenon. (30)   A portion of this study involved the preparation of lithium and silver complexes of [Me(Ph)PN]$_n$, 1, and [Me$_2$PN]$_n$, 23 (eq 16). Each of these complexes (24 - 28) was soluble in CH$_2$Cl$_2$ and had a single resonance in the room temperature ^{31}P NMR spectra. Since this was true even when small amounts of metal were incorporated, it appeared that the single resonance involved a fluxional process. Indeed, this signal was resolved into two resonances in the - 90 °C spectrum of 28. Hence, it appears that the metal ions are mobile at room temperature. When 1 was treated with PtCl$_2$, an insoluble material and trace amounts of soluble polymer were obtained. Presumably the multiple coordination sites in platinum allow for crosslinking of the polymer. Polymers 1 and 23 were also protonated with anhydrous HCl and were recovered unchanged by washing with K$_2$CO$_3$. As expected for the incorporation of the ionic groups, the T_g's of these complexes were higher than that of 1 (37 °C) and 23 (-42 °C) (4) and ranged from -20 °C for 28 to 121 °C for 25. (30)

$$ \underset{\underset{Me}{|}}{\overset{\overset{R}{|}}{\{P=\ddot{N}\}_n}} \xrightarrow[CH_2Cl_2]{M^+BF_4^-} \left\{ \underset{\underset{Me}{|}}{\overset{\overset{R}{|}}{\{P=\ddot{N}\}_x}} \underset{\underset{Me}{|}}{\overset{\overset{R}{|}}{\{P=\ddot{N}\}_y}} \right\} y\, BF_4 \qquad (16) $$

1, R = Ph
23, R = Me

24, R = Ph, x = 0.83, y = 0.17, M = Ag
25, R = Ph, x = 0.70, y = 0.30, M = Ag
26, R = Me, x = 0.85, y = 0.15, M = Ag
27, R = Ph, x = 0.84, y = 0.16, M = Li
28, R = Me, x = 0.80, y = 0.20, M = Li

Conclusion

The reactivity of poly(alkyl/arylphosphazenes) is proving to be surprisingly diverse and provides access to a large number of new poly(phosphazenes). Substitution at the simple methyl and aryl groups allows for the attachment of many functional groups most of which can be used for further derivatization chemistry. Though less well developed than the substitution chemistry of poly(phosphazenes) prepared by ring opening polymerization, the chemistry of these condensation polymers is significantly broad and yields polymers with a vast range of properties. Moreover, the coordinating ability of the backbone nitrogen is enhanced by the electron releasing effects of the directly bonded alkyl and aryl groups and thus offers a relatively new dimension to poly(phosphazene) chemistry.

Acknowledgments. This work has been supported primarily by the the the U. S. Army Research Office and the Robert A. Welch Foundation. Acknowledgement is also made to the donors of the Petroleum Research Fund administered by the American Chemical Society, The Texas Advanced Technology Program administered by the Texas Higher Education Coordinating Board, and Southern Methodist University (e.g., the SMU Research Council) for additional support of this research.

Literature Cited

1. Mark, J. E.; Allcock, H. R.; West, R. *Inorganic Polymers*; Prentice Hall: Englewood Cliffs, NJ, 1992, Chapter 3.
2. Allcock, H. R. *ACS Symposium Series* **1988**, *360*, 250.
3. Allcock, H. R. *Angew. Chem., Int. Ed. Engl.* **1977**, *16*, 147.
4. Neilson, R. H.; Hani, R.; Wisian-Neilson, P.; Meister, J. J.; Roy, A. K. *Macromolecules* **1987**, *20*, 910.
5. Wisian-Neilson, P.; Neilson, R. H. *Inorg. Synth.* **1989**, *25*, 69.
6. Neilson, R. H.; Wisian-Neilson, P. *Chem. Rev.* **1988**, *88*, 541.
7. Montague; R. A.; Matyjaszewski, K. *J. Am. Chem. Soc.* **1990**, *112*, 6721.
8. D'Hallum, G.; De Jaeger, R.; Chambrette, J. P.; Potin, P. *Macromolecules* **1992**, *25*,1254.
9. Franz, U.; Nuyken, O.; Matyjaszewski, K. *Macromolecules* **1993**, *26*, 3723.
10. Neilson, R. H.; Jinkerson, D. L.; Karthikeyan, S.; Samuel, R.; Wood, C. E. *Polym. Prepr., (Am. Chem. Soc. Div. Polym. Chem.)* **1991**, *32(3)*, 483.
11. Wisian-Neilson, P.; Ford, R. R.; Roy, A. K.; Neilson, R. H. *Macromolecules* **1986**, *19*, 2089.
12. Wisian-Neilson, P.; Ford, R. R. *Organometallics* **1987**, *6*, 2258.
13. Wisian-Neilson, P.; Ford, R. R. *Macromolecules* **1989**, *22*, 72.
14. Wisian-Neilson, P.; Chrusciel, J.; Bahadur, M., unpublished results.
15. Bailey, L. M. S. Thesis, Southern Methodist University, December, 1993.
16. Wisian-Neilson, P.; Huang, L.; Quamrul Islam, M.; Crane, R. A. *Polymer*, in press.
17. Allcock, H. R.; Connolly, M. S.; Sisko, J. T.; Al-Shali, S. *Macromolecules* **1988**, *21*, 323.
18. Wang, T. M. S. Thesis, Southern Methodist University, August, 1990.
19. Crumbliss, A. L.; Cooke, D.; Castillo, J.; Wisian-Neilson, P. *Inorg. Chem.* **1993**, *32*, 6088.
20. Wisian-Neilson, P.; Islam, M. S.; Ganapathiappan, S.; Scott, D. L.; Raghuveer, K. S.; Ford, R. R. *Macromolecules* **1989**, *22*, 4382.
21. Hoyle, C. E.; Wisian-Neilson, P.; Chatterton, P. M.; Trapp, M. A. *Macromolecules* **1991**, *24*, 2194.
22. Wisian-Neilson, P.; Schaefer, M. A. *Macromolecules* **1989**, *22*, 2003.
23. Wisian-Neilson, P. Dubois, D.; York, G., unpublished results.
24. Wisian-Neilson, P.; Garcia-Alonso, F. J.; Crane, R. A., unpublished results.
25. Wisian-Neilson, P.; Islam, M. S.; Schaefer, M. A. *Phosphorus, Sulfur, and Silicon* **1989**, *41*, 135.
26. Wisian-Neilson, P.; Islam, M. S. *Macromolecules* **1989**, *22*, 2026.
27. Claypool, C.; Wisian-Neilson, P., unpublished results.
28. Crane, R. A., Castillo, J.; Huang, L.; Wisian-Neilson, P., unpublished results.
29. Wisian-Neilson, P.; Bahadur, M.; Iriarte, J. M.; Wood, C. E. *Macromolecules*, in press.
30. Wisian-Neilson, P.; Garcia-Alonso, F. J. *Macromolecules* **1993**, 26, 7156.

RECEIVED June 24, 1994

Chapter 20

Polyphosphazene Molecular Composites

In Situ Polymerizations of Silicon, Titanium, Zirconium, and Aluminum Alkoxides

W. T. Ferrar, B. K. Coltrain, C. J. T. Landry, V. K. Long, T. R. Molaire, and D. E. Schildkraut

Research Laboratories, Eastman Kodak Company, Rochester, NY 14650–2129

Metal alkoxides of silicon, titanium, zirconium, and aluminum have been used to prepare optically transparent composite films with the etheric phosphazene polymer poly[bis(2-(2-methoxyethoxy)ethoxy)-phosphazene] (MEEP). Dynamic mechanical and stress-strain techniques are used to evaluate and compare the mechanical properties of the different composites. Despite strong interactions with the inorganic oxide network, the glass transition temperature (T_g) of MEEP does not shift to higher temperature upon addition of the inorganics, but the polymer is reinforced by the oxides. The addition of potassium triflate to the composites resulted in materials with both excellent mechanical properties and increased conductivity.

Polymer electrolytes based on polyethylene oxide (PEO)-salt complexation are capable of transporting cations by pseudo-crown ether coordination of low lattice energy salts. In particular, complexes of the etheric phosphazene polymers with salts such as lithium triflate are known to display high electrical conductivity when compared with other polymeric electrolytes (1). Although attachment of the PEO side group moieties to high molecular weight phosphazene backbones improves the physical properties relative to low molecular weight PEO, these properties remain unacceptable for many applications. A persistent goal of several research groups has been to increase the mechanical properties of this amorphous material (2,3).

We reported previously that the mechanical properties of the amorphous etheric poly[bis(methoxyethoxyethoxy)-phosphazene] (MEEP) were significantly improved by in situ polymerization of tetraethoxysilane (TEOS) (4). This work on etheric phosphazene composites was consistent with reports from the research groups of Mark (5) and Wilkes (6), where metal alkoxides were added to poly(dimethylsiloxane) and poly(tetramethylene oxide). We now extend the phosphazene sol-gel composites to include the more reactive metal alkoxides of titanium, zirconium, and aluminum, and we report that when polymerized in situ, each is compatible with the etheric phosphazene polymer. Dynamic mechanical and stress-strain techniques are utilized to evaluate and compare the mechanical properties of the different composites. Thermal and

0097–6156/94/0572–0258$08.00/0

mechanical data of the composites, with and without salt, are presented. Comparisons are also made between these composites and MEEP blends with organic polymers.

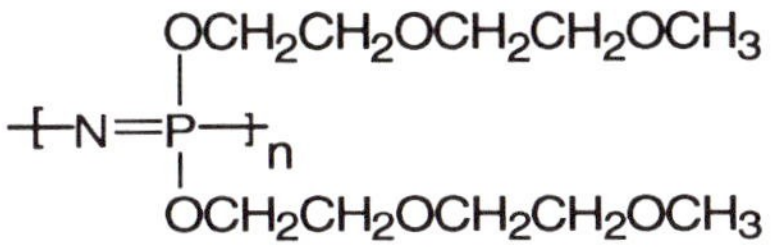

$$OCH_2CH_2OCH_2CH_2OCH_3$$

$$\left[N{=}P \right]_n$$

$$OCH_2CH_2OCH_2CH_2OCH_3$$

MEEP

Sample Preparation

Composites were prepared from four metal alkoxides (Scheme 1) and the etheric phosphazene polymer MEEP. The four types of composites are referred to throughout the report as: MEEP/Si, MEEP/tetraethoxysilane; MEEP/Ti, MEEP/tetraisopropoxyorthotitanate; MEEP/Zr, MEEP/zirconium n-butoxide.butanol complex; MEEP/Al, MEEP/aluminum tri-sec-butoxide.

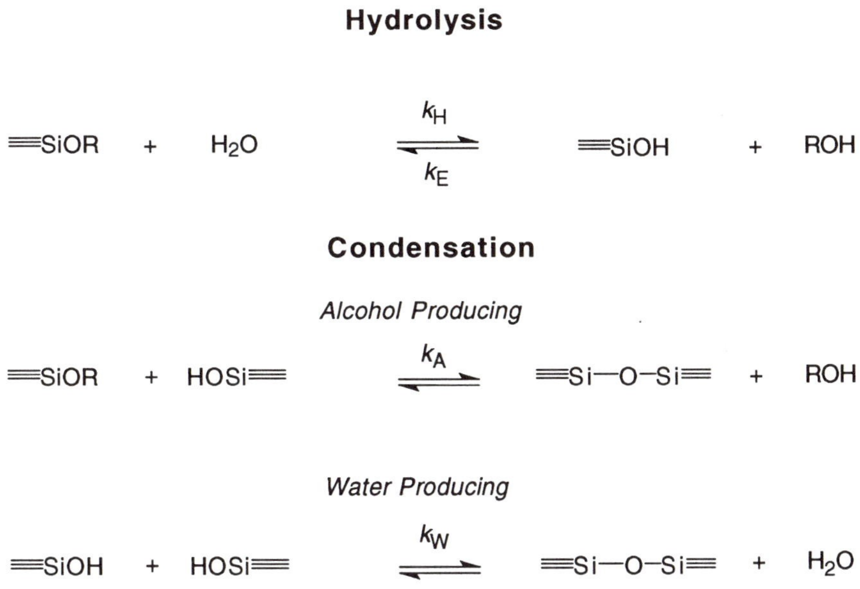

Hydrolysis

$$\equiv\!\text{SiOR} \quad + \quad H_2O \quad \underset{k_E}{\overset{k_H}{\rightleftharpoons}} \quad \equiv\!\text{SiOH} \quad + \quad \text{ROH}$$

Condensation

Alcohol Producing

$$\equiv\!\text{SiOR} \quad + \quad \text{HOSi}\!\equiv \quad \overset{k_A}{\rightleftharpoons} \quad \equiv\!\text{Si}{-}\text{O}{-}\text{Si}\!\equiv \quad + \quad \text{ROH}$$

Water Producing

$$\equiv\!\text{SiOH} \quad + \quad \text{HOSi}\!\equiv \quad \overset{k_W}{\rightleftharpoons} \quad \equiv\!\text{Si}{-}\text{O}{-}\text{Si}\!\equiv \quad + \quad H_2O$$

Scheme 1

The preparation of MEEP has been described previously (*4*). Inorganic alkoxides were obtained from Kodak, Aldrich, and Fluka. A general procedure for the films and castings was as follows. A stock solution of MEEP was prepared by dissolving 2.0 g of MEEP in 80 g of a 1/1 mixture of THF/ethanol. Twelve grams of MEEP stock solution (0.3 g, 1 mmol) was added to a 50 mL Erlenmeyer flask equipped with a magnetic stirrer. The desired alkoxide was then added to the stirred solution (0.3 g if a 1/1 wt ratio is desired), followed by the addition of the triflate salt if so desired. The solution was then cast in a Teflon mold, which was covered to reduce the evaporation rate of the solvent. The thickness of the films used for dynamic mechanical analysis was usually ca. 75 μm. These were dried under vacuum at ambient temperature for several days, then at 60°C for at least 16 h.

Dynamic Mechanical Measurements

Dynamic mechanical measurements [obtained using a Rheovibron DDV-II dynamic tensile tester (Tokyo Measuring Instruments, Ltd., Japan) automated by IMASS, Inc. at 110 Hz] of MEEP/Si, MEEP/Ti, MEEP/Zr, and MEEP/Al (1/1 by weight) are shown in Figure 1. The storage modulus (E′) decreased at ca. -75°C because of the glass transition of MEEP. The storage modulus reached a minimum at -20°C and then increased gradually as the samples were heated. The increase in E′ with temperature has been attributed to the continuation of network formation of the silicate during the experiment (7). This increase in modulus with temperature was not as evident in this study (with Ti, Zr, and Al) as in the earlier work with MEEP/Si (4). This suggests that in situ polymerization of the inorganics in these composites was nearly complete during the processing, consistent with the known higher reactivity of these inorganics. The presence of a plateau in E′ above the T_g is thought to reflect a continuous, load bearing, inorganic oxide network in these composites.

The storage modulus in the rubber plateau region (above T_g) for MEEP/Ti (1/1) was the highest of all the composites. A factor that would influence the modulus is the higher reactivity of the titanium alkoxide. Thus the titanium sol-gel would be more fully densified under the conditions employed here. At 240°C, we have found the moduli of the silicon (not shown) and titanium polymer composites to be similar. Curing rate differences are not as large when comparing titanium with zirconium or aluminum alkoxides, as reflected in the smaller differences in the plateau moduli of the corresponding composites. Another factor, which is important to consider when comparing the observed physical properties of the composites, is that the mole fraction and volume fraction of oxide in these composites is not equal. For MEEP/Ti, the ratio of the two represents both a weight and a mole ratio because the molecular weight of a MEEP repeat unit and titanium isopropoxide are fortuitously close. The zirconium alkoxide has a much higher equivalent weight and thus fewer moles of oxide are produced per unit weight of the alkoxide.

Composites of MEEP/Al were more difficult to prepare than other composites because of the extreme moisture sensitivity of the aluminum alkoxides. Nonetheless, translucent, tough films were prepared from equal weights of aluminum tri-sec-butoxide and polymer. As with the composites discussed above, differential scanning calorimetry (DSC) revealed the glass transition at -79°C, indicating the backbone motion of a large portion of polymer segments was not perturbed by the metal oxide.

Stress-Strain

The ambient temperature stress-strain results for the composites with a 1/1 ratio by weight of MEEP to metal alkoxide show that the nature of the metal alkoxide had a significant effect on the physical properties. The stress-strain curves for these samples are shown in Figure 2. None of the curves is characteristic of a rubber material, where the stress increases gradually until the sample breaks. All of the curves are more characteristic of a ductile, glassy polymer showing various degrees of brittleness. In the Al and Si sample curves the stress increases to a yield point, however, neither the Zr nor Ti samples have a distinct yield point. Whereas the Ti composite was brittle, the MEEP/Zr was extremely ductile.

The magnitude of the tensile modulus for each of the composites was on the order of 30-220 MPa. This was much larger than expected for a low T_g rubber (*8*). As an example the modulus of a filled, crosslinked polydimethylsiloxane (PDMS), T_g = -120°C, is approximately 3-4 MPa (*9*). Therefore the metal alkoxides have a significant effect in reinforcing the MEEP. The Ti and Al samples had much greater tensile moduli than the Zr and Si samples. This is consistent with the dynamic mechanical results, where the Ti and Al samples have greater rubbery plateau moduli. Once again, the relative differences in the molar amounts likely affect the resulting morphology, and this, along with the structure of the oxide, may be significant in determining the ultimate chemical response. The maximum stress was similar for all the samples and is similar to that of a filled, crosslinked low T_g rubber (*9*).

Addition of Low Lattice Energy Salts

The preceding analysis of the mechanical properties of the MEEP/inorganic oxide composites shows that regions of the composites retain the low glass transition characteristics of MEEP polymer. Thus it is not unexpected that should the composites be formed in the presence of low lattice energy salts, they should display similar charge carrying properties to MEEP solid electrolytes. The addition of potassium triflate to the polymer/alkoxide solutions in THF/ethanol had no noticeable effect on the composite formation. The solutions remained clear with no evidence of precipitation and the composites were again optically clear. The samples containing the salt did show a slightly extended glassy region in the dynamic mechanical spectra (an increase in T_g), but this appeared to be a minor perturbation of the composite. Overall the spectra showed the same general features as the composites in Figure 1, low glass transition temperatures and extended plateau moduli to above 200°C.

The amount of salt added was 17 wt % of potassium triflate for resistivity measurements. This value was chosen to be slightly higher than the level determined by Shriver to be the optimum for conductivity. Shriver et al. (*1*) have reported that the maximum conductivity in a MEEP-salt electrolyte is achieved at a molar ratio of 0.25 salt to 1.0 repeat unit of polymer. This corresponds to 16 oxygen atoms from the PEO side groups for each potassium atom from the salt. A slight excess of salt was used in the composite to compensate for consumption (e. g. adsorption) by the oxide. Thus the molar ratio of salt to polymer repeat unit in the composites was 0.29, which corresponds to 17 wt % of potassium triflate to MEEP.

Impedance Measurements. The resistivities of the films were examined using complex impedance measurements. The composite films were sandwiched between ion blocking electrodes of glassy carbon. The measurements were performed under

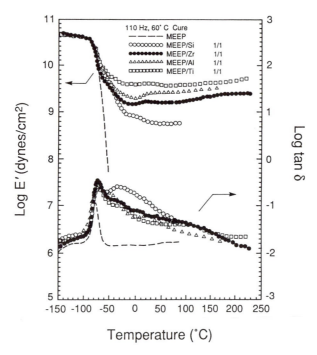

Figure 1. Dynamic mechanical spectra of MEEP and MEEP composites.

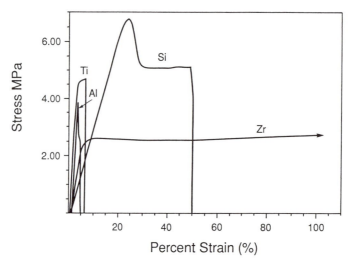

Figure 2. Stress-strain curves of MEEP composites (1/1).

ambient conditions over a frequency range of 10 to 30,000 Hz. Impedance measurements under ambient conditions were made on free standing films to give bulk conductivity values (*10*). It is important to note that the conductivity values are probably biased by the lack of environmental control of the samples, resulting primarily from water absorption. Thus the results reported here are useful only for comparing samples with and without salt and are not meant to be absolute measurements.

Complex impedance plots typical of ionic conduction were obtained for all of the samples, including the composites that lacked potassium triflate. The plots were characterized by an arc at high frequencies and a spur as the frequency was decreased(*11*). Conductivity was calculated using the x-axis intercept of the lower frequency side of the arc. The composites with salts displayed similar conductivity behavior to polymeric electrolytes reported for MEEP and lithium triflate (*1*). Comparison of films made with and without salt showed that although the general shape of the curves did not change, the conductivity of the composites containing salts were several orders of magnitude higher than those without salts deliberately added (*12,13*). For example, the MEEP/Ti composite to which no salt was added displayed a well-defined arc when plotted using megohms for the axes (Figure 3a) and had a conductivity of log -8.7 (ohm cm)$^{-1}$. The plot for the more conductive MEEP/Ti/K$^+$CF$_3$SO$_3^-$ composite is shown in Figure 3b using kiloohms on the axes. This sample had a more well-defined spur and gave a conductivity of log -5.0 (ohm cm)$^{-1}$. Although not absolute in value, these numbers are consistent with measurements made by others on MEEP itself (*1*).

Polymer Blends Compared with MEEP/Si Composites

The formation of homogeneous MEEP/Si composites described above is dependent upon thorough mixing and compatibility of the polymer with the forming inorganic oxide. Because the sol-gel process is a combination of hydrolysis and condensation with complex equilibria between many different products (Scheme 1), it is difficult to know which species are compatible with the polymer. Hydrolysis of the inorganic alkoxides to the sols under acidic conditions leads to various forms of silicic acid, Si(OR)$_x$(OH)$_{4-x}$ (*14*), followed rapidly by condensation to various suboxides. The complex equilibria that accompany the sol-gel process prevent direct examination of the proposed acid-base pairing between the silicic acid moieties and the MEEP. The tendency of the etheric side groups to act as a pseudo-crown ether could stabilize the inorganic alkoxides (*1*). Etheric alcohols such as methoxyethanol or methoxyethoxyethanol are known to modify the reactivity of the inorganic alkoxide precursors (*15,16*). Alternatively, the polymer may strongly interact with the condensed oxide, as MEEP has been shown to be an excellent dispersant of 0.4 micron alumina, probably caused by adsorption of the side groups onto the particle surface (*17*), or interact with the growing silicate network via hydrogen bonding.

In an attempt to picture the interaction of MEEP with the hydrolyzed growing silicate, a cartoon (Figure 4) shows how the inorganic network of the sol-gel might appear and compares it with the carbon backbone polymer of poly(vinylphenol) (PVPh). Both the growing silicate and the PVPh have hydroxy moieties capable of hydrogen bonding to a suitable base. We have recently reported the preparation of the fully

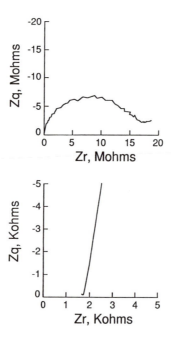

Figure 3. Impedance plot using ion blocking electrodes (10 Hz- 32 KHz) for (a) MEEP/Ti and (b) MEEP/Ti/$K^+CF_3SO_3^-$.

miscible polymer blends of PVPh with MEEP (*18*). Figure 5 shows a plot of the glass transition temperature of the polymer blend as a function of composition. Only one T_g is observed for each composite of this blend, indicating that no regions of phase separation are present. Infrared studies suggest that mixing in these materials is a result of hydrogen bonding of the hydroxy functionality of the PVPh to the ethylene oxy side groups of the MEEP. This dispersing of the phenolic materials throughout the MEEP network may thus be viewed as similar to the dispersal of the partially condensed inorganic sol throughout the polymer matrix in the early stages of composite formation.

PVPh provides an interesting comparison with the partially condensed silicate network. Strong interactions between the growing silicate and MEEP are important in retarding phase separation (via hetero-interactions), resulting in a homogeneous MEEP/Si composite, similar to the hydrogen bonding interactions leading to miscible MEEP/PVPh blends. Although similarities exist between MEEP/PVPh and MEEP/Si "blends", an important distinction should be made. The T_g of the polymer blend increases as the amount of PVPh increases (Figure 5). PVPh has a T_g of 183° C (DSC on set). This behavior is generally observed for miscible polymer blends, and is also observed when low lattice energy salts are added to MEEP to form solid electrolytes (*1*). However, ultimately MEEP is not thermodynamically miscible with the condensed silicate. With further condensation the oxide forms its own phase (perhaps entrapping some MEEP chains within it and some at the surface). Macroscopic phase separation is prevented since mobility of the components is restricted due to vitrification and the composite remains optically transparent. Thus, a two-phase system results in which the dimensions of the phases are extremely small (< 100 Å) and two T_g values were observed, one corresponding to pure MEEP , and the other to the other component or mixed phase. The latter would have a T_g that is too high to be observable thermally, without degradation of the polymer. This is supported by the dynamic mechanical spectra of the MEEP/sol-gel composites that show the T_g of the polymer is not affected by the cured sol-gel (see above). In fact, two separate relaxations have been observed in the solid state phosphorus NMR spectra of the MEEP/Si composites, consistent with microphase separation in these composites (*4*). One relaxation corresponds to pure MEEP, and the other slower relaxation is attributed to an interfacial region where MEEP is in intimate contact with the partially condensed oxide. It would be these "pure MEEP" regions of the composites that carry the charge upon doping with potassium triflate. This discussion was primarily focused on MEEP/Si composites because the silicate network is known to contain many silanols in acidic conditions, however, similar mechanisms could be occurring with the MEEP/Ti, MEEP/Al, and to a lesser extent, MEEP/Zr composites.

CONCLUSIONS

The in situ polymerization of metal alkoxides in the presence of the etheric phosphazene polymer MEEP results in clear composite films with improved physical properties. Four metal alkoxides were examined in terms of mechanical properties and ionic conductivities. Many factors such as the amount of hydrolysis and condensation of the metal alkoxides are important in determining the composite properties. It is important to note that in none of the composites did the T_g increase above the -75°C detected for

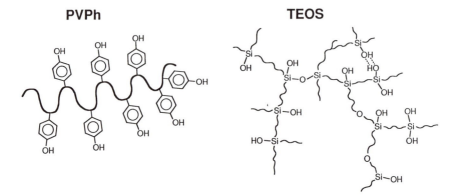

Figure 4. PVPh and partially condensed TEOS both have hydroxy functionality that promote miscibility in polymer blends and control phase separation in sol-gel composites.

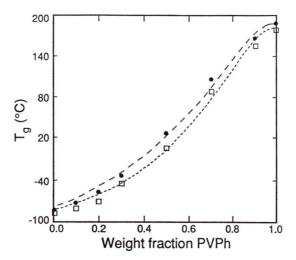

Figure 5. Composition dependence of the glass transition temperature [DSC, 20°C/min] for blends of PVPh with MEEP [(□) onset and (●) midpoint]. (Adapted from ref. 18.)

the pure polymer. The addition of low lattice energy salts to the solution of the polymer with the metal alkoxide resulted in polymeric electrolytes with good mechanical integrity. Complex impedance plots typical for ionic conductivity were obtained. Comparisons were made between the composites and MEEP blends with the organic polymer PVPh.

Literature Cited

1. Blonsky, P. M.; Shriver, D. F.; Austin, P.; Allcock, H. R. *J. Am. Chem. Soc.* **1984**, *106*, 6854; *Solid State Ionics* **1989**, *18&19*, 258.
2. Tonge J. S; Shriver, D. F. *J. Electrochem. Soc.: Accel. Brief Commun.* January 1987, 270.
3. Semkow, K. W.; Sammells, A. F. *J. Electrochem. Soc.* **1987**, *134*, 767.
4. Coltrain, B. K.; Ferrar, W. T.; Landry, C. J. T.; Molaire, T. R.; Zumbulyadis, N. *Chem. Mater.* **1992**, *4*, 358.
5. Mark, J. E. *J. Inorg. Org. Polym.* **1991**, *1*, 431.
6. Brennen, A. B.; Wang, B.; Rodrigues, D. E.; Wilkes, G. L. *J. Inorg. Org. Polym.* **1991**, *1*, 167.
7. Fitzgerald, J. J.; Landry, C. J. T.; Pochan, J. M. *Macromolecules* **1992**, *25*, 3715.
8. Ward, I. M. *Mechanical Properties of Solid Polymers*, Wiley-Interscience New York, 1971.
9. Long, V. K. Unpublished Results.
10. Jacobs, P. W. M.; Lorimer, J. W.; Russer, A.; Wasiucionek, M. *J. Power Sources* **1989**, *26*, 483.
11. Hardy, L. C.; Shriver, D. F. *J. Am. Chem. Soc.* **1985**, *107*, 3823.
12. Tonge, J. S.; Shriver, D. F. in *Polymers for Electronic Applications*, Lai, J. H., Ed.; CRC Press: Cleveland, Ohio 1989, p 157.
13. Cowie, J. M. G.; Cree, S. H. *Annu. Rev. Phys. Chem.* **1989**, *40*, 85.
14. Iler, R. K. *The Chemistry of Silica;* John Wiley and Sons: New York, 1979.
15. Hayashi, Y.; Blum, J. B. *J. Mater. Sci.* **1987**, *22*, 2655.
16. Schwartz, R. W.; Payne, D. A. *Mater. Res. Soc. Symp.* **1988**, *121*, 199.
17. Exashos, G. J.; Ferris, K. F.; Friedrich, D. M.; Samuels, W. D. *J. Am. Ceram. Soc.* **1988**, *71*, C-406.
18. Landry, C. J. T.; Ferrar, W. T.; Teegarden, D. M.; Coltrain, B. K. *Macromolecules* **1993**, *26*, 35.

RECEIVED July 1, 1994

Chapter 21

Oxygen Gas Permeability and the Mechanical Properties of Poly(*n*-butylamino)(di-*n*-hexylamino)phosphazene Membranes

M. Kajiwara and T. Kimura

Department of Applied Chemistry, School of Engineering, Nagoya University, Furo-cho, Chikusa-ku, Nagoya, 464–01 Japan

Poly(n-butylamino)(di-n-hexylamino)phosphazene was prepared by the reaction of poly(dichlorophosphazene), di-n-hexylamine and n-butylamine using THF solvent and triethylamine at room temperature. To increase the mechanical properties of the polymer films, cross-linking agents such as neopentyl glycol diglycidyl ether and trimethylol propane triglycidyl ether were added to the original polymer. The original and cross-linked polymer films were prepared by solution casting. The films with cross linking agents were heated at 60 °C and 120 °C, giving transparent films. The oxygen gas permeability and the mechanical properties such as Young's modulus of the cross-linked film were higher than those of the original film.

Linear poly(organophosphazene) are typical inorganic-organic polymers. A variety of organic side-groups bond to each phosphorous can be easily introduced and modified. For example, poly(organophosphazenes) bearing suitable side-groups can be used as membrane materials in the field of gas permeation (1), organic liquid separation pervaporation (2), ion separation (3), ultrafiltration (4), and microfiltration (5). After Krieev (6) described gas permeation and separation of oxygen, nitrogen and argon using poly(diamyloxyphosphazene) membrane, patents and reports were described by many companies (7) and investigators (8) . However, only one report (9) described the oxygen gas permeability of poly(organophosphazene) membranes under wet condition. This paper describes oxygen gas permeability in water and the mechanical properties of poly(n-butylamino)(di-n-hexylamino)phosphazene membranes and the membranes cured with cross-linking agents.

0097–6156/94/0572–0268$08.00/0

Experimental

Preparation of Hexachlorocyclotriphosphazene and Polydichlorophosphazene.

Hexachlorocyclotriphosphazene $(NPCl_2)_3$ was synthesized by the method of Saito (10). That is, $(NPCl_2)_3$ was prepared by the reaction of 500 g of phosphorus pentachloride PCl_5 and 200 g of ammonium chloride NH_4Cl using 1500 mL of tetrachloroethane or chlorobenzene. Pure trimer was obtained by repeated fractional crystallization from light petroleum ether.

Linear polydichlorophosphazene $(NPCl_2)_n$ was prepared by the method of Kajiwara (11). That is, 5 g of trimer and 0.1 g of sodium dibutyldithiocarbamate were placed in a three-necked flask equipped with a stirrer and condenser, which was then evacuated to 10 Torr for 1 h. Then 5 g of dichlorobenzene was added to the flask, and the mixture was heated to 190 °C for 24 h in a flow of dry nitrogen gas. After the reaction was over, the polymer-oligomer mixture was dissolved in 200 mL of dry tetrahydrofuran (THF). To remove the oligomer, the THF solution was added to 500 mL of n-heptane. After the polymer was precipitated, the precipitate was separated by decantation, followed by dissolution in THF and reprecipitated. This purification procedure was repeated several times and the purified polymer was then dissolved in 100 mL of THF.

Preparation of Poly(n-butylamino)(di-n-hexylamino)phosphazene.

Linear $(NPCl_2)n$ that is soluble in THF or benzene was prepared with solution polymerization of $(NPCl_2)_3$. Also, poly(n-butylamino)(di-n-hexylamino)phosphazene was prepared by the reaction of $(NPCl_2)_n$, di-n-hexylamine, and n-butylamine using triethylamine in THF at room temperature and bubbling dry nitrogen gas. After the reaction was over, triethylamine hydrochloride was removed by filtration. When the filtrate was added to ethanol, white colored polymer was precipitated. The polymer was dissolved in THF, and then the solution was added to ethanol. This purification procedure of the polymer was repeated several times.

Preparation of the Membranes.

The purified polymer was dissolved in THF and the solution was filtered using a 0.5 mm filter to remove the cross-linked polymers. The filtrate was added to a Petri dish and kept at room temperature until the membrane was formed. Also, to increase the mechanical properties of the membrane, cross-linking agents such as neopentyl glycol diglycidyl ether (NPG) and trimethylol propane triglycidyl ether (TMP) were added to the polymer solution. The membranes prepared in this manner were heated in the oven at 60 °C or 120 °C. The cross-linking agents such as (NPG) and (TMP) were selected because they are transparent liquids and are easily handled.

Determination of Oxygen Gas Permeability of the Membranes.

Measurement of the oxygen gas permeability (D_k) value in water is determined with an oxygen permeater of Seikanken type at 25 - 35°C using the equation [1]. That is,

$$D_k = K \times i \times L \times 760/P' \qquad (1)$$

P' = oxygen pressure (mmHg), i = current of the sample(mA), L = thickness of the sample (mm). Also, K is determined with equation [2]. That is,

$$K = 14.2 \times p/i \times L \times 760 \qquad (2)$$

p = oxygen pressure of standard plate (Teflon^R), i = current of standard plate, L = thickness of standard plate.

Characterization of Poly(n-butylamino)(di-n-hexylamino)phosphazene and the Cross-Linked Polymers Prepared with NPG and TMP. Poly(n-butylamino)(di-n-hexylamino)phosphazene and the cross-linked polymers were characterized by ^{1}H-NMR, ^{31}P-NMR, FTIR and GPC.

Mechanical Properties of Poly(n-butylamino)(di-n-hexylamino)phosphazene and the Cross-Linked Polymer. Mechanical properties (Young's modulus, tensile strength or elongation) of the original and cured polymer films were determined with Toyo Baldmin Co Ltd UM II-20 (Tensilon). The test film size was 30 mm x (0.05 ~ 0.15) mm and the rate of tension was 10 mm/min using 20 mm as the distance between bench markers. After three test films were measured, the average value was estimated from the measurement. Also, the tests were done after keeping the films at 20 °C, 65%RH for 4h. Young's modulus(YM), tensile strength(TS), breaking strength(BS), elongation yield(YP) and elongation breaking(BP) were estimated with following the equations [3], [4], [5], [6] and [7], respectively. That is,

$$YM(kg/cm^2) = a/0.02 \times S \qquad (3)$$

$$TS(kg/cm^2) = b/S \qquad (4)$$

$$BS(kg/cm^2) = c/S \qquad (5)$$

$$YP(\%) = (e \times 1/10/2.0) \times 100 \qquad (6)$$

$$BP(\%) = (d \times 1/10/2.0) \times 100 \qquad (7)$$

where S is cross section of the polymer film, a is load produced 2 % of strain, b max load, c load breaking, d elongation at breaking point and e elongation owing to max load, respectively.

Results and Discussion

Characterization of Poly(n-butylamino)(di-n-hexylamino)phosphazene and the Cross-Linked Polymer Films. The molecular weight of poly(n-butylamino)(di-n-hexylamino)phosphazene was determined with GPC using THF solvent and calibration curves prepared with polyethylene. It is found that the polymer molecular weight was about 5 x 10^4. The ^{1}H-NMR of the polymer is shown in Figure 1. The peak near 2.5 ppm belongs to the primary and secondary amine protons. The peaks near 1.2 ppm and 0.8 ppm are assigned to -CH$_2$ and -CH$_3$. Furthermore, the proton ratio (NH + N-CH$_2$: CH$_2$: CH$_3$) estimated from Figure 1 is similar to 4 : 10 : 5. ^{31}P-NMR of the polymer is given in Figure 2. The broad peak appears near 5 ppm. However,

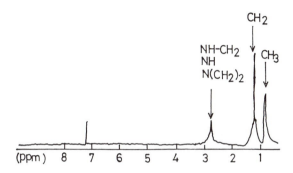

Figure 1. ^{1}H-NMR of poly(n-butylamino)(di-n-hexylamino)phosphazene.

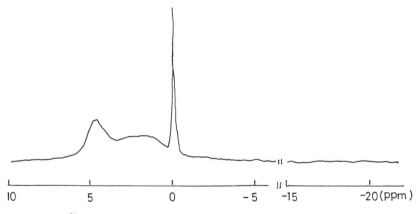

Figure 2. ^{31}P-NMR of poly(n-butylamino)(di-n-hexylamino)phosphazene.

unsubstituted P-Cl bonds occur at higher field (-19.0 ppm) are not observed in Figure 2. Consequently, it is assumed that almost all the chlorine atoms in $(NPCl_2)_n$ polymer are substituted with two kinds of amines and the polymer has the structure shown in Scheme I. That is, they are non-geminal type.

The polymers cured with NPG and TMP do not dissolve in THF or other organic solvents. FTIR of original and the cured polymer films are given in Figure 3. It is found from FTIR that -NH groups near 3,470 cm^{-1} and 3,370 cm^{-1} decrease with increasing -CH(OH) groups appearing near 1,350 cm^{-1} ~ 1,260 cm^{-1} and 1,100 cm^{-1}. It is assumed that the cross-linked reaction occurs as given in Scheme II. That is, the epoxy ring of NPG and TMP is cleaved with a proton in NHBu- on the polymer and the intermolecular cross-linked reaction is caused by heating. The polymer films cured with NPG and TMP are transparent.

Mechanical Properties of the Original and Cross-linked Polymer Films. Young's modulus and tensile strength of the original and the cross-linked polymer films with 0.05 wt.%, 0.1 wt.% of NPG and TMP are given in Figures 4, 5, 6 and 7, respectively. As for Figures 4 and 5, it is found that Young's modulus and tensile strength of the films cured with 0.05 wt.% or 0.1 wt.% of NPG are decreased with increasing curing time compared with that of the original film. Also, Young's modulus and tensile strength of the film cured with TMP as shown in Figures 6 and 7 are increased with increasing curing time. Furthermore, the results of elongation and breaking strength of the film are summarized in Tables 1 and 2. It is found that the value of elongation yield (YP) and breaking point (BP) of the cured films are lower than that of the original film. On the other hand, breaking strength of the film cured with 0.1 wt.% of NPG at 60 °C and 120 °C is higher than that of the original film. As for the film cured with TMP, elongation yield (YP) and breaking point (BP) are lower than that of the original film regardless of curing times and temperatures. Even if the film having the highest Young's modulus is prepared with curing agents such as NPG or TMP, the modulus value is lower than that of PVA or polyethylene films.

Oxygen Gas Permeability of the Films Cured with NPG and TMP. Oxygen gas permeability (D_k) of polymer films cured with NPG and TMP are determined and the results are summarized in Table 3. It is found from Table 3 that D_k values of the cured films are higher than that of the original film. The film cured with 0.05 wt.% of TMP has the highest D_k value. This value is about three times higher compared with that of the original film. Generally, it is said that the films having higher cross-linking densities give lower D_k values. However, the highest D_k value is higher than that of a silicone polymer film such as poly(dimethylsiloxane) ($D_k = 59.8 \times 10^{-11}$ ml cm/cm^2 s mmHg) (12).

Summary

NPG and TMP are a good cross-linking agents for poly(n-butyl-amino)(di-n-hexylamino)phosphazene. Oxygen gas permeability and Young's modulus of the film cured with NPG and TMP are higher than that of the original polymer film.

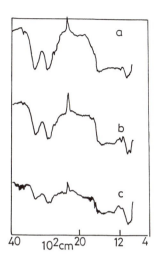

Figure 3. FTIR of the original (a) and cured films (b) or (c).
(b) : TMP 0.05 wt% 120 ℃/ 24 hr,   (c) : TMP 0.1 wt% 120 ℃/ 24 hr.

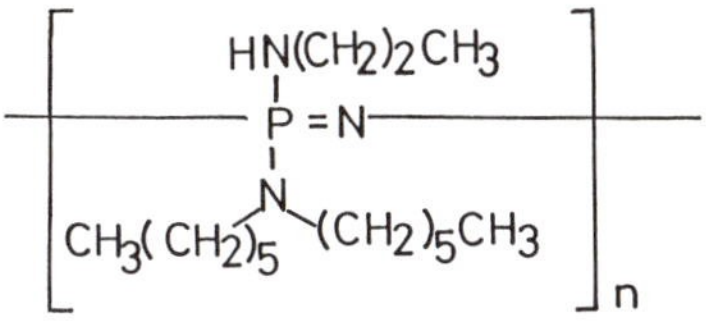

$$\left[\begin{array}{c} HN(CH_2)_2CH_3 \\ | \\ -P=N- \\ | \\ N \\ CH_3(CH_2)_5 \diagdown (CH_2)_5CH_3 \end{array} \right]_n$$

$$\overline{Mw} \approx 5 \times 10^4$$

Scheme I

$$\left[\begin{array}{c} NR_2 \\ | \\ -N=P- \\ | \\ HNBu\text{-}n \end{array} \right]_n \quad + \quad TMP \text{ or } NPG$$

$$\begin{array}{c} Bu\text{-}n \qquad OH \; n\text{-}Bu \\ | \qquad\quad | \\ P\text{-}NCH_2CH\text{-}R\text{-}CHCH_2N\text{-}P \\ \| \qquad\quad OH \qquad\quad \| \\ N \qquad\qquad\qquad\qquad N \\ | \\ CH_2 \\ | \end{array}$$

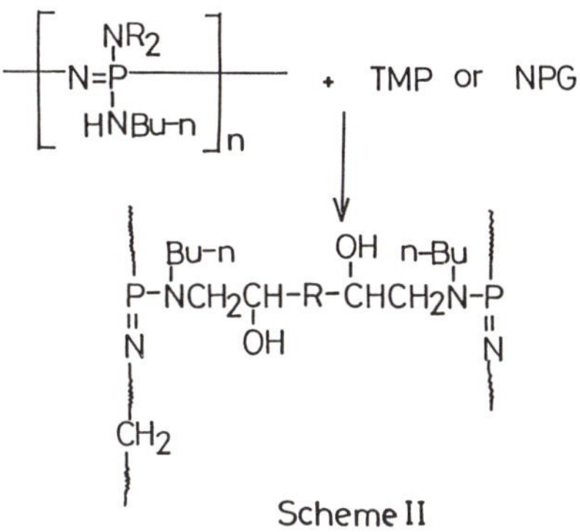

Scheme II

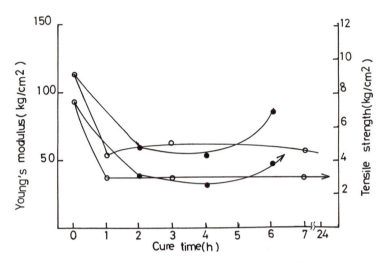

Figure 4. Mechanical properties of poly(n-butylamino)(di-n-hexylamino)-phosphazene films cured with 0.05 wt% of NPG. ◯: 50 ℃, ●: 100 ℃.

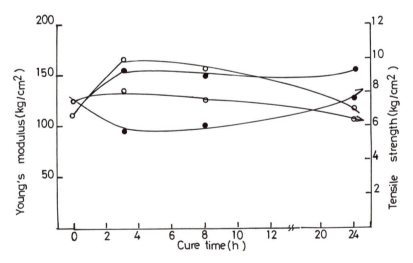

Figure 5. Mechanical properties of poly(n-butylamino)(di-n-hexylamino)-phosphazene films cured with 0.1 wt% of NPG. ◯: 60 ℃, ●: 120 ℃.

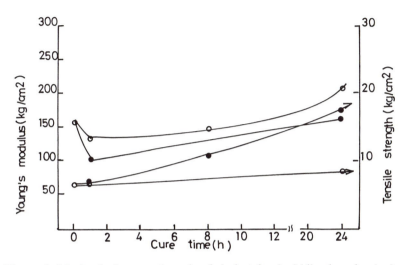

Figure 6. Mechanical properties of poly(n-butylamino)(di-n-hexylamino)-phosphazene films cured with 0.05 wt% of TMP. ○: 70 ℃, ●: 120 ℃.

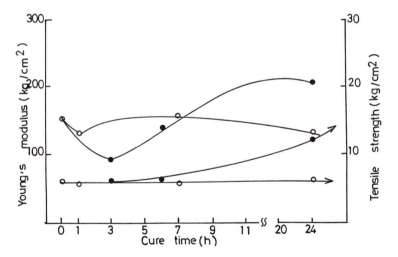

Figure 7. Mechanical properties of poly(n-butylamino)(di-n-hexylamino)-phosphazene films cured with 0.1 wt% of TMP. ○: 70 ℃, ●: 120 ℃.

Table 1. Elongation (yield and breaking point) and breaking strength of
poly(n-butylamine)(di-n-hyxylamine)phosphazene films cured with NPG.

NPG wt%	Cure Temp. (°C)	Time (hr)	Film thickness (mm)	Elongation (%) YP	BP	Breaking strength (kg/cm²)
0			0.47	16	58	3.1
0.05	50	1	0.26	14	20	2.4
		3	0.23	13	18	2.7
		7	0.31	16	23	2.8
	100	2	0.29	14	16	3.0
		4	0.22	13	15	2.6
		6	0.27	12	14	3.4
0.1	60	3	0.18	12	34	5.0
		8	0.18	13	38	3.7
		24	0.24	13	23	5.9
	120	3	0.31	12	29	5.0
		8	0.24	14	20	5.9
		24	0.18	14	30	5.7

NPG : Neopentyl glycol diglycidyl ether

Table 2. Elongation (yield and breaking point) and breaking strength of
poly(n-butylamine)(di-n-hyxylamine)phosphazene films cured with TMP.

TMP wt%	Cure Temp. (°C)	Time (hr)	Film thickness (mm)	Elongation (%) YP	BP	Breaking strength (kg/cm²)
0			0.47	16	58	3.1
0.05	70	1	0.20	16	36.5	4.8
		24	0.19	15.8	40.2	5.0
	120	1	0.26	7.8	32.2	4.3
		8	0.22	10.8	44.2	7.0
		24	0.25	17.6	44.3	15.1
0.1	70	1	0.69	17.8	30.5	4.0
		7	0.69	12.2	20.8	4.4
		24	0.60	17.3	30.3	3.8
	120	3	0.13	16.5	23.0	4.5
		6	0.13	11.0	25.5	5.2
		24	0.13	14.8	31.5	5.4

TMP : Trimethylol propane triglycidyl ether

Table 3. Oxygen gas permeability (in water) of poly(n-butylamine)(di-n-hyxylamine)phosphazene films cured with NPG and TMP.

Cure agent wt%	Cure Temp. (°C)	Time (hr)	Film thickness (mm)	$D_k \times 10^{11}$ (ml $\cdot$ cm/cm^{2}s^{-1}mmHg^{-1})	YM (kg/cm^2)
NPG 0.1	60	24	0.290	102.3	135
TMP 0.1	120	24	0.784	140.4	207
0.05	120	24	0.979	153.9	209
0			0.132	52.0	126

NPG : Neopentyl glycol diglycidyl ether, epoxy equivalent 135 - 165
TMP : Trimethylol propane triglycidyl ether, epoxy equivalent 130 - 150

References

1. Drioli, E.; Zhang, S. M.; Basile, A.; Goleme, G.; Gaeta, S. N.; Zhang, H. C. *Gas Separation and Purification* **1991**, *5*, 252.
2. (a) McAtee, R. E.; McCaffrey, R. R.; Grey, A. E.; Allen, C. A.; Gummings, D. G.; Appelhaus, A. D. *J. Membrane. Sci.* **1986**, *28*, 4. *Sepa. Sci. Technol.* **1987**, *22*, 873. (b) Suzuki, F.; Onozato, K.; Taegashi, H.; Masuko, T., J. *Appl. Polym. Sci.* **1987**, *35*, 2187. U. S. Pat., 4783202.
3. Allen, C. A.; Gumming, D. G.; Grey, A. E.; McAtee, R. E.; McCaffrey, R. R. *J. Membrane Sci.* **1987**, *33*, 181.
4. Gaeta, S. N.; Zhang, H. C.; Drioli, E., *Proc. 1st Conference Inorganic Membranes (ICIM 89)* France, **1989**, 65.
5. Zhang, S. M.; Basile, A.; Gaeta, S. N.; Drioli, E. *3rd Eur. Polym. Fed. Symp. Poly. Mater.* Sorrento, **1990**, 1.
6. Bittrava, F. A.; Krieev, V. V.; Mikitaev, A. K. *Vsykmol. Soe. Ser.* **1981**, *B23*, 30.
7. Japan Pat., 1984, 59-15140: 60-87828: 62-25118: U. S. Pat., 1981, 4710204.
8. Kajiwara, M. *J. Mater. Sci.* **1988**, *23*, 1360.
9. Kajiwara, M. *Polymer* **1989**, *30*, 1536.
10. Saito, H.; Kajiwara, M. *Kogyo Kagaku Zashi* **1963**, *66*, 618.
11. Kajiwara, M.; Shiomoto, K. *Polym. Comm.* **1984**, *25*, 93.
12. Mizutani, Y.; Miwa, Y.; Kuriaki, M.; Nozaki, S. *J. Japan Contact Lens* **1980**, *22*, 2.

RECEIVED July 5, 1994

Chapter 22

Grafting Reactions onto Poly(organophosphazenes)

Mario Gleria[1], Francesco Minto[1], Pietro Bortolus[1], Giacomo Facchin[2], and Roberta Bertani[2]

[1]Istituto di Fotochimica e Radiazioni d'Alta Energia del Consiglio Nazionale delle Ricerche, Sezione di Legnaro, Via Romea 4, 35020 Legnaro, Padova, Italy
[2]Centro di Chimica e Tecnologia dei Composti Metallorganici degli Elementi di Transizione del Consiglio Nazaionale delle Ricerche, c/o Istituto di Chimica Industriale, Facoltà di Ingegneria dell'Università di Padova, Via Marzolo 9, 35100 Padova, Italy

An overview of the grafting reactions of organic substrates onto poly(organophosphazenes) is presented. It was found that the type of phosphazene substrate used in the reactions, the swelling properties of the polymeric films, the characteristics of the photoinitiators and the composition of the reaction media are the major factors that influence the light-induced grafting of organic macromolecules onto polyphosphazene matrices. When the thermal grafting of maleic anhydride onto poly(organophosphazenes) is considered, the nature of the polymer, the reaction temperature, and the peroxide concentration are of importance.

Since the middle of sixties, when H.R.Allcock first reported the synthesis of poly(organophosphazenes), (POPs) (1-3), the scientific, technological and industrial relevance of these materials has increased tremendously (4-7). During the last decade our interest in this field was mainly focused on the photochemistry and photophysics of POPs, and on their possible practical exploitation in the photochemical field (8,9). A few years ago, however, we became interested also in the functionalization of POPs as obtained by the introduction of suitable chemical functionalities in selected macromolecules. These investigations led us to the synthesis and characterization of water-soluble phosphazenes (10-14), double-bond-containing substrates (15,16), and polymers bearing epoxide moieties (16).

Expanding this research, we recently explored another strategy that permits the chemical modification of POPs and leads to the preparation of new phosphazene materials, i.e. the grafting reaction of organic substrates onto POP matrices (17-23). Grafted copolymers, in fact, are of importance in blending processes (24), adhesion improvements (25) and biocompatibilization (26-28) of polymeric materials.

0097–6156/94/0572–0279$08.00/0

For these processes alkoxy- and aryloxy- substituted POPs were taken into consideration, such as poly[bis(4-*iso*propylphenoxy)phosphazene] (PiPP), poly[bis(4-*sec*butylphenoxy)phosphazene] (PsBPP), poly[bis(4-benzylphenoxy)phosphazene] (PBPP), poly[bis(2,2',2"-trifluoroethoxy)phosphazene] (PTFEP), poly[bis(4-methylphenoxy)phosphazene] (PMPP) and poly[bis(4-*tert*butylphenoxy)phosphazene] (PtBPP),

PiPP   PsBPP   PBPP

PTFEP   PMPP   PtBPP

in which tertiary, secondary and primary mobile hydrogen atoms respectively are present in the side groups.

In general, the grafting reactions of organic substrates onto polyphosphazene matrices can be achieved according to the scheme reported in Figure 1, using an ionic approach *(29-32)* or free radical reactions *(17-19,21-23,26,27,33,34)*.

P.Wisian-Neilson reported the grafting reactions of polystyrene *(30,31)*, polydimethylsiloxane *(29)* and polymethylmethacrylate *(32)* onto poly(methylphenylphosphazene), using alkyl lithium reagents; the phosphazene macroanions produced by heterolytic splitting of the labile hydrogens present in this polymer (Figure 1, Part A) are sufficiently reactive to bring about the anionic growth of the polystyrene, polydimethylsiloxane or polymethylmethacrylate chains onto the polyphosphazene matrix.

The free radical method, preferentially adopted in our laboratory, implies the homolytic removal of mobile hydrogens present in the side substituents of a selected POP, for instance poly[bis(4-*iso*propylphenoxy)phosphazene] **(1)** (Figure 1, Part B) using chemical methods *(17,18,20,33)*, UV light *(19,21-23,34)* or γ-rays *(26,27,35)* to give the phosphazene macroradicals **(2)**. These intermediates are very reactive species, able to induce the radical grafting of different types of unsaturated monomers, to form the species **(3)** and **(5)** in which the active center is now located in the grafted monomer sites. When the unsaturated species used as grafting substrate is maleic anhydride (MA), which has very low tendency to homopolymerize *(36,37)*, the macroradical **(3)** can be stabilized by abstracting hydrogens from the reaction medium, impurities, etc., forming the non reactive species **(4)**, thus stopping the grafting process at monomer level. Alternatively, when the species **(2)** reacts with unsaturated vinyl monomers originates new materials **(6)** with a comb-like

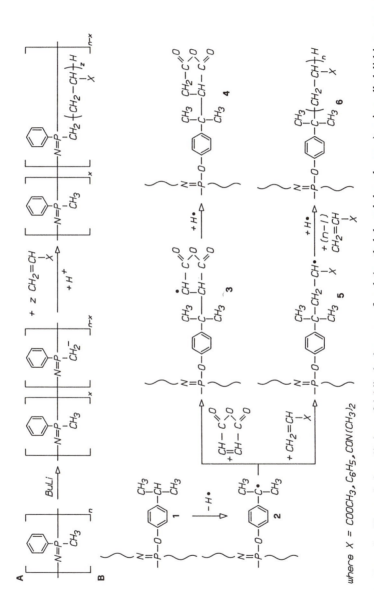

Figure 1. Part A: Heterolytic splitting of labile hydrogens of poly(methylphenylphosphazene) using alkyl lithium reagents.
Part B: Homolytic removal of mobile hydrogens of poly[bis(4-isopropylphenoxy)phosphazene].

structure in which side organic branches grow from the main polyphosphazene matrix *(38)*.

In this paper we will provide examples of both these types of processes.

Experimental Section

Materials. PiPP *(18)*, PsBPP *(20)*, PBPP *(39)*, PTFEP *(40)*, PMPP *(41)* and PtBPP *(42)* were prepared according to the literature. Methylmethacrylate (MMAM) and N,N'-dimethyl-acrylamide (NNDMAAM) monomers were Aldrich products purified from the stabilizers by vacuum distillation. Maleic anhydride (MA), 9,10-anthraquinone (AQ) and benzoyl peroxide (BPer) were purchased from Aldrich. 1-benzoyl-cyclohexanol (Irgacure 184) (IRG) was a Ciba Geigy product. All these chemical were used as received. The solvents were C.Erba analytical grade; when necessary they were purified or dried with standard techniques *(43)*.

Equipments. IR (both ATR and Transmission) spectra were obtained with a Bruker spectrometer Mod IFS 88. Contact angle determinations were carried out with a RAME'-HART goniometer at 25°C. UV spectra were measured with a Perkin Elmer spectrophotometer Mod 320. ^{1}H, ^{13}C and ^{31}P NMR spectra were performed using a Bruker AC 200 and a Varian FT 80 spectrometers. T_g were measured with a Mettler DSC Mod TA 72. Adhesion determinations were carried out with an Instrom 1195. Weight Average Molecular Weight measurements (MW) were obtained by light scattering technique using a Sophica spectrogoniodiffusimeter.

Grafting Procedures: a)vinyl polymers. Films of selected POPs obtained by casting technique were immersed in degassed (10^{-6} Torr) methanol solutions containing various amounts of vinyl monomers and photoinitiators, allowed to swell for 20 mins, and eventually irradiated with an HBO 150 W high pressure mercury lamp equipped with cut off filters at 340 nm. The final grafted copolymers (POP-*g*-Organic Polymers) obtained are thoroughly washed with suitable solvents in order to eliminate residual monomers, photoinitiators and organic homopolymers always formed in these conditions, and eventually dried in vacuum.

The grafted percentage *(44)*,

$$G_g(\%) = (\text{weight of grafts/weight of POP film}) \times 100$$

grafting efficiency *(44)*,

$$G_e(\%) = (\text{weight of grafts/total weight of homopolymer and grafts}) \times 100$$

swelling degree *(21)*,

$$G_s(\%) = [(\text{weight of the swollen film - weight of the film})/\text{weight of the film}] \times 100$$

and the amount of organic homopolymer formed during irradiation were determined gravimetrically.

Grafting procedures: b)maleic anhydride. 50 mg of selected POPs dissolved in 1 mL of xylene were treated with a fixed amount of maleic anhydride (MA) and benzoyl peroxide (BPer), at temperatures ranging between 95 and 139 °C, for different periods of time. In these conditions the grafting reactions took place, and succinic anhydride (SA) residues were grafted onto the POP matrices in percentages ranging between 0.1 and 10 % w/w, depending on the nature of POPs and on the adopted experimental conditions. The functionalized polymers, recovered by precipitation in chilled methanol or in *n*-heptane, were dried in vacuum and characterized by IR and NMR spectroscopy, and by molecular weight determinations.

Results and Discussion

Light-Induced Grafting Reactions of Organic Polymers onto POPs. The light-induced grafting of vinyl polymers onto phosphazene matrices is a reaction of paramount importance, potentially able to produce new materials of high scientific and technologic impact.

In spite of the very simple methodology used for the preparation of these substrates (see Experimental Section), the grafting of vinyl organic macromolecules onto polyphosphazene substrates appears to be a rather complicated phenomenon that depends on several different experimental parameters.

It was found, in fact, that the type of phosphazene polymer used in the grafting processes has a deep influence in determining the overall percentage of organic macromolecules grafted onto the polyphosphazene matrix. Thus, in the grafting reaction of PNNDMAAM onto PiPP, PsBPP, PBPP and PTFEP, PiPP and PsBPP proved to be the most reactive substrates, as easily abstractable tertiary hydrogens are present in the side phosphorus substituents of these polymers; by contrast, PBPP and PTFEP, that bear secondary hydrogen atoms in the side groups, show reduced reactivity. It may be pointed out, moreover, that steric factors, such as the spatial accessibility of the reactive hydrogens in the substituents, may also condition the final reactivity order of the PiPP-PsBPP and PBPP-PTFEP couples. These results are reported in Figure 2, where the grafting percentage of PNNDMAA onto the above mentioned POPs is reported vs. irradiation time. From these plots it is evident also that $G_g(\%)$ increases linearly with the irradiation time, at a fixed methanol/monomer ratio and for the same type of photoinitiator.

Another factor that influences strongly the final percentage of organic polymers grafted onto POPs , i.e. $G_g(\%)$, is the concentration of monomer present in the methanol/monomer mixtures used as reaction medium for the grafting experiments. This is reported in Figures 3 and 4 that refer to the grafting reaction of PMMA onto PiPP *(19)* and PNNDMAA onto PBPP *(21)* respectively. The Figures show that the variation of the monomer concentration in the reaction media has opposite consequences on the grafting processes of PMMA onto PiPP and of PNNDMAA onto PBPP respectively. In the first case an increase of the MMAM percentage in the system leads to a decrease of $G_g(\%)$, while in the second one an increase of NNDMAAM concentration results in an enhancement of the formed PBPP-*g*-PNNDMAA grafted copolymer.

These contrasting results prompted us to start investigations on the role played by the monomer on the overall grafting process of organic macromolecules onto POPs *(23)* by varying its percentage in the reaction mixture over a large range of concentrations. The obtained results are summarized in Figures 5 and 6. In the case of the grafting reaction of PMMA onto PsBPP, by using AQ as a photoinitiator (Figure 5), it was found that the value of $G_g(\%)$ measured after 1 h of irradiation increases with increasing the percentage of MMAM until a maximum is reached for the percentage of 20-25 % of monomer in the system. After this point, further addition of monomer results in the decrease of $G_g(\%)$. Similar findings were obtained also when the same experiments were carried out for PsBPP-*g*-PMMA copolymers, using IRG as a photoinitiator (Figure 6). In this case, however, the maximum $G_g(\%)$ value was found to be considerably higher than that observed for the system in which AQ is

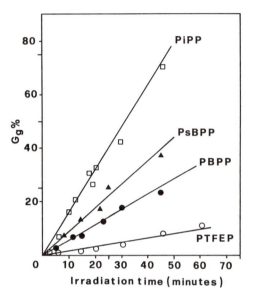

Figure 2: Percentage of PNNDMAA grafted onto different POP films. Methanol/monomer ratio = 1.7. Reproduced with permission from reference 21. Copyright 1994 Pergamon Press.

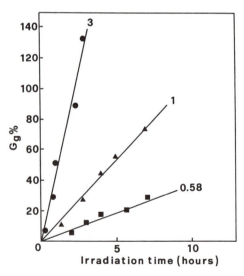

Figure 3: Percentage of PMMA grafted onto PiPP at various methanol/monomer ratios (3.0, 1.7, 0.58). Reproduced with permission from reference 19. Copyright 1992 Pergamon Press.

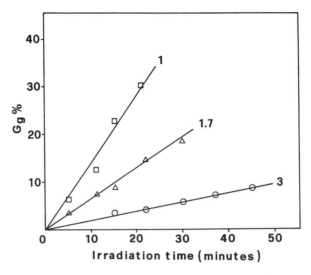

Figure 4: Percentage of PNNDMAA grafted onto PBPP at various methanol/monomer ratios (1.0, 1.7, 3.0). Reproduced with permission from reference 21. Copyright 1994 Pergamon Press.

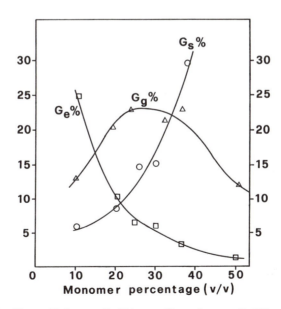

Figure 5: Grafting efficiency $G_e(\%)$, swelling degree $G_s(\%)$ and grafting percentage $G_g(\%)$ for the grafting of PMMA onto PsBPP using 9-10 anthraquinone photoinitiator.

utilized as photoinitiator. This is not unexpected, as AQ is known to be a photoinitiator that produces radicals under illumination by hydrogen abstraction reactions, while IRG is operative with a photofragmentation process *(45)*.

These experiments seem to indicate that the type of photoinitiator used to induce the grafting processes involving polyphosphazene substrates is very important in determining the percentage of organic macromolecules grafted onto POPs. They suggest also that the amount of PMMA grafted onto PsBPP increases with the increasing amount of MMAM in the reaction medium. This is likely due to the fact that MMAM acts as a solvent *(21)* for the phosphazene polymer films (as shown in Figures 5 and 6 by the increase of $G_s(\%)$ increasing the amount of MMAM). This favours the penetration of the monomer and the photoinitiator in the bulk of the phosphazene films and eventually facilitates the overall grafting process. A further increase of the monomer concentration, however, leads to quenching of the photoinitiator excited triplet state, as demonstrated by flash photolysis experiments *(23,46)*, and depresses the photoreactivity of the photoinitiator and the final value of $G_e(\%)$. The best compromise between these two processes is observed when 20-25 % v/v of monomer is present in the reaction medium, as in these conditions, the maximum of $G_g(\%)$ is observed for all the exploited photoinitiators.

The final samples of POP-*g*-PMMA and POP-*g*-PNNDMAA grafted copolymers obtained according to the above described experimental procedure, were characterized by DSC, IR spectroscopy and contact angle techniques

DSC measurements carried out on the PiPP-g-PMMM system *(19)*, showed the presence in the thermograms of two distinct transitions at 0 °C and 120-130 °C attributed to the glass transition temperature of PiPP and PMMA respectively; this demonstrates that PMMA is successfully attached to the polyphosphazene matrix during the light-induced grafting reaction.

Further support to this fact comes from IR spectroscopy. In fact, FTIR measurements, both ATR and transmission, carried out on POP-g-PMMA or POP-g-PNNDMAA showed, in the first case, a sharp peak at 1730 cm^{-1} attributed to the vibration of the -COOCH$_3$ ester functionalities, and, in the second case, a new band in the spectral range of the carbonyl groups (1700-1600 cm^{-1}), assigned to the vibration of the -CO-N(CH$_3$)$_2$ moiety *(47)*. The usual peaks at 1200-1100 cm^{-1} (v - P=N-) and at 970-940 cm^{-1} (v P-O-Ph) attributed to the polyphosphazene skeleton and to the phosphazene substituents respectively, are also observed *(48)*.

Portions of these spectra, that refer to the IR-ATR and to IR-transmission measurements carried out on PsBPP-g-PMMA grafted copolymers, are reported in Figures 7 and 8, respectively. The absorbance measured by transmission spectroscopy is remarkable higher than that obtained by ATR superficial measurements. This seems to indicate that the grafting process of the organic polymer onto the polyphosphazene support takes place predominantly in the bulk of the polyphosphazene matrix and that only a minor part of PNNDMAA is grafted onto the surface of the POP films. This is in agreement with the above mentioned abilities of MMAM to swell the polyphosphazene substrates.

As a consequence, it may be expected that the bulk properties of the POPs are remarkably modified with respect to those of the starting materials. Preliminary results of investigations in this direction are promising. In spite of the above mentioned facts, however, the PNNDMAA grafted onto the surface of the POP films

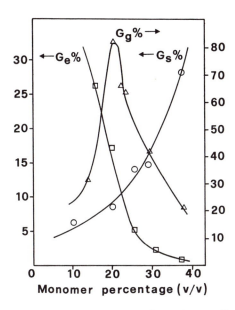

Figure 6: Grafting efficiency G_e(%), swelling degree G_s(%) and grafting percentage G_g(%) for the grafting of PMMA onto PsBPP using Irgacure 194 photoinitiator.

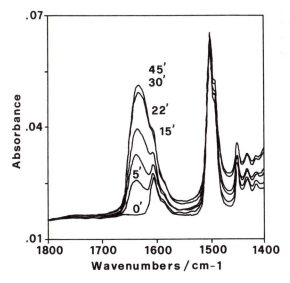

Figure 7: FTIR-ATR spectra of PNNDMAA grafted onto PBPP films at various irradiation times (0, 5, 15, 22, 30, 45 mins). Reproduced with permission from reference 21. Copyright 1994 Pergamon Press.

is able to impart considerable hydrophilicity to the system, as demonstrated by the contact angle measurements reported in the Table 1.

Table 1. Contact Angle Measurements for POP-g-PNNDMAA Grafted Copolymers[***]

Time	PiPP		PsBPP		PBPP		PTFEP[**]	
t(min)	Θ_a	%	Θ_a	%	Θ_a	%	Θ_a	%
0	86	0	75	0	84	0	99	0
2			66	2				
4.5			67	3.84				
5					53	3.93		
6	59	7					63	0.6
7			65	7.2				
10	67	16.7						
11			63	10.2				
12.5	33[*]	21.1						
15	33[*]	26.7	60	14.2	40[*]	8.3	67	2.3
20	32[*]	31.6						
22			59	16.8	33[*]	15.5		
30	30[*]	42.1			27[*]	18.5		
45							69	8.1
60			58	16.2			65	10.7

[*] In these values　swelling of POP-g-PNNDMAA grafted copolymer films　was observed during the contact angle measurements.
[**] In the case of PTFEP-g-PNNDMAA grafted copolymers the films were found not rigorously planar during the contact angle measurements.
[***]Reproduced with permission from reference 21. Copyright 1994.

As can be seen for all the exploited POPs, the increase of PNNDMAA grafted onto the polyphosphazene films brings about a concurrent decrease of the contact angles measured, to suggest an increased surface hydrophilicity of the system.

In conclusion, the light-induced grafting reactions of organic macromolecules onto phosphazene polymer films is a process that depends on several factors, i.e. the type of POP used as grafting support, the characteristics and the concentration of the exploited monomers, the type of photoinitiator, the swelling of the phosphazene substrate, etc. The accurate control of these parameters is mandatory in order to prepare phosphazene grafted copolymers of predictable characteristics.

Grafting of Maleic Anhydride onto POPs. The field of polymer blends is very important in the modern macromolecular chemistry both from the point of view of the scientific understanding and of the technologic interest *(24)*. The formation of blends appears to be particularly desirable in the case of POPs because of the poor attitude of these polymers to compatibilize with organic macromolecules *(49,50)* and also with polyphosphazenes bearing different phosphorus substituents *(51)*. A suitable way to form blends is to reduce interfacial tension between the different polymers using compatibilizing agents *(52)*. In this respect polymers containing anhydride residues lend themselves nicely because they are excellent starting materials for the preparation of grafted copolymers able to act as compatibilizing agents *(53)*.

In an attempt to promote the formation of blends between POPs and macromolecules of different nature, we started investigations on the grafting reactions of maleic anhydride onto polyphosphazene matrices *(20)*. The grafting reaction was carried out onto PiPP, PsBPP, PBPP, PMPP and PtBPP, by dissolving these polymers in boiling xylene, in the presence of variable amounts of MA and BPer, and heating the system for different periods of time. In these conditions, the opening of the MA double bond allows the grafting of succinic anhydride (SA) residues onto the POP substrates, according to the reaction scheme reported in Figure 9. The radical nature of this reaction and the poor capability of MA to homopolymerize (so that only monomer anhydride groups are eventually attached to the polyphosphazene system) are well known facts, widely discussed in other publications *(20,36,37)*.

The obtained POP-*g*-SA were characterized by IR and NMR (^{1}H, ^{13}C, ^{31}P) spectroscopy, and by Weight Average Molecular Weight (MW) measurements.

All the IR spectra of these materials show the presence of the usual polyphosphazene absorptions (skeleton and substituents), and of two additional peaks located at 1865 and 1785 cm^{-1}, assigned to the carbonyl groups of the anhydride moiety; these signals support the presence of SA groups in the functionalized POPs *(20)*.

NMR characterization data confirm this conclusion. It was found in fact, that ^{1}H and ^{13}C NMR spectra of the original POPs and those of the phosphazenes containing SA groups are almost superimposable, with the exception of low-intensity peaks located in the spectral range of 175-165 ppm attributed to the -CH$_2$-CO- and -CH-CO- carbonyl groups of the anhydride moieties. Moreover, ^{31}P NMR spectra show sharp peaks in the spectral range of -18/-19 ppm for the original POPs, that broaden considerably after grafting few percent of the SA groups. This is reasonable due to the fact that the grafted SA moieties lie quite apart from the polymer chain and should have only minor influence on the phosphorus atoms of the phosphazene skeleton.

Weight Average Molecular Weight measurements carried out on PsBPP-*g*-SA copolymers *(20)* showed that the grafting of MA onto PsBPP occurs with the decrease of molecular weight, to suggest the occurrence of degradation phenomena during the process. The MW of the original POP, in fact, passed from 2.3x10^6 to 9.5x10^5 by treating 50 mg of PsBPP in xylene with 5 mg of BPer at 125 °C for 20 min. Preliminary results concerning MW determinations carried out on the other POPs used in this work seem to indicate that the same phenomena are operative also for these macromolecules.

The percentages of succinic anhydride grafted onto the five selected polyphosphazenes, vs. the amount of benzoyl peroxide used in the reaction and vs. the temperature of the grafting process are reported in the Figures 10 and 11 respectively. These pictures show that the grafting of MA onto the selected POPs is an easily controllable process, that allows to insert predictable amounts (from 0.1 up to 10%) of SA groups in the phosphazene polymers depending on the type of POP used for the process. Moreover it can be seen that PiPP and PsBPP, i.e. substrates bearing tertiary

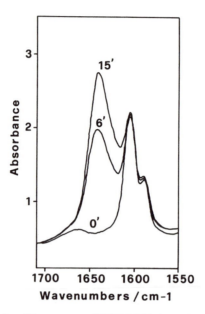

Figure 8: Transmission IR spectra of PNNDMAA grafted onto PBPP films at various irradiation times (0, 6, 15, mins). Reproduced with permission from reference 21. Copyright 1994 Pergamon Press.

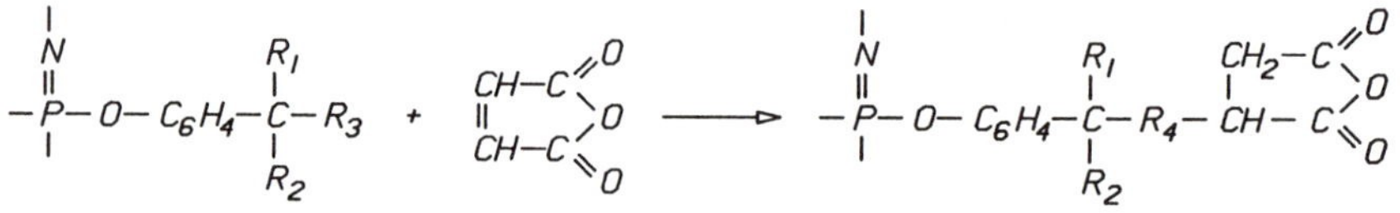

	R_1	R_2	R_3	R_4
PIPP	CH_3	CH_3	H	–
PsBPP	CH_3	C_2H_5	H	–
PBPP	H	C_6H_5	H	–
PMPP	H	CH_3	H	–
PtBPP	CH_3	CH_3	CH_3	CH_2

Figure 9. Grafting of succinic anhydride onto polyphosphazene matrices.

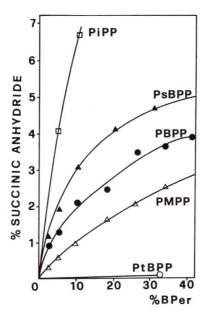

Figure 10: Weight percentage of SA grafted onto different POPs at various BPer percentages.

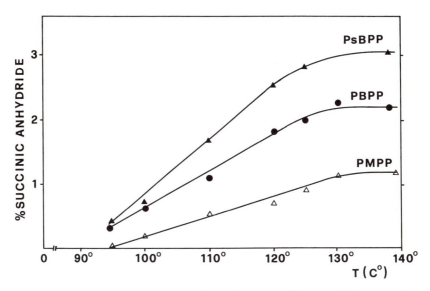

Figure 11: Weight percentage of SA grafted onto different POPs at various reaction temperatures.

hydrogens in the side phenoxy substituents, are more reactive than PBPP, a polymer that possesses secondary hydrogens in the side groups; this macromolecule, in turn, reacts with MA easier than PMPP and PtPBB, in which only primary hydrogens are present in the phenoxy moieties. The trend of the reactivities confirms the radical character of the grafting reaction of MA onto polymeric matrices (53), whose primary step implies the homolytic cleavage of the carbon-hydrogen bonds present in the phosphazene substituents. Tertiary hydrogen, in fact, are more easily abstractable than secondary ones, that, in turn, can be split easier than the primary H. This is in agreement also with the increasing values of the C-H bond dissociation energies (54).

It may be observed, moreover, that the same order of reactivity reported for the grafting of MA onto POPs has been found also during the light-induced grafting reactions of organic macromolecules onto these materials (see Figure 2), to confirm the strong similarities existing between these two processes.

The SA functionalized POPs show very different physical and chemical characteristics as compared to those of the unmodified materials. The metal-polymer adhesion properties in the melt state of these polymers are considerably increased by grafting increasingly high amounts of SA onto POPs, as shown in the case of PsBPP-g-SA (20). The initial adhesion strength of PsBPP (73 N/m) increases up to 2153 N/m when 2.4 % w/w of SA groups are inserted in the polymer. This value is of the same order of magnitude of that of the commercially available polypropylene containing 0.4 % w/w of SA groups (3537 N/m). Moreover, the glass transition temperature, Tg, of the same substrate was found to increase increasing the SA content in the polymer, to indicate a progressive rigidity of the polyphosphazene backbone when bulky groups (SA) are attached to the polymer substituents. From a chemical of viewpoint, moreover, the reactivity of phosphazene polymers containing variable amounts of SA groups is dramatically different compared to that of the pristine polymers. In fact, anhydride moieties are known to react easily with -OH or -NH$_2$ terminated organic substrates. This fact may open remarkable perspectives in the preparation of grafted copolymers to be used as interfacial agents able to favour the formation of polymer blends. In this context several options are available. POP-g-SA copolymers, in fact, can react with -NH$_2$ end-capped nylon oligomers, or with -OH terminated polyesters to form grafted copolymers that may facilitate blending processes among POPs and polyamides or polyesters, respectively. Hydroxyl-group-containing macromolecules, like polyvinyl alcohol, polyHEMA or poly[bis(4-hydroxyphenoxy)phosphazene], can be reacted with POP-g-SA to form new crosslinked materials. Very promising results, for instance are obtained with PsBPP-g-SA and OH group containing siloxane derivatives. The research on this new topic is still under way in our Institute and will be reported in forthcoming papers.

Acknowledgements
The Authors are indebted to the Progetto Finalizzato Chimica Fine e Secondaria II del Consiglio Nazionale delle Ricerche for financial support.

Literature Cited
1.   Allcock, H. R.; Kugel, R. L. *J. Am. Chem. Soc.* **1965**, *87*, 4216.
2.   Allcock, H. R.; Kugel, R. L.; Valan, K. J. *Inorg. Chem.* **1966**, *5*, 1709.
3.   Allcock, H. R.; Kugel, R. L. *Inorg. Chem.* **1966**, *5*, 1716.
4.   Allcock, H. R. *Chem. Eng. News* **1985**, *63*, 22.
5.   Potin, P.; De Jaeger, R. *Eur. Polym. J.* **1991**, *37*, 341.

6. Allcock, H. R. *Science* **1992**, *255*, 1106.
7. Allcock, H. R. In *Inorganic and Organometallic Polymers with Special Properties*; Laine, R. M., Ed.; Nato Asi Series E: Applied Sciences: 1992; Vol. 206; p 43.
8. Gleria, M.; Bortolus, P.; Flamigni, L.; Minto, F. In *Inorganic and Organometallic Polymers with Special Properties*; Laine, R. M., Ed.; Nato ASI Series E: Applied Sciences: 1992; Vol. 206; p 375.
9. Bortolus, P.; Gleria, M. *J. Inorg. Organomet. Polym.* **1994**, in press.
10. Montoneri, E.; Gleria, M.; Ricca, G.; Pappalardo, G. C. *Makromol. Chem.* **1989**, *190*, 191.
11. Montoneri, E.; Gleria, M.; Ricca, G.; Pappalardo, G. C. *J. Macromol. Sci., Chem.* **1989**, *A26*, 645.
12. Montoneri, E.; Ricca, G.; Gleria, M.; Gallazzi, M. C. *Inorg. Chem.* **1991**, *30*, 150.
13. Montoneri, E.; Savarino, P.; Viscardi, G.; Gallazzi, M. C.; Gleria, M.; Minto, F. *J. Inorg. Organomet. Polym.* **1992**, *2*, 421.
14. Medici, A.; Fantin, G.; Pedrini, P.; Gleria, M.; Minto, F. *Macromolecules* **1992**, *25*, 2569.
15. Facchin, G.; Bertani, R.; Berton, A.; Gleria, M. *Inorg. Chim. Acta* **1988**, *147*, 165.
16. Fantin, G.; Medici, A.; Fogagnolo, M.; Pedrini, P.; Gleria, M.; Bertani, R.; Facchin, G. *Eur. Polym. J.* **1993**, *29*, 1571.
17. Audisio, G.; Bolognesi, A.; Catellani, M.; Destri, S.; Porzio, W.; Gleria, M. *Proceedings of the International Conference on Functional Polymers and Biopolymers*, Oxford, U.K., September 19-19th, 1986, p 119.
18. Gleria, M.; Bolognesi, A.; Porzio, W.; Catellani, M.; Destri, S.; Audisio, G. *Macromolecules* **1987**, *20*, 469.
19. Minto, F.; Scoponi, M.; Fambri, L.; Gleria, M.; Bortolus, P.; Pradella, F. *Eur. Polym. J.* **1992**, *28*, 167.
20. Gleria, M.; Minto, F.; Scoponi, M.; Pradella, F.; Carassiti, V. *Chem. Mater.* **1992**, *4*, 1027.
21. Minto, F.; Scoponi, M.; Gleria, M.; Pradella, F.; Bortolus, P. *Eur. Polym. J.* **1994**, *30*, 375.
22. Gleria, M.; Minto, F.; Bortolus, P.; Facchin, G.; Bertani, R.; Scoponi, M.; Pradella, F. *ACS Polym. Prep.* **1993**, *34(1)*, 270.
23. Minto, F.; Gleria, M.; Bortolus, P.; Scoponi, M.; Pradella, F. *Proceedings of the 3rd International Congress on Polymer Photochemistry*, Sestri Levante (Genoa), Italy, September 5-10th, 1993, p 162.
24. *Polymer Blends: Processing Morphology and Properties*; Martuscelli, E.; Palumbo, R.; Kryszewski, M., Eds.; Plenum Press: New York, 1980.
25. Bratawidjaja, A. S.; I, G.; Watanabe, Y.; Hatakeyama, T. *J. Appl. Polym. Sci.* **1989**, *37*, 1141.
26. Lora, S.; Carenza, M.; Palma, G.; Caliceti, P. *Radiat. Phys. Chem.* **1990**, *35*, 117.
27. Lora, S.; Carenza, M.; Palma, G.; Pezzin, G.; Caliceti, P.; Battaglia, P.; Lora, A. *Biomaterials* **1991**, *12*, 275.
28. Lora, S.; Palma, G.; Bozio, R.; Caliceti, P.; Pezzin, G. *Biomaterials* **1993**, *14*, 430.

29.   Wisian-Neilson, P.; Islam, M. S. *Macromolecules* **1989**, *22*, 2026.
30.   Wisian-Neilson, P.; Schaefer, M. A.; Islam, M. S. *ACS Polym. Prep.* **1989**, *30(1)*, 185.
31.   Wisian-Neilson, P.; Schaefer, M. A. *Macromolecules* **1989**, *22*, 2003.
32.   Garcia-Alonso, F. J.; Wisian-Neilson, P. *Proceedings of the Workshop on the Status and Future of Polyphosphazenes*, Raleigh, NC, USA, March 15-18th, 1992, Poster No. 10.
33.   Brossas, J.; Clouet, G.; Tissot, M. *French Patent* 2 510 128, **1983**.
34.   Minto, F.; Gleria, M.; Bortolus, P.; Scoponi, M.; Pradella, F.; Fambri, L. In *Polymer Photochemistry*; Allen, N. S., Edge, M., Bellobono, I. R., Selli, E., Eds.; Ellis Horwood Ltd: 1994, in press.
35.   Pezzin, G.; Palma, G.; Lora, S.; Carenza, M. *Chem. Ind. (Milan)* **1991**, *73*, 185.
36.   Smeller, J. A. *J. Mod. Plast. Int.* **1985**, 42.
37.   Borsig, E.; Klimova, M.; Hrckova, L.; Szöcs, F. *J. Macromol. Sci., Pure Appl. Chem.* **1993**, *A30*, 105.
38.   Chapiro, A. *Radiation Chemistry of Polymeric Systems*; Wiley: New York, 1962.
39.   Allcock, H. R.; Mang, M. N.; Dembek, A. A.; Wynne, K. J. *Macromolecules* **1989**, *22*, 4179.
40.   Ferrar, W. T.; Marshall, A. S.; Whitefield, J. *Macromolecules* **1987**, *20*, 317.
41.   Singler, R. E.; Hagnauer, G. L.; Schneider, N. S.; LaLiberte, B. R.; Sacher, R. E.; Matton, R. W. *J. Polym. Sci., Polym. Chem. Ed.* **1974**, *12*, 433.
42.   Quinn, E. J.; Dieck, R. L. *J. Fire & Flammability* **1976**, *7*, 5.
43.   Vogel, A. J. *A Textbook of Practical Organic Chemistry*; Longman: London, 1970.
44.   Ogiwara, Y.; Kanda, M.; Takumi, M.; Kubota, H. *J. Polym. Sci., Polym. Lett. Ed.* **1981**, *19*, 457.
45.   Pappas, S. P. *UV Curing: Science and Technology*; Stamford Technology Marketing: Norwalk, 1978; Volume 1 and 2.
46.   Schnabel, W. *Photogr. Sci. Eng.* **1979**, *23*, 154.
47.   Silverstein, R. M.; Bassler, C. G.; Morrill, T. C. *Spectrophotometric Identification of Organic Compounds*; Wiley: New York, 1974.
48.   Thomas, L. C. *Interpretation of the Infrared Spectra of Organophosphorus Compounds*; Heyden: London, UK, 1974.
49.   Beres, J. J.; Schneider, N. S.; Desper, C. R.; Singler, R. E. *Macromolecules* **1979**, *12*, 566.
50.   Allcock, H. R.; Visscher, K. B. *Chem. Mater.* **1992**, *4*, 1182.
51.   Gleria, M.; Minto, F.; Bortolus, P.; Porzio, W.; Meille, S. V. *Eur. Polym. J.* **1990**, *26*, 315.
52.   Datta, S.; Lohse, D. J. *Macromolecules* **1993**, *26*, 2064.
53.   De Vito, G.; Lanzetta, N.; Maglio, G.; Malinconico, M.; Musto, P.; Palumbo, R. *J. Polym. Sci., Polym. Chem. Ed.* **1984**, *22*, 1335.
54.   *Handbook of Chemistry and Physics*; Weast, R. C.; Selby, S. M., Eds.; CRC Press: Cleveland, OH, 1974.

RECEIVED April 8, 1994

Chapter 23

Poly[bis(p-chlorophenoxy)phosphazene]– Polystyrene Blends

Preparation, Compatibility, and Properties

Y. W. Chen-Yang[1], H. F. Lee[1], and T. T. Wu[2]

[1]Department of Chemistry, Chung Yuan Christian University, Chung-Li,
Taiwan 320, Republic of China
[2]Man-made Fiber Research Department, Union Chemical Laboratories,
Industrial Technology Research Institute, Hsinchu, Taiwan 300,
Republic of China

Blends with different compositions of PCPP and PS have been prepared.
The combination of the observation of optical clarity, DSC measurements
and SEM micrographs show that PCPP and PS are compatible in the
solution casting blend films with the compositions of PCPP less than 50%
and are partially compatible or incompatible with higher content of PCPP.
FTIR spectroscopy indicates a weak interaction between the component
polymers in the blends. The blends are phase separated when they are
heated above T(1) mainly due to repacking of the PCPP crystalline
regions. The thermal properties and the flammability of the blends were
also studied.

Polyphosphazenes are some of the most diverse inorganic type polymers known. The presence of phosphorus and nitrogen atoms in the backbone and the unusual method of synthesis, with which an almost infinite variety of side groups and combinations of different side groups can be linked to the backbone, confer to the polymers a variety of interesting properties (1-3). Therefore, the applications of polyphophazenes can be very broad, and includes flame retardent materials, high performance elastomers, optical and electronic materials, biomedical materials etc.(3-8). The application potentials of some elastomers included aerospace, marine, oil exploration and industrial fields. However, despite to the hundreds of the polyphosphazenes synthesized, they are still expensive to produce and very few of them have been commercialized (6).

Up to now, the preparation of polyphosphazenes with unique properties have mainly involved chemical reactions on existing polymers, developing new methods and techniques of polymerization, and combining existing monomers in such a way that the resulting materials have certain superior properties. The first is the most popular and traditional two-step process pioneered by Allcock, in which the ring-opening polymerization of hexachlorocyclotriphosphazene is followed by macro-molecular substitution (9-11). In fact, the largest variety of polyphosphazene molecules have been obtained by this method, and this is still the main method for polyphosphazene synthesis today. The second approach is a ring-opening of partially substituted cyclics followed substitution. It is carried out by three-step synthesis (12). Since one kind of functional group is substituted in the first step before ring-opening step is carried out and the other functional group is substituted in the last step,

the order of the functional group on backbone may be controlled in this method. The third approach is the condensation method, developed by Neilson and Wisian-Neilson, which provides a method to generate polymers with all the side groups linked to the skeleton P-C bonds (13). The copolymerization and grafting method are the examples of the fourth approach. These have been used to prepare polyphosphazene-organic polymer hybrids and to improve some properties, such as flame retardency and thermal stability, of the commercially valuable organic polymers by the incorporation of polyphosphazene (14-18).

All the approaches described above involve the formation of covalent bonds between polyphosphazenes and the interesting organic entity. On the other hand, the method of physically combining two existing polymers, without forming covalent bonds, to achieve new and useful properties (such as blending, interpenetrating polymer network, composite etc.), is one of the most commercially important areas for development of new polymer materials and has received much attention by the organic polymer researchers and industrial companies (19). Moreover, the approach was applied to prepare new polyphosphazene-organic polymer materials just recently (20-26). Among them, Borisenkova et.al. reported an improved melt process comprising mesophase poly[bis(trifluoroethoxy)phosphazene] and ultrahigh molecular weight polyethylene (20). Yamamoto et.al. patented work on fire-resistant polyolefin-phosphazene polymer copolymer compositions for electrical insulation (21). Ferrar et.al. filed a patent on the miscible blends of polyphosphazenes and acidic functional polymers and used FTIR spectroscopy to show the existence of the hydrogen bonding in the blend of poly(bis(methoxyethoxyethoxy))phosphazene and polyvinylphenol (22). Using the sol-gel method Ferrar et.al. also studied the preparation of composite materials by combining polyphosphazene with a silicon oxide precursor tetraethoxy-silane, TEOS, which was then polymerized in situ (23). Allcock et.al. reported the synthesis and characterization of several new interpenetrating polymer networks composed of polyphosphazenes with organic polymers (24,25). In addition, they also studied the polymer blends derived from polyphosphazenes and organic polymers (26).

Since both miscible and immiscible phosphazene blends are of considerable interest as membrances, biomaterials, or flame retardant materials, it is worthwhile to study the compatibility of these kinds of blends in more detail in order to understand more about the interaction between the polymer pair and the stability of the blends. Our goal in this study is to prepare PCPP/PS blends and investigate the compatibility and the properties of the blends by optical clarity, DSC, SEM, FTIR, TGA and LOI. PCPP/PS is chosen in this study because (a) PCPP is known to be flame retardant (5), (b) polystyrene is a well-known versatile organic polymer and has been selected to graft or blend with phosphazene polymers in most of the polyphosphazene-organic polymer hybrids systems (14,15), and (c) the similarity of aromatic side groups in both of the polymers. In addition, since PCPP has a special thermotropic transition temperature, T(1), and in order to further understand the stability of the blends, the compatibility influenced by temperature is also studied before and after T(1) transition by DSC.

Experimental

Materials. PCPP was synthesized from soluble poly(dichlorophosphazene), $(PNCl_2)_n$, which was prepared by the thermal ring opening of hexachlorocyclo-triphosphazene, $N_3P_3Cl_6$ (9). The poly(dichlorophosphazene) dissolved in benzene solution was reacted with the sodium salt of p-chlorophenol, which was prepared by reaction of sodium and p-chlorophenol in a 2-ethoxyethyl ether / 1,4-dioxane co-solvent system, to form a fully substituted homopolymer, $[PN(OC_6H_5Cl)_2]_n$, PCPP, as reported by Reynard et al.(27). The product, PCPP, was then isolated and purified

by precipitation from THF solution into methanol and dried in vacuum for 1 day. The polymerization yield was about 80%. The ^{13}C NMR spectrum exhibited sharp resonances near 148.7, 129.2, 128.4 and 120.9 ppm for aromatic carbons. ^{31}P NMR exhibited a singlet at -20.1 ppm and confirmed that all chlorines were replaced by OC_6H_5Cl group. The intrinsic viscosity of PCPP was determined to be 1.5 dL/g in THF at 40°C. GPC measurements indicated a broad molecular weight distribution with an Mn value of 1.2×10^4 and an Mw value of 1.5×10^5 vs. Toyo Soda PS standards. PS was supplied by Taida Chemical Co. Ltd. (951 LL), Taiwan, R.O.C.. The Mn and Mw value of the PS, measured by GPC, were 1×10^5 and 3.2×10^5, respectively.

Preparation of Samples. A PS solution in THF was prepared by dissolving the polymer at 50 °C under continuous stirring over a period of complete dissolution. The PCPP solution in THF was prepared by the same method. The two solutions, thus separately prepared, were mixed at 70 °C in the desired proportions, so that the relative composition of the two polymers in the mixed solution ranged from 20/80 to 80/20 of weight percent. After stirring for about one day, the mixed solutions were used to prepare blend samples by the coagulation method and the solution casting method.

The coagulation method reported by Y. Nishio (28) was carried out following the procedure for the Limiting Oxygen Index (LOI) test. Each blended solution prepared as above was poured into a rectangular teflon tray. An appropriate amount of methanol was then carefully added into the solution, where the PCPP/PS blends were precipitated as a gelatinous thick film. The blends were steeped in methanol overnight and finally allowed to dry gradually at room temperature. An additional 24 hrs of drying at room temperature under high vacuum was then carried out on the solid films. The PS and the PCPP homopolymer films were also obtained by the same method.

On the other hand, the conventional solution casting method was used to prepare the films for measuring other properties. The blends prepared as above were cast on glass plates using gardner knife and heated to remove solvent in a 50 °C oven for about 40 min. The blends were then removed from the glass by steeping in 4 °C ice water and dried in vacuum to remove the remaining solvent at room temperature.

Measurements. The instruments used were Mettler DSC-30 for Differential Scanning Calorimetry (DSC), Mettler TG-50 for Thermogravimetric Analysis (TGA), JEOL JSM-6300 for Scanning Electron Microscopy (SEM), BIO-RAD FTS7 and a GRASEBY-SPECAC P-N11080 attenuated total reflectance (ATR) accessory for Fourier Transform Infrared Spectroscopy and Stanton Redcroft COI-710 for Limiting Oxygen Index (LOI) measurement. DSC was performed under an atmosphere of nitrogen. All the samples were heated to remove the remaining solvent before measurement. Two scans were carried out for the blends and the homopolymers. The first scan was carried out at a rate of 10 °C/min from -50 °C up to 200 °C. The second scan followed after quick cooling (200°C/min) from the first scan and was done at the same rate from -50 °C up to 250 °C. The temperature readings were calibrated with an indium standard. All the glass transition temperatures were measured by the midpoint method. The morphology of the solution cast films was examined by SEM at 20 kV accelerating voltage after gold sputter coating. For FTIR spectra sixty four scans at a resolution of 2-4 cm^{-1} were signal averaged and stored on a magnetic disc system. TGA measurements were carried out at a heating rate of 20 °C/min from 50 °C to 750 °C in nitrogen atmosphere at a flow rate of 200 cm^3/min. The flammabilities of the samples were determined in terms of Limiting Oxygen Index(LOI) according to ASTM D2863-77 on specimens with 6"x1/4"x1/8" molded under pressure.

Results and Discussion

Compatibility. It is known that the compatibility, or miscibility of the polymer pair
in a blend depends on the affinity between the two mixing polymers. When they do not
have enough affinity, two distinct or immiscible phases remain, an incompatible
polyblend is obtained. Due to the presence of the two distinct phases, usually two
glass transition temperatures of the respective parent polymers are detected and
opaquness or cloudness is observed. When the affinity exists but only partial mixing
of the two polymers takes place at the molecular level, two phases with different
compositions may be obtained and the components are considered to be partially
compatible. The blends of this kind have two glass transition temperatures between the
two Tg of the parent polymers and are usually translucent. On the other hand, when
the affinity between the two polymers in the blend is strong enough and a
homogeneous phase is formed, the two polymers are, then, compatible. The blend
with a compatible polymer pair is characterized by a single glass transition temperature
and is usually transparent. The observation of the optical clarities of polyblends
prepared is, therefore, the simplest method providing valuable information for the
compatibilities of polyblends. On the other hand, DSC measurement, detecting the
glass transition temperatures, is convenient and one of the most important techniques
for studying the compatibilities of polyblends.

The optical clarities of the PCPP/PS blends with different compositions
prepared in this study are listed in Table I. It indicates that the blend with 80/20 wt %
of PCPP/PS composition are not compatible and the others are possibly compatible.
Figure 1 displays the thermograms obtained in the first heating scan for the samples
and the data obtained through analysis of the DSC curves are summarized in Table II.
The glass transition temperatures of the pure PS and the pure PCPP samples are
estimated to be 100 °C and 13 °C, respectively. As can be seen from the other curves,
when PCPP is blended into PS up to 50%, only one Tg is observed in the blend and
the Tg is shifted to the lower temperature region, sensitively depending on the blend
composition.

Table I. The optical clarity of PCPP/PS blends.

PCPP/PS	Optical clarity
100/0	transparent
80/20	opaque
60/40	hazy
50/50	translucent
40/60	transparent
20/80	transparent
0/100	transparent

Table II. The glass transition temperature Tg, and the first transition temperature T(1) of PCPP/PS blends.

PCPP/PS	1st heating			2nd heating		
(w/w)	$Tg_{,pcpp}$	$Tg_{,ps}$	$T(1)_{,pcpp}$	$Tg_{,pcpp}$	$Tg_{,ps}$	$T(1)_{,pcpp}$
100/0	13.0	----	148.2	13.0	---	170.0
80/20	14.6	98.0	148.0	17.9	97.2	167.6
60/40	13.5	71.0	146.5	14.0	95.4	168.7
50/50		61.2	147.0		97.5	168.1
40/60		68.6	143.0		92.3	165.4
20/80		74.0	145.3		91.4	163.9
0/100	---	100	---	---	100	---

Two conventional methods are usually used to show the dependence of Tg on the composition of the miscible blend systems and compare with the experimental values. (a) The simple equation:

$$Tg = w_1 Tg_1 + w_2 Tg_2 \tag{1}$$

where w_i is the weight fraction of the homopolymer i with Tg_i (29).
(b) The Couchman relation:

$$\ln Tg = \frac{w_1 \Delta Cp_1 \ln Tg_1 + w_2 \Delta Cp_2 \ln Tg_2}{w_1 \Delta Cp_1 + w_2 \Delta Cp_2} \tag{2}$$

where ΔCpi is the difference in molar heat capacity at the corresponding Tgi (30). As Table III and Figure 2 show, the Tg values detected are nearly consistent with the calculated values. Together with the evidence of the optical clarity, the blends contain less than 50% of PCPP are considered to be compatible. This implies that there is a certain affinity between PCPP and PS molecules in these blends.

Table III. Comparison between experimental and calculated values of PCPP/PS blend.

PCPP/PS	$Tg_{exp.}(^0C)$		$Tg_{calc.}(^0C)$		pcpp-rich[c]		ps-rich[c]	
	$Tg_{1,pcpp}$	$Tg_{2,ps}$	a	b	w'$_1$	w'$_2$	w"$_1$	w"$_2$
100/0	13.0	---	13.0	13.0				
80/20	14.6	98.0	30.4	27.7	.9799	.0201	.0465	.9535
60/40	13.5	71.0	47.8	45.2	.9861	.0139	.5085	.4915
50/50		61.2	56.5	54.4				
40/60		68.6	65.2	63.8				
20/80		74.0	82.6	82.2				
0/100	---	100	100	100				

a. Calculated by eq.(1)
b. Calculated by eq.(2)
c. All w's calculated from eq.(3), the values of ΔCp were 0.089 and 0.21 for PCPP and PS, respectively.

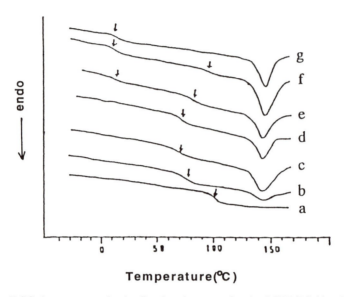

Fig 1. DSC thermograms in the first heating scan for the PCPP/PS blends.
PCPP/PS wt%: (a) 0/100 (b) 20/80 (c) 40/60 (d) 50/50 (e) 60/40 (f) 80/20
(g) 100/0

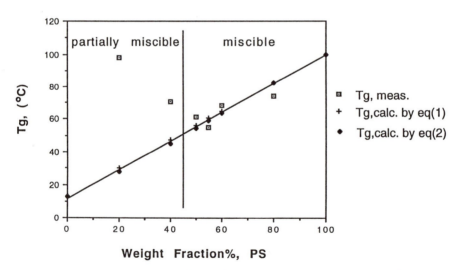

Fig 2. Experimental and calculated glass transition temperature (Tg) for
PCPP/PS blends.

However, as shown in Figure 1, when the PCPP content increases to 60% or higher, the blends exhibits two Tg's. The locations of the two Tg's suggest the coexistence of a PCPP-rich phase and a PS-rich phase in the blend showing partially compatibility for these blends. For these blends Eq(2) may be rearranged to

$$w'_1 = \frac{\Delta Cp_2(\ln Tg_{1,b} - \ln Tg_2)}{\Delta Cp_1(\ln Tg_1 - \ln Tg_{1,b}) + \Delta Cp_2(\ln Tg_{1,b} - \ln Tg_2)} \tag{3}$$

where w'_1 is the apparent weight fraction of polymer 1 in the polymer 1-rich phase, and $Tg_{1,b}$ is the observed Tg of polymer 1 in the blend. According to the DSC results and eq(3), the apparent weight fraction of PCPP and PS in the PCPP-rich phase and in the PS-rich phase are calculated and listed in Table III. It reveals that the PCPP dissolved more in the PS phase than did the PS in the PCPP phase. This may be ascribed to the ease of formation of the crystals of PCPP.

As we know, in addition to the glass transition at Tg, which exhibits the thermal behavior of the amorphous part of the polymer, the semicrystalline polyphosphazenes like PCPP also undergo the thermotropic transition from crystal to mesophase at the so-called T(1) temperature for the crystalline part. The location and the magnitude of T(1) depend on how easy the folded morphology of the monoclinic form is unfolded during the first DSC run, i.e. the temperature and size of the thermotropic transition is dependent on the thermal history of the sample. Therefore, all of the thermograms in this study were obtained in the same manner in order to eliminate the effect of the thermal history on T(1) values. From Figure 1, it is seen that T(1) of the blend is shifted downward with increase of PS content up to 40% and gradually shifted back to 148 ⁰C , the T(1) of PCPP. This implies that similar to the amorphous part, the phase transition of the crystalline part of PCPP is also influenced by the presence of PS molecules indicating the existence of an interaction between PS and PCPP in molecular level, which makes the unfolding the monoclinic form a little bit easier.

The results of the second heating scan are shown in Figure 3. Obviously, after first heating through T(1), the Tg's of all the blends are shifted close to the glass transition temperature of the pure PS polymer and the thermotropic transition temperatures, T(1)'s, are also shifted close to 170 ⁰C, the T(1) of pure PCPP. On account of this, we can conclude that the compatibility of PCPP/PS blend is unstable on heating and a phase separation was occured during the heating process. This may be ascribed to the fact that the extending and the repacking of the crystalline part occurred in the T(1) transition of PCPP, which was found in the phase transition of halogenated phenoxypolyphosphazenes by Magill et.al.(31). The weak affinity between PS and PCPP was destroyed when the thermotropic transition was proceeded during heating through T(1).

As can be seen in Figure 4, the morphology of PCPP/PS blend determined by SEM is changed with the composition. Figure 4-a shows almost a homogeneous dispersion of PCPP in PS matrix in the 20/80 wt% of PCPP/PS blend. For the 40/60 blend, PCPP is still well dispersed in the PS matrix with about 1μm of the diameter of the spherical domain as shown in Figure 4-b. A regular two-phase structure with a periodic distance of a few micrometer and a phase connectivity are found in Figure4-c for the 50/50 blend, indicating an early stage of phase separation. From Figure 4-d, we can see that phase separation between the PS-rich phase and the PCPP-rich phase is more pronounced in the case of 60/40 blend. This confirms that PS and PCPP are compatible up to about 50% of PCPP content and is only partially compatible in 60/40 blend.

In addition to the appearence of the optical clarity, the DSC thermograms and

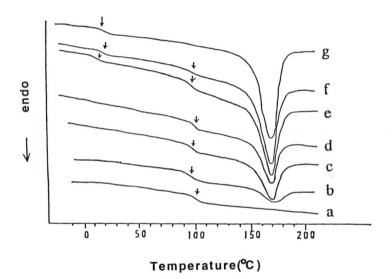

Fig 3. DSC thermograms in the second heating scan for the PCPP/PS blends PCPP/PS wt%: (a) 0/100 (b) 20/80 (c) 40/60 (d) 50/50 (e) 60/40 (f) 80/20 (g) 100/0

Fig 4. SEM micrographys of PCPP/PS blends with various contents.
PCPP/PS wt%: (a) 20/80 (b) 40/60 (c) 50/50 (d) 60/40

Continued on next page

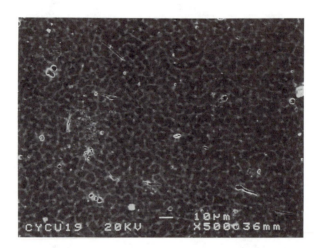

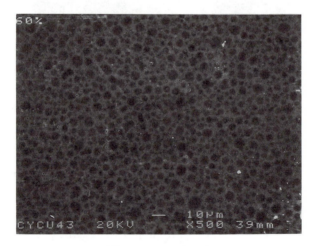

Figure 4. *Continued*

the morphology by SEM micrographs, composition dependent frequency shifts and band broadening of FTIR spectrum has been used to probe the possible interactions or conformation changes in the polymer blends (32).

Figure 5 shows the FT-IR spectra in the range 650-1600 cm^{-1} of pure PS, pure PCPP, and the series of the PCPP/PS blends with various composition. Comparing the spectrum of pure PCPP and that of PS, it is reasonable to choose the range of 1350-800 cm^{-1} to study the change of PCPP in the blends, while the band at 696 cm^{-1} was chosen to examine the change of PS in the blends. As shown in the figure, except the 80/20 wt% PCPP/PS blend, the spectra of the blending system are significantly different from that of the spectra of pure PCPP and pure PS at certain characteristic absorbances. We consider these spectra changes evidence for chemical interactions or conformational changes in the blends.

In order to further examine the differences of the FTIR spectra observed, the digital absorbance subtraction technique to obtain difference spectrum described by Garcia (33) was used in the analysis. Figure 6 shows the infrared spectra of pure PCPP and the difference spectrum obtained by subtracting the PS spectrum from that of the various composition blends. It can be seen that the difference spectrum of 80/20 wt% PCPP/PS is almost identical to that of pure PCPP. This suggests that the PCPP in this blend was unaffected by the presence of PS, i.e. no interaction between two polymers, and confirms the existence of phase separation. It is also found that the characteristic bands of PCPP are all narrower in the compatible blends implying that the vibrations are restricted owing to the presence of PS. In addition, considerable composition-dependent frequency shifts of the characteristic bands of PCPP are observed and listed in Table IV. As large as 12, 19 and 10 cm^{-1} shifts are found for the P-O-Ar stretching at 1186 cm^{-1} and 910 cm^{-1} and the C-H out of plane bending on Cl-phenoxy group at 822 cm^{-1}, respectively. These are evident for some small but distinct interactions and conformation changes involving the polymer backbone and the side groups of PCPP due to the presence of PS in the blends.

Table IV. FTIR band shifts of PCPP/PS blend - PS

PCPP/PS (w/w)	P-O-C stretching(cm^{-1}) (PCPP)		C-H out of plane bending of (C6H4Cl) (PCPP)	C-H out of plane bending of (C6H5) (PS)
100/0	1186	910	822	-
80/20	+0	+0	+0	0.0
60/40	+7	+13	+4	-3.0
50/50	+8	+14	+6	-2.3
40/60	+11	+17	+9	-1.5
20/80	+12	+19	+10	-1.0
0/100	-	-	-	696.0

Figure 7 shows the infrared spectra of pure PS and the difference spectra obtained by subtracting the pure PCPP spectrum from that of the various composition blends in a range 650-750 cm^{-1}. Due to the contribution involved by PCPP in this region, a direct correlation between the structural change and the spectrum difference is difficult to ascertain. Nevertheless, the frequency shifts and the change in band shape of the bands at 696 cm^{-1}, representing the C-H out of plane bending on benzene ring, do exsist in the blends. This indicates the existence of interaction between polymers. Yet, the perturbation of PS by the presence of PCPP is not as much as that of PCPP

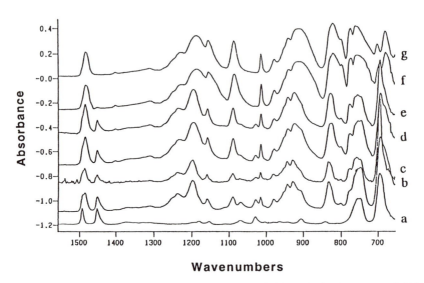

Fig 5. The FT-IR spectra of PS, PCPP and PCPP/PS blends in the range of 650 - 1600 cm^{-1}. PCPP/PS wt%: (a) 0/100 (b) 20/80 (c) 40/60 (d) 50/50 (e) 60/40 (f) 80/20 (g) 100/0

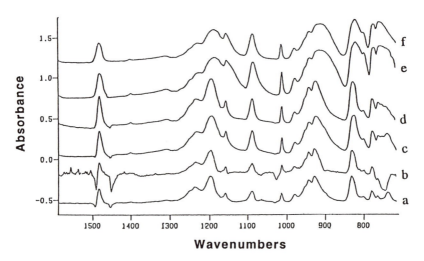

Fig 6. The difference spectra of PCPP/PS blends - PS in the range of 650 - 1600 cm^{-1} PCPP/PS wt%: (a) 20/80 (b) 40/60 (c) 50/50 (d) 60/40 (e) 80/20 (f) 100/0

by PS. This may also relate to the previous result, in which we found that PCPP dissolves more in the PS-rich phase than PS in the PCPP-rich phase.

Thermal Stability. The thermal stability and the degradation of polyblends were studied by TGA, which is a continuous process that involves the measurement of sample weight as the reaction temperature is changed by means of a programmed rate of heating. From Figure 8, it can be seen that with increasing the composition of PCPP, the temperature with 50% weight loss, (50%DT), is increased from about 400 to about 500 °C, the percentage residue (char yield) at 750 °C is increased from 0 to about 30%, and the decomposition rate is decreased considerably. It indicates that the thermal stability is increased with increasing the content of PCPP.

As listed in Table V, 50%DT is not linearly increased with the content of PCPP. It is raised significantly from pure PS to 20/80 wt% of PCPP/PS blend which is compatible, while it is dropped significantly from pure PCPP to 80/20 wt% of PCPP/PS blend which is almost completely incompatible. This reveals that the thermal stability of the blend is improved due to hybridization with PCPP, however, it is dependent on the miscibility between PCPP and PS rather than the amount of PCPP mixed. The char yield that remained at 750 °C, on the other hand, is almost linearly increased with the content of PCPP as shown in Figure 9. This is understandable since after heating to 750 °C no residues is left for PS as indicated in Table V.

Table V. The char yield, LOI and 50%DT of PCPP/PS blends

PCPP/PS	Char Yield(%)	LOI(%)	50%DT(°C)
100/0	32.3	41.2	508
80/20	27.9	34.6	490
60/40	18.4	27.8	452
50/50	17.2	25.9	450
40/20	12.0	24.6	443
20/80	4.3	----	431
0/100	0.0	17.0	399

Flammability. The flammabilities of PCPP/PS blends were measured in terms of the LOI value which describes the minimum volume fraction of oxygen in nitrogen necessary to sustain candle-like burning. The values of LOI obtained for PCPP/PS blends are given in Table V. It shows that the LOI of the blends increase linearly with PCPP weight fraction and almost proportionally to the char yield. This indicates that the flame retardant ability of PS was improved by blending with PCPP.

Conclusions

In this study, the optical clarities observed, the glass transition temperatures found in DSC thermograms and the morphologies from SEM micrographs all show that the solution casting blends of PCPP/PS are compatible when the content of PCPP is less than about 50%, while those with higher PCPP content are only partially compatible or incompatible. The compatibility of the blends is also influenced by heating through T(1). FTIR spectroscopies indicate a weak interaction between the component polymers in the blends. The TGA and the LOI measurements indicate that the thermal stability and the flame resistance of PS are improved by blending with PCPP.

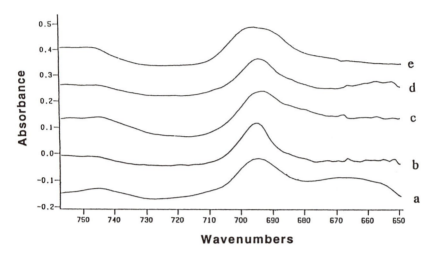

Fig 7. The difference spectra of PCPP/PS blends - PCPP in the range of 650 -
800 cm⁻¹ PCPP/PS wt%: (a) 20/80 (b) 40/60 (c) 50/50 (d) 60/40 (e) 0/100

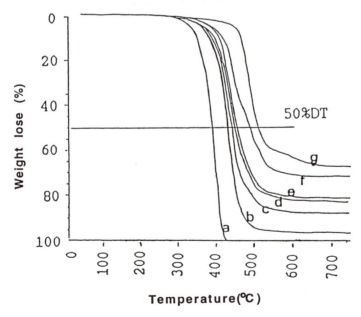

Fig 8. The TGA curves of the PCPP/PS blends.
PCPP/PS wt%: (a) 0/100 (b) 20/80 (c) 40/60 (d) 50/50 (e) 60/40 (f) 80/20
(g) 100/0

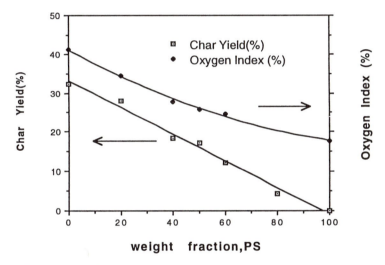

Fig 9. The effect of the blend composition on the LOI value and the char yield at 750°C the PCPP/PS blends.

Acknowledgments We thank the National Science Council of R.O.C. for the support of this work under project NSC 82-0208-M-033-018, and Nippon Fine Chemical Co. Japan, for providing the hexachlorocyclotriphosphazene.

Literature Cited

1. Allcock, H. R. *Phosphorus-Nitrogen Compounds* Academic Press, New York, **1972**.
2. Tate, D. P.; Antowiak, T. A. *Kirk-Othmer Encycl. Chem. Technol.* 3rd. Ed.. **1980**, *10*, 939.
3. Singler, R. W.; Hagnauer, G. L.; Sicka, R. W. In *Polymers for Fibers and Elastomers* Arthur, J.C., Ed.; *ACS Symposium Series,* No. 260. American Chemical Society, Washiongtonm, D.C., **1984**, p.143.
4. Allcock, H. R. *Chem. & Eng. News* **1985**, March 8, p.22.
5. Tate, D. P.; *Rubber World* **1975**, Sep.,41.
6. Wilson, A.; *J. Coated Fab.* **1978**, *7*, 233.
7. Lohr, D. F.; Beckman, J. A. *Rubber and Plastics News* **1982**, *16*
8. Penton, H. R.; In *Inorganic and Organometallic Polymers* Zeldin, M.; Wynne, K. J.; Allcock, H. R., Ed.; *ACS Symposium Series*, No. 360. American Chemical Society, Washiongtonm, D.C., **1988**, p.277.
9. Allcock, H. R.; Kugel, R. L. *J. Am. Chem. Soc.* **1965**, *87*, 4216.
10. Allcock, H. R.; Kugel, R. L. *Inorg. Chem.* **1966**, *5*, 1709.
11. Allcock, H. R.; Kugel, R. L. *Inorg. Chem.* **1966**, *5*, 1716.
12. Allcock, H. R. J. *Inorg. Organometallic Polym.* **1992**, *2*, 197.
13. Neilson, R. H.; Wisian-Neilson, P. *Chem. Rev.* **1988**, *88*, 541.
14. Wisian-Neilson, P.; Schaefer M. A. *Macromolecules* **1989**, *22*, 2003.
15. Gleria, M.; Bolognesi, A.; Porzio, W.; Catellani, M.; Destri, S.; Audisio, G. *Macromolecules* **1987**, *20*, 469.
16. Allcock, H. R.; Fitzpatrick, R. J.; Visscher, K. B. *Chem. Mater.* **1992**, *4*, 775.
17. Allen, C. W.; Shaw, J. C.; Brown, D. E. *Macromolecules*, **1988**, *21*, 2655.
18. Allen, C. W. *The Chemistry of Inorganic Ring System,* Steudel, R. Ed., Elsevier Science Publishers B. V., **1992**, Vol 14, pp 171-191.
19. Manson, J. A. Sperling, L. H. *Polymer Blends and Composites*, Plenum Press, New York and London, **1976**. p. 270.
20. Borisenkova, E. K.; Kulichikhin, V. G.; Tur, D. R.; Litvinov, I. A.; Plate, N. A.; *Otkrytiya Izobret* **1991**, *36*, 112. *Chemical Abstract*: **1992**, *117*, 70940q.
21. Yamamoto, Y.; Tonamachi, M.; Takahata, N. *Jpn. Kokai Tokyo Koho* Jp 04154852. *Chemical Abstracts*: **1992**, *117*, 173135k.
22. Landry, C. J. T.; Ferrar, W. T.; Teegarden, D. M. *Eur. Pat. Appl.* EP 485033. *Chemical Abstracts*: **1992**, *117*, 192999m.
23. Coltrain, B. K.; Ferrar, W. T.; Landry, C. J. T.; Molaire, T. R.; Zumbulyadis, N. *Chem. Mater.* **1992**, *4*, 358.
24. Visscher, K B.; Manners, I.; Allcock, H. R. *Macromolecules* **1990**, *23*, 4885.
25. Allcock, H. R.; Visscher, K. B.; Manners, I. *Chem. Mater.* **1992**, *4*, 1188.
26. Allcock, H. R.; Visscher, K. B. *Chem. Mater.* **1992**, *4*, 1182.
27. Reynard, K. A.; Gerber, A. H.; Rose, S H. Final report, Contract DAGG-46-71-C-0103, *AAMMRC* CTR 72-29, **1972**.
28. Nishio, Y.; Manley, R. J. *Macromolecules* **1988**, *21*, 1270.
29. Manson, J. A. Sperling, L. H. *Polymer Blends and Composites*, Plenum Press, New York and London, **1976**. p.290.
30. Couchman, P. R. *Macromolecules*, **1978**,*11*, 1156.
31. Young, S. G.; Kojima, M.; Magill, J. H.; Lin, F. T. *Polymer* **1992**, *33*, 3215.
32. Lu, F. J.; Benedetti, E.; Hsu, S. L. *Macromolecules*, **1983**, *16*, 1525.
33. Garcia, E. J. D. *J. Polym. Sci. Polym. Phys Ed.* **1984**, *22*, 107 .

RECEIVED May 2, 1994

Chapter 24

Polyphosphazene Random and Block Copolymers with Alkoxyalkoxy and Trifluoroethoxy Side Groups

Michael L. White and Krzysztof Matyjaszewski

Carnegie Mellon University, 4400 Fifth Avenue, Pittsburgh, PA 15213

The synthesis of N-silyl-phosphoranimines bearing mixed alkoxyalkoxy and trifluoroethoxy substituents is reported. The polymerization kinetics and the proposed anionic mechanism are also discussed. These monomers are utilized for the preparation of polyphosphazene random copolymers by the simultaneous polymerization of two phosphoranimines. Block copolymers have also been synthesized by addition of a second phosphoranimine after conversion of the first. Evidence for copolymer formation includes ^{1}H and ^{31}P NMR, SEC, solubility and DSC data. The differences between analogous random and block copolymers are discussed.

Over the past fifteen years, a number of phosphoranimines have been prepared as monomers for the synthesis of polyphosphazenes. (1-3) Polymerization of these compounds has allowed access to new materials that had not been previously prepared by the traditional Allcock ring opening polymerization method. (4) An example was the use of dimethyl-2,2,2-trifluoroethoxy-N-trimethylsilylphosphoranimine for the synthesis of the first reported dialkyl substituted polyphosphazene in 1980. (2) However, the thermal process suffers from the limitation that 2-12 days at temperatures in the neighborhood of 200 °C are required to obtain complete conversion. It has recently been shown that a variety of nucleophilic and anionic species will accelerate the polymerization of tris-(2,2,2-trifluoroethoxy)-N-trimethylsilylphosphoranimine. (5-7) The conversion of this monomer is complete after 4 hours at ≈ 100 °C using tetra-(n butyl)ammonium fluoride as an initiator/catalyst. Additional advantages to this method include the ability to control the molecular weight of the resulting polymers and the possibility of producing N-functionalized polyphosphazenes by the use of chain terminating phosphoranimines. (8)

In order to study the generality of this method, it was desirable to synthesize a number of other phosphoranimines. Additionally, it was of interest to utilize these monomers to produce random copolymers by crosspropagation and block copolymers by the addition of the second monomer after conversion of the first so that the properties of these polymers could be systematically controlled. Although random copolymers have been synthesized by the thermal polymerization of phosphoranimines (1) and by the ring opening polymerization method, (4) polyphosphazene AB block copolymers have not been previously reported.

The Synthesis and Homopolymerization of Mixed Phosphoranimines

Monomer Synthesis. Phosphoranimines bearing various combinations of 2-methoxyethoxy, 2-(2-methoxyethoxy)ethoxy and 2,2,2-trifluoroethoxy groups have been synthesized by the Staudinger reaction (*9*) between azidotrimethylsilane and the appropriately substituted phosphites. (*11-14*) Phosphites can be prepared via the reaction of the corresponding sodium alkoxide with tris-(2,2,2-trifluoroethyl)phosphite or from PCl_3 and the appropriate alcohols in the presence of $(CH_3CH_2)_3N$.

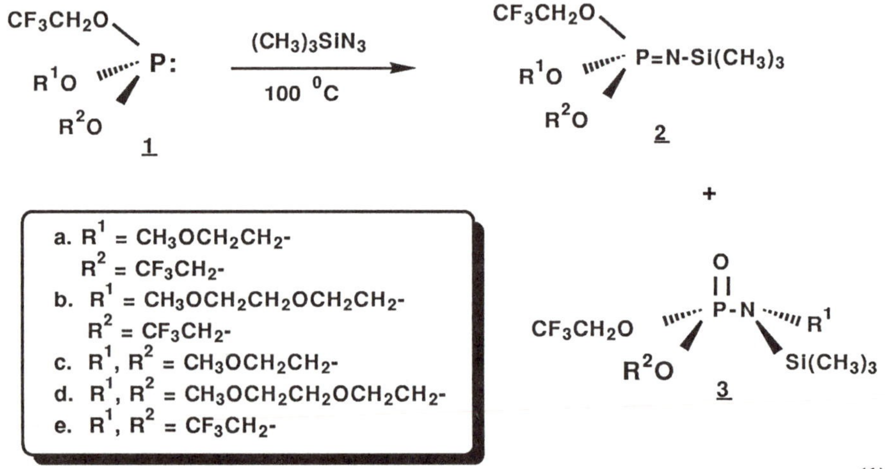
CF3CH2O
R1O
P:
R2O
1
(CH3)3SiN3
100 °C
CF3CH2O
R1O
P=N-Si(CH3)3
R2O
2
+
a. R1 = CH3OCH2CH2-
R2 = CF3CH2-
b. R1 = CH3OCH2CH2OCH2CH2-
R2 = CF3CH2-
c. R1, R2 = CH3OCH2CH2-
d. R1, R2 = CH3OCH2CH2OCH2CH2-
e. R1, R2 = CF3CH2-
O
P-N
CF3CH2O
R1
R2O
Si(CH3)3
3

(1)

In addition to the desired phosphoranimine, phosphoramidate isomer is also obtained as a product of this reaction. The phosphoramidate yield varies from 8% **3a** when phosphite **1a** is utilized to 32% **3d** with **1d** and is apparently dependent upon inductive effects in which electron withdrawing substituents tend to minimize the formation of isomer. Isomer **3e** has not been observed.

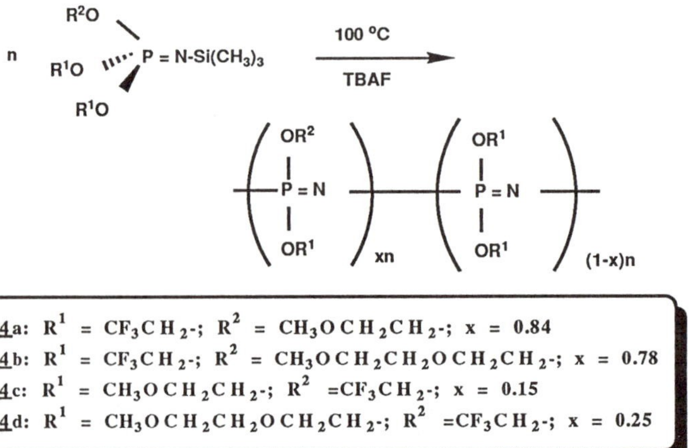
n
R2O
R1O
P = N-Si(CH3)3
R1O
100 °C
TBAF
OR2
P = N
OR1
xn
OR1
P = N
OR1
(1-x)n
4a: R1 = CF3CH2-; R2 = CH3OCH2CH2-; x = 0.84
4b: R1 = CF3CH2-; R2 = CH3OCH2CH2OCH2CH2-; x = 0.78
4c: R1 = CH3OCH2CH2-; R2 =CF3CH2-; x = 0.15
4d: R1 = CH3OCH2CH2OCH2CH2-; R2 =CF3CH2-; x = 0.25

(2)

Polymerization of Mixed Phosphoranimines. Although numerous anionic and nucleophilic reagents will accelerate the polymerization of phosphoranimines, tetrabutylammonium fluoride (TBAF) has proven to be particularly effective due to the "silylphilic" nature of the fluoride anion. Both 2,2,2-trifluoroethoxy and alkoxyalkoxy moieties can act as leaving groups in the polymerization of mixed phosphoranimines **2a-2d** (although the former is favored). *(14)* This leads to polymers with more than one repeating unit in the backbone.

Kinetics of Polymerization. The kinetics of polymerization have been extensively investigated for phosphoranimines **2a**, **2b** *(14)* and **2e** *(6)* and it has been shown that the following equation is valid:

$$R_p = -\frac{d[M]}{dt} = k_p^{app} [M][I]_o^n, \qquad (3)$$

where [M] is the monomer concentration at a given time, k_p^{app} is the apparent rate constant calculated based on the assumption that the initiator quantitatively produces active sites and $[I]_o$ is the initial initiator concentration. The polymerization of **2a-2e** are first order in monomer and n, the order in initiator, is 1 for the bulk polymerization of **2a** . *(14,15)* Typical kinetic plots are shown in Figure 1.

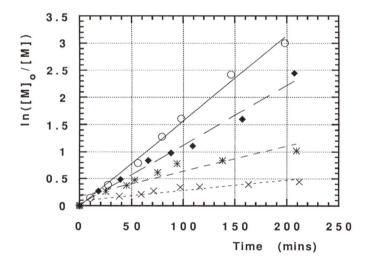

Figure 1. Kinetics of the polymerization of **2a** with various concentrations of TBAF at 130 °C. $[I]_o$ = (O) 2.50×10^{-2} M, ($\blacklozenge$)1.43×10^{-2} M, (*) 9.10×10^{-3} M, (X) 3.2×10^{-3} M.

Arrhenius plots (Figure 2) indicate that $\Delta S^{\ddagger} = -157$ J/mol °K and $\Delta H^{\ddagger} = 47.4$ kJ/mol for the bulk polymerization of **2a** using 1% TBAF. The low value obtained for the entropy of activation may indicate that the polymerization involves a transition state in which relatively few components are involved.

There are significant solvent effects for the initiated polymerization of phosphoranimines. The rates of polymerization for **2a** in bulk are 400 times higher than in toluene using N-methylimidazole (N-MI) as a catalyst and the TBAF and N-MI initiated polymerization rates are greatly enhanced in DMSO. *(14)* This is evidence for ionic intermediates. In addition, the presence of water in the reaction affects the polymerization rate. It has been demonstrated that water acts as both an accelerator and a chain transfer agent in the polymerization of phosphoranimines. (*16*)

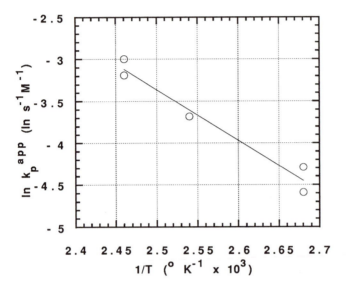

Figure 2. Arrhenius plots for **2a** using [M] = 2.32 mol/L and [I] = 0.023 mol/L.

Mechanism. The initiation mechanism is thought to involve the abstraction of the phosphoranimine silyl group by a "silylphilic" initiator followed by the nucleophilic attack of the resulting "phosphazene anion" on another monomer resulting in a phosphazene dimer.

$$(5)$$

Evidence for this initiation mechanism includes GC/MS identification of the volatile Me_3SiF by-product of the reaction. (*7*) Propagation may occur by the attack of the alkoxide obtained from the initiation of the reaction on the silyl group on the end of the growing chain. The true nature of the active site is not known.

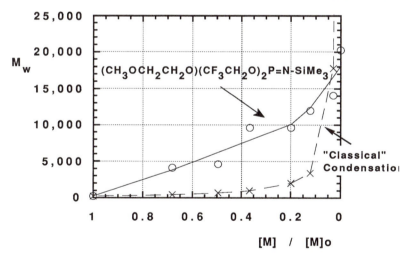

$$(6)$$

Molecular weight vs. conversion curves as followed by GPC vs. polystyrene standards indicate that relatively high molecular weights are obtained at low conversions ($M_n \approx$ 20,000 for the polymerization of **2e**; 4000 for **2a**, 3500 for **2b**). This supports a chain growth mechanism. Figure 3 shows a comparison of the M_w vs. time curve for **2a** compared to the expected molecular weights calculated for a step growth process.

Figure 3. M_w vs. conversion curve for **2a** with 1% TBAF at 133 °C compared to the theorical curve calculated assuming a step growth process.

However, the molecular weights obtained in this initial chain growth stage are apparently limited by chain transfer to monomer which involves the attack of the active site on a monomer silyl group resulting in the temporary termination of one growing chain and the initiation of a new chain.

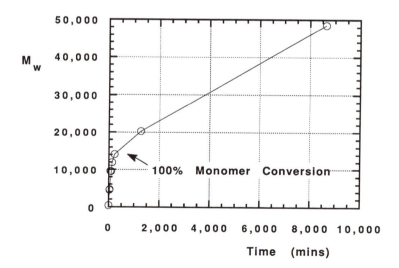

(7)

Alternatively, an alkoxide residue can attack a monomer molecule, thus starting a new chain. One possible explanation for the chain growth nature of this polymerization is that the N-Si bond on the monomer is stronger than the analogous bond on the growing chain. This theory is supported by MNDO calculations (7) and polymerization kinetics which show that the apparent propagation rate constants for **2e** are 6 times higher when polymerizing off of the end of an existing **3e** polymer chain than during homopolymerizion. (15) The molecular weights obtained during the chain growth phase vary depending upon the substituents. This is presumably because of inductive effects which alter the strength of the N-Si bond.

Another phenomenon which is noted is that the molecular weights of these materials continue to grow even after 100% monomer conversion as shown in Figure 4. (14,17)

Figure 4. M_W vs. time curve for the polymerization of **2a** with 1% TBAF at 133 °C.

This is attributed to a slow macrocondenstion process which involves the attack of a growing chain on the P-terminal end of an existing chain. Since reactive end groups are present on the polymer, this process continues after the monomer is consumed. (*15*)

$$\text{Macrocondensation} \qquad (8)$$

Polyphosphazene Random Copolymers

Synthesis. Random copolymers have also been produced by the simultaneous polymerization of various ratios of phosphoranimines **2a-d** with **2e** using TBAF at 133 °C for 13 hours. (*18*) These materials were purified by dissolution in THF, followed by precipitation from $CHCl_3$.

a. $R^1 = CF_3CH_2-$
     $R^2 = CH_3OCH_2CH_2-$

b. $R^1 = CF_3CH_2-$
     $R^2 = CH_3OCH_2CH_2OCH_2CH_2-$

c. $R^1 = CH_3OCH_2CH_2-$
     $R^2 = CH_3OCH_2CH_2-$

d. $R^1 = CH_3OCH_2CH_2OCH_2CH_2-$
     $R^2 = CH_3OCH_2CH_2OCH_2CH_2-$

$$(9)$$

The GPC, DSC, yield and the repeating unit ratio data are sumarized in Table 1.

Table I. Tabulated Results for the Random Copolymerization of Various Phosphoranimines

Monomer Feed Ratio	n	Yield	M_n (x 10^{-3})	M_w /M_n	T(1) (^{0}C)	ΔH (J/g)	T_i (^{0}C)
2a only	0.827	-	15.5	1.60	N/A	N/A	N/A
3 2a :7 2e	0.166	32.4%	44.9	1.41	54	7	155
2 2a : 8 2e	0.101	48.5%	43.9	1.55	57	20	161
1 2a : 9 2e	0.053	56.1%	59.6	1.59	64	33	187
1 2a : 19 E2e	0.019	82.3%	37.7	2.34	67	40	193
2e Only	0	85.5%	81.5	1.87	71	39	208
2b Only	0.995	-	8.1	1.56	N/A	N/A	
3 2b : 7 2e	0.118	32.0%	9.2	1.71	55	12	
2 2b : 8 2e	0.051	76.7%	18.7	1.61	59	28	152
1 2b : 9 2e	0.026	83.4%	17.5	1.45	68	37	163
1 2b : 19 2e	0.013	97.0%	17.3	1.40	68	38	166
3 2c : 7 2e	0.096	21.5%	8.0	1.35	71	23	173
2 2c : 8 2e	0.062	43.4%	9.7	1.36	75	21	180
1 2c : 9 2e	0.022	67.1%	12.8	1.38	73	27	190
1 2c : 19 2e	0.016	69.6%	16.5	1.39	74	36	192
17 2d : 83 2e	0.089	56.1%	16.6	1.30	46	5	
8 2d : 92 2e	0.025	56.1%	25.3	1.44	67	22	172
4 2d : 96 2e	0.007	56.1%	28.1	1,.58	64	28	166

All samples were copolymerized in bulk at 133 ^{0}C for 13 hours. Yields were determined gravimetrically, molecular weights by SEC vs. polystyrene standards and thermal properties via DSC. n = the fraction of the pendant side groups which are alkoxyethoxy moieties.

There is a body of evidence that suggests that copolymers are being produced in lieu of two homopolymers. First of all, the materials formed from phosporanimines **2a-d** bearing alkoxyalkoxy groups are soluble in $CHCl_3$, whereas poly(bis-trifluoroethoxyphosphazene) is not. The copolymers **4a-d** are insoluble in chloroform until 30% of the repeating units bear alkoxyalkoxy groups. ^{1}H NMR shows the presence of alkoxyalkoxy groups in the the precipitated materials. In order to prove that two homopolymers are not being prepared and then coprecipitated, equal masses of polymers **2a** and **2e** were dissolved in THF and precipitated from $CHCl_3$. No 2-methoxyethoxy groups were detected by ^{1}H NMR. Other data includes in situ ^{31}P NMR studies which show simultaneous monomer consumption and GPC traces which are monomodal and have molecular weights which are in between the molecular weights obtained from independently homopolymerizing the two phosphoranimines. Typical GPC traces are shown in Figure 5. Intermediate molecular weights are expected due to chain transfer to monomer. The refractive index polarity difference in the figure above is due to the concentration of trifluoroethoxy side groups.

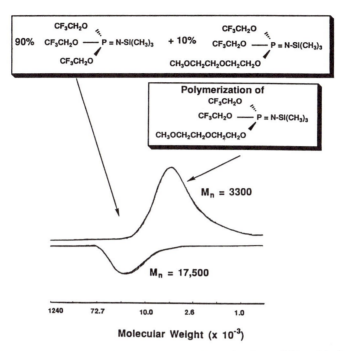

Figure 5. GPC traces from the random copolymerization of **2b** with **2e** and the homopolymerization of **2b**.

Another notable observation regarding these random copolymers is that the thermal properties vary with the repeating unit ratio. As in the case of poly(bis-trifluoroethoxyphosphazene), (*19*) these copolymers exhibit a T_g, a transition to a hexagonal columnar mesophase (T(1)) and an isotropization temperature (T_i). Polymers **3a-d** are completely amorphous and exibit only T_g. The DSC traces of the T(1) transitions for several random copolymers is displayed in Figure 6.

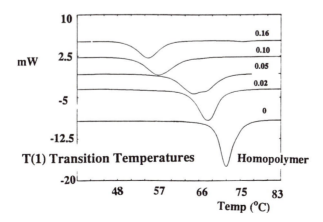

Figure 6. DSC data on the random copolymers produced from $(CF_3CH_2O)_3P{=}NSiMe_3$ and $(CH_3OCH_2CH_2O)(CF_3CH_2O)_2P{=}NSiMe_3$.

In addition to molecular weight effects, the T(1) transition temperatures and the enthalpies of those transitions depend on the percentage of alkoxyalkoxy functional groups pendant to the backbone. When approximately 30% of the repeating units bear alkoxyalkoxy groups, the random copolymers become completely amorphous and show only a T_g transition. As in the trend noted for poly(bis(halophenoxy)phosphazenes), (20) the T(1) transition temperature is higher for polymers at a given M_n for polymers bearing bulkier 2-(2-methoxyethoxy)ethoxy) groups than for those with 2-methoxyethoxy groups.

Polyphosphazene Block Copolymers

It is also feasible to synthesize polyphosphazene block copolymers by the addition of a second phosphoranimine after complete conversion of the first. This is possible due to the existence of silyl end groups present on the polymer chain. These moieties can be spectroscopically observed by ^{1}H NMR. (15) Molecular weight calculations based on the integration of these signals gives good agreement (to within about 20%) with the GPC data. Block copolymers have been prepared by the addition of monomer **2e** after the conversion of **2a** and **2b**. Conversion was determined by ^{31}P NMR. The tabulated data for these polymers can be viewed in Table II.

$$6a. \; R = CH_3OCH_2CH_2-$$
$$6b. \; R = CH_3OCH_2CH_2OCH_2CH_2-$$

(10)

End capping studies in which the N terminal end of the polymer chain is blocked with non-labile t-butyl groups before the addition of the second monomer have shown that the majority of the chain growth occurs from the silyl end group as opposed a macrocondensation or branching mechanism.

Table II. Block Copolymer data

Monomer Feed Ratio	n	Yield	M_n $(x\ 10^{-3})$	M_w/M_n	T(1) (^0C)	ΔH (mJ/mg)
1 **2a** : 19 **2e**	0.019	82.3%	37.7	2.34	67	35
45 **2a** : 55 **2e**	0.294	15.9%	28.8	1.54	69	21
36 **2a** : 64 **2e**	0.197	65.1%	35.1	1.52	71	21
22 **2a** : 78 **2e**	0.113	76.3%	36.2	1.51	75	32
8 **2a** : 92 **2e**	0.043	79.6%	43.0	1.44	73	40
5 **2a** : 95 **2e**	0.028	81.8%	41.2	1.44	74	40
22 **2b**: 78 **2e**	0.050	67.3%	22.7	1.56	66	32
12 **2b**: 88 **2e**	0.019	74.9%	21.1	1.54	67	32
6 **2b** : 94 **2e**	0.012	81.1%	22.7	1.67	71	40

Molecular weight data is determined by SEC vs. polystryrene standards and the thermal properties are quoted for the 2nd heating cycle.

As in the case of the random copolymers, there is considerable evidence that a block copolymer is being formed as opposed to two homopolymers. The block copolymers are insoluble in chloroform. Thus, if the two monomers were simply homopolymerizing, then the two homopolymers would be separated during the purification process. ^{1}H NMR of the precipitated polymer shows the presence of both types of repeating units. GPC traces show an increase in molecular weight after the addition of the second monomer and the refractive index becomes negative as a result of the inclusion of a greater number of trifluoroethoxy groups. All GPC traces are monomodal as in the case of the sample shown in Figure 7. This is additional evidence that copolymers are being formed in favor of two homopolymers.

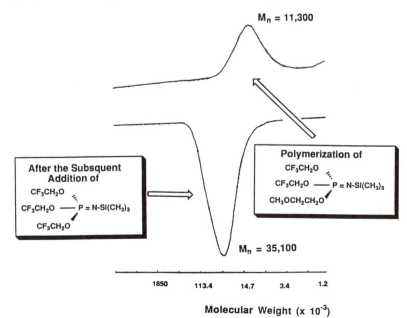

Figure 7. The GPC traces for a block copolymerization before and after the addition of **2e** to the products of the polymerization of **2a** at 133 °C using 1% TBAF.

As in the case of the random copolymers, the thermal properties of the block copolymers are dependent upon the repeating unit ratios as well as molecular weight. However, as shown in Figure 8, the T(1) transition temperatures of the block copolymers are higher than those for the random materials and do not decrease nearly as rapidly with molecular weight and repeating unit ratio in comparison to their random analogs. In fact, the random materials become completely amorphous when more than 25% of the alkoxyalkoxy bearing repeating unit are incorporated into the polymer backbone. The blocky polymers remain crystalline to much higher alkoxyalkoxy ratios.

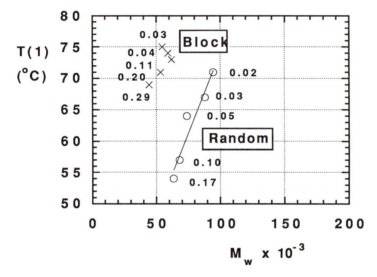

Figure 8. The effect of M_W on T(1) for the copolymers made from **1a** and **1d**. o = random copolymers; x = block copolymers. The numbers next to the data points represent the molecular weight fraction of the repeating units bearing 2-methoxyethoxy moieties as calculated from ^{1}H NMR.

One possible explanation for this behavior is that the block copolymers have relatively long runs of poly(bis-trifluoroethoxyphosphazene) which can orient into a hexagonal mesophase and therefore the systems remain fairly ordered and are not greatly affected by the length of the amorphous segment. In contrast, the random materials can tolerate a certain number of alkoxyalkoxy defects in the backbone, but, in general, the incorporation of disrupting side groups causes disorder in proportion to the number of those groups present and thus the ability of these polymers to order decreases rapidly.

In addition, the physical properties of the random and blocky materials are quite a bit different. The random copolymers vary from crystalline powdery polymers to putty-like materials, depending upon the percentage of repeating units bearing alkoxyalkoxy groups. The blocky materials form much tougher films at the same molecular weights and repeating unit ratios when cast from THF.

Conclusions

It has been demonstrated that it is possible to extend the anionically initiated polymerization of phosphoranimines to monomers bearing alkoxyalkoxy groups. These polymerization reactions obey first order kinetics in monomer and partial to first order kinetics in initiator, depending upon the initiating system and solvent. An anionic intermediate has been proposed to account for the kinetic data. Additionally, random and block copolymers have been synthesized between these monomers and tris(2,2,2-trifluoroethyl-N-trimethylsilylphosphoranimine). It is possible to control the mechanical and thermal properties of these materials by adjusting the monomer feed ratios and therefore the repeating unit ratios of the resulting polymers or by use of simultaneous or sequential polymerization techniques.

Acknowledgments. We would like to thank the US Army Research Office, the National Science Foundation via the Presidential Young Investigator Award, Eastman Kodak, PPG Industries and the Xerox Corporation for their financial support.

References

1. Neilson, R. H.; Wisian-Neilson, P. *Chem. Rev* . **1988**, *88* , 541.
2. Wisian-Neilson, P.; Neilson, R. H. *J. Am. Chem. Soc.* **1980**, *102*, 2848.
3. Flindt, E. P.; Rose, H. *Z. Anorg. Allg. Chem.* **1977**, *428*, 204.
4. Allcock,. H. R. *J. Inorg. and Organomet. Polym.* **1992**, *2*, 197.
5. Montague, R. A.; Matyjaszewski, K. *ACS Polymer Preprints* **1990**, *31* (2), 679.
6. Montague, R. A.; Matyjaszewski, K. *J. Am. Chem. Soc.* **1990**, *112*, 6721.
7. Matyjaszewski, K.; Cypryk, M.; Dauth,J.; Montague, R. A.; White, M. L. *Macromol. Chem., Macromol. Symp.* ,**54/55**, 13-30 (1992).
8. Montague, R. A.; Burkus, F. II.; Matyjaszewski, K. *ACS Polymer Preprints* **1992**, *33(2)*, 316.
9. Staudinger, H.; Meyer, J. *Helv. Chim. Acta* **1919**, *2*, 635.
10. Neilson, R. H.; Wisian-Neilson, P. *J. Macro. Sci, Chem.* **1981**, *A16*, 425.
11. Matyjaszewski, K.; Lindenberg, M. S.; White, M. L. *ACS Polym. Prepr.*, **1992**, <u>33(1)</u>, 305.
12. Matyjaszewski, K.; Lindenberg, M. S.; Moore, M. K.; White, M. L. *ACS Polym. Prepr.*, **1993**, <u>34(1)</u> .
13. Matyjaszewski, K.; Lindenberg, M. S.; Spearman, J. L.; White, M. L. *ACS Polym. Prepr.*, **1992**, <u>33(2)</u>, 176.
14. Matyjaszewski, K.; Lindenberg, M. S.; Moore, M. K.; White, M. L. *Journal of Polymer Sci, Polym. Chem. Ed.*, **1994**, *32(3)*, 465.
15. Matyjaszewski, K.; Moore, M. K.; White, M. L. *Macromolecules*, **1993**, *26*, 6741.
16. Montague, R. A.; Green, J.; Matyjaszewski, K. *ACS Polymer Preprints,* **1992**, *33*(2).
17. Matyjaszewski, K.; Franz, U.; Montague, R. A.; White, M. L. *Polymer*, in press.
18. Matyjaszewski, K.; Lindenberg, M. S.; Moore, M. K.; White, M. L. Kojima, M. *Journal of Inorg. and Organomet. Polymers*, **1993**, *3(4)*, 317.
19. Schneider, N. S.; Desper, C. K.; Singler, R. C.; *J. Appl. Polym. Sci.* **1976**, *20*, 3087.
20. Young, G. S.; Kojima, M.; Magill, J. H. *Polymer*, **1992**, *33(21)*, 4539.

RECEIVED June 20, 1994

Chapter 25

Photophysics and Photochemistry of Poly(methylphenylphosphazene) and Poly(methylphenylphosphazene)-*graft*-polystyrene Copolymers

C. E. Hoyle[1], D. Creed[2], P. Subramanian[2], I. B. Rufus[1], P. Chatterton[1], M. Bahadur[3], and P. Wisian-Neilson[3]

Departments of [1]Polymer Science and [2]Chemistry and Biochemistry, University of Southern Mississippi, Hattiesburg, MS 39406
[3]Department of Chemistry, Southern Methodist University, Dallas, TX 75275

The photophysical and photochemical behavior of poly(methylphenylphosphazene), PMPP, and poly(methylphenyl-phosphazene)-*graft*-polystyrene copolymers have been studied. In both types of polymers a broad red-shifted emission at about 430 nm is attributed to fluorescence from a singlet with a significant geometry change in the excited state relative to the ground state. UV irradiation of PMPP results in chain scission in air-saturated methylene chloride, whereas in the absence of molecular oxygen, a bimolecular process, leading to extensive crosslinking and gel formation, predominates. Biphenyl has been identified as a major photoproduct from PMPP both in the presence and absence of molecular oxygen. The degradation of PMPP is not inhibited by triplet state quenchers, and hence the photoinduced reactions of PMPP are proposed to originate from the singlet state.

Since 1965, when Allcock (*1*) first prepared linear, soluble poly(organophosphazenes), interest in inorganic polymers with a backbone of alternating phosphorus and nitrogen atoms has rapidly increased. Many such polymers with a variety of substituents at phosphorus have been prepared and have been found to be suitable for a variety of technological applications (*2-5*). Since poly(phosphazenes), in general, have high resistance to combustion (*6*), there has been considerable interest in commercial application of these materials as fire-resistant coatings and flame-retardant additives (*7*). Poly(phosphazenes) containing biologically and catalytically active side groups

have also been investigated (*8*).   The thermal stability of several
poly(phosphazenes) has been investigated both in solution and in the bulk phase
(*9-11*) owing to their potential use in a variety of situations requiring exposure
to higher temperatures.   Since they may also be used for applications requiring
exposure to UV radiation, a series of investigations on the photochemical
behavior of poly(organophosphazenes) has been carried out (*12-20*).   To date,
photochemical investigations have been performed on poly(phosphazenes)
bearing alkoxy/aryloxy or amino substituents on the phosphorus atom.   There
have been no reports of the photochemical behavior of poly(phosphazenes) with
an aliphatic or aromatic chromophore directly linked to the phosphazene
backbone.   In   this   article,   the   photochemical   behavior   of
poly(methylphenylphosphazene)   (PMPP)   and   poly(phosphazene)-*graft*-
polystyrene copolymers will be described.   It is also demonstrated that the
photophysics of arylphosphazenes is reflective of the attachment of the phenyl
ring directly to phosphorus resulting in a very unique red-shifted emission with
the characteristics of a severely distorted excited state geometry.

Experimental

All solvents were obtained from Burdick and Jackson and used as received. The
poly(phosphazenes) and the poly(phosphazene)-*graft*-polystyrene copolymers
were prepared and purified by the methods described earlier (*21,22*).   Polymer
film and solution irradiations were conducted with either a Rayonet RPR-100
photochemical reactor (Southern New England Ultraviolet) equipped with a full
complement of sixteen 300 nm lamps, a 254 nm Spectroline (model XX-15F)
low intensity mercury lamp, or a 450 W Canrad-Hanovia medium pressure
mercury lamp.   All irradiations were carried out at room temperature in the
presence of air or nitrogen, as appropriate.   A DuPont 9000 thermal analyzer
was used for the thermogravimetric analysis (TGA) and DSC scans.   DSC and
dynamic TGA scans were obtained using 10 °C/min temperature ramps.
During the TGA experiments a continuous flow of nitrogen gas was maintained
throughout the sample compartment.   Polymer molecular weights were
determined by size-exclusion chromatography (SEC) on a Waters Associates
GPC II instrument with a Waters Maxima data handling system using 500-, 10^4-,
10^5- and 10^6-Å μStyragel columns in series.   The SEC operating conditions
consisted of a mobile phase of THF containing 0.1% (n-Bu)$_4$N$^+$Br$^-$, a flow rate
of 1.5 mℓ/min, a temperature of 30 °C and a sample size of 0.05 mℓ of 0.1%
solution.   The system was calibrated with a series of narrow molecular weight
polystyrene standards in the range of 10^3 - 10^6.   The products formed on
irradiation were identified by gas chromatography (Hewlett-Packard HP 5890A)
using a Hewlett-Packard HP-1 capillary column (crosslinked methyl silicone
gum, L = 25 m, ID = 0.32 mm) and a flame ionization detector.   Retention
times were compared with those of authentic samples.
    A Perkin-Elmer 1600 Fourier transform infrared (FT-IR) spectrometer
was used for recording infrared spectra.   UV spectra were obtained on a Perkin-
Elmer Lambda 6 UV-Vis spectrophotometer.   Corrected fluorescence spectra
were obtained on a SPEX Fluorolog-2 fluorimeter with 3.5 nm bandpass

excitation and emission slits. Phosphorescence spectra were recorded on a
SPEX Phosphorimeter model 1934D. Fluorescence decay lifetimes were
measured by a single-photon counting apparatus from Photochemical Research
Associates International Inc. which is interfaced with a Digital PDP-11
computer. Triplet quenching experiments were carried out with an Applied
Photophysics laser flash photolysis apparatus described elsewhere (23). For
quantum yield measurements, 254 nm band pass filter (Omega Optics, Inc.) was
used to isolate the specific line from a 450 W Canrad-Hanovia medium-pressure
mercury lamp. Samples were irradiated in a 1-cm path-length cell equipped with
a septum (Wilmad Glass), and the light intensities were determined by
ferrioxalate actinometry (24). Product formation was followed by comparison
of gas chromatography retention times with those of authentic samples as
reported elsewhere (25).

Results and Discussion

The polymers and model compound which are the subject of this paper are
given below along with appropriate acronyms for rapid reference. Before
discussing the photochemistry of PMPP and the corresponding graft copolymers,
it is first necessary to consider their photophysical properties. As will be
described, the excited state of PMPP is quite unusual in its own right and may
have implications for designing additional phosphorus centered polymers. The
photophysical results described in the next section are excerpted in large part
from a paper from our laboratory which details in greater depth the
photophysics of PMPP (26).

Photophysical Properties.
As the UV spectra in Figure 1 clearly shows, the basic phosphazene repeat unit
exhibits no absorbance, either for PMPP, poly(dimethylphosphazene) (PDMP),
or the methylphenyl-phosphazene cyclic oligomer (MPCO) model, at
wavelengths which might suggest any type of unusual extended conjugation. The
copolymers (Figure 2) clearly show an overlap between the phenyl groups in the
polystyrene grafts and the phosphazene backbone phenyl groups. This can be
best seen by comparison of the graft copolymer spectra (Figures 2a and 2b) to
those of the ungrafted parent polymer (PMPP in Figure 1a) and pure
polystyrene (Figure 2c). A more thorough description of the absorption of
PMPP and PDMP can be found in reference 26.

   The emission spectrum resulting from the basic methylphenylphosphazene
unit is markedly red-shifted as shown in Figure 3. The excitation spectra of
PMPP or MPCO are identical to the corresponding absorption spectra in
Figure 1. As discussed thoroughly in reference 26, this emission is attributed to
an excited state which has a distinct and quite unique geometry distortion
compared to the ground state. This assignment is based on the estimated small
(2-5 kcal mol^{-1}) energy difference in the singlet and triplet energies (determined

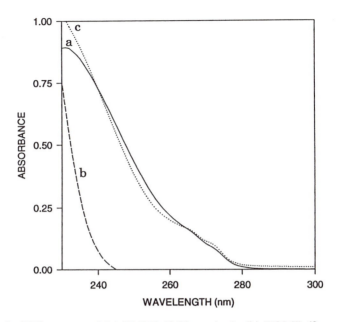

Figure 1. UV spectra of (a) PMPP (0.03 mg/mℓ), (b) PDMP (2 mg/mℓ), and (c) MPCO (0.03 mg/mℓ) in CH$_2$Cl$_2$.

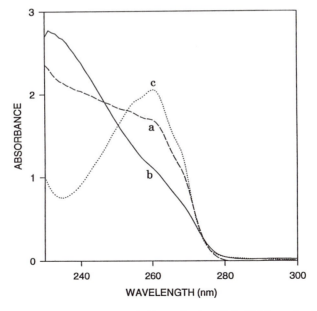

Figure 2. UV spectra of (a) **1** (0.02 mg/mℓ), (b) **2** (0.02 mg/mℓ), and (c) polystyrene (0.14 mg/mℓ) in CH$_2$Cl$_2$.

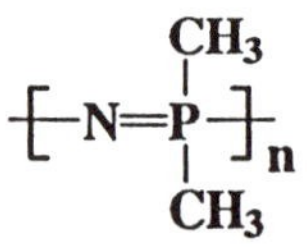

PDMP

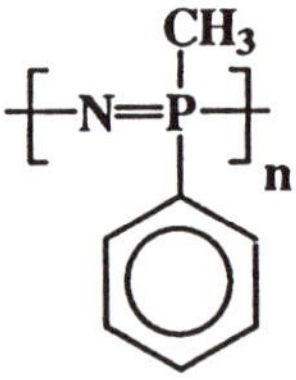

PMPP

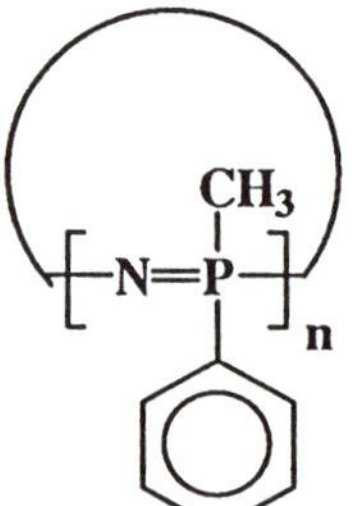

MPCO (n = 3,4)

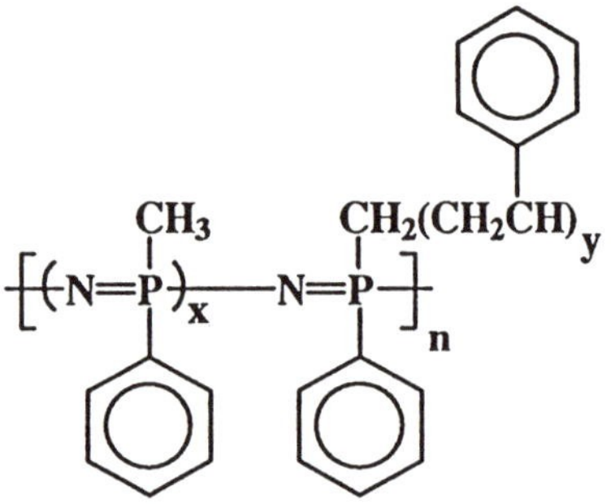

1: x = 9.0; y = 153
2: x = 6.3; y = 20

by fluorescence/phosphorescence spectral analysis), the lack of response of the fluorescence energy maximum to solvent polarity (this rules out emission from an excited charge-transfer state), and the large red-shift for onset of the fluorescence (compared to the absorption).   A full description of the "anomalous" emission of PMPP and a comparison of the similarities and differences between PMPP and triphenylphosphine (27) is provided in reference 26 and will not be elaborated upon further in this paper.   Here we simply mention that triphenylphosphine, like PMPP, also shows an "anomalous" red-shifted emission (see Figure 4) whose excitation spectrum matches its absorption spectrum.   Interestingly, the emission of triphenylphosphine is characterized by a marked solvent polarity dependence (arising from n,π^* character in the excited state) whereas, the position of the fluorescence of PMPP or MPCO is not affected to any great extent by solvent polarity.   This suggests that the emission of the phenylphosphazene unit results from a predominantly π,π^* singlet state, perhaps formed, as suggested in reference 26, by excitation of a π bonding electron from the phenyl ring to the π^* orbital of the delocalized three-center molecular orbital on the phosphazene backbone.   The poor overlap between these orbitals to a π^* phenyl orbital.   This, incidently, would account for the small singlet-triplet energy gap for PMPP.

The fluorescence spectra of the graft copolymers 1 and 2 show two distinct maxima when excited at 270 nm (Figure 5).   Accordingly, the fluorescence spectra of the graft copolymers are comprised of fluorescence from the polystyrene excimer with a maximum at 330 nm (the fluorescence spectrum of pure polystyrene is also shown in Figure 5 for comparison) and the fluorescence from the methylphenylphosphazene repeat unit with a maximum at 430 nm.   If the excitation spectrum for emission at 430 nm of graft copolymer 1 is compared to the excitation spectrum for the emission at 330 nm (Figure 6), it is evident that there is marked difference.   The emission at 330 nm can be attributed to absorption by the styrene repeat units only, since for excitation at wavelengths less than about 260 nm, where the absorption is dominated by the methylphenylphosphazene unit, there is relatively little emission at 330 nm attributable to the polystyrene excimer emission.   This suggests that excitation into the methylphenylphosphazene unit does not lead to energy transfer to the phenyl groups in the polystyrene side chain graft.   This is probably due to rapid vibrational relaxation leading to formation of the relaxed, low energy singlet state of the methylphenylphosphazene unit from whence energy transfer (energetically unfavorable) cannot occur.

Emission from the graft copolymer 1 at both 330 nm and 430 nm was quenched by carbon tetrachloride (CCl_4) in tetrahydrofuran solvent.   Linear Stern-Volmer plots of the fluorescence intensity *versus* CCl_4 concentration were obtained when the fluorescence was monitored at either 330 nm or 430 nm. The emission at 430 nm gave a Stern-Volmer quenching constant (K_{SV}) of 66 L mol^{-1}, while the emission at 330 nm gave a somewhat higher value (K_{SV} = 111 L mol^{-1}) indicating that the polystyrene graft fluorescence (around 330 nm) is

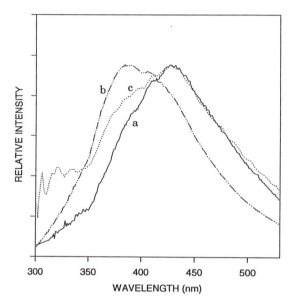

Figure 3.   Fluorescence spectra of (a) PMPP in CH$_2$Cl$_2$ (λ_{ex} 270 nm); (b) PMPP film (λ_{ex} 270 nm); and (c) MPCO (0.03 mg/mℓ) in CH$_2$Cl$_2$ (λ_{ex} 290 nm).

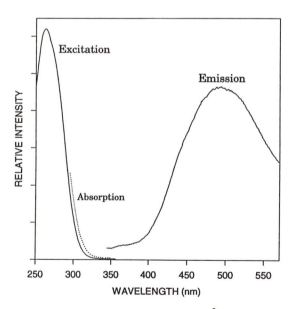

Figure 4.   Excitation and emission spectra of 10^{-3}M of triphenylphosphine in cyclohexane (λ_{ex} 290 nm; λ_{em} 460 nm).   The UV spectrum is shown for comparison with excitation spectrum.

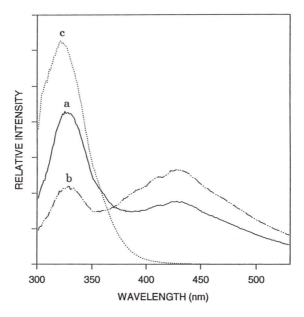

Figure 5. Fluorescence spectra of (a) **1** (0.02 mg/mℓ), (b) **2** (0.02 mg/mℓ), and (c) polystyrene (0.14 mg/mℓ) in CH$_2$Cl$_2$ (λ_{ex} 270 nm).

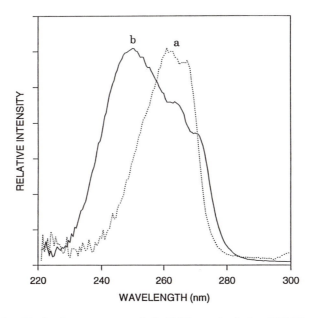

Figure 6. Excitation spectra of **1** (0.02 mg/mℓ) in CH$_2$Cl$_2$: (a) λ_{em} 330 nm; (b) λ_{em} 430 nm.

quenched somewhat more effectively than that of the poly(phosphazene) main chain. Nonetheless, these results demonstrate that both emissions are readily quenched at very rapid rates (since the emission is comprised of multiple components, exact fluorescence quenching constants could not be obtained). In the case of PMPP, the emission at 430 nm gave a K_{SV} value of 34 L mol^{-1}, which is half the value obtained for the graft polymer **1** at 430 nm. The fluorescence decay of PMPP is single exponential with a lifetime of 1.9 ns and from the K_{SV} value a bimolecular quenching rate constant of 1.8 x 10^{10} L mol^{-1} s^{-1}, typical of a dynamic quenching of a singlet excited state at a diffusion controlled rate, was obtained. In order to confirm the steady state quenching data, we have determined that the fluorescence decay lifetimes of the polymers (PMPP, **1**, and **2**) in the presence of CCl$_4$ were also quenched. Though the exact fluorescence lifetime quenching data could not be obtained due to the contribution of several decaying species in the graft copolymers as well as competitive absorption by CCl$_4$ at 270 nm, the quenching data indicates that both the 330 nm and 430 nm emissions are effectively quenched.

Photochemical Properties.
Since poly(phosphazenes) have potential use in outdoor environments, an investigation of their photochemical stability is imperative. The UV absorption spectra of PMPP in air-saturated methylene chloride before and after irradiation with 300 nm Rayonet lamps are shown in Figure 7. The absorption band around 240 - 260 nm, due to the perturbed phenyl groups on the phosphazene backbone, decreases as the irradiation is carried out, whereas the band around 270 nm increases. The decrease in 240 - 260 nm absorption indicates loss of phenyl groups attached to the -P=N- backbone of the polymer, while the new absorption band appearing around 270 nm is due to the photoproducts.

In order to define the impact of irradiation on film thermal properties, poly(phosphazene) films were photolyzed for periods up to 12 h with the unfiltered output of a medium pressure mercury lamp (polystyrene films were also photolyzed as a control experiment). Table I shows T_g and TGA onset and mid-point decomposition temperatures before and after irradiation of the films. Invariably, all of the polymers photolyzed, including polystyrene, exhibit measurable changes in glass transition(s) and onset/midpoint temperatures for thermal decomposition. As an example of other changes which occur during photolysis, we note that the decrease in the thermal stability of the PMPP polymer is also accompanied by a drop in the molecular weight after irradiation: As seen from the results in Table II, irradiation of PMPP in solution or film in the presence of oxygen (air) leads to a pronounced decrease in the polymer molecular weight as exemplified by the decrease in the peak maximum molecular weight (based on polystyrene standards). Interestingly, the polymer film, after irradiation in air, remained completely soluble in methylene chloride indicating that a very efficient chain cleavage process with little crosslinking occurs during the photodegradation of the PMPP polymer. In contrast, when irradiation of the film was conducted in a nitrogen saturated atmosphere, extensive crosslinking and gel formation were observed (Table II), i.e., the photolyzed film was completely insoluble in methylene chloride (the casting solvent) indicating that appreciable crosslinking takes place. Irradiation of the graft copolymers **1** and **2** in air-saturated THF solution leads to a decrease in the molecular weight indicating photodegradation, but to a lesser extent. Photolysis of the graft copolymers in the absence of oxygen was not conducted.

Table I.   Thermogravimetric Analysis and Glass Transitions of Polymers Before and After 12 h Irradiation.[a]

Polymer	T_g(°C)		TGA (onset, mid-point) (°C)	
	Before	After	Before	After
PMPP	42	41	424, 463	318, 350
1	42, 98	37, 90	412, 441	385, 430
2	46, 90	36, 90	415, 444	355, 434
Polystyrene	105	89	401, 434	385, 429

[a]450 W medium pressure mercury lamp.

Table II.   Peak Maximum Molecular Weight Changes of PMPP Before and After Irradiation[b]

Sample	Medium	Before Irradiation	After Irradiation
THF Solution	Air	73,300	45,300[c]
Film	Air	73,300	56,300[d]
Film	N_2	73,300	gel[e]

[b]254 nm Spectroline lamp; [c]3 h irradiation; [d]7 h irradiation; [e]insoluble

IR spectra of films irradiated in the presence and in the absence of molecular oxygen were also recorded. Figure 8 shows the IR spectra before and after irradiation (254 nm Spectroline lamp) of a thin film of PMPP in air for 2 h. A significant and uniform decrease in most of the infra-red bands occurs with increasing irradiation time: The bands at 1300-1296 cm^{-1} and 1195-1175 cm^{-1}, assigned to the P-N-P asymmetric vibration or a degenerate ring stretching mode (*29*), and the band around 875 cm^{-1} assigned to the P-N-P symmetric stretching, decrease upon irradiation, indicating that degradation of the polymer main chain occurs in accordance with the molecular weight changes

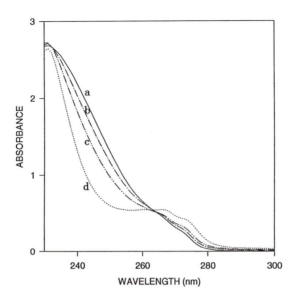

Figure 7.   UV spectra of PMPP (0.1 mg/mℓ) in air-saturated CH$_2$Cl$_2$ before and after irradiation using 300 nm Rayonet lamps: (a) 0 min; (b) 30 min; (c) 1 h; (d) 3 h.

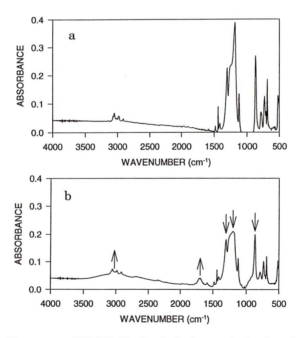

Figure 8. IR spectra of PMPP film in air, before and after irradiation with a 254 nm mercury lamp: (a) 0 h; (b) 2 h.

previously noted in Table II. An increase in intensity observed between 3400-2600 cm^{-1} is attributed to the formation of hydroperoxide groups on the surface of the polymer film: A similar observation was reported earlier (17) for poly[bis(4-isopropylphenoxy)phosphazene]. A new absorption around 1700 cm^{-1} is attributed to carbonyl groups formed on the film surface (17), probably derived by thermal and/or photochemical decomposition of hydroperoxides, since the irradiation is conducted in the presence of molecular oxygen. In the case of the graft copolymers, similar changes in the IR regions of methylphenylphosphazene units were noticed. In the presence of nitrogen, no new absorption was observed upon photolysis of PMPP, and only a small decrease in the P=N stretching (bands around 1300, 1187 and 875 cm^{-1}) occurred (Figure 9). As we will discuss below, when molecular oxygen is excluded, the photodegradation process of PMPP is quite different.

Irradiations were also carried out in the presence and absence of oxygen with the goal of identifying small molecule photoproducts. During the photolysis of PMPP, biphenyl (determined by gas chromatography) was found to be the major product, with benzene and phenol as minor products (it may be that these products are themselves photolyzed). Biphenyl can also be detected by fluorescence spectroscopy. The fluorescence spectra of PMPP in air-saturated methylene chloride recorded upon excitation at 270 nm, both before and after irradiation (300 nm Rayonet lamps), are shown in Figure 10. The disappearance of the red-shifted fluorescence around 430 nm and a gradual increase in the fluorescence intensity around 320 nm are indicative of the photodegradation process. The fluorescence spectra of a PMPP film (cast from methylene chloride) in air before and after irradiation are shown in Figure 11. As in the case of solution irradiation, the structured emission centered around 320 nm increases as the irradiation proceeds. For both irradiation in solution (Figure 10) and film (Figure 11) the 320 nm emission is inferred to be from biphenyl by comparison with an authentic sample: The characteristic vibrational fine structure provides strong evidence for biphenyl formation.

In order to provide a quantitative description of the product formation from the photolysis of PMPP, the quantum yield of biphenyl formation in methylene chloride was determined. Photolysis of PMPP in methylene chloride was therefore conducted with a medium-pressure mercury lamp using 254 nm band-pass filter. The light intensity was measured with ferrioxalate actinometry and the product yield was determined by gas chromatographic analysis of the samples photolyzed in methylene chloride. The quantum yield for biphenyl formation is *ca.* 5 x 10^{-3}. No change in the quantum yield for biphenyl was noticed in the presence and in absence of molecular oxygen. Since only a trace amounts of benzene and phenol were detected by gas chromatography, their formation quantum yields could not be determined accurately.

The effect of a triplet quencher on the formation of photoproducts was also investigated. The PMPP triplet, readily characterized by laser flash photolysis (26), has a triplet lifetime of 9.1 μs in nitrogen saturated methylene

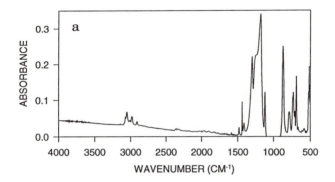

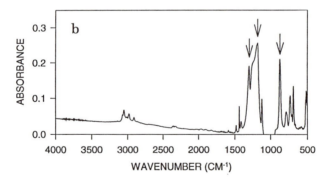

Figure 9.   IR spectra of PMPP film in nitrogen, before and after irradiation with a 254 nm mercury lamp: (a) 0 h; (b) 2 h.

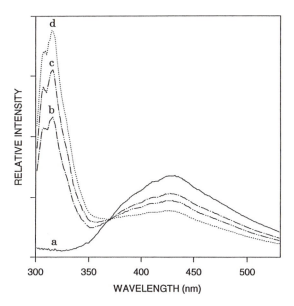

Figure 10. Fluorescence spectra of PMPP (0.1 mg/ml) in air-saturated CH$_2$Cl$_2$ (λ_{ex} 270 nm) before and after irradiation using 300 nm Rayonet lamps: (a) 0 min; (b) 30 min; (c) 45 min; (d) 1 h.

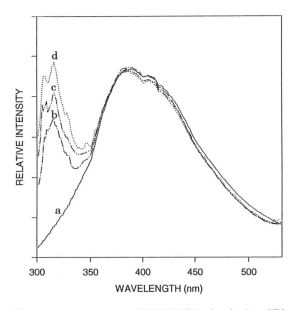

Figure 11. Fluorescence spectra of PMPP film in air (λ_{ex} 270 nm), before and after irradiation, using 300 nm Rayonet lamps: (a) 0 min; (b) 10 min; (c) 20 min; (d) 30 min.

chloride. Oxygen (air saturated solvent) and 1,3-cyclohexadiene quench the triplet of PMPP with bimolecular quenching rate constants of 1.8 x 10^9 L mol^{-1} s^{-1} and 5.6 x 10^9 L mol^{-1} s^{-1} respectively, as determined by laser flash photolysis (26). The formation of biphenyl upon irradiation of PMPP in solution was also monitored by both GC and fluorescence spectroscopy in the presence and absence of piperylene, a quencher with a triplet energy, E_T, of 59 kcal mol^{-1}. (For the product analysis, piperylene was employed as the triplet quencher since cyclohexadiene has a red extended absorption and undergoes photolysis when excited with either the 254 nm or the 300 nm lamp). The amount of biphenyl formed was followed by evaluation of the GC peak area. The build-up of the fluorescence intensity at 320 nm was also recorded. No change in the concentration of biphenyl was observed by either method upon irradiation in the absence or in the presence of piperylene (10^{-3}M) indicating that piperylene did not quench the formation of photoproduct. The lack of quenching of product formation, therefore, provides evidence that the lowest triplet state of PMPP is not involved in the photochemistry of PMPP. Hence, we conclude that the photoinduced reaction must proceed from an excited singlet state, a short-lived upper triplet state, or a hot ground state leading to the formation of biphenyl and other products.

In order to explain the photochemical behavior of PMPP in the presence and absence of oxygen, we propose a mechanism (Scheme I) similar to that proposed earlier (13,14,17-20) for photochemical reactions of other poly(phosphazenes) with alkoxy/aryloxy or amino substituents. We note that biphenyl formation has also been postulated by Allcock (20) for poly(bisphenoxyphosphazene) photolysis. According to Scheme I (part a), the primary act in the photolysis of the polymer is the homolytic scission of the P-C_6H_5 bond in the excited singlet state of PMPP to yield an aryl radical ($C_6H_5\cdot$) and a macroradical. We postulate that the direct dissociation of a -P=N- skeletal bond via a primary photochemical process does not occur, in accordance with previous reports (20). Apparently the -P=N- backbone has an unusual capacity to dissipate the absorbed energy by processes other than bond cleavage (20). The phenyl radical, once formed, can abstract a hydrogen atom, e.g., from the solvent, giving benzene, or it can react with oxygen to ultimately yield phenol, as shown in Scheme I, part b. In accordance with the mechanism in part b, we could identify trace amounts of benzene and phenol by GC. We envision that the formation of biphenyl can occur from two different pathways as depicted in part b of Scheme I: (a) phenyl radical phenyl radical combination and/or (b) reaction of the phenyl radical with a substituted phenyl group on the phosphazene chain forming the cyclohexadienyl type radical and thereby giving biphenyl. Indeed, cyclohexadienyl type radicals have been reported to occur from hydrogen abstraction of biphenyl (30,31). Although we did not detect a cyclohexadienyl type radical by laser flash photolysis (26), this may well be due to the rapid formation and loss, and/or low extinction coefficient, of the

Scheme I

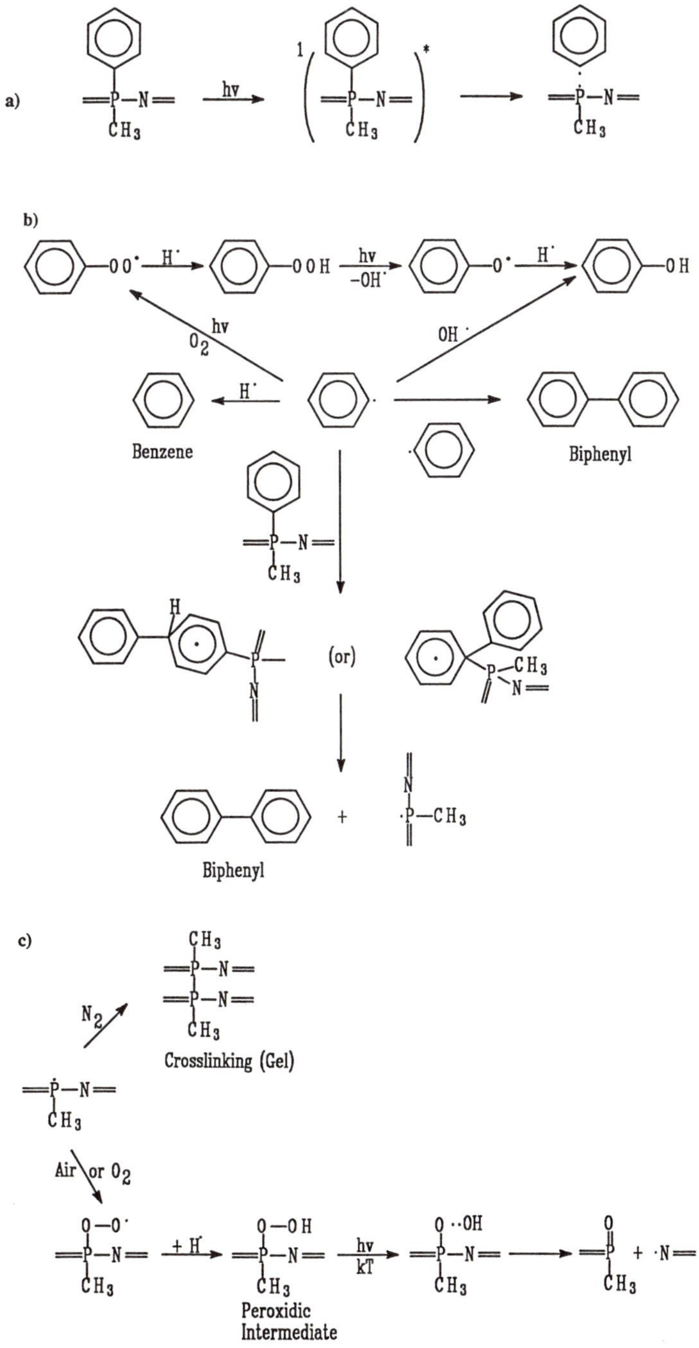
a)
hv
b)
H·
hv
-OH·
O2
OH·
Benzene
Biphenyl
(or)
Biphenyl
c)
N2
Crosslinking (Gel)
Air or O2
+ H·
hv
kT
Peroxidic
Intermediate

cyclohexadienyl radical making it impossible to detect within the timescale of our laser flash system (about 15 ns).

Finally, we project that the polymer macroradical may react by two pathways as depicted in part c of Scheme I, depending on the presence or absence of oxygen. If oxygen is present (in air-saturated solvent or environment), a peroxidic intermediate on phosphorus can form. The existence of this intermediate is supported by IR studies that show bands due to O-H stretching. Such bands do not appear in the absence of oxygen. It is quite reasonable that the breaking of the peroxidic O-OH bond induces chain scission of the polymer with a decrease in molecular weight in accordance with a similar mechanism reported by Gleria *et al.* (*13,16-19*). Unfortunately, the formation of P=O groups could not be confirmed by IR spectroscopy since the P=O group absorb in the same spectral region (1200-1300 cm^{-1}) as the asymmetric vibration of the P-N-P group (*32*). In the absence of oxygen, the formation of inter-chain linkages between two macroradicals, as shown in part c (Scheme I), is probably the dominant process leading to an increase in molecular weight and gel formation. The existence of P-P bonding in phosphazene systems is known (*33*) and the possible existence of such a linkage has been suggested by Gleria *et al.* (*19*) for other phosphazene systems.

Conclusions

In summary, we have reported the photophysical and photochemical behavior of poly(methylphenylphosphazene) and the corresponding polystyrene graft copolymers. The fluorescence of the methylphenylphosphazene chromophore is red-shifted with a maximum around 430 nm, which probably results from a large geometry change in the excited state. The electronic excited state energies of PMPP indicate a very small S_1 - T_1 energy gap and support the supposition of a large geometry change in the excited state (*26*). The emission of the graft copolymers is comprised of both the polystyrene excimer fluorescence and the inherent PMPP fluorescence. The singlet state of the PMPP and polystyrene graft copolymers are readily quenched by CCl_4. Quenching experiments indicate that photoreactions most probably occur from a first excited singlet state, a short-lived upper triplet state, or a hot ground state of PMPP. During the irradiation of PMPP, homolytic cleavage of the P-C_6H_5 bond generates biphenyl as the major product ($\phi_{biphenyl}$ = 5 x 10^{-3}) with only trace amounts of benzene and phenol being detected. The macroradical formed as a result of photolysis results in efficient crosslinking and gel formation in the absence of oxygen. In contrast, in the presence of molecular oxygen, degradation of PMPP results in chain-cleavage and a marked reduction in polymer molecular weight. This process results in a severe decrease in the thermal stability of PMPP.

Acknowledgements

This work was supported, in part, by The Office of Naval Research (Contract No. NOO14-89J-1048), U.S. Army Research Office, and The Robert A. Welch Foundation. The authors also acknowledge the financial support of the National Science Foundation (Grant No.EHR-9108767), the State of Mississippi, and the University of Southern Mississippi.

Literature Cited

(1) Allcock, H.R.; Kugel, R.L. *J. Am. Chem. Soc.* **1965**, *87*, 4216.
(2) Allcock, H.R. *Angew. Chem. Int. Ed. Engl.* **1977**, *16*, 147.
(3) Allcock, H.R. *Phosphorus-Nitrogen Compounds; Cyclic, Linear, and High Polymeric Systems*; Academic Press: New York, NY, 1972.
(4) Allcock, H.R.; Allen, R.W.; O'Brien, J.P. *J. Chem. Soc., Chem. Comm.* **1976**, 717.
(5) Allcock, H.R.; Allen, R.W.; O;Brien, J.P. *J. Am. Chem. Soc.* **1977**, *99*, 3984.
(6) Quinn, E.J.; Dieck, R.L. *J. Fire Flammability* **1976**, *7*, 358.
(7) Singler, R.E.; Schneider, N.S.; Hagnauer, G.L. *Polym. Eng. Sci.* **1975**, *15*, 321.
(8) Allcock, H.R. *ACS Symp. Ser.* **1983**, *49*, 232.
(9) Allcock, H.R.; Cook, W.J. *Macromolecules* **1974**, *7*, 284.
(10) Allcock, H.R.; Moore, G.Y.; Cook, W.J. *Macromolecules* **1974**, *7*, 571.
(11) Hagnauer, G.L.; LaLiberte, B.R. *J. Appl. Polym. Sci.* **1976**, *20*, 3073.
(12) Bortolus, P.; Minto, F.; Beggiato, G.; Lora, S. *J. Appl. Polym. Sci.* **1979**, *24*, 285.
(13) Gleria, M.; Minto, F.; Lora, S.; Bortolus, P. *Eur. Polym. J.* **1979**, *15*, 671.
(14) Gleria, M.; Minto, F.; Lora, S.; Bortolus, P.; Ballardini, R. *Macromolecules* **1981**, *14*, 687.
(15) Gleria, M.; Minto, F.; Lora, S.; Busulini, L.; Bortolus, P. *Macromolecules* **1986**, *19*, 574.
(16) Gleria, M.; Minto, F.; Flamigni, L.; Bortolus, P. *Macromolecules* **1987**, *20*, 1766.
(17) Gleria, M.; Minto, F.; Bortolus, P.; Porzio, W.; Bolognesi, A. *Eur. Polym. J.* **1989**, *25*, 1039.
(18) Gleria, M.; Minto, F.; Bortolus, P.; Porzio, W.; Meille, S.V. *Eur. Polym. J.* **1990**, *26*, 315.
(19) Bortolus, P.; Gleria, M.J. , *Inorg. Organomet. Polym.* **1994**, 4, 1.
(20) O'Brien, J.P.; Ferrar, W.T.; Allcock, H.R. *Macromolecules* **1979**, *12*, 108.
(21) Wisian-Neilson, P.; Neilson, R.H. *J. Am. Chem. Soc.* **1980**, *102*, 2848.
(22) Wisian-Neilson, P.; Schaefer, M.A. *Macromolecules* **1989**, *22*, 2003.
(23) Hoyle, C.E.; Anzures, E.T.; Subramanian, P.; Nagarajan, R.; Creed, D. *Macromolecules* **1992**, *25*, 6551.

(24) Murov, S.L. *Handbook of Photochemistry*; Marcel Dekker: New York, NY, 1973, p 119.

(25) Hoyle, C.E.; Creed, D.; Nagarajan, R.; Subramanian, P.; Anzures, E.T. *Polymer* **1992**, *33*, 3162.

(26) Hoyle, C.E.; Creed, D.; Subramanian, P.; Rufus, I.B.; Chatterton, P.; Bahadur, M.; Wisian-Neilson, P. *Macromolecules* (accepted for publication).

(27) Fife, D.J.; Morse, K.W.; Moore, W.M. *J. Photochem.* **1984**, *24*, 249.

(28) Mark, J.E.; Allcock, H.R.; West, R. *Inorganic Polymers*; Prentice Hall: Englewood Cliffs, NJ, 1992, p 86.

(29) Allcock, H.R. *Phosphorus-Nitrogen Compounds: Cyclic, Linear, and High Polymeric Systems*; Academic Press: New York, NY, 1972, p 51.

(30) Fuller, J.; Petelski, N; Ruppel, D. *J. Phys. Chem.* **1970**, *74*, 3066.

(31) Sawai, T; Hamill, W.H. *J. Phys. Chem.* **1969**, *73*, 2750.

(32) Singler, R.E.; Schneider, N.A.; Hagnauer, G.L. *Polym. Eng. Sci.* **1975**, *15*, 321.

(33) Allcock, H.R.; Connolly, Mar S.; Whittle, Robert R. *Organometalics*, **1983**, 2, 1514.

RECEIVED June 14, 1994

MAIN GROUP
ELEMENT POLYMERS

Chapter 26

Poly(alkyl/aryloxothiazenes), $[N=S(O)R]_n$

New Direction in Inorganic Polymers

Aroop K. Roy[1], Gary T. Burns, Stelian Grigoras, and George C. Lie[2]

Mail Stop 500, Dow Corning Corporation, Midland, MI 48640–0995

The thermally-induced condensation polymerization of appropriately substituted N-silylsulfonimidates and free sulfonimidates affords the new sulfur(VI)-nitrogen backbone inorganic polymers, poly(alkyl/aryloxothiazenes), $[N=S(O)R]_n$. A novel synthetic approach, involving a facile restructuring of sulfonamides without affecting the oxidation state of sulfur, was utilized in the preparation of the polymer precursor N-trimethylsilylsulfonimidates, $Me_3SiN=S(O)R^1(OR^2)$. These N-silylated sulfonimidates readily undergo desilylation with methanol to yield free sulfonimidates, $HN=S(O)R^1(OR^2)$. In some respects, the free sulfonimidates appear to be superior "monomers" for the synthesis of the title polymers. Both homopolymers and copolymers are accessible from the condensation polymerization. The first examples of this new class of polymers have been characterized by GPC, DSC, TGA, NMR, and elemental microanalysis. Theoretical calculations have also been performed on poly(methyloxothiazene), $[N=S(O)CH_3]_n$.

Over the past two-and-a-half decades, efforts to develop new inorganic backbone polymers have steadily intensified. Two primary reasons account for this rising interest. First, materials requirements for advancing technologies have precipitated the need to look beyond primarily carbon-based polymers. Second, unique characteristics of the commercially successful poly(siloxanes) and the tremendous versatility and significant commercial potential of the poly(phosphazenes) have heightened awareness of the potential of inorganic polymers as sources for new materials.

For precisely the first reason mentioned above, we became interested in developing a new polymer system based on an inorganic backbone. A literature search indicated that polymers with an alternating sulfur(VI)-nitrogen backbone, consisting of the repeat unit —N=S(O)X—, would be excellent candidates for exploratory research and development. Of several reasons supporting this choice, two were critical from a synthetic perspective. First, during the early 1960s polymers where X = F, NH_2 were

---

[1]Corresponding author
[2]Current address: 1014 Cumberland Court, Vernon Hills, IL 60061

0097–6156/94/0572–0344$08.00/0

patented and briefly described (*1-3*), and the synthesis of a polymer with X = Ph was inferred from condensation reactions (*4,5*). Second, the sulfur-nitrogen repeat unit bears strong structural resemblances to the phosphorus-nitrogen repeat unit in poly-(phosphazenes). These suggested the possibility of using guidelines from phosphazene chemistry for synthesizing the S-N polymers that we have chosen to call poly-(oxothiazenes), in analogy with poly(phosphazenes). For our studies, we decided to initially focus on the polymers $[N=S(O)R]_n$ **1** with alkyl and aryl substituents attached to sulfur through direct sulfur-carbon bonds.

Since polymers **1** would be structurally analogous to the poly(alkyl/arylphosphazenes), $[N=PR_2]_n$, which are synthesized by a condensation polymerization route (*6,7*), it appeared reasonable that an analogous polycondensation process might allow access to the S-N polymers **1**. Thus, we adopted a retrosynthetic approach for the synthesis of **1**. It was obvious that the polycondensation precursor would have to be

Polymer **1** ← polycondensation precursor ← suitable sulfur(VI) intermediates

a N-silylsulfonimidate **2**, since a 1,2-elimination of silyl ether (Me_3SiOR^2) was envisaged, analogous to that undergone by N-silylphosphoranimines in poly(phosphazene) synthesis. Further, since **2** is an ester (of sulfonimidic acid), a logical synthesis of **2** would be from the corresponding acid chloride **3** which are known as sulfonimidoyl chlorides. Hence, the first objective was to explore viable methods for the synthesis of N-silylsulfonimidoyl chlorides **3**.

$$Me_3SiN{=}\overset{\displaystyle O}{\underset{\displaystyle R^1}{\overset{\|}{S}}}{-}OR^2 \qquad\qquad Me_3SiN{=}\overset{\displaystyle O}{\underset{\displaystyle R^1}{\overset{\|}{S}}}{-}Cl$$

2　　　　　　　　　　**3**

Synthesis of Polymer Precursors

Development of a New Route to Sulfonimidoyl Halides. Even though sulfonimidoyl chlorides are the aza analogs of the well-known sulfonyl chlorides, RSO_2Cl, the former have received little attention as useful synthetic intermediates. We believe one reason for this has been the lack of convenient methods for their synthesis. The two primary routes to sulfonimidoyl chlorides (*8,9*), prior to our work, were based on sulfur(IV) starting materials, many of which are still not readily available or are unstable to storage or distillation. In addition, among the oxidizing agents used to convert sulfur(IV) to sulfur(VI) in these methods several are either not readily accessible or are themselves unstable compounds.

In order to circumvent the difficulties associated with the above syntheses, we decided to examine the possibility of synthesizing sulfonimidoyl chlorides from readily accessible sulfur(VI) compounds such as sulfonic acids, sulfonyl chlorides, and sulfonamides. For our purposes, of particular utility as starting materials would be N-silylated sulfonamides **4** and **5**, since either of these provides a significant portion of the framework needed in the polymer precursor **2**, as well as sulfur in the oxidation state VI. N-silylated sulfonamides are readily synthesized in good to near-quantitative yields from commercially available sulfonyl chlorides, sulfonic anhydrides or sulfonamides (*10-12*).

$$
\underset{\textbf{4}}{\underset{\displaystyle \overset{\displaystyle O}{\underset{\displaystyle O}{R^1-\overset{\|}{\underset{\|}{S}}-N}}\overset{/SiMe_3}{\diagdown SiMe_3}}
\qquad
\underset{\textbf{5}}{\underset{\displaystyle \overset{\displaystyle O}{\underset{\displaystyle O}{R^1-\overset{\|}{\underset{\|}{S}}-N}}\overset{/H}{\diagdown SiMe_3}}
$$

A major hurdle to surmount, however, was the conversion of **4** or **5** to **3**. Sulfone-type structures are known to be virtually indestructible without reduction at sulfur. Yet, removal of an oxygen from sulfur and restructuring to a sulfonimidoyl moiety was necessary to arrive at **3** without going through several redox steps. From bond energy considerations, and from known reactions between sulfur(VI) compounds (halides and acids) and phosphorus halides, it appeared that appropriate halophosphoranes might be useful for the transformation that we sought.

We have found that, indeed, under appropriate conditions dihalophosphoranes of the type Y_3PX_2 (Y = Ph, X = Cl, Br) can bring about an extremely facile transformation of N-silylated sulfonamides to N-silylsulfonimidoyl halides. The nature of the product(s) is dependent on the polarity of the solvent used, the steric bulk of the phosphorane, and on the electronic effect of the substituent on sulfur. If PCl_5 is used as the phosphorane with **4**, either a phosphoranylidene product **6** (equation 1) or a mixture of **6** and sulfonimidoyl chloride **3** is formed depending on whether the solvent is nonpolar CCl_4 or polar $CHCl_3$, respectively. With the bulkier Ph_3PCl_2 as the reagent, however, only sulfonimidoyl chloride is produced in $CHCl_3$ solution with **4** and, more surprisingly, also with **5** (equations 2, 3). The success of the reaction with the monosilylsulfonamides **5** is particularly important since these are much more readily prepared, in higher yields and in wider variety, than the disilylsulfonamides **4**. Additionally, the use of Ph_3PBr_2 with **4** (R^1 = Me) produced the first detectable sulfonimidoyl bromide, which as a class of compounds had only been postulated as intermediates prior to our work. We have reported the details of this new area of sulfur chemistry in recent publications and patents (*13-15*).

$$
\underset{\textbf{4a}}{\underset{\displaystyle \overset{\displaystyle O}{\underset{\displaystyle O}{CH_3-\overset{\|}{\underset{\|}{S}}-N}}\overset{/SiMe_3}{\diagdown SiMe_3}}
\quad
\xrightarrow[\text{- 2 Me}_3\text{SiCl}]{\text{PCl}_5 \text{ in CCl}_4}
\quad
\underset{\textbf{6a}}{\underset{\displaystyle \overset{\displaystyle O}{\underset{\displaystyle O}{CH_3-\overset{\|}{\underset{\|}{S}}-N}}{=PCl_3}}
\qquad (1)
$$

$$
\underset{\textbf{4a}}{\underset{\displaystyle \overset{\displaystyle O}{\underset{\displaystyle O}{CH_3-\overset{\|}{\underset{\|}{S}}-N}}\overset{/SiMe_3}{\diagdown SiMe_3}}
\quad
\xrightarrow[\substack{\text{- Me}_3\text{SiCl}\\ \text{- Ph}_3\text{P=O}}]{\text{Ph}_3\text{PCl}_2 \text{ in CHCl}_3}
\quad
\underset{\textbf{3a}}{Me_3SiN=\underset{\displaystyle CH_3}{\overset{\displaystyle O}{\overset{\|}{S}}}-Cl}
\qquad (2)
$$

$$R^1-\overset{\overset{O}{\|}}{\underset{\underset{O}{\|}}{S}}-N\overset{\diagup H}{\diagdown SiMe_3} \xrightarrow[\substack{-\ Et_3NH^+Cl^- \\ -\ Ph_3P=O}]{Et_3N\text{-}Ph_3PCl_2 \text{ in } CHCl_3} Me_3SiN=\overset{\overset{O}{\|}}{\underset{\underset{R^1}{|}}{S}}-Cl \qquad (3)$$

5   **3**

3a: R^1 = Me
3b: R^1 = Et
3c: R^1 = $CH_2CH_2CH_2Cl$
3d: R^1 = CH=CH–Ph
3e: R^1 = Ph
3f: R^1 = C_6H_4–p–F

The formation of the sulfonimidoyl chlorides **3** is readily followed by ^{1}H NMR spectroscopy for those compounds containing at least one α-hydrogen on the S-alkyl substituent. A significant downfield shift of the signal for this proton occurs in going from the sulfonamide structure to the sulfonimidoyl chloride structure (*13*). The transformation of **5** to **3** occurs in a near-quantitative fashion for the compounds that we have examined so far. The only exception that we have noted is for N-silyl-triflamides, which do not yield sulfonimidoyl chloride. Instead, the phosphoran-ylidene product, $CF_3SO_2N=PPh_3$, is formed in almost quantitative yield. This suggests that at least our current set of phosphorus reagent and reaction conditions is inadequate for sulfonimidoyl halide synthesis from strongly electronegative sulfon-amides.

The sulfonimidoyl chlorides **3a-f** can not be isolated since they are unstable to any solvent removal process that we have attempted. They appear to cleanly decompose to oxothiazene oligomers or low polymers (broad ^{1}H NMR signals for S-alkyl group) via elimination of Me_3SiCl. This condensation may be catalyzed by the amine hydrochloride salts (chloride ion) present in the reaction mixture. A noteworthy observation is that no cyclic oxothiazene species has been detected (by NMR) in this condensation.

Even though the compounds **3** are unstable for extended periods in solution at room temperature they, however, are stable for several hours at 0 °C, and this is a vital feature of these compounds for the synthesis of polymer precursor sulfonimidates, which is described next.

Synthesis of N-silylsulfonimidates. It is known in poly(alkyl/arylphosphazene) synthesis that only those N-silylphosphoranimines which contain either a 2,2,2-tri-fluoroethoxy or an aryloxy group on phosphorus, yield polymer on 1,2-elimination of the corresponding silyl ether (*7*). Hence, directly applying the phosphazene analogy, we prepared several 2,2,2-trifluoroethyl- and phenyl sulfonimidates **2** from the sulfon-imidoyl chlorides **3a-f** by addition of a benzene solution containing the alcohol (or phenol) and triethyl amine to the sulfonimidoyl chloride solution at 0 °C (equation 4) (*13*). In additon, one ethyl sulfonimidate was also synthesized to compare its thermal decomposition behavior with those of the fluoroalkyl- and aryl sulfonimidates.

$$\underset{\underset{R^1}{|}}{\overset{\overset{O}{\|}}{Me_3SiN{=}S}}-Cl + R^2OH \quad \xrightarrow[- Et_3NH^+Cl^-]{Et_3N} \quad \underset{\underset{R^1}{|}}{\overset{\overset{O}{\|}}{Me_3SiN{=}S}}-OR^2 \qquad (4)$$

3a-f $R^2O =$ **2a-k**
           $CF_3CH_2O,$

           $C_6H_5O,$ **2a:** $R^1 = Me,$ $R^2 = CH_2CF_3$

           C_2H_5O **2b:** $R^1 = Me,$ $R^2 = Ph$

                               2c: $R^1 = Et,$ $R^2 = Ph$

                               2d: $R^1 = (CH_2)_3Cl,$ $R^2 = CH_2CF_3$

                               2e: $R^1 = (CH_2)_3Cl,$ $R^2 = Ph$

                               2f: $R^1 = CH{=}CHPh,$ $R^2 = CH_2CF_3$

                               2g: $R^1 = CH{=}CHPh,$ $R^2 = Ph$

                               2h: $R^1 = Ph,$ $R^2 = CH_2CF_3$

                               2i: $R^1 = Ph,$ $R^2 = Ph$

                               2j: $R^1 = C_6H_4\text{-F-p},$ $R^2 = Ph$

                               2k: $R^1 = Me,$ $R^2 = Et$

The N-silylsulfonimidates were obtained in distilled yields of ca. 20-80%. They were characterized by ^1H and ^{13}C NMR spectroscopy and, in most cases, also by elemental microanalysis (13). A characteristic feature of the 2,2,2-trifluoroethyl sulfonimidates is a complex multiplet pattern of the $CF_3C\underline{H}_2O$ ^1H NMR signal (in benzene solution) due to the nonequivalency of the two protons.

Synthesis of Free Sulfonimidates. During investigations of the reactivity of N-silyl-sulfonimidates **2** at the Si-N bond for derivatization purposes, it was discovered that these silylsulfonimidates undergo alcoholysis of the Si-N bond to yield free sulfon-imidates. On closer examination, this reaction proved to be extremely valuable. Compounds **2** were found to react with methanol (used in excess) under mild con-ditions to cleanly, and almost quantitatively, produce free sulfonimidates **6**, without transesterification occurring at sulfur (16,17) (equation 5). More importantly, these free sulfonimidates, with one exception, showed signs of polycondensation at room temperature. Thus, somewhat fortuitously, a new type of polymer precursor was born. In our initial explorations, only a few of these have been made for selective polymerization studies, but this facile synthesis of free sulfonimidates carries significant potential for entry into newer areas of functionalized sulfonimidates.

$$Me_3SiN{=}\overset{\displaystyle O}{\underset{\displaystyle R^1}{S}}{-}OR^2 \quad \xrightarrow[{- Me_3SiOMe}]{MeOH, 20\text{-}55\ ^\circ C} \quad HN{=}\overset{\displaystyle O}{\underset{\displaystyle R^1}{S}}{-}OR^2 \qquad (5)$$

2	**6**

6a: R^1 = Me, R^2 = CH$_2$CF$_3$

6b: R^1 = Me, R^2 = Ph

6c: R^1 = (CH$_2$)$_3$Cl, R^2 = CH$_2$CF$_3$

6d: R^1 = CH=CHPh, R^2 = CH$_2$CF$_3$

6e: R^1 = Ph, R^2 = Ph

6f: R^1 = Me, R^2 = CH$_2$CH$_3$

Condensation Synthesis of Poly(oxothiazenes)

Thermal Polycondensation of N-silylsulfonimidates 2. In order to keep the first polycondensation studies with the new S-N polymer precursors manageable, we focused primarily on the thermal condensation of **2a**, **2b**, and **2i** to obtain preliminary information on the effect of temperature and time of heating on the condensation behavior and the nature of the products.

As with poly(phosphazene) precursors, thermolysis of **2** was carried out in evacuated, sealed, heavy-walled glass ampules or in stainless steel Parr reactors. All the N-silylsulfonimidates that were examined, condensed betwen 120 and 170 °C, usually to produce a two-phase mixture of solid polymer and liquid silyl ether and any uncondensed precursor (*16,18*) (equation 6). The extent of condensation was determined primarily by ^{1}H NMR spectroscopy of the liquid content and also by weighing the volatile silyl ether where possible (usually for Me$_3$SiOCH$_2$CF$_3$).

Even though both 2,2,2-trifluoroethyl and phenyl N-silylsulfonimidates eliminate silyl ether on thermolysis, the condensation behavior of the fluoroalkyl sulfonimidates is sometimes erratic in that little polymer is produced. We have not yet determined the cause of this behavior. Instead, we have shifted focus to the phenyl N-silylsulfonimidates which always and reproducibly yield polymer and silyl ether on condensation.

A noteworthy exception to the polycondensation of N-silylsulfonimidates was the ethyl sulfonimidate **2k**. Even after extended heating at 160 °C, this ethyl ester failed to condense to any detectible extent. Instead, complete rearrangement to the isomeric N-ethyl-N-silylsulfonamide was observed(*16*). This provides strong evidence that the leaving group OR^2 on sulfur must be fairly electronegative and resistant to C-O bond cleavage, so that the drive toward the stable sulfone (-SO$_2$-) structure is totally suppressed during heating and the reaction proceeds in the direction of polycondensation.

$$\underset{\underset{R^1}{|}}{Me_3SiN{=}\overset{\overset{O}{||}}{S}} - OPh\,(OCH_2CF_3) \xrightarrow[\substack{-\,Me_3SiOPh \\ or \\ -\,Me_3SiOCH_2CF_3}]{120\text{-}170\ ^\circ C,\ 4\text{-}12\ days} \left[N{=}\underset{\underset{R^1}{|}}{\overset{\overset{O}{||}}{S}} \right]_n \quad (6)$$

2   **1**

2a: $R^1 = Me$, $OR^2 = OCH_2CF_3$ **1a:** $R^1 = Me$

2b: $R^1 = Me$, $OR^2 = OPh$ **1b:** $R^1 = Et$

2c: $R^1 = Et$, $OR^2 = OPh$ **1c:** $R^1 = Ph$

2h: $R^1 = Ph$, $OR^2 = OCH_2CF_3$ **1d:** $R^1 = C_6H_4\text{-F-p}$

2i: $R^1 = Ph$, $OR^2 = OPh$

2j: $R^1 = C_6H_4\text{-F-p}$, $OR^2 = OPh$

    The ability to copolymerize two or more monomers plays a titanic role in property modification in the realm of carbon-based polymers, silicones, and poly(phosphazenes). We have also found that the co-condensation of two N-silylsulfonimidates with different R^1 substituents, affords oxothiazene copolymers (equation 7).

$$\mathbf{2b} + \mathbf{2i} \xrightarrow[-\,Me_3SiOPh]{\Delta} \left[(N{=}\underset{\underset{Me}{|}}{\overset{\overset{O}{||}}{S}})_x (N{=}\underset{\underset{Ph}{|}}{\overset{\overset{O}{||}}{S}})_y \right]_n \quad (7)$$

7 (x = y)

    The polycondensation of the N-silylsulfonimidates is rather slow, taking between 4-12 days for greater than 90% condensation. Hence, several Lewis acid and Lewis base compounds were examined as potential catalysts (*16,18*). These included $BF_3{\cdot}Et_2O$, $AlCl_3$, WCl_6, $Sn(octoate)_2$, KF/18-crown-6, LiOPh, and others. Though both Lewis acids and bases accelerated condensation, only moderate enhancements in rate were observed with most of the examined compounds. The best combination of rate, molecular weight, and polydispersity of polymer was obtained with $BF_3{\cdot}Et_2O$. The catalyst study was by no means exhaustive, and it is likely that other species can be found for much faster polycondensation with **2** under appropriate conditions (e.g., solution polymerization).

Polycondensation of Free Sulfonimidates 6. Our initial observations that the free sulfonimidates appeared to undergo condensation at room temperature, led us to more closely examine this behavior. In fact, all the free sulfonimidates, with the exception of **6f**, showed significant condensation overnight when left in flasks on the bench top. This condensation, however, could be totally suppressed for several weeks to months if the samples were stored at -20 °C or below.

When heated under nitrogen in a flask, the free sulfonimidates **6a-e**, condensed extremely rapidly when the temperature reached about 85-95 °C (*16,17*) (equation 8). In no case was heating above 120 °C necessary for more than 2-3 hours to bring about near-quantitative condensation. The polycondensations were usually carried out under reduced pressure to facilitate removal of the alcohol (or phenol) byproduct. A solution polymerization in DMF was also found to proceed quite smoothly. Significantly, *both* 2,2,2-trifluoroethyl- and phenyl free sulfonimidates underwent polycondensation in a consistent and reproducible fashion unlike the N-silylsulfonimidates. As with the N-silylsulfonimidates, a copolymerization was also successful with the free sulfonimidates (equation 9).

Thus, it appears at this point that the condensation of free sulfonimidates is an easier and more efficient route to oxothiazene polymers than the condensation based on N-silylsulfonimidates. Although the free sulfonimidate route has yielded polymers with lower molecular weight in most cases, it may be possible to overcome this apparent shortcoming by varying condensation conditions. Some representative molecular weight (GPC) data for polymers from both N-silyl- and free sulfonimidates are provided later (Table I) in the section on polymer characterization and properties.

$$HN{=}S{-}OPh\,(OCH_2CF_3) \quad \xrightarrow[\substack{-\,PhOH \\ or \\ -\,CF_3CH_2OH}]{85\text{-}120\ ^\circ C,\ 2\text{-}3\ hours} \quad {\left[N{=}S\right]}_n \qquad (8)$$

6   **1**

6a: $R^1 = Me,$ $R^2 = CH_2CF_3$ **1a:** $R^1 = Me$
6b: $R^1 = Me,$ $R^2 = Ph$ **1c:** $R^1 = Ph$
6c: $R^1 = (CH_2)_3Cl,$ $R^2 = CH_2CF_3$ **1e:** $R^1 = CH_2CH_2CH_2Cl$
6d: $R^1 = CH{=}CHPh,$ $R^2 = CH_2CF_3$ **1f:** $R^1 = CH{=}CHPh$
6e: $R^1 = Ph,$ $R^2 = Ph$

$$\mathbf{6b}\ +\ \mathbf{6e} \quad \xrightarrow[-\,PhOH]{\Delta} \quad {\left[{(}N{=}S{)}_x{(}N{=}S{)}_y\right]}_n \qquad (9)$$

7 (x = y)

It is of interest to note that while analogy with phosphazenes has been extremely useful in the synthesis of the oxothiazene polymers, the free sulfonimidate method has no analogous precedent in phosphazene polymer synthesis. Thus, the oxothi-azenes may be exhibiting natural chemical differences with phosphorus compounds, and one may anticipate that specific differences between the chemistries of phosphorus and sulfur will also lead to differences in properties at the macromolecular level between the two systems.

Polymer Characterization and Properties

Solubility and Surface tension. The oxothiazene polymers **1** exhibit solubility behavior that is indicative of a very polar structure. This can, ofcourse, be anticipated from the strongly dipolar characteristic of the S=O ($S^{\partial+}$–$O^{\partial-}$) bond at every other link along the backbone. Thus, polymers **1a** and **1d** are soluble in DMF, DMSO and MeNO$_2$, but insoluble in other common organic solvents. The other homopolymers, and copolymer **7**, are soluble in CH$_2$Cl$_2$ and CHCl$_3$. Quite surprisingly, the S-aryl polymers **1c** and **1d** (as well as copolymer **7**) are insoluble in common aromatic hydrocarbons as well as in THF. The assertion that the polymer chains are quite polar is corroborated by contact angle measurements on **1a**, which show the total solid surface tension for this polymer to be in the range 44.8-50.5 mN/m (*16*).

NMR Spectra. Only [N=S(O)CH$_3$]$_n$ has so far been examined by both ^{1}H and ^{13}C NMR spectroscopy. The proton chemical shift of the relatively broad CH$_3$ signal at 3.45-3.50 ppm (relative to d$_6$-DMSO) is about 0.2 ppm downfield from the corresponding sulfonimidate signal. The ^{13}C NMR spectrum of this polymer shows a broad triad (centered at 46.4 ppm), with other outer peaks apparent. This is suggestive of a primarily atactic structure for this polymer. The relatively high polydispersity observed with these polymers certainly contributes to loss in definition of the NMR signals. It is quite interesting from a synthetic perspective that because of the inherent chirality at sulfur, the possibility exists for synthesizing stereoregular polymers. And, because of the unusual, polar S=O substituent, this could lead to a number of interesting properties. Obviously, NMR will play an important diagnostic role in ventures in this direction. For polymer **1e** with a chloropropyl group on sulfur, the ^{1}H NMR signals are even broader because of coupling between the neigboring CH$_2$ protons, but this polymer also exhibits a downfield shift for the S-CH$_2$ protons in going from the sulfonimidate to the polymer.

Molecular Weight Determination. Gel permeation chromatography (GPC) has been used to obtain rough estimates of molecular weight for the polymers **1** and **7** in our initial studies with the oxothiazene polymer system, as is often the case with the first studies of a new polymer system. Because of the insolubility of several of the polymers in traditional GPC solvents, DMF, which is a common solvent for all the oxothiazene polymers examined, was used to obtain molecular weights against polystyrene standards (*16*). In light of the quite different structural make up of the oxothiazenes and polystyrene, our current GPC molecular weights are best interpreted in terms of a comparative scenario within the series **1a-f** and **7**. Obviously, true molecular weights will need to be determined for future studies using primary methods. Additionally, methods such as light scattering will provide valuable information on chain dimensions which is particularly needed to address some puzzling questions on chain flexibility as described later.

The very approximate nature of the molecular weights by GPC not withstanding, some general, but interesting observations have been made. With N-silylsulfonimidates **2** as polymer precursors, molecular weights are approximately an order of magnitude lower when OCH$_2$CF$_3$ is the leaving group on sulfur compared with OPh. There appears to be some steric effect of the S-alkyl/aryl substituent on polymerization, with lower molecular weights for larger substituents. A significant observation is the much lower molecular weight of the S-Me polymer **1a** obtained from free sulfonimidates, irrespective of the leaving group on sulfur. Yet, virtually no change is observed in this respect for the S-Ph polymer **1c**. Solution polymerization with a free sulfonimidate (**6a**) indicates the possibility of obtaining high molecular weight polymers from free sulfonimidates. Further detailed studies encompassing a larger variety of precursors as well as greater variation of polycondensation conditions are needed to obtain a clearer picture of the polymerization process. Some selected molecular weight data are provided in Table I.

Table I. Molecular Weights (by GPC) for polymers 1a-f and 7

Sulfonimidate	Polymer	M_w	M_n
		mol wt	
2a	**1a**	52 000	6 000
2b	**1a**	481 000	35 000
2b	**1a**[a]	479 000	53 000
2c	**1b**[a]	208 000	28 000
2h	**1c**	10 000	4 000
2i	**1c**[b]	195 000	138 000
		14 000	12 000
2j	**1d**[b]	539 000	389 000
		43 000	35 000
2b + 2i	**7**	256 000	37 000
6a	**1a**	45 000	10 000
6b	**1a**	50 000	12 000
6b	**1a**[c]	87 000	27 000
6c	**1e**	44 000	6 000
6e	**1c**[b]	299 000	230 000
		24 000	18 000
6b + 6e	**7**	55 000	15 000

Adapted from ref. 16, copyright 1993, American Chemical Society.
[a]$BF_3 \cdot Et_2O$ catalyst.   [b]Bimodal molecular weight distribution.   [c]Polymerization in DMF solution.

Thermal Analysis. The thermal behavior of the polymers prepared in this study has been examined by differential scanning calorimetry (DSC) and by thermogravimetric analysis (TGA) (*16*). Some striking observations have been made from the DSC curves for the polymers. First, the S-Me polymer **1a** exhibits an unexpectedly high glass transition temperature (T_g) at 55-65 °C, considering the small substituents on sulfur and the low T_g (-46 °C) of the analogous poly(phosphazene), $[N=PMe_2]_n$. Second, and even more perplexing is the apparent trend of decreasing T_g with increasing length of the alkyl group on sulfur (S-ethyl, 29 °C; S-3-chloropropyl, -1 °C). Since polymers of comparable molecular weight show this difference, the trend appears to be real. Based on the closely similar atomic sizes of phosphorus and sulfur, and the comparable P-N and S-N bond lengths in cyclic phosphazene and oxothiazene compounds, a flexible oxothiazene backbone would be anticipated.

At present, the only explanation for the anomalous glass transition behavior lies in the highly polar nature of the S=O bond. It is conceivable that for a "small" Me group on sulfur, the interchain dipolar interactions are strong enough to drastically reduce backbone mobility. With the ethyl and 3-chloropropyl substituents, the chains are forced further apart and, as a consequence, the interchain dipolar attractive force drops significantly allowing the "natural" flexibility of the chain to break through. This postulation is supported by a recent report by Manners on poly(thionylphosphazenes) (19), which by our proposed nomenclature (16) would be poly(phosphazene- -co-oxothiazenes). Very recently, the dramatic effect of eliminating interchain attractive forces in nylon 66 and some rigid and semi-rigid polymers on the T_g of these polymers was reported by Jenekhe (20). By removing H-bond interactions or van der Waal interactions via Lewis acid-base complexation, very large decreases in T_g were reported in going from the pure polymers to the complexed polymers. We suspect that similar experiments on the poly(oxothiazenes) and poly(thionylphosphazenes) are likely to shed more light on the glass transition behavior of these polymers.

Thermogravimetric analysis of polymers 1a-f and 7 also shows interesting degradation behavior. First, virtually all the polymers decompose within the narrow temperature range of roughly 270-300 °C. Second, decomposition temperatures in air are only 6-10 °C lower than in helium. Third, the polymers do not appear to depolymerize to cyclic oxothiazenes. Instead, redox decompositon appears to be operative. It must be mentioned that detailed thermoxidative studies with highly purified polymer samples are warranted for an indepth understanding of the thermal degradation behavior of this new macromolecular system.

Theoretical Calculations

As part of our initial studies with the new S-N polymer system, theoretical calculations on selected aspects of polymer structure were also carried out. These reveal some interesting similarities and differences with theoretical as well as experimental information available on some phosphazene systems.

Conformation of Poly(methyloxothiazene). A series of Hartree-Fock calculations was carried out on a three-and-a-half repeat unit segment of the polymer in order to determine the most stable conformation (16). A 3-21G** basis set (21) was used in the calculations. With certain restrictions, as described in our recent publication (16), to keep the system manageable, only three different general backbone conformations needed to be analyzed to arrive at the most stable conformation. Among these three, **Conformation A**, with *trans* arrangement of the S—N single bonds and *cis* arrangement of the S=N double bonds along the backbone, was found to be the most stable. In this conformation, the most energetically favorable orientation of the side groups (oxygen, for example) was found to be "up-down-up" at alternate sulfur atoms. Note, that the conformation drawn depicts the less favorable "up-up-up" orientation, because no accepted pictorial notation exists for indicating "up" or "down" for a double bond. The other two conformations considered consisted of all *trans* skeletal bonds in one, and *cis* single bonds with *trans* double bonds in the other.

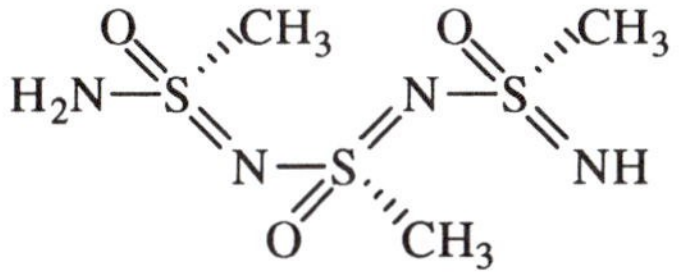

Conformation A

With skeletal **Conformation A** calculated to be the most stable, the effects of rotation around S-N single and S=N double bonds were also studied. For rotation around the central single S-N bond, the energy minimum was located at 33° off the planar conformation for the longest distance between carbon atoms on neighboring methyl groups. Thus, a slightly helical conformation is predicted for this polymer. The calculated IR spectrum that provides the best fit with the experimental IR spectrum is also based on a helical conformation. A Raman spectrum for the polymer has also been recently calculated (*22*). The rotational barrier at the energy maximum was 20 kcal/mol. It was also found that rotation around the S=N double bond is about four times more difficult than rotation around the S-N single bond. Based on these calculations, there appears to be higher substituent-substituent steric interactions in this system compared with the siloxanes and the phosphazenes (*16*).

Bonding. Sulfur-nitrogen "single" and "double" bond lengths were calculated at 1.60 Å and 1.56 Å, respectively. The alternating long and short skeletal bonding characteristics are probably maintained to a large degree since rotation around a short bond was found to be four times more difficult compared with that around a long bond. This is also supported by recent computations (*22*) on skeletal bond orders and net atomic charges (Chart I), using the MOPAC software package (*23*) and Gopinathan's procedure (*24*) based on the concept of molecular orbital valency.

Chart I

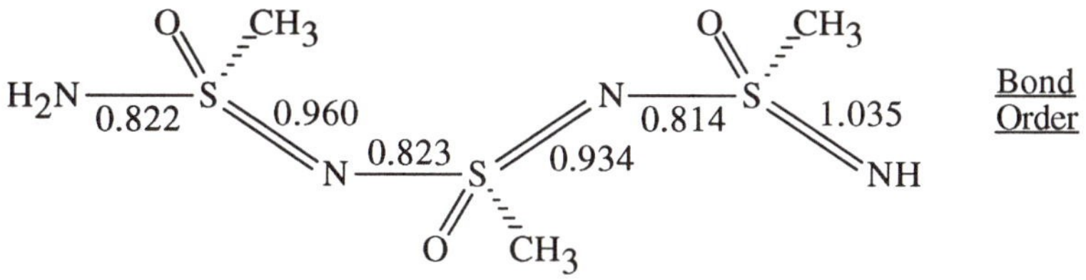

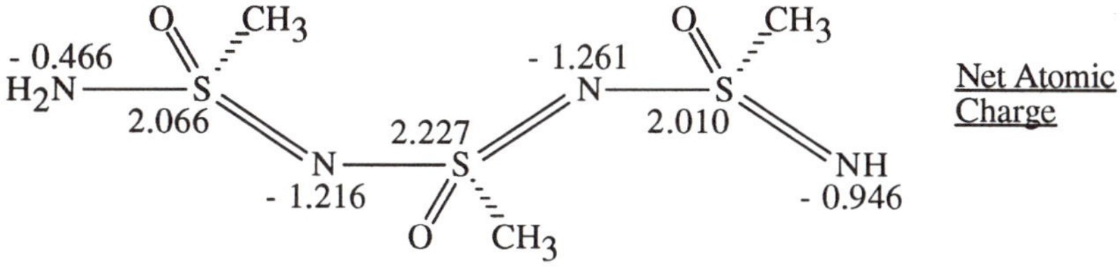

It can be seen from Chart I that all internal S-N bond orders are less than one, with large net atomic charges on the skeletal atoms. This implies a significant ionic character of the skeletal bonding. Since this method of calculating bond order and atomic charges has not been widely tested, comparisons of the oxothiazene system were made with similar calculations on poly(acetylene), poly(dimethylsiloxane), poly(dichlorophosphazene), and poly(dimethylphosphazene), all polymers on which significantly more information exists on structure and bonding. Since the calculations on these polymers provided values consistent with existing knowledge, our conclusions on the nature of skeletal bonding in poly(methyloxothiazene) are very likely a fair representation of actual bonding in poly(oxothiazenes). It should be mentioned that the nature of skeletal bonding in the two poly(phosphazenes) examined and in poly(methyloxothiazene) were very similar with respect to bond orders and net atomic charges, but π-contribution to bonding was found to be the least with the oxothiazene polymer.

The most significant difference with phosphazenes was found in the area of skeletal bond angles, as we reported earlier (16). Even though the skeletal ∠SNS at ca. 123.5° compares quite well with ∠PNP in many cyclic and poly(phosphazenes), the ∠NSN at 103° is substantially smaller than the average ∠NPN at 120° in poly-(phosphazenes). This lower bond angle at sulfur could play a significant role in backbone mobility, since the smaller angle would be expected to result in increased substituent-backbone steric interactions. However, the fact that lowering of glass transition temperature is experimentally observed with increasing length of the S-alkyl substituent suggests that a combination of several factors is at play in bonding, conformation, and skeletal mobility of the oxothiazene system.

Conclusions

A new inorganic polymer system, based on a backbone of alternating sulfur(VI) and nitrogen atoms, with oxygen as a "fixed" substituent and alkyl or aryl groups as the variable substituent, has been developed. The overall strategy for polymer synthesis was based on the analogous condensation synthesis of poly(alkyl/arylphosphazenes). The design and development of appropriate precursors to the S-N polymers was then accomplished through a fundamentally new reaction in sulfur chemistry, namely, the facile transformation of a sulfonamide group to a sulfonimidoyl moiety using a common class of phosphorus reagents. We anticipate that this new reaction of sulfur(VI) compounds will open up synthetic opportunities in several areas of organosulfur chemistry.

The new inorganic polymers, the poly(alkyl/aryloxothiazenes), have shown a number of very interesting properties and characteristics, even from these early studies. The polymers are quite polar, resulting in very selective solubility behavior. The polar characteristic and limited solubility suggest potential applications in several areas, once tailored polymers with appropriate substituents have been made. The issue of chain flexibility is also expected have supreme bearing on potential applications in the area of elastomers. From our own observations so far, and from those of Manners (19) with poly(thionylphosphazenes) and of Allcock (25) with poly(thiophosphazenes), the oxothiazene skeletal system appears to be flexible. A number of factors are known to affect chain flexibility in polymers. Allcock has discussed some of these in the light of a comparative picture between phosphorus-nitrogen and sulfur-nitrogen skeletal systems (25). Further studies are needed to help provide more definitive answers on this key question with S-N skeletal systems. For oxothiazenes, the polar S=O substituent appears to play a significant role in the properties and characteristics observed so far. Only continued studies on these polymers with this unique substituent will tell whether this can be used to an advantage to generate useful materials for commercial application.

Acknowledgments

The early exploratory phases of this research were carried out at Michigan Molecular Institute, Midland, Michigan, under a Dow Corning Corporation Visiting Scientist grant. AKR especially thanks Dr. Dale Meier of this institute for many stimulating discussions, and for fostering an invigorating research atmosphere in his group. We also thank Dow Corning Corporation for financial support of this work.

Literature Cited

1. Seel, F.; Simon, G. *Angew. Chem.* **1960**, *72*, 709.
2. Parshall, G. W.; Cramer, R.; Foster, R. E. *Inorg. Chem.* **1962**, *1*, 677.
3. Cramer, R. D. U.S. Patent 3,017,240.
4. Levchenko, E. S.; Kozlov, E. S.; Kirsanov, A. V. *Zh. Obshch. Khim.* **1962**, *32*, 2585.
5. Levchenko, E. S.; Kozlov, E. S.; Kirsanov, A. V. *Zh. Obshch. Khim.* **1963**, *33*, 565.
6. Wisian-Neilson, P.; Neilson, R. H. *J. Am. Chem. Soc.* **1980**, *102*, 2848.
7. Neilson, R. H.; Wisian-Neilson, P. *Chem. Rev.* **1988**, *88*, 541.
8. Levchenko, E. S.; Markovskii, L. N.; Kirsanov, A. V. *Zh. Org. Khim.* **1967**, *3*, 1273 and references therein.
9. Johnson, C. R.; Jonsson, E. U.; Bacon, C. C. *J. Org. Chem.* **1979**, *44*, 2055 and references therein.
10. Derkach, N. Ya.; Smetankina, N. P. *Zh. Obshch. Khim..* **1964**, *34*, 3613.
11. Krebs, K. -W.; et al. German Patent 2,002,065, 1971.
12. Golebiowski, L.; Lasocki, Z. *Bulletin De L'Academie Polonaise Des Sciences* **1976**, *24*, 439.
13. Roy, A. K. *J. Am. Chem. Soc.* **1993**, *115*, 2598.
14. Roy, A. K. *Polymer Preprints* **1993**, *34(1)*, 332.
15. Roy, A. K. U.S. Patent 5,068,379, 1991(to Dow Corning Corp.).
16. Roy, A. K.; Burns, G. T.; Lie, G. C.; Grigoras, S. *J. Am. Chem. Soc.* **1993**, *115*, 2604.
17. Roy, A. K. U.S. Patent 5,233,019, 1993 (to Dow Corning Corp.).
18. Roy, A. K.; Burns, G.T. U.S. Patent 5,194,557, 1993 (to Dow Corning Corp.).
19. Manners, I. et al. *Macromolecules* **1992**, *25*, 7119.
20. Jenekhe, S. A.; Roberts, M. F. *Macromolecules* **1993**, *26*, 4981.
21. Hehre, W. J.; Radom, L.; Schleyer, P. v. R. *Ab Initio Molecular Orbital Theory*; John Wiley: New York, 1986.
22. Lie, G. C.; Grigoras, S.; Roy, A. K.; Burns, G. T. *Dow Corning Corporation Internal Technical Report* **1993**.
23. Stewart, J. J. P.; Quantum Chemistry Exchange, Program No. 455, Bloomington, Indiana.
24. Gopinathan, M. S.; Siddarth, P.; Ravimohan, C. *Theo. Chim. Acta* **1986**, *70*, 303.
25. Allcock, H. R.; Dodge, J. A.; Manners, I. *Macromolecules* **1993**, *26*, 11.

RECEIVED April 25, 1994

Chapter 27

Recent Developments in Borazine-Based Polymers

R. T. Paine[1] and Larry G. Sneddon[2]

[1]Department of Chemistry, University of New Mexico, Albuquerque,
New Mexico 87131
[2]Department of Chemistry, University of Pennsylvania, Philadelphia,
Pennsylvania 19104–6323

Although they have tremendous applications potential, inorganic
polymers have received relatively little attention compared to organic
polymers. Recent demands for entirely new hybrid materials and
advanced ceramics, however, are driving efforts to develop new
inorganic polymer systems. This paper summarizes some historical
background along with recent results from our two laboratories on the
formation of boron-nitrogen polymers based upon borazine monomer
chemistry that will be useful in the realization of advanced materials
goals.

A wide spectrum of organic polymer chemistry has been developed over the last fifty
years, and the synthesis, characterization and processing of organic polymers has been
one of the most thoroughly studied topics in modern chemistry. Surprisingly, in spite
of their abundance in the periodic table and their flexibility in structure and reactivity,
the polymer chemistry of main group elements has not enjoyed parallel, active
development. This fact was first recognized in the 1950's and 60's, and serious
attempts were made at that time to synthesize selected inorganic polymers, including
polyphosphazenes, polysilicones, polysilicates, polyborates, and polysulfurimines.[1]
Some initial successes were realized, and even some commercial development, e.g.,
polysilicones and polyphosphazenes, occurred, but, in general, interest in these
polymers declined due to synthetic roadblocks, property shortcomings and synthesis
and processing economics.

In the 1980's and 90's the advent of new technological needs in materials
science has reactivated interest in inorganic polymers, and a number of exciting
developments have occurred. In particular, it was found that controlled hydrolysis of
tetraethyl orthosilicate (TEOS) leads initially to soluble silicate polymers that form
processible silicate gels.[2] This observation spawned the now rich field of sol-gel
chemistry that encompasses not only silicate polymers, but also oxo polymers of a
variety of metals. In a different arena, Verbeek[3] and Yajima[4] reported the synthesis of
processible polycarbosilanes that have found use as precursors for the nonoxide
compositions SiC and SiC/Si$_3$N$_4$. These findings and pressing practical needs for
more useful forms of existing and new advanced high-temperature materials are
encouraging additional studies of inorganic polymers and inorganic/organic hybrid
polymers.[5-10]

   Boron nitride is one of several commercially significant high temperature ceramics that is easily obtained in powder form from pyrolysis of simple inexpensive reagents.[11] Unfortunately, it is difficult or impossible to obtain BN in some desired forms (e.g., fibers, coatings) from powders. Requirements for more processable boron nitride have encouraged several research groups to seek new approaches to obtain these forms, and boron-nitrogen containing polymers offer an attractive vehicle.

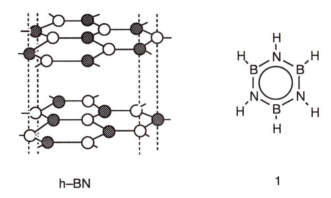

h–BN   1

   One appropriately constituted molecular species upon which polymeric boron-nitride precursors could be based is the ring-compound borazine, $B_3N_3H_6$ (**1**). The compound is easily made and has both the correct boron/nitrogen ratio and a 6-membered ring structure analogous to the solid state structure of h-BN. Furthermore, the similarity of the structures and physical properties of benzene and borazine, suggest that the borazine analogues of many benzene-derived polymers might be possible. In the following sections we outline recent advances in the development of two general classes of borazine-based polymers in which the borazine ring is either a member of the polymer backbone or a pendant on a polymer having an organic or inorganic backbone.

Borazine Backbone Polymers

In our work two general approaches to forming borazine-backbone polymers are being explored. These involve formation of polyphenylene analogues in which the borazine rings are joined either by direct B-B (**2**) or B-N (**3**) linkages or borazine-bridged frameworks with the rings linked via an intervening atom or group, X (**4, 5**).

2   3

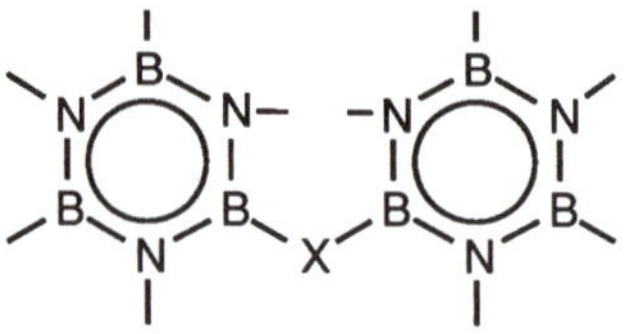

4 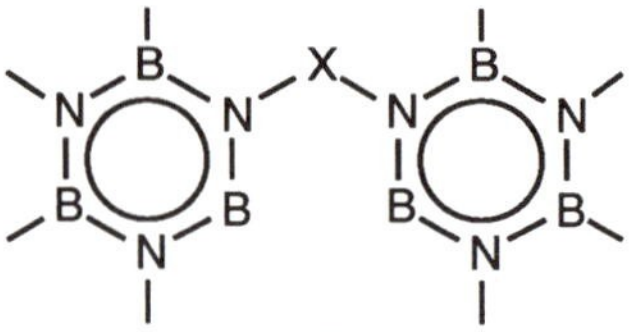   **5**

Polyphenylene Analogues. Early work aimed at the development of linked-ring borazine polymers with structures (**2-3**) has been outlined in a previous review.[11] In particular, it is worth noting that the metathetical reaction of N-lithio pentamethyl borazine with B-chloropentamethyl borazine (eq. 1) led to LiCl metathesis and production of decamethyl-N,B-diborazine.[12] Extensions of this chemistry resulted in formation of a polyborazinyl chains, estimated to contain ten monomer rings, which melted over a broad temperature range. Unfortunately, little follow up on these polymers has been published perhaps because the polymers are not likely to be efficient ceramic precursors due to their expected low char yields and the inclusion of carbon in their ceramic residues.

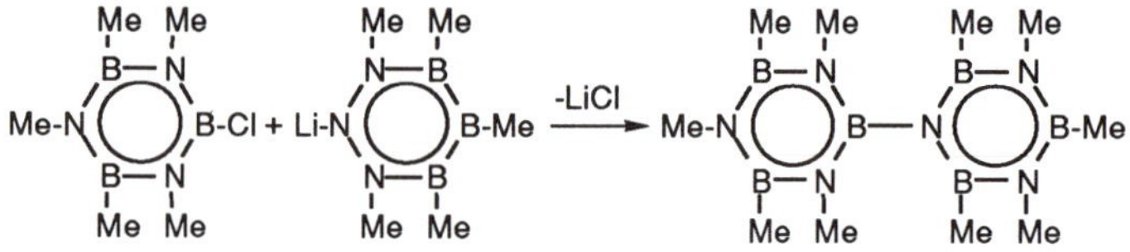

$$(1)$$

The use of other types of elimination reactions for the formation of B-N linked systems has also been used to yield alkylated diborazines.[13] Recently, Niedenzu has reported that the reaction of $Et_3B_3N_3H_2(SiMe_3)$ with B-haloborazines results in the elimination of trimethylsilyl halide and the formation of several new alkylated borazine dimers and trimers including (**6-8**).[14] It can be expected that additional studies of this elimination chemistry may produce interesting polymers.

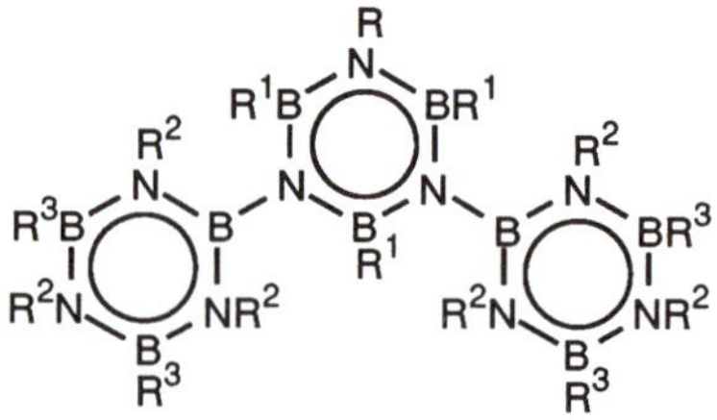

6

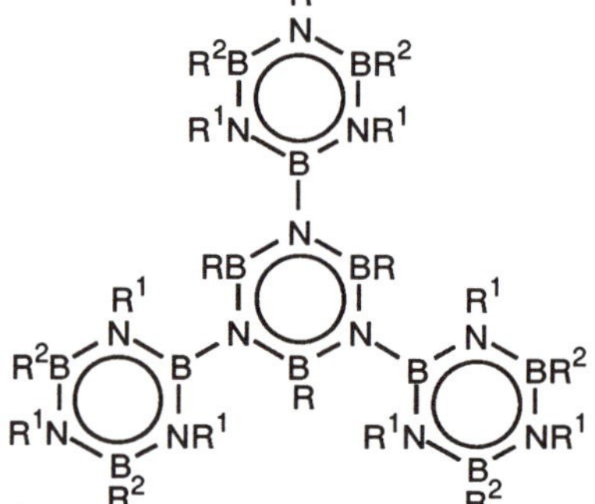

7

8

The use of reductive (Wurtz) coupling of borazine rings by combination of B-haloborazines with Group 1 metals has also been explored.[15] This approach has recently been reexamined by Shore[16] who reported that trichloroborazine and Cs metal react explosively at 125°C with formation of a insoluble, perhaps polymeric, solid some of which has a tubular morphology. Most significantly, the tubular structure is retained in the final boron nitride pyrolysis product.

Perhaps the earliest examples of backbone polymers derived from the parent, $B_3N_3H_6$, borazine are those mentioned by Stock and Pohland.[17] They found that pyrolysis of borazine at 500°C produced an undefined insoluble solid with an approximate empirical composition BNH. Somewhat later, Wiberg[18] proposed that the BNH solid consisted of a network structure. Subsequently, Laubengayer[19] specifically proposed that the biphenyl, $B_6N_6H_{10}$, and naphthalene, $B_5N_5H_8$, analogues formed during the gas phase pyrolysis of borazine. These species were then identified by Margrave[20] in gas phase mass spectrometric analysis. In related studies, Porter[21] found that the same species, as well as an insoluble polymeric solid, are formed by photoinduced decomposition of borazine. These early findings clearly suggest that higher linked- or fused-ring polymers based on the parent borazine should be accessible, although no tractable polymeric species were characterized. Processable dehydrocoupled polymers derived from the parent borazine would, of course, be of particular interest as polymeric precursors to boron nitride, owing to their potentially high ceramic yields (95%) and their close structural relationship to h-BN. These considerations stimulated new investigations at Penn of borazine dehydro-genation-polymerization.

Initial investigations focused on metal catalyzed synthetic routes; however, during the course of these studies it was discovered[22] that controlled thermolysis of liquid borazine in vacuo at moderate temperatures results in nearly quantitative dehydropolymerizations to form *soluble* polyborazylenes. In a typical reaction (eq. 2), borazine is heated in vacuo at 70°C with periodic degassing until the liquid becomes sufficiently viscous that stirring cannot continue. The volatile contents of the flask are then vacuum evaporated, leaving a white solid (**9**) in >90% yield.

$$R = H \text{ or borazinyl} \quad (2)$$

The polymer prepared in this manner is highly soluble in polar solvents and, a typical sample has, according to size exclusion chromatography (SEC)/low angle laser light scattering (LALLS) analysis, Mw 7,600 g/mol and Mn 3,400 g/mol. Polymerizations carried out for shorter times can be used to produce viscous liquid polymers with correspondingly lower molecular weight averages. For example, when the polymerization is carried out for only 24 h, Mw = 2100 and Mn = 980. Elemental analyses of the polymer are consistent with the empirical formula of ~$B_3N_3H_3$, (linear polyborazylene = $B_3N_3H_4$) suggesting the formation of a branched-chain or partially crosslinked structure. Evidence of chain branching has also been found in the SEC/LALLS/UV studies. The detailed structure of the polymer has not yet been established; however, since small amounts of both the N:B coupled dimer (1:2'-$(B_3N_3H_5)_2$) and borazanaphthalene ($B_5N_5H_8$) were isolated in the volatile materials from the reaction, the polymer may have a complex structure, having linear, branched, and fused chain segments.

Because of its composition, high yield synthesis, and excellent solubilities, polyborazylene would appear to be an ideal chemical precursor to boron nitride. Indeed, bulk pyrolyses of the polymer under either argon or ammonia to 1200 °C have been found to result in the formation of white boron nitride powders in excellent purities and ceramic yields 85-93% (theoretical ceramic yield, 95%) (eq. 3).

$$\xrightarrow{1200°C} \quad BN \quad (3)$$

Thermogravimetric studies of the polyborazylene/boron-nitride ceramic conversion reaction showed that the polymer undergoes an initial (2%) weight loss in a narrow range between 125-300°C, followed by a gradual 4% loss ending by 1100°C. In an effort to determine the steps involved in the ceramic-conversion process, a sample of the polymer heated to 400°C was investigated in more detail. This material proved to be an insoluble white powder having an empirical formula of ~$B_3N_3H_2$, suggesting that polymer crosslinking had occurred. This conclusion is supported by diffuse reflectance infrared studies[22d] which show corresponding decreases in the B-H and N-H stretching intensities as the polymer is heated to 400°C and then 1200°C.

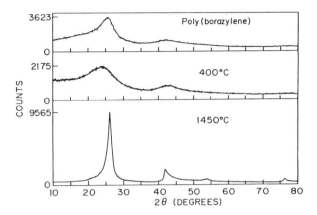

Figure 1. XRD for polyborazylene and pyrolyzates

X-ray diffraction studies (Fig. 1) demonstrate that both polyborazylene and the intermediate material obtained by heating to 400°C exhibit patterns similar to those observed for turbostratic boron nitride, indicating that the layer structure of boron nitride is already present in the polymer. These studies suggest the simple process shown in Scheme I for the conversion of the polymer to boron nitride. This involves a dehydrocoupling reaction between the B-H and N-H groups on adjacent polymer strands resulting in a 2-dimensional-crosslinking of the polymer. Because the polymer has a more complex structure than the idealized linear structure depicted in the scheme, the final high-temperature processing step is probably necessary in order to remove residual hydrogen from the B-H and N-H groups that are not in the proper orientation for the initial low-temperature crosslinking reaction.

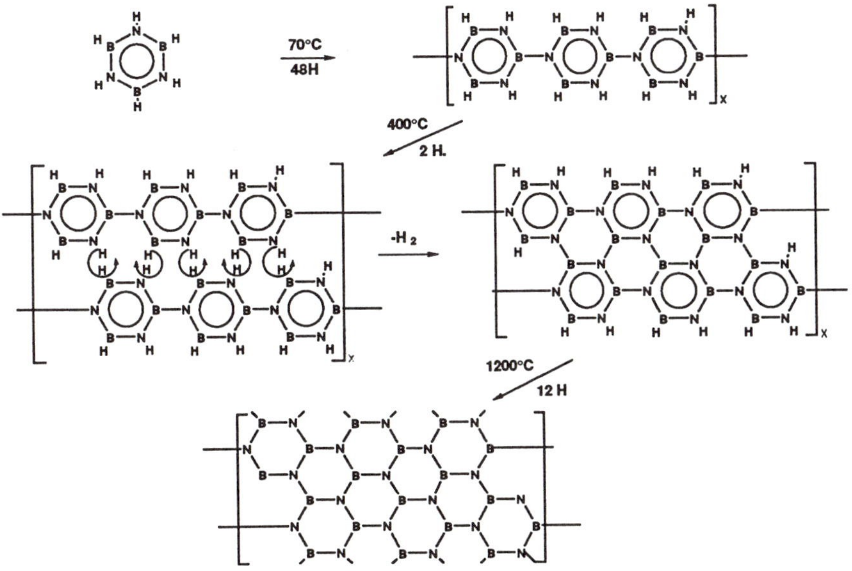

Scheme I

Because of its solubility, low-temperature decomposition, and high ceramic and chemical yields, polyborazylene shows excellent potential for many technological applications requiring a processable BN precursor. Indeed, high quality boron nitride coatings have now been achieved on alumina, silicon nitride, silicon carbide and carbon fibers using polyborazylene. Others have also recently demonstrated that this polymer may be used to generate BN matrices in carbon composites that significantly increase the oxidation stability of the composite.[23]

In addition to being of use as a boron nitride precursor, polyborazylene has proven to serve as a useful reagent for the formation of new composite metal-nitride/metal-boride materials that have improved properties over the individual pure phase metal boride or nitride. For example, titanium-boride/titanium-nitride composites have been made (eq 4) by dispersing titanium powder into polyborazylene

$$(B_3N_3H_4)_x + Ti \xrightarrow{\;\Delta\;} TiN/TiB_2 + H_2 \qquad (4)$$

followed by pyrolysis under inert atmosphere.[24] Ceramic bars with excellent shape retention in observed ceramic yields of 91-96% are achieved by heating shaped green bodies to 1450°C. X-ray diffraction and TEM studies of the ceramic conversion reaction show that the material goes through several changes in composition and crystallinity as the processing temperature is increased. The final ceramic contains intimately mixed polycrystals of 500 nm size grains with residual amorphous phase at triple points and along grain boundaries. The dependence of the crystal size and distribution on the reaction/sinter temperature, the evolution of microstructure and the properties of the final composite material, such as conductivity and hardness, are currently under investigation. The reactions of polyborazylene with other metals and soluble metal complexes are also being explored with the goal of developing a range of new metal boride/nitride precursor systems that will allow the formation of both shaped bodies and coatings.

Linked-Ring Polymers. Yet another potential route to form borazine backbone polymers involves deamination of B-aminoborazines. Early work on this chemistry by Gerrard and coworkers,[13b] and Toeniskoetter and Hall[25] demonstrated that deamination may be used to form bis-borazinyl amines of structure type (4) with X = NR. Extending that work, Taniguchi[26] described the result of heating the monomer, [(H_2N)BNPh]_3, at 250°C. This gave a meltable polymer that was drawn into fibers. Paciorek[27] also studied the thermally promoted deamination of several organo-substituted B-amino borazines and apparently obtained polyborazines. Although these polymers have not been characterized in detail, it appears that they probably contain networks of bridged-ring units such as represented by (4), and that the polymers may be processible and potentially useful in some applications.

In seeking to simplify and utilize the deamination-condensation chemistry and obtain polymers with a minimum of organic substituent groups, at New Mexico we reinvestigated the aminolysis of B-tris(dimethylamino)borazine (10).[28] It was initially found that when excess NH_3 is added to a cold toluene solution of [(Me_2N)BNH]_3 an organic solvent insoluble polymer (11) is obtained. It is assumed that this polymer is highly crosslinked as illustrated in (eq. 5). Subsequently, Kimura[29] reported aminolysis of [(Et_2N)BNH]_3 (12) in liquid NH_3 at 25°C (eq. 6) from which was

$$(5)$$

$$(6)$$

14

obtained a crystalline solid, $[(H_2N)BNH]_3$ (**13**). This compound undergoes rapid deamination at 120°C forming a slightly soluble polymer whose structure has been proposed to resemble that illustrated in (**14**). Further examination of this chemistry has shown that a small fraction (~10%) of the polymer (**11**) is also soluble in toluene. Further, if the insoluble fraction is redissolved in liquid NH_3 and briefly digested in that solvent at -40°C, the resulting polymeric evaporate is soluble in aromatic hydrocarbons. Full characterization of this soluble polymer continues and practical applications for the solutions have been accomplished.

The assembly of borazine rings via bridging groups X as schematically shown by (**15-18**) offers a particularly flexible approach to formation of both nonfunctionalized and functionalized borazine polymers. The terminology two-point (**15, 16**) and three-point (**17, 18**) is employed to designate the number of bridging groups in the polymer units. Nearly all efforts to date to make representatives of these polymers has occurred with linkages through the boron atom (**15, 17**), since that substitution chemistry is more facile and precedented in the literature.[30,31]

15   **16**

17   **18**

Lappert[32] first reported coupling of borazine rings via a bridging group, X = NR, in 1959. That work was followed by numerous related studies in the period 1960 - 70,[11] but few of these polymers were adequately characterized due to their generally low solubility, and it is unclear if practical applications were realized. In 1978, Meller and Füllgrabe[33] briefly reported simple 1:1 reactions between hexamethyldisilazane (HMDS) and 2-alkyl-4,6-dichloro-1,3,5-trimethylborazines (eq. 7). Based upon limited solution NMR data and mass spectra, it was proposed that cyclic two-point oligomers were formed having four and five borazine rings in the cyclic structure.

$$ (7) $$

That result eventually stimulated further studies of the HMDS/borazine systems by Paciorek[27] and at New Mexico,[34] and the reader is referred a previous review[11] for a detailed summary. In general, it is found that introduction of organic groups on the borazine nitrogen atoms and/or on one boron atom lead in some cases to polymers that have modest to very good organic solubility. Unfortunately most of these polymers have not been extensively characterized since they are generally poor as ceramic precursors. Nonetheless, as interest in organic modified polyborazine systems has broadened, more effort is being given to developing this chemistry.

The reaction of HMDS with B-trichloroborazine is quite complex. With excess HMDS, the reaction is reported to give an organic solvent soluble polymer that has been used to produce fibers and coatings.[11,27] Reactions performed at New Mexico with 2:1 to 1:1 (HMDS/borazine) mole ratios, (eq. 8) on the other hand, form polymers that are initially insoluble in organic solvents.[35] However, if these polymers are briefly dissolved in liquid NH_3 and then dried, they redissolve in organic solvents. It is reasoned that the initial polymer formed is highly cross-linked, but NH_3 solvolysis likely breaks down the extended structure leaving more soluble fragments. Indeed it is possible then to process these polymer solutions to form fibers,[36] coatings,[37] dense or porous bodies, xerogels,[38] aerogels,[39] and aerosols[40] all of which are converted to BN in these forms.

$$+ \; xs \; (Me_3Si)_2NH \longrightarrow \qquad\qquad\qquad (8)$$

The pyrolytic decomposition of the three point poly(borazinylamine) has been followed by TGA, thermal desorption mass spectrometry and isotope distribution analysis.[41] Although the full details of the solid state decomposition and lamellar structure assembly are not yet revealed, it can be concluded that borazine ring opening is a crucial reaction at low temperatures and some degree of solid state lamellar order is found for samples heated between 400° C and 500°C. In fact, these samples show x-ray patterns similar to those shown in Fig. 1. High temperature pyrolysis is required to remove the last hydrogen atoms in this structure as well. The early stage of the decomposition is illustrated schematically in Scheme II.

Some additional attempts have recently been made to prepare hybrid and copolymer materials containing boron and nitrogen by utilizing inherent reactivity of borazine monomers or chemistry that likely transits through borazine ring formation. That work includes blending of vinylic polysilane (VPS) with polyborazinylamine and reaction of $(EtNH)_3B$ with Et_2AlNH_2 and $(Me_2N)_4Ti$[42] to give viscous polymers. At New Mexico, we have also found that the structure and reactivity of two-point polyborazinylamines (**19**) can be modified by varying the nature of the substituent group, X, e.g. X = H, Me_2N, NH_2, N_3. Further, the reactive sites are ideal for introducing chemical constituents for the formation of new copolymers and eventually composites. As an example, it was found that reaction of (**19**) (X = NH_2) with metal amides, e.g., $(Me_2N)_4Ti$, followed by treatment with NH_3 produced a hybrid titanyl borazinylamine polymer. This polymer can be pyrolyzed at modest temperatures and

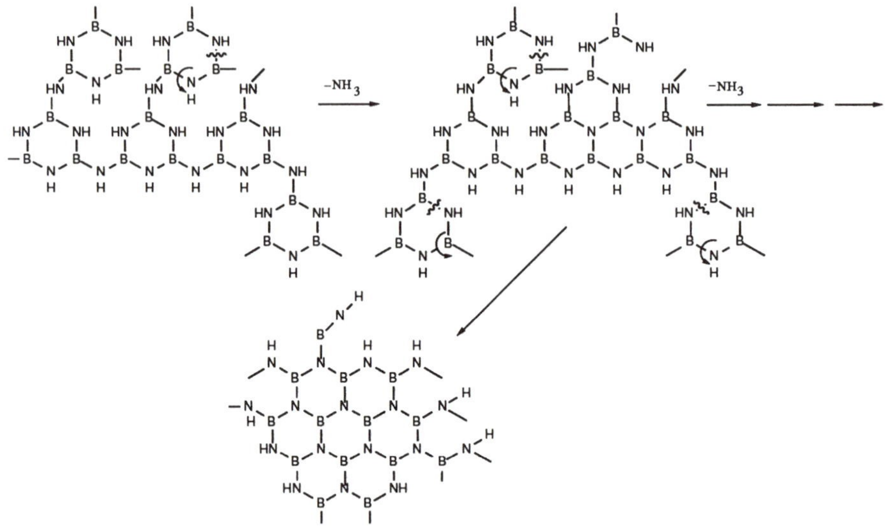

Scheme II

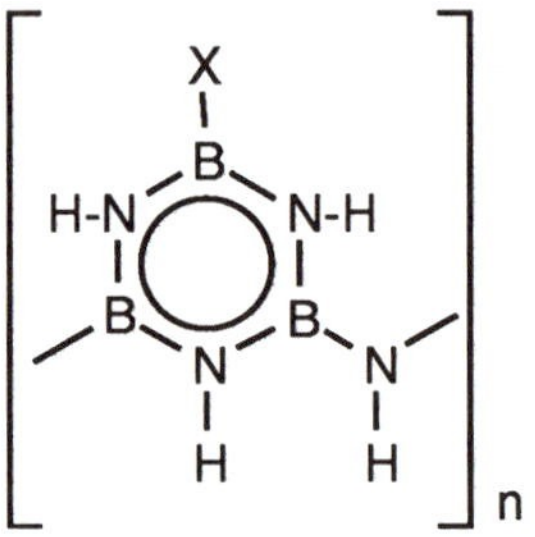

19

nano-composites of TiN/BN are obtained. Similar metal nitride dispersions in BN have been achieved with V, Nb, Ta Zr and Al. It is also possible to add non-metals to this polymer, e.g., SiCl$_4$, that lead to growth of a polysilaborazinylamine that converts to Si$_3$N$_4$/BN composites. Obviously this is also a rich area for continued development, and we are currently engaged in the extensions of the work.

Borazine Pendant Group Polymers

As the field of organic polymer chemistry matured, many investigators turned their attention to construction of more complex organic hydrid and copolymer systems. This has proven to be a very fruitful venture as some of the organic materials have proven to have unusual and enhanced properties such as mechanical strength, shock resistance, chemical resistance and unique opto-electric behavior. With this background, it is logical to expect that hybrid inorganic/organic and inorganic/inorganic polymers may be prepared, and these materials should display special properties with unlimited applications. Indeed, in the last few years there have been a number of reports of the synthesis of silicate/organic copolymer systems (ormsils, ceramers, etc),[43,44] as well as a host of systems involving polycarbosilanes/organic polymers, polysiloxane/organic polymers and polycarbosilane/polysilazane copolymers. The limited results available clearly indicate that this field deserves serious attention.

Two general approaches can be envisioned to attain a polymer containing a borazine pendant group: (1) polymerization or copolymerization of a suitable borazine monomer or (2) controlled functionalization by a borazine reagent of a preformed organic or inorganic polymer. Both of these approaches are currently being explored as illustrated by the examples discussed below.

Poly(vinylborazine). Initial efforts at Penn to generate borazine-substituted polymers focused on the construction of polystyrene analogues in which the borazine would be bound as a pendant group on a conventional polyethylene polymer backbone. Because of their similarity to conventional organic vinyl polymers, these types of borazine polymers might be both soluble and processable. Vinylborazine, $CH_2=CH-B_3N_3H_5$, which could be considered the inorganic analogue of styrene, could serve as the repeating unit in the construction of these new polystyrene-type polymers. Vinyl derivatives of alkylated borazines have been prepared;[45] however, the presence of the ring substituents was found to inhibit polymerization. Furthermore, even if polymers based on these compounds could be made, their high carbon contents would make them less desirable as boron nitride precursors. The parent B-trivinylborazine has also been synthesized.[46] Although the polymerization reactions of this compound have not been reported, it would be expected, due to the three vinyl substituents, to form only insoluble crosslinked polymers and have little preceramic utility.

Therefore, at the outset of this work there were no general synthetic routes for the formation of suitable B-alkenylborazine monomers and our initial goal was to develop an efficient synthesis of the parent B-vinylborazine. Since our previous work[47] has demonstrated that transition metal reagents, similar to those widely employed in organic and organometallic chemistry, may be used to catalyze a variety of transformations involving B-H activation reactions, we investigated the use transition metals to catalyze the alkyne-addition and olefin-substitution reactions of borazine. We found (eq. 9) that B-vinylborazine may be readily obtained[48] via the $RhH(CO)(PPh_3)_3$ catalyzed reaction of borazine with acetylene and this reaction has allowed the large scale production of this monomer.

$$B_3N_3H_6 + HC \equiv CH \xrightarrow{RhH(CO)(PPh_3)_3} \text{(borazine–CH=CH}_2\text{)} \tag{9}$$

Insoluble polymeric materials result from thermal polymerizations of neat liquid B-vinylborazine, but soluble poly(B-vinylborazine) homopolymers are readily obtained by solution polymerization with the free radical initiator AIBN (eq. 10).[49] Thus, when the polymerization of B-vinylborazine is carried out in benzene solution

$$(10)$$

for 20 h in the presence of 1.6 mol% AIBN, a white polymeric solid is obtained that is completely soluble in solvents such as benzene or ethers. Molecular weight studies using SEC/LALLS show Mw = 18,000 and Mn = 10,700 for the polymer prepared in this manner. The molecular weight distribution of these polymers is generally skewed in the high molecular weight end, suggesting that some crosslinking or branching may occur, perhaps by the formation of B-N linkages such as found in the poly(borazylene) polymer. The molecular weight of the polymer may also be controlled by varying the AIBN concentrations.

These new polyvinylborazine polymers have now been used to produce, depending upon pyrolysis conditions, a variety of ceramic materials, ranging from black, high carbon content materials to white h-BN. In each case, however, the polymer/ceramic conversions are found to take place with both high ceramic and chemical yields and give materials with B/N ratios of 1.0. Best results for the production of pure h-BN are obtained from pyrolyses under an ammonia atmosphere, which produce white, crystalline h-BN with a composition of $B_{1.00}N_{1.01}C_{0.006}H_{0.04}$.

Poly(styrene-co-B-vinylborazine) Copolymers. The discovery that soluble poly(B-vinylborazine) homopolymers could be readily formed in solution at 80 °C with the aid of AIBN initiation has enabled the systematic production of new soluble organic/inorganic hybrid polymers. In initial studies at Penn, we demonstrated the production of a new family of poly(styrene-co-B-vinylborazine) copolymers (eq.11).[49]

$$(11)$$

Elemental analyses and the spectroscopic data are consistent with the styrene-co-B-vinylborazine formulations. By adjusting the monomer feed ratios and the initiator concentrations, copolymers ranging in composition from 10 to 90% styrene and molecular weights from ~1,000 to >1,000,000 can be produced. The use of these AIBN initiated olefin copolymerization reactions are now being explored as routes to a wide variety of borazine-containing hybrid organic polymers.

Borazine-Modified Hydridopolysilazanes. Polysilazanes have been shown to be excellent polymeric precursors to silicon nitride or SiNC ceramic materials. Chemical modification of polysilazanes has also been proposed as a means of modifying and enhancing the properties of the polymer and/or the final ceramic materials. For example, the incorporation of boron in the polysilazane has been claimed to decrease the crystallinity of the silicon nitride derived from these polymers and thereby extend the effective use-temperature of these ceramics. Given the potential importance of SiNCB ceramics, investigations of the generation of new classes of hybrid polymers that might serve as processable precursors to such composites were initiated.

It was found[50] that the high yield synthesis of hybrid borazine-substituted hydridopolysilazanes can be achieved by the reaction of the preformed silicon-based preceramic polymer, hydridopolysilazane (**HPZ**), with liquid borazine at moderate temperatures (eq.12). The amounts of borazine incorporated into the hydridopolysilazane can be readily controlled. For example, polymers of compositions $(B_3N_3H_5)_{0.02}(HSi)_{0.30}(Me_3Si)_{0.19}(NH)_{0.26}N_{0.23}$ and $(B_3N_3H_5)_{0.07}(HSi)_{0.34}(Me_3Si)_{0.18}(NH)_{0.13}N_{0.28}$ were prepared by heating **HPZ** in excess borazine at 73°C for 2.2 h and 7.0 h, respectively. The spectroscopic data for the new polymers indicate that they retain their hydridopolysilazane backbones, but are substituted with pendant borazines by means of a borazine-boron to polymer-nitrogen linkage. Such linkages may form by either hydrogen or trimethylsilane elimination reactions.

$$(12)$$

Studies of the thermolytic reactions of these polymers demonstrate that they are converted in high ceramic and chemical yields to composite SiNCB ceramic materials. The ceramic yields (~70%) to 1400°C are increased over that observed for the unmodified **HPZ** (~57%) due to both the retention of boron and nitrogen in the final ceramic, and a borazine-crosslinking reaction which retards the loss of polymer backbone components. It is also very significant that, in contrast to the ceramic materials obtained from unmodified **HPZ**, the borazine-modified polymers yield ceramics that are amorphous to 1400°C. Pyrolyses carried out to higher temperatures also demonstrate that the boron-modified ceramics are more stable with respect to nitrogen loss and exhibit significant differences in the onset, phases and extent of crystallization in comparison to the ceramic derived from the unmodified **HPZ**.

Because of their ease of synthesis and processability, as well as the increased thermal stabilities and reduced crystallinity of their derived ceramics, these new hybrid borazine-modified hydridopolysilazanes polymers may prove to be reasonable precursors for enhanced SiNCB fibers, matrices for ceramic-ceramic or carbon-carbon composites and/or binders for nitrogen based ceramic powders. These applications are presently under investigation.

Conclusions

The diverse collection of boron-nitrogen polymer chemistry summarized here points out that efforts in this area are fruitful and worthy of continuing attention. Furthermore, the considerable progress made in processing some of the polymers into useful ceramic compositions and forms demonstrates that fundamental inorganic polymer chemistry has practical and, perhaps in the near future, commercial utility.

Acknowledgements. At the University of New Mexico we thank the NSF Center for Microengineered Ceramics for their support. At the University of Pennsylvania we thank the U.S. Department of Energy, Division of Chemical Sciences, Office of Basic Energy Sciences, Office of Energy Research, and the National Science Foundation Materials Research Laboratory at the University of Pennsylvania for their support of this research. We also thank the many coworkers and collaborators that are cited in the references for their contributions to the work described in this chapter.

References

1.   (a) Allcock, H. R. "Heteroatom Ring Systems and Polymers"; Academic Press: New York, 1967; (b) Lappert, M. F. "Developments in Inorganic Polymer Chemistry"; (c) Lappert, M. F., Leigh, G. J., Eds.; Elsevier: Amsterdam, 1962.
2.   Brinker, C. J.; Scherer, G. W. "Sol-Gel Science: The Physics and Chemistry of Sol-Gel Processing"; Academic Press: San Diego, 1990.
3.   Verbeek, W. U.S. Pat, 3,853,567, 1974.
4.   Yajima, S.; Hasegawa, Y.; Okamura, K.; Matsuzawa, T. *Nature* **1978**, *273*, 525.
5.   Wynne, K. J.; Rice, R. W., *Ann Rev. Mater Sci.* **1984**, *14*, 297.
6.   Mark, J. E.; Allcock, H. R.; West. R. "Inorganic Polymers"; Prentice Hall: Englewood Cliffs, New Jersey, 1992.
7.   Zeldin, M.; Wynne, K. J.; Allcock, H. R. (eds.) "Inorganic and Organometallic Polymers, ACS Symp Ser 360, Amer. Chem. Soc., Washington D.C., 1988.
8.   Laine, R. M. (ed.) "Inorganic and Organometallic Polymers with Special Properties"; NATO ASI Ser. Appl. Sci, Vol. 206, Kluver: Dordrecht, 1992.
9.   McNally, L.; Allen, C. W. *Heteroatom. Chem.* **1993**, *4*, 159.
10.   Laine, R. M. (ed.) "Transformation of Organometallics into Common and Exotic Materials: Design and Activation"; Niyhoff: Dordrecht, 1988.
11.   Paine, R. T.; Narula, C. J. *Chem. Rev.* **1990**, *90*, 73.
12.   (a) Melcher, L. A.; Adcock, J. L.; Lagowski, J. J. *Inorg. Chem.* **1972**, *11*, 1247; (b) Steinberg, H.; Brotherton, R. J. "Organoboron Chemistry"; Vol. 2, Wiley: New York, 1966.
13.   (a) Harris, J. J. *J. Org. Chem.* **1961**, *26*, 2155. (b) Gerrard, W.; Hudson, H. R.; Mooney, E. F. *J. Chem. Soc.* **1962**, 113.
14.   Bai, J.; Niedenzu, K.; Serwatowska, J.; Serwatowski, J. *Inorg. Chem.* **1992**, *31*, 228.
15.   (a) Brotherton, R. J.; McCloskey, A. L. U.S. Patent 3,101,369, 1963; *Chem. Abstr.* **1964**, *60*, 547; (b) Gutmann, V.; Meller, V.; Schelegel, R. *Monatsh. Chem.* **1964**, *95*, 314.

16. Hamilton, E. J. M.; Dolan, S. E.; Mann, C. M.; Colijin, H. O.; McDonald, C. A.; Shore, S. G. *Science* **1993**, *260*, 659.
17. Stock, A.; Pohland, E. *Chem. Ber.* **1926**, *59*, 2215.
18. Wiberg, E. *Naturwissenschaften* **1948**, *35*, 182, 212.
19. Laubengayer, A. W.; Moews, P. C., Jr.; Porter, R. F. *J. Am. Chem. Soc.* **1961**, *83*, 1337.
20. Manatov, G.; Margrave, J. L. *J. Inorg. Nucl. Chem.* **1961**, *20*, 348.
21. (a) Neiss, M. A.; Porter, R. F. *J. Am. Chem. Soc.* **1972**, *94*, 1438. (b) Porter, R. F.; Turbini, L. J. *Top. Curr. Chem.* **1981**, *96*, 1.
22. (a) Fazen, P. J.; Beck, J. S.; Lynch, A. T. *Chem. of Mater.* **1990**, *2*, 96-97. (b) Sneddon, L. G.; Su, K.; Fazen, P. J.; Lynch, A. T.; Beck, J. S. in *Inorganic and Organometallic Oligomers and Polymers*; Laine, R. M.; Harrod, J. F. eds. Kluwer: Dorrdrecht (Netherlands) **1991**. (c) Sneddon, L. G.; Mirabelli, M. G. L.; Lynch, A. T.; Fazen, P. J.; Su, K.; Beck, J. S. *Pure and Applied Chem.* **1991**, *63*, 407-410. (d) Fazen, P. J.; Remsen, E. E.; Sneddon, L. G. *Polymer Preprints,* **1991**, *32*, 544-545.
23. Kim, D.; Economy, J. *Chem. of Mater.* **1993**, *5*, 1216-1220.
24. (a) Su, K.; Nowakowski, M.; Bonnell, D.; Sneddon, L. G. *Chem. of Mater.* **1992**, *4*, 1139-1141. (b) Nowakowski, M.; Su, K.; Sneddon, L. G.; Bonnell, D.; *Mat. Res. Soc. Sym. Proc.* **1993**, *286*, 425-430. (c) Szabo, V.; Nowakowski, M.; Su, K.; Sneddon, L. G.; Ruhle, M.; Bonnell, D. *Mat. Res. Soc. Sym. Proc.* **1993**, *286*, 431-437.
25. Toeniskoetter, R. H.; Hall, F. R. *Inorg. Chem.* **1963**, *2*, 29.
26. Taniguchi, I.; Harada, K.; Maeda, T. *Jpn. Kokai Tokkyo* 76,53,000, 1976; *Chem Abstr* **1976**, *85*, 96582v.
27. (a) Paciorek, K. J. L.; Krone-Schmidt, W.; Harris, D. H.; Kratzer, R. H. *ACS Symp. Ser.* **1988**, *360*, 392; (b) Paciorek, K. J. L.; Masuda, S. R.; Kratzer, R. H. *Chem of Mater.* **1991**, *3*, 88.
28. Narula, C. K.; Schaeffer, R.; Datye, A.; Paine, R. T. *Inorg. Chem.* **1989**, *28*, 4053.
29. Kimura, Y.; Kubo, Y.; Hayashi, N. *J. Inorg. Organomet. Polymer* **1992**, *2*, 231.
30. "Gemelin Handbuch der Anorganischen Chemie". Vol. 13 (Part 1); Springer-Verlag: New York, 1974.
31. Maringgele, W. "The Chemistry of Inorganic Homo- and Hetero-cycles"; Haiduc, I. (ed.); Academic Press: London, 1987.
32. Lappert, M. F.; Majunder, M. F. *Proc. Chem. Soc.* **1961**, 425.
33. Meller, A.; Füllgrabe, H. J. *Z. Naturforsch B* **1978**, *33B*, 156.
34. Paine, R. T. *J. Inorg. Organomet. Polym.* **1992**, *2*, 183.
35. Narula, C. K.; Schaeffer, R.; Paine, R. T. *J. Am. Chem. Soc.* **1987**, *109*, 5556.
36. Lindquist, D. A.; Janik, J. F.; Datye, A. K.; Paine, R. T.; Rothman, B. *Chem. of Mater.* **1992**, *4*, 17.
37. (a) Narula, C. K.; Datye, A. K.; Schaeffer, R.; Paine, R. T. Chem. Mater. **1989**, *1*, 486. (b) Datye, A. K.; Xiaomei, Q.; Borek, T. T.; Paine, R. T.; Allard, L. F. *Proc. Mat. Res. Soc.* **1990**, *180*, 807. (c) Allard, L. F.; Datye, A. K.; Nolan, T. A.; Mahan, S. L.; Paine, R. T. *Ultramicroscopy* **1990**, *37*, 153.
38. Borek, T. T.; Ackerman, W.; Hua, D. W.; Paine, R. T.; Smith, D. M. *Langmuir* **1991**, *7*, 2844.
39. Lindquist, D. A.; Borek, T. T.; Kramer, S.J.; Narula, C. K.; Johnston, G.; Schaeffer, R.; Smith, D. M.; Paine, R. T. *J. Amer. Ceram. Soc.* **1990**, *73*, 757.
40. Lindquist, D. A.; Kodas, T. T.; Smith, D. M.; Xiaomei, Q.; Hietala, S.; Paine, R. T. *J. Amer. Ceram. Soc.* **1991**, *74*, 3126.

41   Rye, R. R.; Tallant, D. R.; Borek, T. T.; Lindquist, D. A.; Paine, R. T. *Chem. of Mater.*, **1991**, *3*, 286.
42.   (a) Interrante, L. V.; Schmidt, W. R.; Marchetti, P. S.; Maciel, G. E.*Mat. Res. Soc. Symp. Proc.* **1992**, *249*, 31. (b) Kwon, D.; Schmidt, W. R.; Interrante, L. V.; Marchetti, P. J., Maciel, G. E. "Inorganic and Organometallic Oligomers and Polymers"; Harrod, J. F.; Laine, R. M. (eds.); Kluver, Netherlands, 1991.
43.   Wilkes, G. L.; Huang, H.; Glaser, R. H. In "Silicon Based Polymer Science", Zeigler, J. M.; Fearon, F. W. G., Eds., Advances in Chemistry 224; American Chemical Society: Washington, D.C., 1990.
44.   Shea, K. J.; Loy, D. A.; Webster, O. *J. Am. Chem. Soc.* **1992**, *114*, 6700 and references therein.
45.   (a) Klancia, A. J.; Faust, J. P.; King, C. S. *Inorg. Chem.* **1967**, *6*, 840. (b) Klancia, A. J.; Faust, J. P. *Inorg. Chem.* **1968**, *7*, 1037. (c) Jackson, L. A.; Allen, C. W. *J. Chem. Soc., Dalton Trans.* **1989**, 2423. (d) Pellon, J.; Deichert, W. G.; Thomas, W. M. *J. Polym. Sci.* **1961**, *55*, 153. (e) Jackson, L. A.; Allen, C. W. *Phosphorus, Sulfur, Silicon* **1989**, *41*, 341.
46.   Fritz, P.; Niedenzu, K.; Dawson, J. W. *Inorg. Chem.* **1964**, *3*, 626.
47.   Sneddon, L. G. *Pure and Appl. Chem.* **1987**, *59*, 837-846 and references therein.
48.   (a) Lynch, A. T.; Sneddon, L. G. *J. Am. Chem. Soc.* **1987**, *109*, 5867-8. (b) Lynch, A. T.; Sneddon, L. G. *J. Am. Chem. Soc.* **1989**, *111*, 6201-9.
49.   (a) Su, K.; Remsen, E. E.; Thompson, H.; Sneddon, L. G. *Macromolecules,* **1991**, *24*, 3760-3766. (b) Su, K.; Remsen, E. E.; Thompson, H.; Sneddon, L. G. *Polymer Preprints,* **1991**, *32*, 481-482.
50.   (a) Su, K.; Remsen, E. E.; Zank, G.; Sneddon, L. G. *Polymer Preprints* **1993**, *34*, 334-335. (b) Su, K.; Remsen, E. E.; Zank, G.; Sneddon, L. G. *Chem. of Mater.* **1993**, *5*, 547-556.

RECEIVED April 8, 1994

Chapter 28

Synthesis and Preceramic Applications of Poly(aminoborazinyls)

Yoshiharu Kimura and Yoshiteru Kubo

Department of Polymer Science and Engineering,
Kyoto Institute of Technology, Matsugasaki, Kyoto 606, Japan

Soluble and thermoplastic derivatives of poly(aminoborazinyls) were used as ceramic precursors for fabricating hexagonal boron nitride (h-BN) ceramics. They were readily prepared by solution condensation of B,B,B-tri(methylamino)-borazine (MAB) and bulk co-condensation of MAB and laurylamime (LA). The structure analysis revealed that both the solution and bulk condensates consist of several borazinyl rings which have two- and three-point linkages with other rings as well as a one point-linkage. The bulk co-condensate of MAB-LA was melt-spun and pyrolyzed to fabricate BN fiber, while the solution condensate was used as a ceramic binder in the powder consolidation molding of h-BN and as a matrix material in the fabrication of fiber-ceramic composites (FCC). These examples demonstrated that these BN precursors are useful for developing well-designed super heat-resistant materials that may be essential for future technological innovation.

Among the newly developed high-tech ceramics, hexagonal boron nitride (h-BN) has been occupying an unique position (1), because of its excellent heat resistance, low density, and structural similarities with graphite. In the utilization of h-BN, however, there have been intrinsic problems related with its processability and mechanical properties. Since the processing of h-BN cannot usually be performed by conventional thermal sintering, a high-energy process like hot-pressing has been utilized. Therefore, fabrication of fibers and molding with complicated, fine shape are difficult. Furthermore, mono-lithic h-BN is likely to crack from a surface micro flaw, and reinforcement with ceramic fiber is preferable (2). For these purposes the so-called precursor method using an appropriate inorganic polymer as the ceramic precursor should be effective (3), since the polymeric precursor should be easily processed to a desired shape and then pyrolyzed into the desired ceramics by the ordinary thermal sintering process without using sinter-aiding agent (4, 5). In this method the structure and properties of the precursor have a close relationship with the performance of the ceramics obtained, and poly(aminoborazinyls) consisting of the unit components of h-BN have received much attention for use as the h-BN precursor (1). In particular, Paciorek (6), Paine (7), and Sneddon (8) designed various derivatives of poly(aminoborazinyls) which can be readily converted to h-BN. Recently, we demonstrated that poly(aminoborazinyls) prepared from B,B,B-

0097–6156/94/0572–0375$08.00/0
© 1994 American Chemical Society

triaminoborazine (TAB) and B,B,B-tri(methylamino)borazine (MAB) are potentially useful for the fabrication of BN fibers and molds (9-11). The further applications of this technique, bench-scale preparation of BN fiber, powder consolidation molding of BN ceramics, and fabrication of fiber-ceramic composite (FCC) have been studied in detail. This paper deals with the concepts of these methods as well as a preliminary feasibility study utilizing poly[(methylamino)borazinyl], i.e., the MAB condensate, as the precursor for making BN ceramics.

Synthesis of MAB

MAB was prepared in high yield by the following reaction scheme (12). It was isolated as white waxy material.

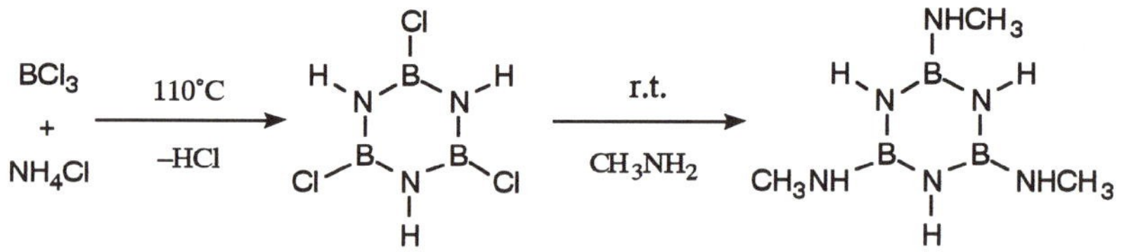

MAB: IR (nujol) 3460 (NH), 1410 cm^{-1} (B$_3$N$_3$ ring), etc.; ^{1}H NMR (C$_6$D$_6$) δ 1.4 ppm (broad, 3H, MeNH), 2.4 (d, 9H, CH_3), 2.6 (s , 3H, NH), ^{11}B NMR (C$_6$D$_6$) δ 26 ppm (the line width at half height = 284 Hz).

Solution polymerization of MAB

MAB polymerizes in the bulk state when heated above 130°C to form a thermoplastic condensate. However, its thermoplasticity is likely to be lost by crosslinking during the bulk condensation (10). In order to control the degree of condensation, the thermal condensation of MAB was carried out in refluxing chlorobenzene solution. With extended reflux an insoluble crosslinked product precipitated. Therefore, the reaction was stopped upon observation of the first trace of the precipitate, and the MAB condensate was isolated from this solution as white solid material.

Table I shows a typical result of the solution condensation of MAB as compared with that of the bulk condensation. The yields of the condensate were almost identical, but the solution condensate was wholly soluble in organic solvents, while the bulk condensate contained some insoluble gel. The average molecular weight of the former was slightly lower than that of the soluble part of the bulk condensate. The softening point of the solution condensate was about 80°C, which was similar to that of the co-condensate of MAB and laurylamine (LA) (90/10 in wt; vide infra).

Table I. Thermal condensation of MAB

Solvent	Temp. (°C)	Time (h)	Condensate (wt%)	Mn
chlorobenzene	130	2	85	500
-	150	2	87	720

The ^{1}H NMR spectrum of the condensate is shown in Figure 1(b). As compared with the spectrum of MAB (Figure 1(a)), the signals at δ 1.4 and 2.4 ppm were reasonably ascribed to the NH and methyl protons of the substituent

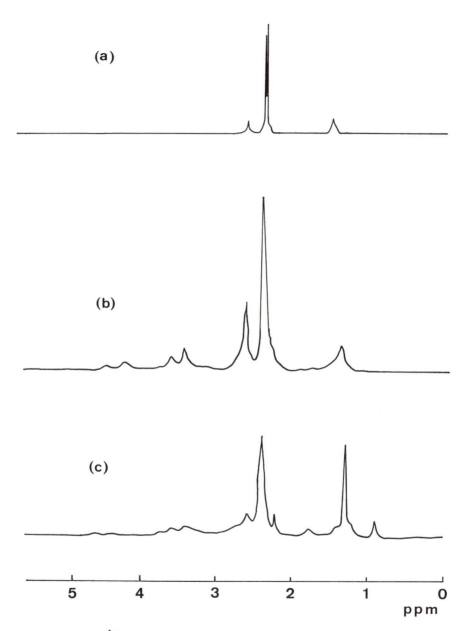

Figure 1. ^{1}H NMR spectra of (a) MAB, (b) the solution condensate of MAB, and (c) the bulk co-condensate of MAB-LA (at 200 MHz, in C_6D_6).

methylamino groups, respectively. In contrast, the NH protons of the borazine ring appeared as three sets of signals around δ 2.6, 3.5, and 4.3 ppm. While the highest-field one is ascribed to the NH protons of the borazine ring that is bonded with only one borazinylamino group (one-point-linkage form). The middle- and the lowest-field signals are reasonably ascribed to those of the borazine rings that are bonded with two and three borazinylamino groups (two- and three-point-linkage forms), respectively. The splittings of the latter signals may be due to the asymmetry of the borazinyl substituents. With this signal assignment combined with the signal intensities and the molecular weight, the structure of the condensate could be represented by **1**, for example, whose average degree of polymerization was five. These facts supported that the homo condensation of MAB is easily controlled in solution state.

1

Bulk Co-Condensation of MAB and Laurylamine

Another effective method for preparing thermoplastic poly(aminoborazinyl) is to make a co-condensate of MAB with such a higher alkylamine as LA. We have shown that the latter can be trapped in the condensate to decrease the crosslinks and work as plasticizer (9).

In the present study, a larger scale synthesis of the co-condensate was studied by heating about 20 g of MAB and 10 wt% of LA at 200-260°C for 1 h in a nitrogen flow. Since in this larger scale reaction the elimination of methylamine was slowed, a relatively higher reaction temperature was necessary to effect the condensation.

Table II. Thermal Co-Condensation[a] of MAB and LA

Run No.	LA (wt%)	Temp. (°C)	Time (h)	Weight Loss (%)	Mn[b] (dalton)	Spinnability
1	10	200	1	10.1	-	good
2	10	220	1	13.4	-	good
3	10	240	1	17.1	470	excellent
4	10	260	1	18.6	-	poor

[a] 20 g of MAB and 2.0 g of LA were reacted.
[b] By GPC relative to PPG standards.

As summarized in Table II, the degree of condensation in terms of weight loss wassuccessfully controlled by changing the reaction temperature, and colorless thermoplastic products were formed when the weight loss due to the elimination of

MA was within 10-20 wt%. M_n of the soluble condensates were in the range of 400-800 as determined by GPC, which corresponds to that of pentamers - decamers of MAB. The typical ^1H NMR spectrum of the condensate is shown in Figure 1(c). The characteristic signals due to the methyl and methylene protons of lauryl group were noted at δ 0.9, 1.28, and 2.2 ppm, respectively. The methyl signal of $NHCH_3$ group at δ 2.4 ppm became broader, and the two signals at δ 1.3 and 1.8 ppm may be ascribed to the pendant NH of the methylamino and laurylamino groups, respectively. The signals of the NH protons of the borazine ring were detected at the region where those of the borazine rings with one- and two-point linkages should appear. Therefore, it appears that the number of three-point-linkage was decreased. The ^{11}B NMR signal of this product was detected as a single peak at the same chemical shift (δ 26 ppm) as observed for MAB. Based on these spectroscopic data, the structure of the MAB-LA condensate is postulated as **2**, in which LA units are incorporated as alkylimino and alkylamino units to interfere with the formation of crosslinks in the product.

2 (R=$C_{12}H_{25}$)

The DTA of these condensates exhibited a broad endothermic peak around 85°C and an exothermic peak ranging from 200 to 430°C which are ascribed to their melting and thermal decomposition, respectively. A large weight loss was observed at 200-500 °C in TGA, and the final weight loss at 900°C was 38 %.

Fabrication of BN Fiber from Bulk Co-Condensate of MAB-LA

As discussed in the Introduction, the poly(aminoborazinyls) can be pyrolyzed into h-BN ceramics and have various potential applications for use as preceramics. In this article, some examples based on the different processabilities of the above MAB condensates are shown.

In the early 1970s, Economy *et al.* invented a chemical conversion method of fabricating BN fiber, in which boron oxide fiber was melt-spun and converted to BN by solid phase nitridation in an ammonia atmosphere (*13*). In this process, however, quality control of the precursor fiber and the process control of the nitridation were not easy. Therefore, an alternative preceramic process was developed (*14, 15*), although the spinnability of the polymer precursors utilized were too poor to fabricate precursor fibers with excellent quality. Recently, the author reported a new method of fabricating high-performance h-BN fiber based on a thermoplastic derivative of poly[(methylamino)borazinyl] as the precursor polymer (*9, 11*). In this process the precursor fiber is spun by melt-spinning techniques, pyrolyzed directly up to 1000°C in an ammonia gas flow, and further sintered up to 1800°C in a nitrogen flow. The strength of the sintered BN fiber was found to reach 1 GPa. In the present study, the bench-scale preparation of BN fiber was investigated by a similar approach using the co-condensate of MAB-LA prepared above.

Melt spinning. Using a micro screw extruder with a nozzle of 500 mm in diameter, the condensate was melt-spun at 160-170°C. The extrudate from the spinneret was taken up by a winder to give a continuous filament of about 50 mm in diameter. The best melt-flow of the extrudate was obtained for the condensate obtained at 240°C (Run No.3 in Table I). The good spinnability of the MAB-LA condensate in spite of its low molecular weight is attributed to the strong intermolecular aggregation between the planar borazine rings. This phenomenon is comparable to that of amorphous carbon pitch (16), which is utilized as the precursor to carbon fiber and has an excellent spinnability. The latter consists of aromatic rings in place of borazine networks with Mn of 800 - 1000. A typical SEM micrograph of the precursor fiber (see Figure 2) indicates that the cross-section of the fiber is circular and that the inner phase is homogeneous.

Pyrolysis of fiber. During the spinning, the fiber surface was slightly hydrolyzed, and the fiber became infusible. Therefore, it was directly pyrolyzed in an electric tube furnace without special thermosetting. The temperature of the furnace was raised to 600-1000°C at a rate of 100°C/h in a flow of ammonia gas (30 ml/min), held at the predetermined temperature for 1 h, and then cooled to room temperature. At this stage the sintered fiber was gray due to the carbonization of the organic residue. It was further sintered up to 1400 and 1800°C in a nitrogen flow. Then, its color turned white above 1400°C. Figure 2 shows the typical SEM photographs of the fibers pyrolyzed at various temperatures. The inner phase of the fibers pyrolyzed at 1000 and 1400°C was homogeneous, while that of the fibers sintered above 1600°C exhibited a slight grain growth.

Mechanical properties. We reported in the former paper (9) that the modulus of the BN fiber prepared by the precursor method increased with increasing pyrolysis temperature, while its tensile strength decreased around 1400°C and increased again at the higher temperature. The average tensile strength and modulus observed were 1 GPa and 78 GPa, respectively, which are lower than those of the common ceramics fibers reported thus far (3). Most of the latter fibers, however, are known to undergo serious deterioration above 1400°C because of the abrupt crystallization and coarse grain growth (17). The present BN fiber is characterized by the exceptional increase in strength at high temperature greater than 1400°C.

Table III. Comparison of the properties of the BN fibers prepared by precursor and chemical conversion methods

	Precursor Method	Chemical Conversion Method[a]
Density	2.05 g/cm^3	1.85-2.10 g/cm^3
Diameter	10 mm	4-6 mm
Tensile strength	0.98 GPa	0.3-1.3 GPa
Tensile modulus	78 GPa	35-67 GPa
Heat resistance	-	2500°C (in nitrogen)
	-	900°C (in air)
Thermal conductivity	-	0.0665cal/sec/cm^2/0°C/cm (500°C)
Electroresistivity	-	10^{13} Ω·cm

[a]See ref. (33).

Table III summarizes the properties of the present BN fiber as compared with those of the BN fiber prepared by the chemical conversion method (13). The mechanical properties of the both fibers are almost identical. We expect that further

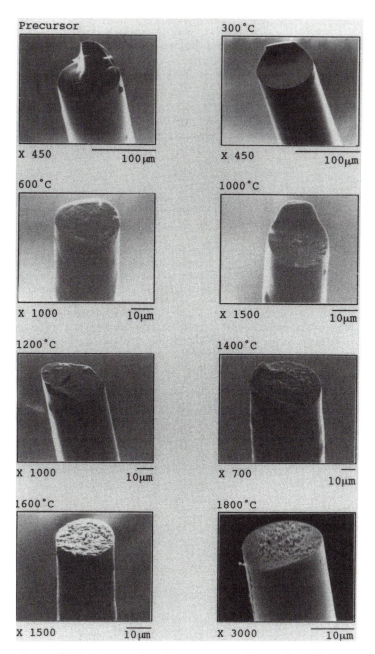

Figure 2.   SEM micrographs of the precursor fiber and the fibers pyrolyzed at the temperatures noted.

increase in strength would be realized by quality control of the precursor fiber, as well as by control of the coarse grain growth in the sintered fiber.

Fabrication of BN Disk with Solution Condensate as Ceramic Binder

For sintering BN powder which has no sintering ability, a high-temperature, high-pressure process of hot-pressing has generally been adopted in the presence of a sinter-aiding agent such as boron trioxide. When the polymer precursor is used as ceramic binder, powder consolidation may be allowed by the conventional sintering technique without loading pressure (18, 19). Here, the solution condensate of MAB was used as the ceramic binder in powder molding of BN ceramics, because it is soluble and contains no higher alkylamine which ought to create large pores and carbon contaminant in the pyrolysate.

A typical experimental procedure is shown in Scheme I. The MAB condensate obtained was dissolved in a least amount of chlorobenzene, and the solution was mixed with h-BN powder in a condensate to powder ratio of 25/75 by weight. After the mixture had been well agitated, chlorobenzene was evaporated. The powder mixture obtained was then compressed to a circular disk of 10.15 mm in diameter and 1.35 mm in thickness. It was first pyrolyzed up to 1000°C in an ammonia flow and was subsequently sintered up to 1800°C in a nitrogen flow. In each step, a hard, well-shaped disk was obtained, indicating the effectiveness of the MAB condensate as the binder. The size, weight, and bulk density of the disk obtained are summarized in Table IV. The disk was found to shrink slightly in diameter with constant thickness when it was sintered up to 600°C. Above 600°C, however, further shrinkage was not observed in spite of the decrease in weight. Therefore, the density of the disk decreased with increasing porosity, with the disk sintered up to 1800°C having a density of 1.58 g/cm^3. Figure 3 shows the morphology change of the disk with sintering temperature, as observed by optical microscopy and SEM. SEM revealed that the BN particles adhered well with increasing temperature. In the disk sintered up to 1800°C the formation of flake like particles was observed. These data indicate that the solution condensate of MAB is an effective ceramic binder.

Table IV. Sintering of BN powder with MAB condensate as the binder[a]

Sintering Temp. (°C)	Diameter (mm)	Thickness (mm)	Weight (mg)	Bulk Density (g/cm^3)
-	10.15	1.35	195	1.79
600	10.00	1.35	181	1.71
1000	10.00	1.35	175	1.65
1400	10.00	1.35	173	1.63
1800	10.00	1.35	167	1.58

[a]A circular disk comprising 75 wt% of BN powder and 25 wt% of the MAB condensate was made by compression molding.

The disk sintered up to 1800°C was then subjected to an impregnation process to decrease porosity as illustrated in Scheme II. The disk was first soaked in a 50 wt% solution of the MAB condensate in chlorobenzene. The whole system was evacuated so that the solution could penetrate small pores inside the disk. The disk was then taken out, thoroughly dried in vacuo, and pyrolyzed up to 1000°C in an ammonia flow. This cycle was repeated several times, and the disk was finally sintered up to 1800°C in a nitrogen flow. Figure 4 shows the increase in bulk density with the number of impregnation cycles. It suggests that the increase in density leveled off beyond the fourth impregnation cycle where the bulk density reached 1.74 g/cm^3.

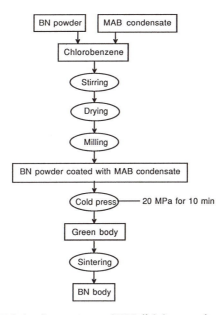

Scheme 1. Fabrication process of BN disk by powder consolidation.

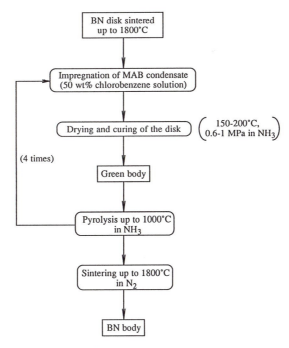

Scheme 2. Impregnation process.

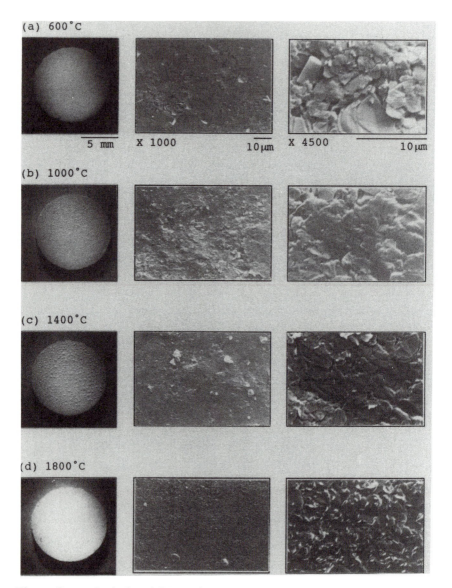

Figure 3. Optical and SEM micrographs of the BN disks pyrolyzed at the temperatures noted.

When this impregnated disk was finally sintered up to 1800°C, the density became 1.72 g/cm^3. This value is much higher than that of a commercially available BN mold fabricated by the ordinary powder sintering (1.4 g/cm^3), but slightly lower than that of a hot-pressed mold (1.9 g/cm^3). Figure 5 shows the SEM of the fracture surfaces of the BN disks before and after the impregnation process. The former has a number of cracks and pores, while in the latter pore sizes and porosity were decreased by impregnation.

Fabrication of Carbon Fiber-BN Matrix Composites

Incorporation of continuous fibers into a ceramic matrix yields a composite material with superior properties (*20*). Here, a BN matrix composite reinforced with carbon cloth was fabricated by using the solution condensate of MAB as the matrix precursor. Scheme III shows the procedure of making a carbon fiber-BN matrix composite. A thick film (thickness : 100 mm) of the MAB condensate was cast by hot-pressing at 70°C. Two sheets of this condensate film were piled up with a piece of carbon cloth in the center, covered with a parting paper from both sides, and hot-pressed at 70°C for impregnation. The prepreg thus obtained was cut into a prescribed size (100 x 100 x 0.25 mm), and twelve piles of this prepreg were laid up on a flat steel plate with a dam. The whole system was put into a Teflon bag and pressed at 70°C at 1.5 MPa by the conventional vacuum bag technique to form a green body. It was then transferred to an tubular furnace and pyrolyzed up to 1000°C in an ammonia flow. The pyrolysate was subjected to the aforementioned impregnation process, in which the soaking and pyrolysis were repeated several times. The composite obtained was finally sintered up to 1800°C in a nitrogen flow. From the weight of the composite, the final volume fraction of carbon fiber (V$_f$) was calculated to be 24%.

After the first pyrolysis of the green body up to 1000°C, the density of the composite was 0.74 g/cm^3. So, the aforementioned impregnation cycle was repeated in order to increase the density of the composite. The increase in density with the number of impregnation cycle is shown in Figure 6. After the sixth impregnation cycle the density reached a maximum of 1.10 g/cm^3. When this composite was finally sintered up to 1800°C in nitrogen flow, the density became 1.05 g/cm^3, indicating that it still had a high porosity. Its mechanical properties were evaluated by the ordinary three point-flexural test. The maximum flexural strength and modulus of the composite was found to be 61 MPa and 23 GPa, respectively. These values are much higher than those of the compacted plate of h-BN ceramics (20-30 MPa in strength), although they should be improved by decreasing the porosity.

Conclusions

Soluble and thermoplastic derivatives of poly(aminoborazinyls) were prepared by solution condensation of MAB and bulk co-condensation of MAB and LA. Both of the condensates consist of several borazinyl rings which have two- and three-point linkages with other rings as well as a one point-linkage.

A bench-scale preparation of BN fiber was successfully carried out by the precursor method with the bulk condensate of MAB-LA. The results were as follows:
1. The precursor fiber could be spun by the conventional melt spinning of the MAB-LA condensate.
2. Because the surface of the fiber was slightly hydrolyzed during the spinning, the precursor fiber was directly pyrolyzed to 1000°C in an ammonia flow and further sintered up to 1800°C in nitrogen flow.
3. The mechanical properties of the BN fiber was found to increase above 1400°C.
4. Further increase in strength would be realized by quality control of the precursor fiber, as well as by control of the coarse grain growth in the sintered fiber.

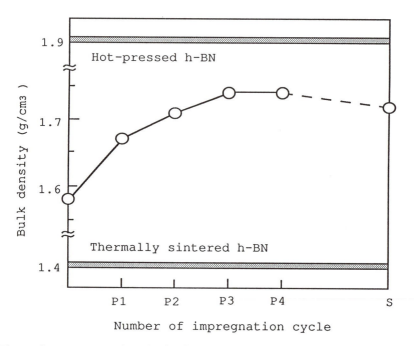

Figure 4. Increase in bulk density of the BN disk with increasing number of impregnation cycle. Each impregnation cycle (shown by P) involves pyrolysis up to 1000°C in ammonia flow. The final sintering up to 1800°C in nitrogen flow is shown by S.

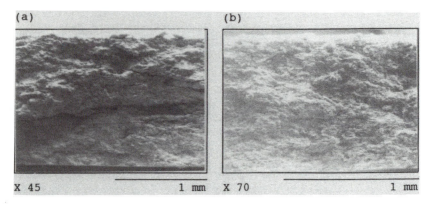

Figure 5. Optical micrographs of the fracture surface of the BN disks (a) before and (b) after the impregnation cycles.

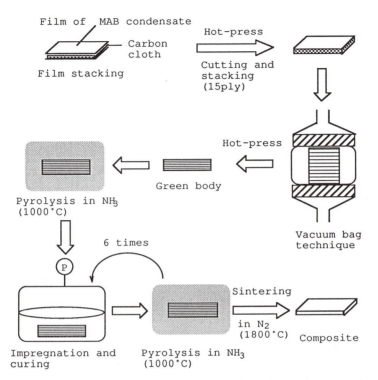

Scheme 3. Fabrication process of carbon fiber-BN matrix composite.

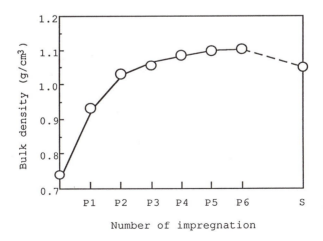

Figure 6. Increase in bulk density of the carbon fiber re-inforced BN matrix composite ($V_f = 24\%$) with increasing number of impregnation cycle. See Figure 4 for the notations.

The solution condensate of MAB was used as ceramic binder in molding BN ceramics from BN powders and as matrix material in fabricating carbon fiber-BN matrix composite. The following results were obtained :

1. With the MAB condensate as ceramic binder, BN powders could be sintered by ordinary thermal sintering to give a well-shaped BN mold.
2. The bulk density of the mold was higher than that of commercially available BN mold, but slightly lower than that of hot-pressed mold.
3. The impregnation procedure with MAB condensate was successful for densification of the mold.
4. The mechanical properties of the composite were superior to those of the h-BN ceramics, but inferior to the recently developed carbon-fiber reinforced composites such as carbon-carbon composites.
5. h-BN matrix composites may be promising as a high temperature structural material in terms of heat resistance.

Literature Cited

(1) Paine, R. T.; Narula, C. K. *Chem. Rev.* **1990,** *90*, 73-91
(2) Prewo, K. M. *Am. Ceram. Soc. Bull.* **1989,** *68*, 395-400
(3) Rice, R. W. *Am. Ceram. Soc. Bull.* **1983,** *62*, 889
(4) Schwab, S. T.; Graef, R. C.; Davidson, D. L.; Pan, Y. *Polym. Prepr. (Am. Chem. Soc., Div. Polym. Chem.)* **1991,** *32 (3)*, 556-558
(5) Sato, K; Suzuki, N; Hunayama, T; Isoda, T. *J. Ceram. Soc. Jap* **1992,** *100 (4)*, 444-447
(6) Paciorek, K. J. L.; Harris, D. H.; Kratzer, R. H. *J. Polym. Sci., Polym. Chem. Ed.,* **1986,** *24*, 173-185
(7) Narula, C. K.; Schaeffer, R.; Datye, A. K.; Borek, T. T.; Rapko, B. M.; Paine, R. T. *Synthesis of Chem. Mater.* **1990,** *2 (4)*, 384-389
(8) Fazen, P. J.; Beck, J. S.; Lynch, A. T.; Remsen, E. E.; Sneddon, L. G. *Chem. Mater.* **1990,** *2 (2)*, 96-97
(9) Kimura, Y.; Hayashi, N.; Yamane, H.; Kitao, T. *Proc. 1st. Japan International SAMPE Symposium*, Nov. 28-Dec. 1 **1989,** *1*, 906
(10) Kimura, Y.; Kubo, Y.; Hayashi, N. *J. Inorg. Organometal Polym.* **1992,** *2*, 231-242
(11) Kimura, Y.; Kubo, Y.; Hayashi, N. *Composites Sci. Tech.*, in print
(12) Brown, C. A. ; Laubengayer, A. W. *J. Chem. Soc.* **1955,** *77*, 3699
(13) Economy, J.; Anderson, R. V. *J. Polym. Sci.* **1967,** *C19*, 283
(14) Paciorek, K. J. L.; Krone-Schmidt, W.; Harris, D. H.; Kratzer, R. H.; Wynne, K. J. In *Inorganic and Organometallic Polymers, ACS Symposium Series 360*; Zeldin, M.; Wynne, K. J.; Allcock, H. R. Ed. Am. Chem. Soc., Washington 1988, pp. 392-406,
(15) Lindquist, D. A.; Janik, J. F.; Datye, A. K.; Paine, R. T. Chem. Mater. **1992,** *4*, 17-19.
(16) Singer, L. S. *Carbon* **1978,** *16*, 408
(17) Douglas, J. P.; Kenneth, C. G.; Robert, S. H. J.; Richard, E. T. *J. Am. Ceram. Soc.* **1989,** *72 (2)*, 284-288
(18) Yajima, S.; Shishido, T.; Okamura, K. *Ceram. Bull.* **1977,** *56*, 1060
(19) Schwartz, K. B.; Rowcliffe, D. J. *J. Am. Ceram. Soc.* **1986,** *69 (5)*, C-106-C-108
(20) Nishikawa, N. *Kagaku Kougyou* **1989,** *40 (9)*, 780, 803-809

RECEIVED April 14, 1994

Chapter 29

Copolymerization Reactions of Inorganic Rings Containing Olefinic Substituents
Quantitative Reactivity Studies

C. W. Allen[1], D. E. Brown[1,2], R. F. Hayes[1], R. Tooze[3], and G. L. Poyser[3]

[1]Department of Chemistry, University of Vermont,
Burlington, VT 05405–0125
[3]Wilton Materials Research Centre, ICI plc, P.O. Box 90, Wilton,
Cleveland TS6 8JE, England

Reactivity ratio determinations for copolymerization reactions of monomers in which a cyclophosphazene or a cycloborazine with attached olefin containing substituents are discussed. Several previously reported systems have been re-examined using improved calculational methodologies. The effect of both the terminal chain unit and the penultimate unit on the mode of chain propagation was detected. The Alfrey Price Q/e values derived from the reactivity ratios are interpreted in terms of the electronic interactions between the olefinic substituent and the inorganic ring system. The role of the analytical methods employed for generation of the copolymer composition data in obtaining reliable reactivity ratios has been explored.

Work in our laboratories has focused on development of a new class of hybrid inorganic-organic polymers which consist of a carbon chain backbone and an inorganic ring system as a substituent (1). The inorganic rings which have been utilized in these polymers include the cyclophosphazenes and cycloborazines. In comparison to traditional inorganic polymers such as poly(siloxanes) and poly(phosphazenes), in which the backbone consists of a thermodynamically stable but kinetically labile inorganic entity protected by organic substituents, the polymers we have developed have the inorganic moiety as a substituent. The

[2]Current address: Eastman Kodak, Kodak Park, Rochester, NY 14650–2132

0097–6156/94/0572–0389$08.00/0
© 1994 American Chemical Society

accessibility of the inorganic substituent allows for it to be exploited synthetically (1,2). The inorganic ring system polymers are a subset of the general class of addition polymers obtained from polymerization of inorganic molecules with olefinic substituents (3). To date, polymerization of the olefin containing inorganic ring systems has been limited to radical chain processes. Attempts at both cationic and anionic initiation failed presumably due to capture of the cationic species by the nucleophilic sites on the ring and capture of the anionic species by the electrophilic sites. The advantage of utilizing inorganic ring systems is that the resulting polymers have a high concentration of inorganic material and, where appropriate such as with the cyclophosphazenes, more reactive sites. A further desirable characteristic of this general class of inorganic-organic polymers is the ability, through copolymerization, to incorporate the inorganic entity into traditional organic polymers (1,4). The existence of authentic copolymerization reactions allows for the application of quantitative kinetic models leading to the determination of monomer reactivity ratios which in turn provide information on electronic effects operative in the olefin site of the inorganic ring system. In this paper, we report the application of this methodology to analyze factors controlling the reactivity of these novel inorganic monomers.

Alkenylfluorophosphazene Copolymers

The first type of of monomers under consideration are those in which the olefin is directly attached to the inorganic ring. Our initial work in this area involved the propenylphosphazene, $N_3P_3F_5C(CH_3)=CH_2$, and its copolymerization with styrene like monomers (5). Reactivity ratios were calculated using linear methods and while they are believed to give an approximate view of the system behavior, it is well known that linear methods are susceptible to a variety of errors (6,7). Reactivity ratios for the $N_3P_3F_5C(CH_3)=CH_2/C_6H_5CH=CH_2$ system have been recalculated using nonlinear minimization techniques with a newly available (7), improved statistical approach (Error-in-Variables method, EVM). It should be noted that in this, and the other copolymerization studies discussed below, experimental design was not fully optimized so the precision of the resulting reactivity ratios has not been maximized (7,8). The values calculated for the propenylphosphazene are significantly different from those calculated using linearization methods (Table I). The new reactivity ratios are more consistent with the nature of the phosphazene derivative and its observed copolymerization behavior (5). Thus r_1 and r_2 set of 1.53,0 is consistent with a radical center on the propenylphosphazene (M_2) which is highly polarized but not resonance stabilized. The data further suggest that the copolymerization proceeds with the incorporation of one phosphazene monomer followed by a polystyrene sequence of variable length and hence phosphazene composition is always less than 50 mole percent. The propenylphosphazene, and the remaining alkenylfluorophosphazene monomers, will be expected to exhibit minimal degrees of homopolymerization in radical processes since the reactions proceed at temperatures greater than the ceiling temperatures typically observed for 1,1-

Table I. Reactivity Ratios for Copolymerization of $N_3P_3X_5R(M_2)$

M_1	M_2 X	M_2 R	Calculation[a] Method	Terminal r_1	Terminal r_2	Model — Penultimate r_1	Penultimate r_1'	Penultimate r_2	Penultimate r_2'	Reference
Styrene	F	C(Me)=CH$_2$	Linearization	2.20	1.35					5
			EVM	1.53	(-0.5)[c]					this work
Styrene	F	C(OEt)=CH$_2$	EVM	3.80	0.85					this work
			PEN			6.63	3.07	0.15	35.11	this work
Styrene	F	p-C$_6$H$_4$C(Me)=CH$_2$	MT	0.41	0.28					11
			EVM	0.42	0.26					this work
			PEN			0.63	0.18	0.63	0.095	11
MMA[b]			PEN			0.70	0.23	0.38	2.90	11
Styrene	F	m-C$_6$H$_4$C(Me)=CH$_2$	MT	0.58	0.28					11
MMA[b]			PEN			1.23	0.32	(-0.037)[c]	39.50	11
N$_4$P$_4$Cl$_7$OCH=CH$_2$	Cl	OCH=CH$_2$	MT	0.60	1.00					this work
MMA	Cl	O(CH$_2$)$_2$OC(O)C(Me)=CH$_2$	MT	1.08	0.46					25
			EVM	1.16	0.55					this work
			PEN			1.44	0.64	(-0.26)[c]	0.85	25
MMA	OCH$_2$CF$_3$	OC(O)C(Me)=CH$_2$	MT	0.29	0.19					this work
			PEN			0.13	1.05	0.11	0.59	this work

a. Error variables (EVM) method reference 7; Mortimer-Tidwell (MT) method, reference 6; Penultimate (PEN) method, reference 10.
b. Methylmethacrylate
c. Value assumed to be zero.

distributed olefins.   The reactivity ratios for the $N_3P_3F_5C(OEt)=CH_2$/
$C_6H_5CH=CH_2$ system have been previously calculated by a variety of methods
(9). Recalculation using the EVM method gives a 95% joint confidence limit
(Figure 1) which suggests that the terminal model may not adequately describe

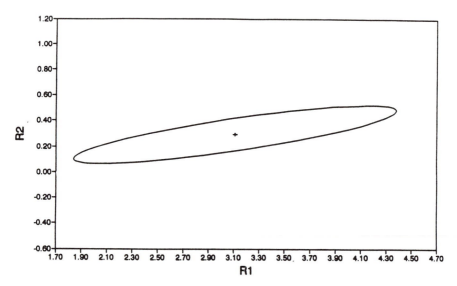

Figure 1.   The 95% joint confidence limit for the styrene/
$N_3P_3F_5C(OEt)=CH_2$ system calculated by the EVM method.

the copolymerization behavior i.e. the value of the reactivity ratio associated with
styrene (r_1) has an enormous range of uncertainty. Reconsideration of the data
using a penultimate model (10) uncovers a large difference between the
phosphazene reactivity ratios, r_2 (k_{222}/k_{221}) and r_2' (k_{122}/k_{121}). The low value of
r_2 compared to r_2' (Table I) suggests that the occurrence of two consecutive
phosphazene units in the growing chain has a strong deactivating effect on the
terminal phosphazene based radical. This interpretation is consistent with the
alternating tendency demonstrated for the propenylphosphazene and is more
reasonable than the alternative which would involve an activating effect of
styrene in the penultimate position.   The higher degree of phosphazene
incorporation and the possibility of more than one phosphazene based monomer
units occurring sequentially can be ascribed to the electron donating effect of the
ethoxy function in the vinyl ether derivatives which partially counterbalances the
phosphazene electron withdrawing effect (9).

The strong sigma polarization of olefins by a directly attached
cyclophosphazene moiety can be mediated, in part, by placement of a molecular

unit between the phosphazene and the olefin to serve an insulating function (1,4,11). Phosphazene derivatives of α-methylstyrene in which the phenyl group

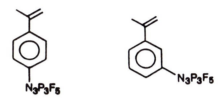

is between the phosphazene and the olefinic center provide an example of this type of monomer (11,12). The recalculated reactivity ratios (Table I) for the $N_3P_3F_5C_6H_4C(Me)=CH_2$/styrene system are in remarkable agreement with those previously obtained using other nonlinear minimization techniques (11) which suggests a high level of confidence in the validity of the data. The Q/e values calculated for the para (0.72,0.72) and meta (0.63,0.58) α-methylstyrene derivatives show that conjugation does not extend outside of the α-methylstyrene unit yet the phosphazene still exerts a strong σ electron withdrawing effect (11). These observations are consistent with the picture of the phosphazene electronic effects derived from systems with olefin directly attached to the phosphorous center.

The reactivity of the α-methylstyrene phosphazene derivatives with methyl methacrylate follows a more complicated pattern than that observed in the copolymerization with styrene. The reactivity of the para-α-methylstyrene derivative is marginally fit by the terminal model but the behavior of the para derivative can only be accounted for by the penultimate model (Table I). The penultimate model is the most widely accepted alternative to terminal behavior (13), however questions remain about the suitability of this approach. Certainly, it can be argued that, on a mathematical basis, a model with addition adjustable parameters will always yield a superior fit to one with fewer parameters. Criteria involving experiment design (8) or sequence distribution determination (14) have been advanced as providing collaborative evidence for the validity of the penultimate effect being operative in specific systems. In the systems in the present investigation, consistency of the mode of operation of the penultimate effect in related systems lends credence to the validity of the application of this model to selected copolymerization reactions of alkenylphosphazenes. Thus in each case, the effect is only manifested in the value of r_2' and is consistent with a strong deactivating effect of sequential sigma electron withdrawing phosphazene units. A spacial proximity effect is also operative in the α-methylstyrene derivatives since strong penultimate behavior is limited to the monomer with a meta placement of the phosphazene. The effect in this case may be steric in nature which would be consistent with the tendency of penultimate behavior being most commonly observed in monomers such as the alkenylphosphazenes that contain bulky and/or polar substituents (15). The intriguing possibility of an electrostatic interaction between a penultimate phosphazene and a terminal methylmethacrylate radical could also be considered.

Alkenylborazine Copolymers

Reactivity studies have not been limited to cyclophosphazene containing monomers. Recent work has been devoted to monomers similar to the alkenylphosphazenes but with a borazine as the inorganic ring system. Sneddon

has reported reactivity ratios for the copolymerization of styrene and the isoelectronic borazine, $N_3B_3H_5CH=CH_2$ (16). Styrene (M_1) is the significantly more reactive monomer of the two ($r_1,r_2=4.02,0.78$). We have calculated Q/e values for this system (Q,e=0.16,0.28) and these data show an olefin with a weak electron withdrawing substituent which has a significantly reduced mesomeric interaction. The values we have calculated are typical of other olefins with weakly interacting substituents; e.g., for $ClCH=CH_2$, Q,e=0.056,0.16. We have previously proposed, based on NMR data, that the vinyl group in $N_3B_3Me_5CH=CH_2$ is out of the borazine plane (17) and it is reasonable to apply similar reasoning to rationalize the decreased copolymerization reactivity of $N_3B_3H_5CH=CH_2$. The lack of resonance stabilization of the propagating radical combined with steric hindrance at the adjacent nitrogen center is sufficient to prevent radical chain polymerization of $N_3B_3Me_5CH=CH_2$. This monomer does, however, undergo a slow thermal polymerization reaction (18). The steric drive to twist aryl substituents out of the B_3N_3 plane is very significant and in the copolymerization of $N_3B_3Ph_3Me_2C_6H_4CH=CH_2$ with styrene, we have shown that the monomer reactivities were indistinguishable within the 95% joint confidence limit for the reactivity ratios (19). The N-methyl derivative, $N_3B_3Me_5C_6H_4CH=CH_2$, undergoes extensive cross-linking in all attempts at radical chain homo- and copolymerization. This behavior apparently is the result of facile radical abstraction at the N-methyl group (19).

Vinyloxyphosphazenes

Recent investigations in our laboratories (1,4) and others (1,20) have focused on vinyl containing substituents connected to the cyclophosphazene through an oxygen atom. The simplest of these substituents is the vinyloxy unit, derivatives of which are available from the reaction of the enolate anion of acetaldehyde with chlorocyclophosphazenes (21,22).

$$(NPCl_2)_n + LiOCH=CH_2 \rightarrow N_nP_nCl_{2n-1}OCH=CH_2$$

$$n=3,4$$

Attempts at copolymerization of $N_3P_3Cl_5OCH=CH_2$ with organic monomers produced mixed results. When styrene is employed as the comonomer, the apparent copolymerization is actually two separate homopolymerization reactions. The ^{13}C NMR spectrum of $N_3P_3Cl_5OCH=CH_2$ demonstrates a close similarity of the electronic structure of the vinyl group in the phosphazene and vinyl acetate (23). If we can model the vinyloxyphosphazene with vinyl acetate, then the styrene copolymerization behavior is understandable since similar behavior is observed in the styrene/vinyl acetate system. Separate homopolymerization of the vinyloxyphosphazene provides materials with molecular weights up to the 500,000 Dalton range. The copolymerization of $N_3P_3Cl_5OCH=CH_2$ with vinyl acetate is accompanied by concomitant decomposition of the type previously identified for the poly(vinyloxyphosphazene) homopolymer (4). Copolymerization of two vinyloxyphosphazenes can be affected. The copolymer composition in the $N_3P_3Cl_5OCH=CH_2/N_4P_4Cl_7OCH=CH_2$ system was established from the integrated intensities of the respective PN ring modes in the IR spectra. The trimer reactivity is significantly greater than that of the tetrameter (Table I). If one approximates the Q/e values for $N_3P_3Cl_5OCH=CH_2$ with those of vinyl acetate (0.026,-0.88), the calculated Q/e parameters for $N_4P_4Cl_7OCH=CH_2$ (0.014,-0.17) show virtually no change in Q and a significant change in the e value. The ^{13}C NMR data indicates a polarity difference in the two monomers but not to the degree suggested by the calculations.

Since NMR studies show a significant perturbation of the olefinic center in vinyloxycyclophosphazenes by phosphazene substituents (1,23), a substituent which would be electronically isolated from the phosphorus center appeared desirable. The hydroxyethylmethacrylate function has been shown to undergo facile polymerization to yield matrix materials in $N_3P_3[O(CH_2)_2OC(O)C(CH_3)=CH_2]_6$ (24), so $N_3P_3Cl_5O(CH_2)_2OC(O)C(CH_3)=CH_2$ was prepared (25). The homo- and copolymerization behavior of this monomer has been examined in detail. The observation that molecular weight decreases with increasing phosphazene content leads to the suggestion of phosphazene based chain transfer events (25). Recalculated reactivity ratios are in good agreement with those previously reported and no indication of penultimate behavior has been detected (Table I). The Q/e parameters for this monomer (0.83,1.07) show a strong polarity effect attributed to the phosphazene. Some caution should be exercised, however since the monomer undergoes a phosphazene/phosphazane rearrangement at a rate which would result in formation of some of the rearranged material over the course of the polymerization reaction (Hayes, R. F.; Allen, C. W., unpublished observations).

A new approach to the synthesis of organooxy cyclophosphazenes has been developed. The reaction of $N_3P_3(OCH_2CF_3)_6$ with strong bases provides the oxyanion $N_3P_3(OCH_2CF_3)_5O^-$ (26) which is allowed reacted with chloroacrylates to yield $N_3P_3(OCH_2CF_3)_5OC(O)C(R)=CH_2$ (R=H,CH$_3$). These monomers undergo facile homo- and copolymerization reactions (Tooze, R.;

Poyser, G.L., unpublished observations). Estimation of the monomer concentrations in the copolymers with methylmethacrylate was investigated by both elemental analysis and gel permeation chromatography. In the latter method a low molecular weight selective column was employed so that the concentrations of both monomers could be followed and quantified. A similar approach to copolymer composition using gas chromatography has previously been employed (27). Based on the range of the 95% joint confidence limit plots of the calculated reactivity ratios, the chromatographic method gave superior results. This is undoubtable due, in part, to the larger number of data points available as demonstrated by narrowing of the confidence limit range with increased number of runs analyzed using the chromatographic approach. The difference in the calculated reactivity ratios in this case was important since in the solution derived from elemental analysis data the phosphazene had the greater r value while the reverse was true when the more accurate data set was employed. No evidence for penultimate behavior was noted and both reactivity ratios are less than one (Table I). The low value of r_1 x r_2 (0.06) indicates a strong tendency towards alternating behavior and as in other systems the phosphazene has a deactivating effect on the olefin although in this case the steric bulk of the penta(trifluoroethoxy)cyclotriphosphazene unit may be more important than any electronic perturbation.

In summary, it is clear that both appropriate calculational methodology and accurate copolymer composition determination play an important part in the generation of reliable reactivity ratios. The cycloborazines function as weak electron withdrawing groups with little resonance stabilization. The cyclophosphazenes exhibit strong sigma electron withdrawing effects but again with little resonance stablilization. The steric bulk and polarity of the phosphazene can also lead to significant penultimate behavior in the polymerization mechanism.

Literature Cited

1. Allen, C. W. in The Chemistry of Inorganic Ring Systems; Steudel, R., Ed., Elsevier: Amsterdam, **1992**, Chapter 10.
2. Allen, C. W., Carter, K. R., Bahadur, M., Brown, D. E. ACS Polymer Preprints, **1991**, 32(3), 479.
3. McNally, L., Allen, C. W. Heteratom. Chem., **1993**, 4, 159.
4. Allen, C. W. ACS Symp. Ser., **1988**, 360, 290.
5. Dupont, J. G., Allen, C. W. Macromolecules, **1979**, 12, 169.
6. (a) Tidwell, P. W., Mortimer, G. A. J. Polym. Sci., Part A, **1965**, 3, 360. (b) Tidwell, P. W., Mortimer, G. A. J. Macromol. Sci., Rev. Macromol. Chem., **1970**, 4, 281.
7. Dube, M., Sanayei, R. A., Penlidis, A., O'Driscoll, K. F., Reilly, P. M. J. Polym. Sci., Polym. Chem. Ed., **1991**, 29, 703.
8. Burke, A. L., Duever, T. A., Penlidis, A. J. Polym. Sci., Polym. Chem. Ed., **1993**, 31, 3065.
9. Allen, C. W., Bright, R. P. Macromolecules, **1986**, 19, 571.
10. Pittman, C. U., Jr., Rounsefell, T. D. J. Polym. Sci., Polym. Chem. Ed., **1973**, 11, 621. Pittman, C. U., Jr., Rounsefell, T. D. Comput. Chem. Instrum., **1977**, 6, 145.
11. Allen, C. W., Shaw, J. C., Brown, D. E. Macromolecules, **1988**, 21, 2653.
12. Shaw, J.C., Allen, C. W. Inorg. Chem., **1986**, 25, 4632.
13. Kuchanov, S. I. Adv. Polym. Sci., **1992**, 103, 1.
14. Hill, D. J. T., O'Donnell, J. H. Makromol. Chem., Makromol. Symp., **1987**, 10/11, 375.
15. Odian, G. Principle of Polymerization, 3rd Ed., Wiley: New York, **1991**, 503.
16. Su, K., Remsen, E. E., Thompson, H. H., Sneddon, L. G. Macromolecules, **1991**, 24, 3760.
17. Jackson, L. A., Allen, C. W. J. Chem. Soc., Dalton Trans., **1989**, 2423.
18. Allen, C. W., Jackson, L. A., Abstract 146, 199th American Chemical Society Meeting, Inorganic Division, Boston, 1990.
19. Jackson, L. A., Allen, C. W. J. Polym. Sci., Polym. Chem. Ed., **1992**, 30, 577.
20. Inoue, K., Nishikawa, Y., Tanigaki, T. J. Am. Chem. Soc., **1991**, 113, 7609.
21. Allen, C. W., Ramachandran, K., Brown, D. E. Inorg. Syn., **1989**, 25, 74.
22. Brown, D. E., Allen, C. W. Inorg. Chem., **1987**, 26, 934.
23. Allen, C. W., Brown, D. E., Carter, K. R. Phosphorus Sulfur, Silicon, **1989**, 41, 311.
24. Yaguchi, A., Mori, S., Kitayama, M., Onda, T., Kurahashi, A., Ando, H. Thin Solid Films, **1992**, 216, 123.
25. Brown, D. E., Allen, C. W. J. Inorg. Organomet. Polymers, **1991**, 1, 189.
26. Ferrar, W. T., Marshall, A. S., Whitefield, J. Macromolecules, **1987**, 20, 317.
27. Pittman, C. U., Jr., Jayaraman, T. V., Prester, R. D., Jr., Spencer, S., Rauch, M. D., Macomber, D. M. Macromolecules, **1981**, 14, 237.

RECEIVED April 14, 1994

Chapter 30

Synthesis of Organoboron Polymers by Hydroboration Polymerization

Yoshiki Chujo

Division of Polymer Chemistry, Graduate School of Engineering, Kyoto Univeristy, Yoshida, Sakyo-ku, Kyoto 606–01 Japan

Hydroboration polymerization is described as a novel methodology for the preparation of organoboron polymers. Polyaddition between dienes and thexylborane produced polymers consisting of carbon-boron bonds in the main chains. The resulting organoboron polymers can be regarded as polymer analogs of trialkylboranes and can be viewed as novel types of reactive polymers. On the other hand, hydroboration polymerization of dicyano compounds with monoalkylboranes or dialkylboranes produced air-stable poly(cyclodiborazane)s having B-N four-membered rings via dimerization of iminoborane species. Allylboration polymerization of dicyano compounds with triallylborane also gave the corresponding poly-(cyclodiborazane)s. In haloboration polymerization between diynes and boron tribromide, poly(organoboron halide)s were obtained and showed characteristic properties as a polymeric Lewis acid.

Organoboron compounds are known as useful reagents or reaction intermediates for the preparation of a wide variety of functional compounds such as alcohols, amines, carboxylic acids, ketones, aldehydes, olefins, and halides (*1-3*). For several decades, Brown and his coworkers have studied this chemistry.

0097–6156/94/0572–0398$08.00/0

It is expected to provide a novel type of reactive polymers by the introduction of organoboron moieties in polymeric chains.   In other words, organoboron polymers can be used as precursor polymers for the preparation of various functional polymers.

Hydroboration (addition of B-H to C=C) is a well-known reaction which takes place under mild conditions almost quantitatively, for the preparation of various alkylborane compounds in organic synthesis.   However, the direct use of this reaction in polymer synthesis has been very limited.

Polymers having olefin groups such as polydienes can be subjected to the hydroboration reaction.   The functionalization of polydienes by means of this hydroboration is known (4-7), especially from the viewpoint of industrial use.   Recently, Chung and his coworkers have examined this method to prepare polymers having organoboron moieties in the branches (8-13). The advantages of this chemistry are (i) the stability of the borane moiety to the transition metal catalyst usually used in Ziegler-Natta or ring-opening metathesis polymerization, (ii) the solubility of borane compounds in hydrocarbon solvents such as hexane or toluene, and (iii) the versatility of borane groups, which can be transformed to a remarkable variety of functionalities.

Previously, Urry et al. reported that an organoboron polymer was formed by pyrolysis of 2,5-dimethyl-2,5-diborahexane (14). Mikhailov et al. also reported the formation of an organoboron polymer by thermal isomerization of triallylborane and triisobutylborane (15).

Among numerous studies reported by Brown and his group, the formation of polymeric materials has been described from the reaction of thexylborane and 1,3-butadiene (16) and from monochloroborane and 1,7-octadiene (17).   However, no details were reported about the yield, molecular weights, and structure of the resulting polymers.   These polymers were formed as intermediates in the formation of boron heterocycles by so-called thermal depolymerization.

Recently, we explored a novel polyaddition between dienes and monoalkylboranes and termed this hydroboration polymerization (18).   Diynes and dicyano compounds can also be used in this hydroboration polymerization.   The organoboron polymers obtained are effectively converted to polyalcohols or polyketones.   In other words, polymers having organoboron units in the main chains can be expected to be reactive polymers.   Here, the scope of these hydroboration polymerizations is viewed as methodologies for the preparation of organoboron polymers.

Hydroboration Polymerization between Dienes and Thexylborane

As a monoalkylborane (bifunctional hydroborane) component in hydroboration polymerization, thexylborane was employed because of its stability after distillation. As a diene monomer, relatively longer chain dienes were used to avoid the competitive cyclization reaction. The general scheme of this polymerization is shown in Scheme 1, which proceeds in THF at 0°C under nitrogen without catalyst (18). For example, the resulting organoboron polymer obtained from the reaction between thexylborane and 1,7-octadiene was isolated as a colorless gum after coagulation and was soluble in common organic solvents such as chloroform, THF and benzene. The structure of the polymer was supported by its IR and ^{1}H-, ^{13}C-, and ^{11}B-NMR spectra. The molecular weight of the polymer increased as the feed ratio approached unity. This effect is taken as a normal behavior of a polyaddition reaction. As summarized in Table I, this hydroboration polymerization can be applied to various combinations of dienes with thexylborane that produce the corresponding organoboron polymers.

These organoboron polymers were stable in protic solvents such as water and alcohol under nitrogen. However, under air, these polymers decomposed as is typical for organoboron compounds. This decomposition was monitored by GPC after bubbling air through a THF solution of the polymer at room temperature. After 3 min of air-bubbling, the peak became broader and moved to the low molecular weight region. At 30 min, the polymer was found to decompose. However, it should be noted that the organoboron polymers were more stable toward air than conventional trialkylboranes.

Organoboron copolymers were prepared by polyaddition of a mixture of dienes with thexylborane (19). When a mixture of dienes such as 1,2-diallyloxyethane and p-diallylbenzene was polymerized with thexylborane, the peaks in the GPC using both UV and RI detectors were shifted to higher molecular weight regions with increase of the amount of thexylborane. The molecular weight distribution of the copolymer obtained by this method was clearly differed from that of a mixture of two homopolymers. This result supports the formation of copolymers. In a similar manner, other copolymers were prepared using various combinations of 1,7-octadiene, 1,2-diallyloxyethane, p-divinylbenzene, and p-diallylbenzene.

Scheme 1

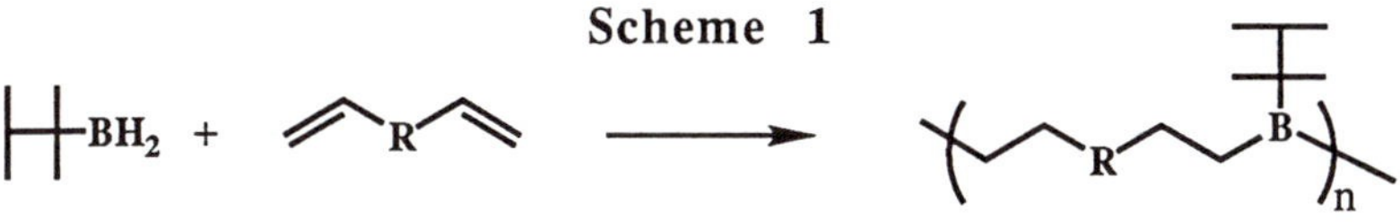

Table I   Hydroboration Polymerization between
Thexylborane and various Dienes[a]

Run	Diene	$\overline{M}_n$[b]	$\overline{M}_w$[b]
1		19,000	27,700
2		18,400	27,400
3		19,000	29,200
4		9,400	16,900
5		1,200	2,600
6		1,900	4,500
7		1,900	3,200
8	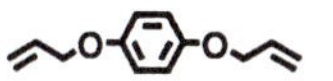	5,100	11,200
9	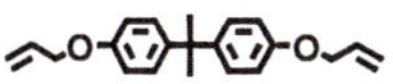	7,600	15,400

a) Polymerizations were carried out by adding small excess of thexylborane
to the 1M THF solution of diene at 0°C.

b) GPC (dry THF), polystyrene standard.

Hydroboration Polymerization of Diynes with Thexylborane

Hydroboration polymerization with monoalkylborane can also be applied to diynes.   Generally, the reactivity of a terminal acetylene group toward hydroboration is quite different from that of an internal acetylene group.   That is, the terminal acetylenes preferentially give dihydroboration products via a further hydroboration of the initially formed vinylborane species.   On the other hand, the internal acetylenes give monohydroboration products regardless of the bulkiness of borane.   Thus, terminal diyne and internal diyne are expected to show the different polymerization behavior due to the steric effect on the second (further) hydroboration reaction.

As a typical example for hydroboration polymerization of internal diyne, the reaction of 3,9-dodecadiyne and thexylborane was carried out in THF at 0°C under nitrogen (20).   The polymerization behavior was quite similar to hydroboration polymerization of dienes.   The organoboron polymer obtained was a colorless wax and soluble in common organic solvents such as benzene, THF or chloroform.   An IR spectrum of the polymer showed the absorption band assignable to C=C bond, which was also supported by ^{1}H- and ^{11}B-NMR spectra.   3,8-Undecadiyne and 3,10-tridecadiyne were also used in hydroboration polymerization with thexylborane to produce the corresponding organoboron polymers.

On the other hand, in the case of terminal diyne such as 1,7-octadiyne, the further hydroboration reaction of olefin of the resulting polymer with thexylborane occurred (20).   In fact, when the ratio of thexylborane to acetylene unit was higher than 1.7, gelation was observed.   Thus, hydroboration of terminal diynes gives organoboron polymers having a branched structure due to the further hydroboration reaction toward the initially formed vinylborane structures.   That is, terminal diynes are taken to have potential as a multifunctional monomer.   When the polymerization between 1,7-octadiene and thexylborane was carried out in the presence of a small amount of 1,7-octadiyne, the molecular weights of organoboron polymers were found to increase when the ratio of diyne/diene was increased (19).

The organoboron polymers prepared from diynes, especially from internal diynes, consisted mainly of divinylborane units.   Thus, different reactivity and stability originating from this structure can be expected in comparison with the organoboron polymers prepared from dienes.

Reactions of Organoboron Polymers

The organoboron polymers prepared by hydroboration polymerization have a new structure consisting of C-B bonds in the main chain.   Generally, these organoboron polymers can be regarded as a polymer homolog of trialkylborane, which is known to be a very versatile compound in organic synthesis.   In other words, the polymers obtained by hydroboration polymerization can be expected to be reactive polymers.

The versatile reactions of organoboron polymers prepared by hydroboration polymerization are summarized in Scheme 2. For example, the polymer prepared from thexylborane and 1,7-octadiene was reacted with carbon monoxide at 120°C followed by treatment with NaOH and H_2O_2 to produce a polyalcohol (*21*). This conversion includes the migrations of polymer chain and thexyl group from a boron atom to carbon as shown in Scheme 3. Table II summarizes the results of preparation of polyalcohols from various organoboron polymers.   This migration reaction is known to proceed intramolecularly.   Thus, it can be mentioned that the molecular weight of the starting organoboron polymers should be higher than (or the same as) that of polyalcohols. When this reaction was carried out under milder condition, polyketone segments were included in the polymer structure due to the incomplete migration of thexyl group.   These conversions may offer a new synthetic method for the preparation of polyalcohols from the corresponding dienes.

For the selective preparation of polyketones, the organoboron polymer prepared by hydroboration polymerization between thexylborane and 1,7-octadiene was subjected to reaction with KCN (*22*).   After oxidation of the reaction mixture, the desired polyketone was obtained.   The polymer formed was a white solid and stable in air.   This reaction also includes the migration of main chains of organoboron polymer from boron to carbon.

As an alternative method for the preparation of polyalcohols from organoboron polymers, dichloromethyl methyl ether (DCME) can be used (*23*).   Various organoboron polymers prepared by hydroboration polymerization of dienes and thexylborane were reacted with DCME at 0°C in THF followed by treatment with lithium triethylmethoxide.   After oxidative treatment with NaOH and H_2O_2, the corresponding poly(alcohol)s were obtained.   The

Scheme 2

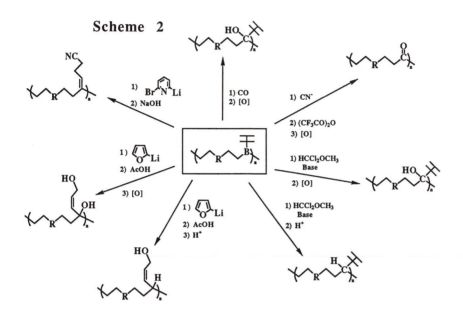

Scheme 3

Table II   Synthesis of Poly(alcohol)s by the Reaction of Organoboron Polymers with Carbon Monoxide.[a]

Run	R	B-Polymers $\bar{M}_n$[b]	Poly(alcohol)s $\bar{M}_n$[c]	$\bar{M}_w$[c]	Yield[d] (%)
1	$-(CH_2)_4-$	19,500	4,080	13,500	8 2
2	$-(CH_2)_6-$	20,500	4,200	9,600	6 0
3	—⬡—	14,600	2,990	7,650	7 4
4	$-CH_2-$⬡$-CH_2-$	9,300	5,400	12,300	7 7
5	$-CH_2O(CH_2)_2OCH_2-$	1,520	1,280	2,200	2 4
6	$-CH_2O(CH_2)_4OCH_2-$	2,280	1,620	2,800	5 9
7	$-CH_2O((CH_2)_2O)_3CH_2-$	3,200	2,200	3,720	4 7
8	$-CH_2O-$⬡$-OCH_2-$	6,800	1,940	3,700	3 9
9	$-CH_2O-$⬡$-$⬡$-OCH_2-$	11,700	2,320	4,400	8 9

a) Reactions were carried out in the presence of 0.1ml ethylene glycol under carbon monoxide (20-25 kg/cm^2) in an autoclave.
b) GPC (dry THF), PS standard.   c) GPC (THF), PS standard.
d) Isolated yield after reprecipitation into methanol/water (v/v=2/1).

structures of the resulting poly(alcohol)s were the same as those prepared by the reaction with carbon monoxide described above.

We also examined the reactions of organoboron polymers via ring-opening of pyridine or furan moieties. The organoboron polymer prepared by hydroboration polymerization between 1,7-octadiene and thexylborane was reacted with 6-bromo-2-lithiopyridine followed by treatment with NaOH to produce a cyano group-branching polymer (24). The reaction of organo-boron polymer with α-furyllithium was followed by treatment with acetic acid and then with NaOH and H_2O_2 to form the polymer having primary and tertiary alcohol groups (25). These conversions involve the ring-opening reactions of pyridine and furan, respectively.

All of the polymers described in this section are difficult to prepare by conventional synthetic methods. These conversions demonstrate the utility and versatility of organoboron polymers as polymeric precursors to functional polymers.

Poly(cyclodiborazane)s by Hydroboration Polymerization

It is known that hydroboration reactions of cyano groups give iminoborane species which dimerize to form boron-nitrogen four-membered rings (cyclodiborazane) (26). When this reaction is used with bifunctional monomers, formation of polymeric materials consisting of cyclodiborazane units can be expected.

As a typical example, isophthalonitrile was reacted with *tert*-butylborane-trimethylamine in diglyme at 100°C under nitrogen to produce a boron-containing polymer as shown in Scheme 4 (27). The white solid polymer was soluble in various organic solvents such as THF, chloroform and benzene. This polymerization involves the dimerization of iminoborane to form the four-membered ring as a key essential step. In this hydroboration polymerization, terephthalonitrile and 1,5-di(4,4'-cyanophenoxy)pentane also gave the corresponding polymers. Aliphatic dicyano compounds, however, resulted in the formation of oligomers with very low molecular weights. The structures of these boron-containing polymers were supported by spectro-scopic data in comparison with the model compound obtained from reaction of benzonitrile and *tert*-butylborane. The boron-containing polymers were sufficiently stable to handle in air. Thermogravimetric analysis (TGA) of this boron-containing polymer shows that the weight loss started at 140°C and was complete at 700°C at which point a black solid remained. This finding suggests that the resulting poly(cyclodiborazane) prepared by hydroboration polymerization of dicyano compound can be used as a polymeric precursor for the preparation of boron-containing inorganic materials.

Thexylborane also produced poly(cyclodiborazane)s by the reaction with aliphatic dicyano compounds, although the molecular weights of the polymers obtained were relatively low (*28*). On the other hand, when polymerization was examined with less hindered monoalkylboranes (i.e., n-BuBH$_2$·NMe$_3$ or i-PrBH$_2$·NMe$_3$), crosslinking reactions took place due to the further hydroboration of iminoboranes (*29*).

Dialkylboranes can also be used as borane monomers in hydroboration polymerization with dicyano compounds. In this case, the substituents at the boron atoms in poly(cyclodiborazane)s should be two alkyl groups. When polymerization between dibutylborane and adiponitrile was carried out in bulk under nitrogen, the reaction was complete within several hours, and the corresponding poly(cyclodiborazane) was obtained in high yield. Other dicyano monomers having different methylene chains also produced the corresponding poly(cyclodiborazane)s. Table III summarizes the results of hydroboration polymerization between dibutylborane and various dicyano compounds. In comparison with monoalkylboranes, the hydroboration polymerization of dialkylboranes produced higher molecular weights polymers even from aliphatic dicyano compounds.

Allylboration   Polymerization

The boron-containing polymer with the cyclodiborazane ring was also prepared by so-called allylboration polymerization between triallylborane and dicyano compounds (Scheme 5) (*30*). Polymerizations were carried out by adding triallylborane to dicyano monomer in bulk at 0°C under nitrogen. When isophthalonitrile was used as a monomer, the polymerization system became homogeneous within a few seconds and became very viscous after a few hours. The glassy polymer obtained after the reaction for 1 day was dissolved in THF and purified by reprecipitation. Both of aromatic and aliphatic dicyano compounds successfully gave polymers in this polymerization with triallylborane as shown in Table IV. Trimethallylborane, allyldialkylboranes and methallyldialkylboranes were also used in this allylboration polymerization with dicyano compounds to produce the corresponding poly(cyclodiborazane)s.

Scheme 4

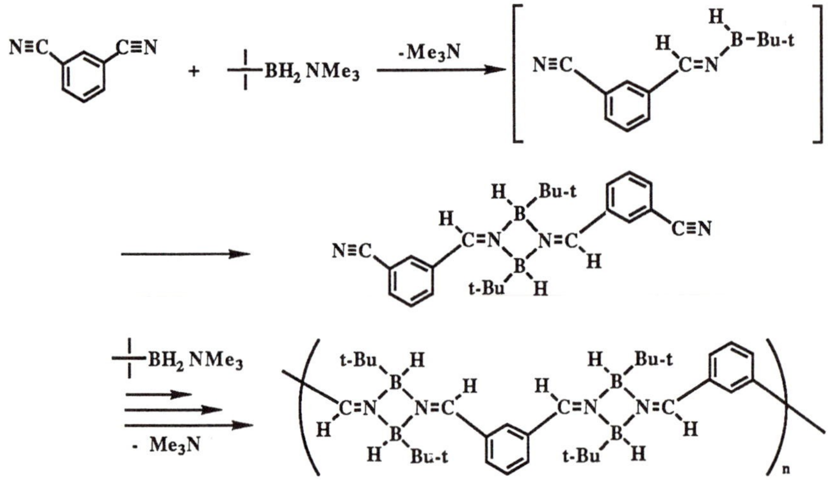

Scheme 5

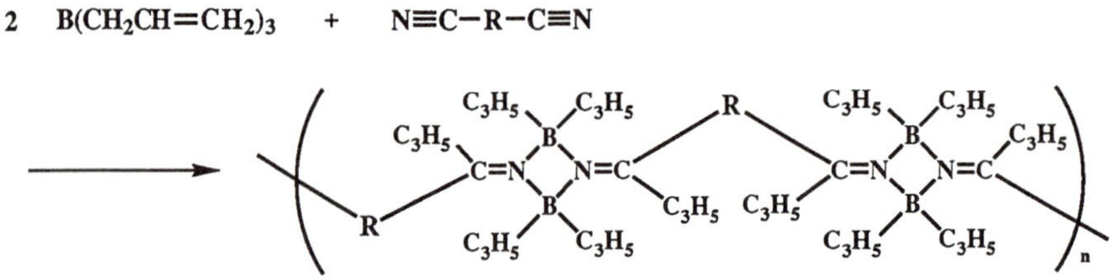

Table III   Hydroboration Polymerization of Various
Dicyano Compounds with Dibutylborane.[a]

Run	Dicyano Compounds	Yield[b] (%)	M_n[c]	M_w[c]
1[d]	$NC(CH_2)_2CN$	88	16,800	49,900
2	$NC(CH_2)_3CN$	90	10,100	23,500
3	$NC(CH_2)_4CN$	78	17,100	51,700
4	$NC(CH_2)_5CN$	77	25,000	69,900
5	$NC(CH_2)_6CN$	76	3,080	5,830
6	$NC(CH_2)_8CN$	88	5,930	9,070
7	$m\text{-}NC\text{-}C_6H_4\text{-}CN$	25	3,540	14,700

a) Polymerizations were carried out in bulk at r.t.
b) Isolated yields after precipitation into EtOH/H$_2$O (1/1).
c) GPC (THF), polystyrene standard.
d) Polymerization was carried out at 70°C for 30min., then at r.t. for 2h.

Table IV   Allylboration Polymerization of Various
Dicyano Compounds with Triallylborane[a]

Run	Dicyano Compounds	Yield[b] (%)	$\overline{M}_n$[c]	$\overline{M}_w$[c]	$\overline{M}_w / \overline{M}_n$
1	$m\text{-}NC\text{-}C_6H_4\text{-}CN$	78	14,000	25,700	1.84
2	$p\text{-}NC\text{-}C_6H_4\text{-}CN$	89	5,900	11,000	1.86
3	$NC(CH_2)_2CN$	79	4,500	9,200	2.04
4	$NC(CH_2)_3CN$	78	5,500	12,000	2.18
5	$NC(CH_2)_4CN$	81	6,300	13,600	2.16
6	$NC(CH_2)_5CN$	93	8,900	18,000	2.02
7	$NC(CH_2)_6CN$	85	9,000	20,400	2.27
8	$NC(CH_2)_8CN$	89	11,900	24,600	2.07

a) Polymerizations were carried out in bulk at 0°C.
b) Isolated yields after precipitation into EtOH/H$_2$O (1/1).
c) GPC (THF), polystyrene standard.

For the preparation of poly(cyclodiborazane)s, a condensation method can also be used. For example, N,N'-bis(trimethylsilyl)terephthalaldehyde diimine was reacted with monochlorodihexylborane to produce a boron-containing polymer with boron-nitrogen four-membered rings (*31*). This polymerization can be applied to various bis(silylimine)s such as N,N'-bis(trimethylsilyl)isophthalaldehyde diimine. However, N,N'-bis(trimethylsilyl)anthraquinone diimine gave only the monomeric iminoborane species and no polymer.

Haloboration Polymerization

Substituted boron halides are known to be useful reagents for ether cleavage (*32*) and selective haloboration reactions (*33-34*) under mild conditions. Polymeric homologs of these materials, therefore, may have a potential to show unique properties as reactive polymers. Recently, we reported a polyaddition between boron tribromide and terminal diynes as a novel methodology for the preparation of poly(organoboron halide)s (*35*). We termed this haloboration polymerization.

As shown in Scheme 6, the polyaddition between 1,7-octadiyne and boron tribromide produced the corresponding poly(organoboron halide) as a brown solid without gelation. This polymer was soluble in common organic solvents such as chloroform. The structure of the polymer was supported by its ^{1}H-, ^{11}B-NMR, IR and UV spectra in comparison with those for the model compound prepared from boron tribromide and 1-hexyne. Table V summarizes the results of haloboration polymerization using various terminal diynes.

The B-Br bonds in poly(organoboron halide)s were found to be replaced by B-OEt bonds during the precipitation of the polymer into ethanol. The reaction of the polymer with diol or with water caused gelation. The B-Br bond in the polymer was also subjected to further haloboration reaction with phenylacetylene. The characteristic property of the polymer as a poly(Lewis acid) was also demonstrated by the reaction with THF to produce 4-bromo-1-butanol after hydrolysis. These reactions are summarized in Scheme 7. Poly(organoboron halide)s obtained by haloboration polymerization are potentially useful and relatively rare polymeric Lewis acids.

Poly(organoboron halide)s were also prepared by hydroboration polymerization between various dienes and monobromoborane-dimethyl sulfide (*36*).

Table V   Haloboration Polymerization between BBr_3 and various Diynes

Run	Diyne	$\dfrac{[diyne]}{[BBr_3]}$	Yield[a] (%)	$\overline{M}_n$[b]	$\overline{M}_w$[b]
1	$HC \equiv C(CH_2)_5 C \equiv CH$	1.01	86	6,740	32,400
2	$HC \equiv C(CH_2)_6 C \equiv CH$	1.01	91	7,230	26,600
3	$HC \equiv C$—⟨benzene⟩—$C \equiv CH$	1.00	92	990	1,660
4	$HC \equiv C$—⟨benzene⟩—$C \equiv CH$	1.00	69	700	1,370

a) Isolated yields after reprecipitation into EtOH.
b) GPC ($CHCl_3$), polystyrene standard.

Scheme 6

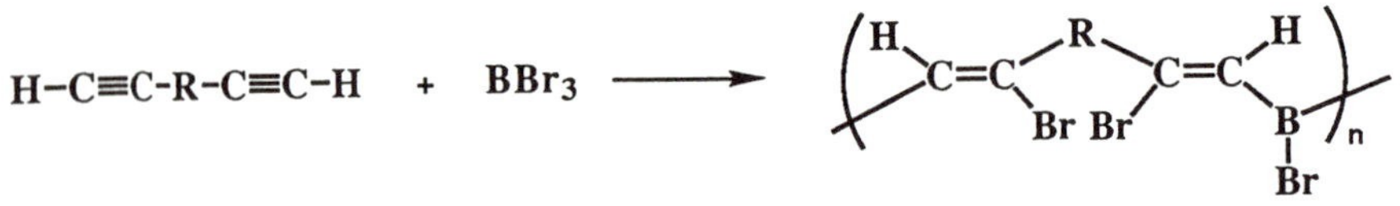

$$H-C\equiv C-R-C\equiv C-H \;+\; BBr_3 \longrightarrow$$

Scheme 7

EtOH

HO-R-OH
(or H_2O)

PhC≡CH

THF

1) H_2O
2) EtOH

$$Br\!\!\!\!\diagdown\!\!\!\!\diagup\!\!\!\!\diagdown\!\!\!\!\diagup OH$$

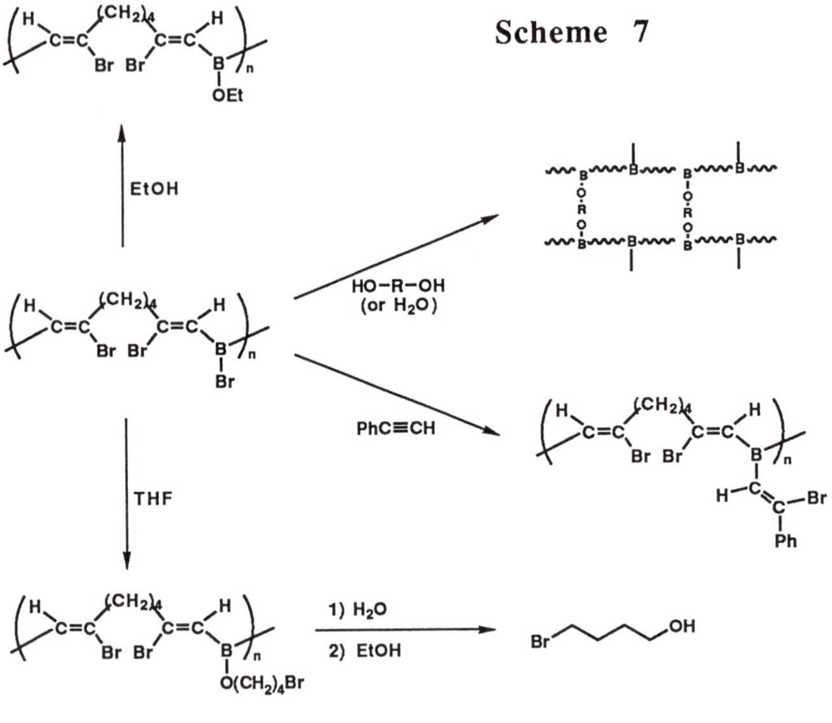

Scheme 8

$$HO-R-OH \;+\; \vert\!\!\vert\text{-}BH_2 \longrightarrow \left(O-R-O-B\right)_n$$

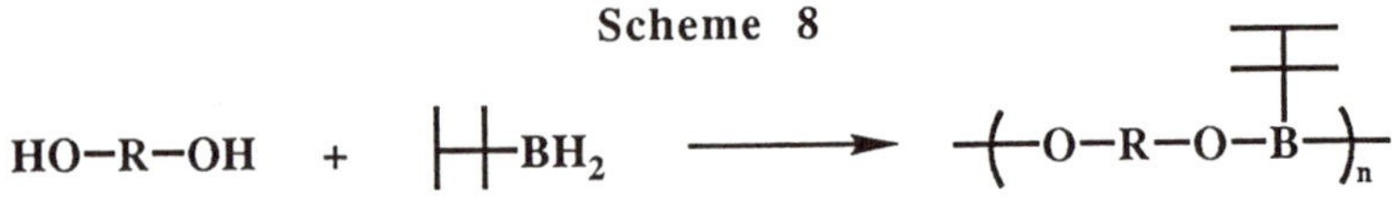

Bóronate Oligomers by Dehydrogenation

Soluble boronate oligomers were prepared by dehydrogenation reactions of diols with thexylborane as shown in Scheme 8 (*37*). Bifunctional thexylborane was employed as a monomer, and dehydrogenation between diols and thexylborane was used to produce a poly(boronic ester).

As a typical example, polymerization of thexylborane and 1,6-hexanediol was carried out at room temperature under nitrogen. Various diols were subjected to this dehydrogenation polymerization with an equimolar amount of thexylborane. The examples of diols were 1,8-octanediol, 1,12-dodecanediol, hydroquinone and bisphenol-A. The boronate oligomers decomposed readily by water, as is usual for boronic esters.

Conclusions

In this article, synthetic methodologies for the preparation of organoboron polymers are described and the reactions of these polymers are discussed. The organoboron polymers themselves are new polymeric materials with interesting properties. In addition, these organoboron polymers can be used as polymeric precursors of functional polymers (*38*). A wide variety of functional groups can be introduced at the main chains, side chains, terminal ends, or even at the center of polymeric chains (*39*) starting from these organoboron polymers.

While the studies on organoboron polymers have just started very recently, there is a rich chemistry for these reactive polymers. The author believes that the study of organoboron polymers will continue to expand rapidly, and that many applications of this chemistry will be developed in the near future.

Literature Cited

1) Brown, H. C. *Hydroboration*; W. A. Benjamin, Inc.: New York, 1962.
2) Brown, H. C. *Organic Synthesis via Boranes*; Wiley-Interscience: New York, 1975.
3) Pelter, A.; Smith, K.; Brown, H. C. *Borane Reagents*; Academic Press: London, 1988.
4) Pinazzi, C.; Brosse, J. C.; Pleurdeau, A.; Reyx, D. *Appl. Polym. Symp.* **1975,** *26,* 73.

5) Pinazzi, C.; Vassort, J.; Reyx, D. *Eur. Polym. J.* **1977**, *13*, 395.

6) Pinazzi, C.; Guillaume, P.; Reyx, D. *Eur. Polym. J.* **1977**, *13*, 711.

7) Yamaguchi, H.; Azuma, K.; Minoura, Y. *Polym. J.* **1972**, *3*, 12.

8) Chung, T. C. *Macromolecules* **1988**, *21*, 865.

9) Chung, T. C.; Roate, M.; Berluche, E.; Schulz, D. N. *Macromolecules* **1988**, *21*, 1903.

10) Ramakrishnan, S.; Berluche, E.; Chung, T. C. *Macromolecules* **1990**, *23*, 378.

11) Ramakrishnan, S.; Chung, T. C. *Macromolecules* **1990**, *23*, 4519.

12) Chung, T. C.; Rhubright, D. *Macromolecules* **1991**, *24*, 970.

13) Chung, T. C.; Ramakrishnan, S.; Kim, M. W. *Macromolecules* **1991**, *24*, 2675.

14) Urry, G.; Kerrigan, J.; Parsons, T. D.; Schlesinger, H. I. *J. Am. Chem. Soc.* **1954**, *76*, 5299.

15) Mikhailov, B. M.; Tutorskaya, F. B. *Izv. Akad. Nauk SSSR* **1959**, 1127.

16) Brown, H. C.; Negishi, E. *J. Am. Chem. Soc.* **1972**, *94*, 3567.

17) Brown, H. C.; Zaidlewicz, M. *J. Am. Chem. Soc.* **1976**, *98*, 4917.

18) Chujo, Y.; Tomita, I.; Hashiguchi, Y.; Tanigawa, H.; Ihara, E.; Saegusa, T. *Macromolecules* **1991**, *24*, 345.

19) Chujo, Y.; Tomita, I.; Saegusa, T. *Polym. Bull.* **1992**, *27*, 375.

20) Chujo, Y.; Tomita, I.; Hashiguchi, Y.; Saegusa, T. *Macromolecules* **1992**, *25*, 33.

21) Chujo, Y.; Tomita, I.; Hashiguchi, Y.; Saegusa, T. *Macromolecules* **1991**, *24*, 3010.

22) Chujo, Y.; Tomita, I.; Hashiguchi, Y.; Saegusa, T. *Polym. Bull.* **1991**, *25*, 1.

23) Chujo, Y.; Tomita, I.; Saegusa, T. *Polym. Bull.* **1991**, *26*, 165.

24) Chujo, Y.; Morimoto, M.; Tomita, I. *Polym. Bull.* **1992**, *29*, 617.

25) Chujo, Y.; Morimoto, M.; Tomita, I. *Polym. J.* **1993**, *25*, 891.

26) Hawthorne, M. F. *Tetrahedron* **1962**, *17*, 112.

27) Chujo, Y.; Tomita, I.; Murata, N.; Mauermann, H.; Saegusa, T. *Macromolecules* **1992**, *25*, 27.

28) Chujo, Y.; Tomita, I.; Saegusa, T. *Polym. Bull.* **1993**, *31*, 553.

29) Chujo, Y.; Tomita, I.; Saegusa, T. *Polym. Bull.* **1993**, *31*, 547.

30) Chujo, Y.; Tomita, I.; Saegusa, T. *Macromolecules* **1992**, *25*, 3005.

31) Chujo, Y.; Tomita, I.; Asano, T.; Saegusa, T. *Polym. J.* **1994**, *26*, 85.

32) Bhatt, M. V. *J. Organomet. Chem.* **1978**, *156*, 221.

33) Hara, S.; Dojo, , H.; Takinami, S.; Suzuki, A. *Tetrahedron Lett.* **1983**, *24*, 731.

34) Hara, S.; Satoh, Y.; Ishiguro, H.; Suzuki, A. *Tetrahedron Lett.* **1983**, *24*, 735.

35) Chujo, Y.; Tomita, I.; Saegusa, T. *Macromolecules* **1990**, *23*, 687.

36) Chujo, Y.; Takizawa, N.; Sakurai, T. *J. Chem. Soc., Chem. Commun.* **1994**, 227.

37) Chujo, Y.; Tomita, I.; Saegusa, T. *Polym. J.* **1991**, *23*, 743.

38) Chujo, Y.; Tomita, I.; Saegusa, T. *Makromol. Chem., Macromol. Symp.* **1993**, *70/71*, 47.

39) Chujo, Y.; Tomita, I.; Asano, T.; Saegusa, T. *Polym. Bull.* **1993**, *30*, 215.

RECEIVED April 14, 1994

Chapter 31

Poly(carborane–siloxane–acetylene) as Precursor to High-Temperature Thermoset and Ceramic

Leslie J. Henderson, Jr. and Teddy M. Keller

Materials Chemistry Branch, Code 6127, Naval Research Laboratory, Washington, DC 20375–5320

A linear carborane-siloxane-acetylenic hybrid polymer has been synthesized and characterized. The poly (carborane-siloxane-acetylene) is a viscous liquid at room temperature and can be easily processed to a high temperature polymeric thermoset under either thermal or photochemical conditions. Thermal curing of the linear polymer to the thermoset is accomplished above 150°C. This novel thermosetting polymer is readily converted into a ceramic material by heating to 1000°C under inert or oxidative conditions. The thermoset and ceramic material are stable in air at elevated temperatures.

The search for polymeric materials that will maintain their properties for extended periods above 300°C in an oxidative environment is being conducted by many investigators over a broad spectrum of disciplines. These polymers are usually composed of aromatic and/or heterocyclic units connected by flexible groups to impart processability and to enhance mechanical properties. Relatively few of the many polymers that have been synthesized have been exploited commercially. Moreover, there is little evidence that any marked improvements in thermal and oxidative stabilities of organic or carbon-based polymers will be forthcoming over currently available resin systems.

An emerging technology that holds promise for extending the temperature stability of polymers is inorganic-organic hybrid polymers. The polymeric hybrid combines the desirable features of inorganics and organics within the same polymeric system such as thermal and oxidative stability and processability. The development of carborane-siloxane polymers in the early 1960s was a major breakthrough in the search for high temperature elastomers (1). Poly(carborane-siloxane) elastomers show superior thermal and oxidative properties and retarded depolymerization (2,3) at elevated temperatures relative to poly(siloxanes). The carborane unit is incorporated into a polymeric chain to impart either high temperature and/or specialized chemical resistance. Patents have been issued for thermally stable elastomers consisting of meta- or para-carborane and either silyl or siloxyl units. Some have been produced commercially. However, they lack acetylenic groups or any other functional group for cross-linking purposes to produce thermosetting polymers. Researchers have successfully incorporated carboranes into the backbone of most of the common types of addition and condensation polymers and as pendant groups or side chains (*4*). It was recognized that the attachment of a carborane side group onto the backbone of a polymer did not enhance the thermal properties. However, if the carborane becomes a part of the polymeric backbone, appreciable improvements in thermal stability were achieved.

In our continuing investigations of high temperature polymers, carborane-siloxane-acetylenic polymers have been synthesized and are being evaluated as high temperature matrix materials for composites and as precursor materials for ceramics. The major advantage of our approach is that the desirable features of inorganics and organics such as high thermal and oxidative stability and processability are incorporated into the same polymeric chain. The siloxane units provide thermal and chain flexibility to polymeric materials. Silylene-acetylenic polymers have also been made but lack the thermally and oxidatively stable carborane units. The chemistry involved in synthesizing poly(carborane-siloxane) has been modified to accommodate the inclusion of an acetylenic unit in the backbone. The presence of the acetylenic linkages within the backbone provides the opportunity to convert the initially formed liquid linear polymer into a thermoset. The viscosity of the linear polymer depends on the molecular weight. The cross-linked density of the poly(carborane-siloxane-acetylene) is easily controlled as a function of the quantity of reactants used in the synthesis. The acetylenic functionality provides many attractive advantages relative to other cross-linking centers. The acetylene group remains inactive or dormant during processing at lower temperatures and will react either thermally or photochemically to form conjugated polymeric cross-links without the evolution of volatiles. This paper is concerned with the synthesis of a carborane-siloxane-acetylenic polymer, its conversion into a high temperature polymer/ceramic, aging under high temperature oxidative conditions, and characterization by FTIR, thermal analyses, optical and scanning electron microscopies, and X-ray photoelectron spectroscopy (XPS).

Experimental

Thermal analyses were performed with a Dupont 2100 thermal analyzer equipped with a thermogravimetric analyzer (TGA, heating rate 10°C/min) and a differential scanning calorimeter (DSC, heating rate of 10°C/min) at a gas flow rate of 50 cc/min. Thermal and oxidative studies were carried out in nitrogen and air, respectively. Infrared spectral studies were performed with a Perkin-Elmer 1800 FTIR spectrophotometer purged with nitrogen gas. FTIR spectra of soluble samples were obtained from thin films solvent-cast on NaCl plates. FTIR spectra of insoluble samples (cured or pyrolyzed) were acquired via KBr pellets. Molecular weights were determined by gel-permeation chromatography (GPC) using a Hewlett-Packard Series 1050 pump equipped with two Altex μ-spherogel columns (size 10^3 and 10^4 Å, respectively) connected in series. *n*-Butyllithium (2.5M in hexane) and hexachlorobutadiene were obtained from Aldrich. The latter was freshly distilled at reduced pressure prior to use. Tetrahydrofuran (THF), purchased from Aldrich, was dried under argon with potassium metal and freshly distilled from potassium/benzophenone prior to use. 1,7-Bis(chlorotetramethyldisiloxyl)-*m*-carborane 1 was purchased from Dexsil Corporation and was used as received. All syntheses were carried out under a dry argon atmosphere.

Synthesis of poly(butadiyne-1,7-bis(tetramethyldisiloxyl)-*m*-carborane) 1. In a typical synthesis, a 2.5M hexane solution of *n*-BuLi (34.2 ml, 85.5 mmol) in 12.0 ml of THF was cooled to -78°C under an argon atmosphere. Hexachlorobutadiene 2 (5.58g, 21.4 mmol) in 2.0 ml THF was added dropwise by cannula. The reaction was allowed to warm to room temperature and stirred for 2 hr. The dilithiobutadiyne 3 in THF was then cooled to -78°C. At this time, an equal-molar amount of 1,7-bis(chlorotetramethyldisiloxyl)-*m*-carborane 4 (10.22 g, 21.4 mmol) in 4.0 ml THF was added dropwise by cannula while stirring. The temperature of the reaction mixture was allowed to slowly rise to room temperature. While stirring the mixture for 1 hour, a copious amount of white solid (LiCl) was formed. The reaction mixture was poured into 100 ml of dilute hydrochloric acid resulting in dissolution of the salt and the separation of a viscous oil. The polymer 1 was extracted into ether. The ethereal layer was washed several times with water until the washing was neutral, separated, and dried over sodium sulfate. The ether was evaporated at reduced pressure leaving a dark-brown viscous polymer 1. A 97% yield (9.50 g) was obtained after drying in vacuo. GPC indicated the presence of low molecular weight species ($\approx$500) as well as higher average molecular weight polymers (Mw$\approx$4900, Mn$\approx$2400). Heating of 1 under vacuum at 150°C removed lower molecular weight volatiles giving a 92% overall yield. Major FTIR peaks (cm^{-1}): 2963 (C-H); 2600 (B-H); 2175 (C$\equiv$C); 1260 (Si-C); and 1080 (Si-O).

Thermosetting Polymer 5A. A sample of 1 (1.5490 g) was quickly heated to 300°C in argon. The polymer 1 was then cured by heating at 300, 350, and 400°C, consecutively, for 2 hours at each temperature. Upon cooling at 1°C/min to ambient conditions, a void-free dark brown solid 5A, which was 96 wt% (1.4840 g) of the starting material 1, was isolated. FTIR (cm^{-1}): 2963 (C-H); 2599 (B-H); 1600 (C=C); 1410 and 1262 (Si-C); and 1080 (Si-O).

An alternate procedure involves heating a sample of 1 (1.51 g) at 300, 350, and 400°C, consecutively, in air for 2 hours at each temperature. During the cure, the sample developed an amber film on the outer surface. The polymer was stable upon aging in air at 500°C for 100 hours.

Photocrosslinking of 1 to Produce 5B. The polymer 1, dissolved in methylene chloride, was coated on a 1 inch square platinum screen, and the solvent evaporated. This process was repeated until a reasonable FTIR spectrum of 1 could be obtained. A mercury UV lamp with a Jarrell-Ash power supply was used without monochromator or filters to irradiate 1. The screen was mounted on an IR cell holder so that 1 could be irradiated and the assembly could be moved without disturbing the relative positions for monitoring by FTIR.

Pyrolysis of Thermoset 5A to Ceramic 6. Pyrolysis of thermoset 5A (1.4251 g) in argon to the ceramic 6 was accomplished by heating directly or at various temperatures in sequence to 1000°C. Alternatively, 5A could be heated in sequence at various temperatures to 1000°C. The sample was then cooled back to ambient conditions at 0.5°C/min producing the black solid ceramic, 6, (1.2045 g, 85% yield). FTIR (cm^{-1}): 3210; 1474; 1402; 1195; 795.

Results and Discussion

The synthesis of 1 is a one pot, two stage reaction (see Scheme 1). Dilithiobutadiyne 3 was prepared by the method of Ijadi-Magshoodi and Barton (5,6). Dilithiobutadiyne 3 was not isolated and was reacted with an equal-molar amount of the 1,7-bis(chlorotetramethyldisiloxyl)-m-carborane 4 to afford a dark-brown viscous polymer 1 (see Scheme 1) in quantitative yield (97-100%). GPC indicated the presence of low molecular weight species (≈500) as well as higher average molecular weight polymers (Mw≈4900, Mn≈2400). Heating under vacuum at 150°C removed the lower molecular weight components leaving a 92-95 % overall yield.

The novel carborane-siloxane-acetylenic polymer 1 has the advantage of being extremely easy to process since it is a liquid at ambient temperature and is soluble in most organic solvents. It is designed as a thermoset polymeric precursor. Cross-linking of 1 can occur by thermal or photochemical means through the triple-bonds of the acetylenic units to afford thermosetting polymers 5A and 5B, respectively (7-9). A shiny void-free dark brown solid 5A was produced by thermally curing 1 at 300, 350, and 400°C, consecutively, for 2 hours at each temperature either under inert condition or in air. Gelation occurred during the initial heat treatment at 300°C. An FTIR spectrum (see Figure 1) of 5A shows the disappearance of the acetylenic absorption at 2175 cm^{-1} and the appearance of a new, weak peak centered at 1600 (C=C) cm^{-1}. A spectrum of 5A cured in air also exhibited an absorption at 1714 cm^{-1}, attributed to a carbonyl group. The other characteristic peaks were still present. Pyrolysis of 5A to 1000°C yielded a black solid ceramic material 6 in 85% yield that retained its shape except for some shrinkage. The characteristic FTIR absorptions of 1 and 5A were now absent.

The thermal polymerization of 1 was studied by differential scanning calorimetry (DSC) from 30 to 400°C under inert conditions (see Figure 2). A small broad exotherm is apparent from about 150 to 225°C and was attributed to the presence of a small amount of primary terminated acetylenic units. This peak was absent when 1 was heated at 150°C for 30 minutes under reduced pressure. These low molecular weight components must be removed to ensure the formation of a void-free thermoset. A larger broad exotherm commencing at 250°C and peaking at 350°C was attributed to the reaction of the acetylene functions to form the cross-links. This exotherm was absent after heat treatment of 1 at 320°C and 375°C, respectively, for 30 minutes. The polymer 1 could be degassed at temperatures up to 150°C without any apparent reaction of the acetylenic units. A fully cured sample of 5A did not exhibit a T$_g$, which enhances its importance for structural applications.

Cross-linking of 1 by photochemical means to produce 5B was achieved in both air and argon atmospheres. Ultraviolet (UV) irradiation of the polymer 1 resulted in a decrease of the intensity of the triple bond absorption (2170 cm^{-1}) as determined from FTIR spectroscopy. Even after irradiating for 4 days, the brown film 5B still showed some absorption due to the triple bond. Comparison of the

Scheme 1

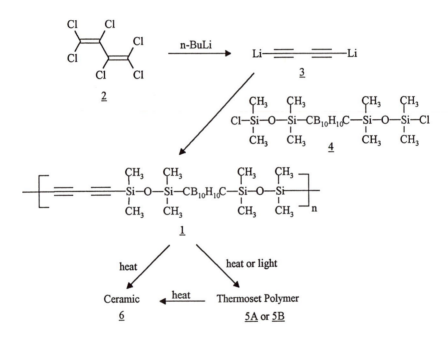

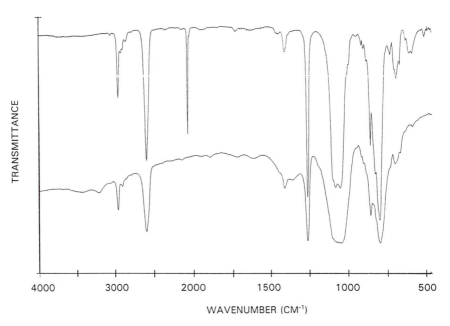

1.   FTIR spectra: film of <u>1</u> on NaCl disk (top) and KBr pellet of <u>5A</u>
      (bottom)

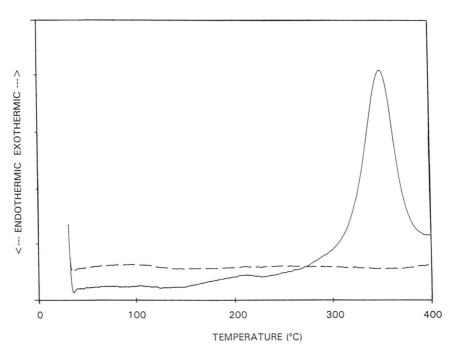

2.   DSC thermograms of: <u>1</u> (solid line) and cured <u>5A</u> (dash line)

FTIR spectra for 5A and 5B disclosed that different polymeric compositions were formed by thermal and photochemical means. Thermoset 5B exhibited an absorption at 1723 cm^{-1} that was absent in 5A.

The thermal and oxidative stability of 1 and its conversion into 5A and 6 were studied by thermogravimetric analysis (TGA). The TGA studies were determined between 50 and 1000°C at a heating rate of 10°C/min in both argon and air atmospheres (see Figures 3 and 4). Upon heating to 1000°C in argon, 6 exhibited a ceramic yield of 85%. Further heat treatment at 1000°C for 12 hours resulted in no additional weight loss. When the ceramic material 6 was cooled back to 50°C and rescanned to 1000°C in air at a flow rate of 50 cc/min, the sample gained weight (≈2%) attributed to oxygen uptake. A TGA thermogram of 1, which was initially converted into 5A and then heated to 1000°C in air, showed less weight loss relative to pyrolysis under inert conditions and exhibited a ceramic yield of 92%. Further TGA studies of the ceramic 6 in air revealed that additional weight loss did not occur and that the sample actually increased in weight as observed previously. Furthermore, a sample that had been heated to 1000°C in air, cooled, and then rerun in argon showed no weight changes. These observations show the stability of 6 under both inert and oxidative conditions.

Isothermal aging studies were performed on both 5A and 6 under inert and oxidative conditions. A sample of 1 was cured by heating in sequence at 300°C and 400°C for 4 hours under an argon atmosphere and lost 8% by weight. Upon exposure to an air atmosphere for a short time, 5A seemed to segregate into different components like cement around irregularly shaped flat stone tiles with the cement having an orangish brown outline at the interfaces. A sample of 5A was further heated in sequence at 500, 650, 800, and 900°C for 4 hours each under an argon atmosphere resulting in a total weight loss of 15%. Upon cooling and aging in an air circulating oven at 500°C for 50 hours, a weight loss of 2.5% was observed. When the sample was aged for an additional 50 hours, 6 did not lose any more weight. A sample of 1 heated under a flow of air at 50 cc/min to 900°C and held at this temperature for 10 hours afforded a ceramic yield of 87%. When the sample 6 was further heated in air at 500°C for 12 hours, no apparent weight changes occurred. Regardless of the heat treatment, the samples retained their structural integrity. Sample 5A showed more erosion or weight loss than 6 when aged in air at 400°C and 500°C, respectively. A similar heat related difference in oxidation protection was observed for carbon fibers and graphite samples coated sequentially with chemical vapor deposition (CVD) of silicon carbide (SiC) and boron oxide (B_2O_3) paste (10). The lack of protection was attributed to insufficient silicon oxide (SiO_2) formation at the lower heat treatment temperature.

The pyrolysis of 1 in a stream of argon to 900°C at 10°C/min afforded a black monolith 6 in 85% ceramic yield that retained its shape except for some shrinkage. Elemental analysis of 1 and 6, which were prepared in argon and were amorphous as determined by X-ray diffraction, is presented in Table I. The major weight percentage changes were at the expense of carbon and hydrogen. Moreover, the ceramic material 6 contains a large excess of carbon (compare calculated values for B_4C and SiC, respectively: 78.26% B, 21.74% C and 70.04% Si, 29.96% C). This was not unexpected since it is known that ceramics formed from pyrolysis of

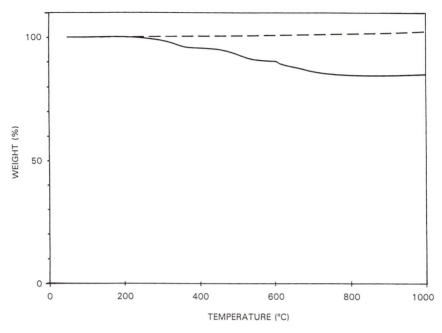

3.   TGA thermograms of: <u>1</u> under nitrogen atmosphere (solid line) and cooled and rerun <u>6</u> in air

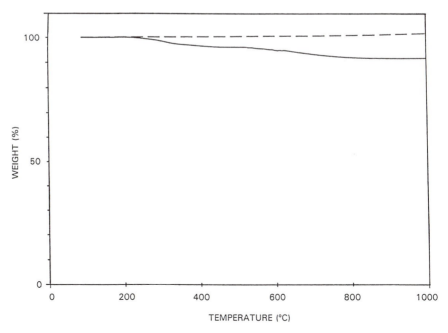

4.   TGA thermograms in air: <u>1</u> (solid line) and cooled and rerun <u>6</u> (dash line)

polymers containing unsaturated organic groups attached to silicon, which makes them less likely to be cleaved, have high carbon levels (*11,12*). Our exploratory experiments show polymer 1 to be potentially useful as a binder or matrix material for advanced composites and in the formation of shaped ceramic bodies.

Table I. Elemental Analysis (%)

Sample	Hydrogen	Boron	Carbon	Silicon	Oxygen[a]
Theory 1	7.53	23.77	36.97	24.70	7.03
1	7.59	23.20	36.86	24.98	7.37
5[b]	7.70	22.55	37.07	23.55	9.13
6[c]	0.43	22.76	35.06	23.70	18.05

[a] Value determined as difference. [b] Sample cured at maximum temperature of 400°C under argon atmosphere. [c] Sample pyrolyzed at 900°C under argon atmosphere.

Optical and scanning electron microscopies were used to examine surface structures of 5A and 6. Samples of 5A cured to 400°C under an argon atmosphere exhibited a uniformly glossy void-free black surface. When 5A was exposed to air and aged at 400°C, a tan layer developed on the surface. A sample of 6 that had been annealed for 12 hours at 900°C and then aged in a flow of air also developed a tan layer. This layer was composed of large numbers of amber crystallites and merged smoothly into the underlying layer. There was no gap between the tan layer and the bulk material.

The composition of the oxidized surface and bulk were examined by XPS. The bulk compositions of 5A and 6 remained unchanged upon oxidation treatments. After curing and aging in air at 400°C, 5A developed a bilayer inorganic surface, which mainly consisted of boron oxide and silicon oxide (see Table II). Prolonged oxidation of 6 at 500°C resulted in the volatilization of boron oxide leaving an almost pure silicon oxide layer, which blended into the bulk boron and silicon concentrations at the interface between the surface layer and the bulk.

In summary, carborane-siloxane-acetylenic-based thermosetting polymers are designed to take advantage of the inorganic-organic hybrid approach to produce high temperature polymers and ceramic-based materials. The major advantage of the hybrid high temperature polymeric approach is that the desirable features of inorganics (high thermal and oxidative stability) and organics (processability) are incorporated into novel polymeric material. Polymer 1 possesses exceptional thermal and oxidative stability, is a viscous liquid that is soluble in most organic solvents, and is easy to process thermally into structural components. Thermal cross-linking occurs through the triple bonds to produce the thermoset 5A. Thermoset 5A, in turn, produces a ceramic material 6 in high yield (85% and 92%, respectively) upon pyrolysis in nonoxidizing and oxidizing environments. When heated in an oxidizing environment, 6 initially gained weight. The potential utility of polymer 1 as a matrix

Table II. Surface Compositions Obtained From XPS Sputter Profiles After >120 Sputtering Time

Sample	C(Atomic) %	O(Atomic)%	Si(Atomic)%	B(Atomic)%
5[a]	45	6	8	31
5[b]	17	48	24	11
5[c]	5	66	20	10
6[d]	3	58	30	9

[a] Cured at maximum temperature of 400°C for 2 hours. [b] Cured at 300°C for 4 hours in argon and at 400°C for 4 hours in air. [c] Cured in air at 320°C, 350°C, and 400°C in sequence for 2 hours and aged for 100 hours in air at 400°C. [d] Cured and heated at 900°C for 4 hours and aged in air at 500°C for 100 hours.

material for advanced composites and further conversion into a shaped ceramic component is due to the ease of processability, high ceramic yield, and oxidative stability at elevated temperatures. Surface analysis studies of samples exposed to air indicate that the surface of both 5A and 6 is enriched with oxide forms of boron and silicon, which apparently protect the interior part against further oxidation. Further studies are underway to fully evaluate the thermal properties of 5A and 6 and to develop new chemistries pertaining to the design of high temperature polymeric materials and ceramics.

Acknowledgement

The authors wish to thank the Office of Naval Research for financial support of this work. The authors are grateful to Dr. Pehr Pehrsson of the Surface Branch of the Chemistry Division for the XPS studies and Dr. Tai Ho of George Mason University for his assistance with the GPC measurements.

Literature Cited

1.   Schroeder, H. A. Inorg. Macromol. Rev. **1970**, 1, 45
2.   Papetti, S.; Schaeffer, B. B.; Gray, A. P.; Heying, T. L. J. Polym. Sci., A-1 **1966**, 4, 1623
3.   Critchley, J. P.; Knight, G. J.; Wright, W. W. Heat-Resistant Polymers, Plenum Press: New York, **1983**, 363
4.   Williams, R. E. J. Pure Appl. Chem. **1972**, 29, 569
5.   Ijadi-Maghsoodi, S.; Pang, Y.; Barton, T. J. J. Polym. Sci.: Part A: Polym. Chem., **1990**, 28, 955
6.   Ijadi-Maghsoodi, S.; Barton, T. J. Macromolecules, **1990**, 23, 4485
7.   Rutherford, D. R.; Stille, J. K. Macromolecules, **1988**, 21, 3530
8.   Neenana, T. X.; Callstrom, M. R.; Scarmoutzos, L. M.; Stewart, K. R.; Whitesides, G. M.; Howes, V. R. Macromolecules **1988**, 21, 3525

9. Callstrom, M. R.; Neenan, T. X.; Whitesides, G. M. Macromolecules **1988**, 21, 3528

10. Gee, S. M.; Little, J. A. J. Mater. Sci., **1991**, 26, 1093

11. White, D. A.; Oleff, S. M.; Fox, J. R. Adv. Ceram. Mater. **1983**, 2, 53

12. Burns, G. T.; Taylor, R. B.; Xu, Y.; Zangvil, A.; Zank, G. A. Chem. Mater. **1992**, 4(6), 1313

RECEIVED April 25, 1994

Chapter 32

Organoaluminum Precursor Polymers for Aluminum Nitride Ceramics

James A. Jensen

Lanxide Corporation, Newark, DE 19714–6077

An overview of synthetic routes to aluminum-nitrogen polymers including recent results describing the use of alkylaluminum imine precursors to aluminum nitride will be presented. Inorganic polymer precursors provide a non-traditional route to ceramic and inorganic composite materials utilizing plastic forming techniques. Polyalazanes or polyaminoalanes, compounds which possess polymer backbones consisting of Al-N bonds, have been shown to be effective precursors to AlN-containing ceramics. The latter sections of the chapter focus on alazane preceramic polymers prepared from alkylaluminum imine feedstocks. The aluminum imine complexes are synthesized from aluminum hydride reagents and nitriles. This synthetic approach is quite versatile due to the commercial availability of a large number of nitriles of various functional group substitution. Consequently, the resulting polymers have a range of properties encompassing curable liquids to thermoplastic solids.

Aluminum nitride, AlN, is a refractory ceramic with low dielectric constant, high intrinsic thermal conductivity, and a low coefficient of thermal expansion very close to that of silicon. In fact, aluminum nitride is one of only few materials which is both a good electrical insulator and a good thermal conductor. Consequently, there is widespread interest in its use as a replacement for alumina and beryllia in electronic packaging applications. Such a high degree of interest in AlN stems from its high thermal conductivity, which has been measured at 319 W/m·K on a single crystal (*1*). The theoretical value is 320 W/m·K.

AlN is also very stable to molten metals (in particular aluminum) and has good corrosion resistance. Additional applications for AlN include parts which come into contact with molten metal or other corrosive materials, and parts which require good heat transfer characteristics, thermal shock resistance or wear resistance.

0097–6156/94/0572–0427$08.00/0
© 1994 American Chemical Society

Consequently, considerable research on a variety of AlN processing techniques has ensued (2-4).

Recently, there has been a great deal of research activity in polymer precursors which convert to ceramic materials upon pyrolysis (5). The use of polymer technology to prepare ceramic materials offers certain processing advantages over conventional ceramic processing routes. The hallmark of preceramic polymers is the ability to fabricate complex shapes using a variety of plastic forming methods including molding, fiber spinning, film casting, coating, and infiltration. Consequently, preferred preceramic polymers are those that are curable liquids or soluble, thermoplastic solids. When used in powder processing, preceramic polymer binders obviate extended binder burnout cycles since the polymer "burns in" to the consolidated ceramic or composite phase.

The desire to prepare oxygen-free AlN requires the use of oxygen-free reagents and precursors. Aluminum alkyls, aluminum hydrides, or anhydrous aluminum halides are thus logical starting materials for aluminum-nitrogen polymers. Since these reagents are typically very air sensitive or pyrophoric, the resulting polymers also tend to be quite oxygen and moisture sensitive. Aluminum nitrogen bonds, in general, are hydrolytically unstable.

This chapter provides an overview of routes and reagents which have been utilized to prepare aminoalane polymers as precursors for AlN ceramics. Throughout the text, the terms polyalazane (analogous to the silicon nomenclature), polyaminoalane, and aluminum-nitrogen polymer will be used interchangeably, and do not represent any structural features other than an aluminum-nitrogen polymer backbone. Several research groups have been very active in the area. The cited references are not intended to be comprehensive, but rather to highlight the various approaches which have been employed. The synthetic descriptions will be divided by type depending on the aluminum reagent used. Generally, the systems undergo successive condensation reactions to form polymeric species upon heating. Crosslinking continues as long as replaceable groups are present on aluminum and nitrogen. Replaceable groups consist of alkyl, aryl or hydride bound to aluminum and hydrogen bound to nitrogen. Thus $[H_2NAlR_2]$ species would be expected to further condense, but species of the type [RNAlCl] and [RNAlR] have been reported to be stable to further condensation (6,7). However, $[RNAlCl]_x$ polymers have recently been converted to AlN and the latter half of this chapter will describe the preparation of $[RNAlR]_n$ polymers and their conversion to AlN. A number of reports have appeared utilizing aminoalane precursors in the preparation of composite materials such as AlN/SiC and AlN/Si$_3$N$_4$ (8-12). These will not be addressed in this chapter.

General Synthetic Methods

Polymers Derived from Trialkylaluminum. The most widely studied approach to alazane polymers utilizes ammonia or amine derivatives of trialkylaluminum reagents. The method, following the steps of equation 1, which was first described over fifty years ago (13), was recently reinvestigated (14-17). Trialkyl and triarylaluminum compounds form 1:1 adducts with Lewis bases such as ammonia or amines. These adducts readily undergo intermolecular condensation with elimination of a

$$\underset{-CH_4}{\overset{i}{Me_3Al\text{-}NH_3 \quad \rightarrow}} \quad \underset{-CH_4}{\overset{ii}{[Me_2AlNH_2]_3 \quad \rightarrow}} \quad \underset{-CH_4}{\overset{iii}{[MeAlNH]_x \quad \rightarrow}} \quad AlN \qquad (1)$$

hydrocarbon molecule as long as the amine has available hydrogen atoms bonded to the nitrogen.

Conversion of $[Me_2AlNH_2]_3$ and $[tert\text{-}Bu_2AlNH_2]_3$, prepared by ammonolysis of Me_3Al and $tert\text{-}Bu_3Al$, respectively, to AlN, has been extensively studied (*15*). These oligomers are crystalline solids which sublime under vacuum, serving as MOCVD precursors to AlN (*16,17*). In the condensed state, the cyclic trimers convert to a polymeric structure on thermolysis between 110-200°C in NH_3 (step ii in equation 1). The resulting aluminum imide polymer is a white, insoluble solid. Pyrolysis occurs above 1000°C to yield AlN, which is reported to contain less than 0.5 wt% residual carbon and oxygen. Utilization of ethylenediamine as the nitrogen source rather than ammonia results in soluble thermoplastic polymers from which films can be cast and fibers spun (*18*).

A melt formable, soluble organoaluminum polymer is formed by thermolysis of $[Et_2AlNH_2]_3$ in the presence of small amounts of Et_3Al (*19-21*). The polymer $[(EtAlNH)_n(Et_2AlNH_2)_m(AlEt_3)_x]$ can be prepared as a colorless viscous fluid, a glassy solid, or an infusible solid by controlling the ratio of reagents and reaction conditions. Fibers could be melt spun from the precursor. Heating the polymer under ammonia affords a "cure" to an infusible polymer which converts to AlN on pyrolysis with retention of shape. High aspect ratio flakes were prepared by freeze-drying a benzene solution of the polymer. This polymer is air sensitive and all manipulations were done in a dry box.

Alkylaluminum reagents have also been treated with aminotriazines, such as melamine, to form precursors whose structure has not been described, but which convert to AlN at 1500°C in nitrogen. The resulting AlN has <2 wt% residual carbon (*22*).

Polymers Derived from $LiAlH_4$ and AlH_3. Due to the high aluminum mass content of these starting materials and the absence of carbon, precursors of the form [HAlNH] have been actively pursued. The reaction of ammonia with AlH_3 was first reported nearly forty years ago. The reaction sequence using two different Al:N ratios and tetrahydrofuran solvent is shown in equations 2 and 3 (*23*). This approach has been

$$\underset{-H_2}{\overset{-30°C}{AlH_3 + 3NH_3 \quad \rightarrow}} \quad \underset{-NH_3}{\overset{20°C}{Al(NH_2)_3 \quad \rightarrow}} \quad \underset{-H_2}{\overset{430°C}{Al(NH)NH_2 \quad \rightarrow}} \quad AlN_{1.27}H_x \qquad (2)$$

$$\underset{-H_2}{\overset{-80°C}{AlH_3 + NH_3 \quad \rightarrow}} \quad \underset{-H_2}{\overset{-45°C}{H_3Al\text{-}NH_3 \quad \rightarrow}} \quad \underset{-H_2}{\overset{20°C}{H_2AlNH_2 \quad \rightarrow}} \quad \overset{150°C}{HAlNH \quad \rightarrow} \quad H_{0.23}AlNH_{0.23} \;(3)$$

reinvestigated by several groups (*24,25*). A soluble gel was formed using the stoichiometry of equation 3. However, on drying and calcination at 1100°C in vacuum, the AlN was contaminated with carbon residues from the tetrahydrofuran solvent (25).

Alternatively, lithium aluminum hydride may be treated with ammonium halides as shown in equation 4 (26). NH_4Br gave better results than NH_4Cl, since

$$LiAlH_4 + NH_4X \rightarrow 1/n\,[HAlNH]_n + 3\,H_2 + LiX \tag{4}$$

the LiBr by-product is soluble in the reaction solvent, Et_2O, and only a trace of salt is retained in the insoluble polymeric aminoalane product. However, the polymer does entrap 5-10 wt% of the diethyl ether solvent. The polyaminoalane of equation 4 converts to AlN in 85-87% yield by TGA. Pyrolysis in nitrogen or under vacuum at 950°C gives crystalline AlN which is black, indicating residual carbon. Pyrolysis in ammonia gave white AlN.

Cage compounds of the form $[HAlNR]_n$ are readily prepared from $LiAlH_4$ and primary aminehydrochlorides according to equation 5 or by use of primary amines, equations 6 and 7, (27). These cage compounds have been converted to AlN (9,28).

$$LiAlH_4 + RNH_3Cl \rightarrow 1/n\,[HAlNR]_n + 3\,H_2 + LiCl \tag{5}$$

$$LiAlH_4 + RNH_2 \rightarrow 1/n\,[HAlNR]_n + 2\,H_2 + LiH \tag{6}$$

$$AlH_3\text{-}OEt_2 + RNH_2 \rightarrow 1/n\,[HAlNR]_n + 2\,H_2 + Et_2O \tag{7}$$

Pyrolysis of $[HAlN(i\text{-}Pr)]_6$ at 1000°C for two hours in argon gave a black product in 43% yield which was amorphous by X-ray diffraction. Heating the black solid to 1600°C resulted in an additional 50% weight loss and formation of crystalline AlN.

Polymers Derived from Aluminum Halides. The compound $[Cl_2AlNHSiMe_3]_2$ is prepared from $AlCl_3$ and hexamethyldisilazane (29). Thermolysis at 170-200°C results in formation of a polymer $[ClAlNH]_n$ and evolution of Me_3SiCl (equation 8). Pyrolysis of $[ClAlNH]_n$ in ammonia or vacuum at 700°C is claimed to give crystalline AlN (30).

$$\overset{\text{heat}}{[Cl_2AlNHSiMe_3]_2 \rightarrow} 2/n\,[ClAlNH]_n + 2\,Me_3SiCl \tag{8}$$

The action of potassium amide on $AlBr_3$ results in an infusible aluminum amideimide $[Al(NH_2)NH]_n$, equation 9, which was also prepared by ammonolysis of $AlH_3\text{-}OEt_2$, (equation 2) (24). Pyrolysis at 800°C under vacuum results in AlN.

$$AlBr_3 + 3\,KNH_2 \rightarrow 1/n\,[Al(NH_2)NH]_n + 3\,KBr + NH_3 \tag{9}$$

Polymers Derived Electrochemically from Elemental Aluminum. Aminoalane polymers have also been prepared by electrolysis of aluminum in a solution consisting of an alkyl amine, a solvent, and a supporting electrolyte (31-33) or in a solution of liquid ammonia (24,34). Anodic dissolution of an aluminum electrode under an inert atmosphere in an electrolyte consisting of a primary amine, acetonitrile, and tetrabutylammonium bromide leads to aluminum-nitrogen polymers, the idealized structure of which is given in equation 10 where R is alkyl or H. The proposed

$$Al + 3\,RNH_2 \quad \xrightarrow{-3/2\,H_2} \quad [Al(NHR)_3] \quad \rightarrow \quad 1/n\,[N(R)Al(NHR)]_n + RNH_2 \qquad (10)$$

$Al(NHR)_3$ intermediate was not isolated or observed spectroscopically. When R is an alkyl group such as n-propylamine, the solution is heated under vacuum on completion of electrolysis to drive off excess amine. At about 80°C, a high viscosity liquid results; at 150°C, an insoluble gel-like solid forms. Pyrolysis at 850°C gives a crystalline AlN product. Pyrolysis in argon results in a product with up to 28 wt% carbon. However, pyrolysis in ammonia gave AlN with about 2% oxygen and 0.8% carbon impurities (*31,32*). SiC fibers were coated by dispersing the fibers in the reaction medium after electrolysis followed by pyrolysis to give AlN-coated SiC (*33*).

A modification of the above procedure which simplifies the process is the use of liquid ammonia as both solvent and nitrogen source. Ammonium bromide is used as the supporting electrolyte, so carbon contamination of the final AlN product should be minimized since none of the starting materials contain carbon (*34*). After electrolysis, the ammonia is allowed to evaporate and the mixture of insoluble aluminum polymer/ammonium bromide is calcined at 1100°C in flowing ammonia to give submicron sized AlN powder with no apparent bromine impurities.

Polymers Derived from Alkylaluminum Hydrides. All of the above synthetic routes utilize ammonia or an amine as the nitrogen source for the polymers. Significantly less effort has been spent on the study of the interaction of aluminum reagents with other nitrogen donor molecules. For example, addition of AlH_3-NR_3 adducts to acetonitrile or aziridine solutions yields polymeric products with imine end groups according to equation 11 (*35,36*). The product is soluble in benzene and has a

$$n\,AlH_3\text{-}NR_3 + n\,CH_3CN \quad \rightarrow \quad [H_2Al[N(Et)Al(H)]_{n-1}N=CHCH_3] + n\,NR_3 \qquad (11)$$

degree of polymerization n = 3 to 35. The conversion of this precursor to AlN was not reported.

Trialkylaluminum reagents form adducts with nitriles, which can be converted thermally into imine complexes as shown in equation 12 (*37,38*). The complexes

$$2\,AlR_3 + 2\,R'CN \quad \rightarrow \quad 2\,R_3Al\text{-}NCR' \quad \rightarrow \quad [RR'C=NAlR_2]_2 \qquad (12)$$

$[PhCH=NAlEt_2]_2$ and $[(CH_3)_3CC(CH_3)=NAl(CH_3)_2]_2$ prepared by the method of equation 12 were not observed to undergo rearrangement to form $[RC(R)_2NAl(R)]_n$ polymers upon heating to 200°C and 280°C, respectively.

Aluminum hydride reagents reduce nitriles directly to aluminum imine complexes, equation 13, without adduct formation as observed with trialkylaluminum reagents (*39*). The aluminum imine, $[C_6H_5CH=NAl(i\text{-}C_4H_9)_2]_2$, prepared from

$$2 R_2AlH + 2 R'CN \rightarrow [R_2AlN=CHR']_2 \qquad (13)$$

benzonitrile and diisobutylaluminum hydride is claimed to lose two equivalents of isobutene upon heating to 280°C to form an aluminum nitrogen polymer of the form $[-(RCH_2)NAl-]_n$ (40). The structural assignment was made based upon the recovery of benzylamine on hydrolysis. No attempt was made to identify the polymer or the inorganic residue on hydrolysis. None of these dimeric or polymeric species derived from nitriles have been investigated as AlN precursors.

The conversion to AlN of alazane polymers derived from nitrile and alkylaluminum hydride reagents has been investigated. The commercial availability of a wide range of nitrile reagents with a variety of molecular weights and functional groups provides a single synthetic route to polyaminoalanes with a spectrum of physical properties. Such flexibility has not been obtained by use of the previously described routes.

Experimental

All manipulations were carried out using standard vacuum line and dry box techniques under inert atmospheres or vacuum. Solvents were dried over molecular sieves and stored under nitrogen. Diisobutylaluminum hydride and the nitrile reagents were used as received. **CAUTION: Diisobutylaluminum hydride (DIBAL-H) is very moisture sensitive in dilute solutions and pyrophoric when neat.** The alkylaluminum imine complexes were prepared by literature methods (41). Thermal polymerization of the aluminum imine complexes was accomplished as previously reported (42-45). Controlled headspace analysis was conducted by holding sealed 0.5 gram samples of the aluminum imine in a bath at the appropriate temperature for 1 hour. The sample was cooled to room temperature and the headspace samples analyzed on an Hewlett Packard 5890 Series II Gas Analysis system using a 50 m x 0.5 mm chrompak column. Pyrolyses were conducted in an Astro Model 1000 furnace using graphite heating elements and crucibles.

Results and Discussion

Synthesis. The diisobutylaluminum imine complexes, $[RCH=NAl(i-C_4H_9)_2]_2$, were prepared by diisobutylaluminum hydride reduction of the appropriate nitrile according to equation 13. The reaction proceeds cleanly at room temperature in quantitative yield. Proton NMR spectra of the reaction products show no side product formation. The product imines are dimeric in structure; $[(C_6H_3-2,6-(CH_3)_2)CH=NAl(i-C_4H_9)_2]_2$ displays a four-membered heterocyclic Al-N core with a *trans* geometry in the solid state (41). Equation 13 is general for nitriles containing a wide range of functional groups. Aluminum imines have been prepared with R= alkyl, aryl, and olefin-containing groups.

DIBAL-H reduction of simple nitriles such as acetonitrile and benzonitrile results in liquid products, while substituted alkyls such as trimethylacetonitrile and 2,6-dimethylbenzonitrile form molecular solids with discrete melting points. Polymerization of the alkylaluminum imine complexes is accomplished thermally by a reductive condensation process. The imine double bond is reduced and alkane or olefin volatiles are eliminated from the aluminum center during the polymerization. In general, the product alazane polymers have the form $[R'CH_2NAl(C_4H_9)]_n$. The alkyl group, R', introduced with the nitrile has a pronounced effect on the polymer properties. For example, simple R' groups such as methyl and phenyl yield liquid or solid thermoplastic polymers; olefin-containing R' groups form liquid or meltable solid polymers which can be thermoset by heating.

Thermolysis of the aluminum imine complexes at temperatures below about 200°C generates alazane polymers substituted at both the Al and N atoms with single alkyl or alkenyl groups, i.e. one aluminum alkyl group has been lost on thermolysis. Prior reports had indicated that polymeric units of this type would not undergo further condensation reactions (*6,7*). However, we find this not to be the case. The "replaceable groups" criteria, as described in the introduction to this chapter is not met in the dimer $[RCH=NAl(i-C_4H_9)_2]_2$ or the polymer $[RNAlR]_n$. The nitrogen atoms of these species do not possess the acidic hydrogen earlier thought to be prerequisite for alkane loss. Yet condensation, ultimately to AlN, proceeds upon additional heating to temperatures above 200°C.

The following examples describe the versatility of this synthetic approach. The olefin-containing aluminum imine, $[CH_2=CHCH_2CH=NAl(i-C_4H_9)_2]_2$, **I**, prepared from 3-butenenitrile and DIBAL-H is a low viscosity liquid (approximately 30 cp) (*43,44*). Thermolysis of **I** below about 160°C generates a viscous liquid alazane polymer, **II**, which retains pendant carbon-carbon double bonds as evidenced by ^{13}C NMR and Raman spectroscopy. Above about 180°C, alazane polymer **II** undergoes additional crosslinking through carbon-carbon double bonds, as well as thermal condensation through aluminum-nitrogen bond formation, to form rigid nonmelting solids, which retain their shape upon pyrolysis. No evidence of carbon-carbon double bonds remain in spectra of the cured polymer.

In contrast, thermolysis of the alkylaluminum imine, $[CH_3CH=NAl(i-C_4H_9)_2]_2$, **III**, at about 200°C results in a viscous liquid polymer $\{[(C_2H_5)NAl(C_4H_9)]_x[(C_2H_5)NAl(R')]_y\}_n$, **IV**. Continued heating of **IV** to 300°C results in a solid thermoplastic alazane polymer which has a softening point near 110°C and is represented by the formula $\{[(C_2H_5)NAl(C_4H_9)]_x[(C_2H_5)NAl(R')]_y[(C_2H_5)NAl]_z\}_n$, **V**, where x+y=0.65, z=0.35, the value of n ranges from 6-20 and R' is an octyl group bridging two aluminum centers. The values of x and y in both **IV** and **V** can be varied as a function of thermolysis time and temperature. There are two types of gross aluminum environments in polymer **V**; those retaining one Al-C bond as in units x and y, and those bonded only to nitrogen as in unit z. Polyalazane **V** shows low moisture and oxygen sensitivity; it can be handled in air for short periods of time without appreciable degradation of properties. Polymer **V** can be processed using typical thermoplastic fabrication techniques.

Thermal Analysis. The char yield, at 1000°C, as measured by thermal gravimetric analysis, of a number of aluminum imine-derived polymers, has been obtained. The values range from 15 to 85 weight % retention depending upon the starting nitrile and the polymerization conditions. Primary weight loss typically occurs between 250 and 500°C (*42*).

Gas chromatography of the gas headspace on polymerization of **III** reveals that isobutane and isobutene are formed in varying ratio depending on temperature. Above 300°C ethylene begins to appear. A compilation of relative gas ratios for the evolved gasses as a function of temperature is shown in Table I. Below 150°C alkane

Table I. Gas Headspace Analysis of $[CH_3CH=NAl(i-C_4H_9)_2]_2$ Thermolysis

Temperature (°C)	Relative Percent in Headspace		
	Isobutane	Isobutene	Ethylene
50	68	32	<1
100	64	36	<1
150	42	58	<1
200	22	78	1
250	18	80	2
300	17	80	3
350	17	80	3

evolution dominates, but above about 150°C the major gaseous byproduct is isobutene. The ethylene observed above 300°C results from decomposition of the N-bound ethyl group and does not result from secondary reactions involving cleavage of butane hydrocarbons. This was confirmed by variation of the nitrogen substituent: precursors containing N-bound benzyl groups showed no ethylene, while imines containing $CD_3CH=N-$ groups evolved deuterium-labeled ethylene. Notably, ethylamine was not observed on pyrolysis indicating N-C and Al-C bond cleavage is favored over Al-N cleavage under the thermolysis conditions.

Polymerization Mechanism. Analysis of headspace gases during polymerization provides indirect evidence for the nature of the polymer itself. Hydrolysis of the polymers in sealed vessels was used to determine aluminum ligation. The resulting evolved gas was analyzed by GC and GC/MS. Three different hydrocarbon products were observed on hydrolysis of alazane polymer **IV**: isobutane, isobutene and octane resulting from isobutyl, isobutenyl, and octyl groups bound to aluminum. The isobutyl groups were introduced during the original imine formation reaction as part of the diisobutylaluminum hydride. However, isobutene, formed upon hydrolysis of the polymer, can only result from an isobutenyl group bound to aluminum. The octane ostensibly results from coupled or crosslinked butyl or butenyl groups. A proposed mechanism for the polymerization of $[CH_3CH=NAl(i-C_4H_9)_2]_2$ is shown in Figure 1. While the proposed mechanism shows an intramolecular process, intermolecular processes cannot be ruled out. The coordinated imine in the starting dimer **1** is reduced in the process of ß-hydride migration from a neighboring isobutyl group as shown in **2**. The imine reduction is accompanied by transfer of an acidic gamma-hydrogen from the isobutyl group involved in imine reduction to the other aluminum-bonded isobutyl group (Path A). Six membered rings are likely intermediates **3**. Loss of alkane in the form of isobutane forms an aluminum bound isobutenyl species **4**. The butenyl group formed in **4** can further react by insertion

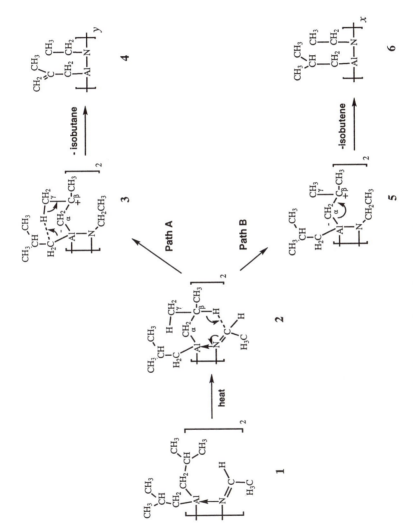

Figure 1. Proposed mechanism for thermal polymerization of $[CH_3CH=NAl(\imath\text{-}C_4H_9)_2]_2$.

into an aluminum-carbon bond to give octyl groups bridging metal centers. The amine substituent formed in Path A is an ethyl group.

Alternatively, following Path B, ß-hydride migration to reduce the coordinated imine can be accompanied by loss of alkene if Al-C_α bonds are broken as shown in **5**.

The second alkyl group bound to aluminum is unaffected in Path B so **6** retains an aluminum alkyl fragment which is unchanged from the starting dimer. The amine substituent formed in Path B is also an ethyl group.

Although the mechanism shown in Figure 1 illustrates an intramolecular process, an intermolecular process involving ß-hydride elimination may also occur. In this process, ß-hydride elimination of the olefin isobutene would generate an aluminum hydride species. The aluminum hydride would further react with another molecule of aluminum imine to reduce the imine with concomitant formation of a new aluminum-nitrogen bond. The only volatile product to be observed during the thermolysis via this mechanism would be the olefin, isobutene. Hydrolysis of the polymer formed by this process would generate only isobutane and ethyl amine products. Since a ß-hydride elimination mechanism can only explain evolution of isobutene during polymer formation, it alone does not explain the observed product distribution. However, it is probable that a combination of intramolecular and intermolecular processes are responsible for polymer formation.

Radical pathways are not anticipated to contribute to the butane formation on imine thermolysis due to the low temperature range where isobutane evolution predominates. Decomposition studies of triisobutylaluminum on Al surfaces show insignificant quantities of butane or butyl radicals in the 450-600 K temperature range studied (*46*).

Pyrolysis Studies. The imine complexes or alazane polymers, **I-V**, can be pyrolyzed in a non-oxidizing atmosphere to crystalline aluminum nitride. Figure 2 shows the X-ray powder diffraction pattern of polyalazane **V** fired from room temperature to 1000°C in ammonia and from 1000°C to 1800°C in nitrogen. Bulk elemental analysis of the powder shows only Al and N present with <0.5 wt% C impurity. Pyrolysis in ammonia without gas purification resulted in fired samples which consistently analyzed low in Al and N suggesting oxygen incorporation. Use of an efficient oxygen and moisture scavenger gave product which analyzed correctly for Al and N. Pyrolysis of **V** in nitrogen from room temperature to 1600°C results in a 40% ceramic yield of a black powder. Elemental analysis of the pyrolyzed sample confirms residual carbon on firing in nitrogen and argon. This result is consistent with previously reported conversion of carbon-containing polymer precursors to nitride ceramics; reducing atmospheres are necessary for complete carbon removal. The crystallization temperature is partially atmosphere dependent (*42*). In ammonia, individual (but not fully separated) peaks are evident in the XRD pattern at 1000°C. But in nitrogen or argon, a temperature of 1400°C is required to obtain a similar degree of crystallinity.

Conclusion

An overview of synthetic routes to aluminum-nitrogen polymer precursors for AlN ceramics has been presented. The most useful systems give precursors which can be

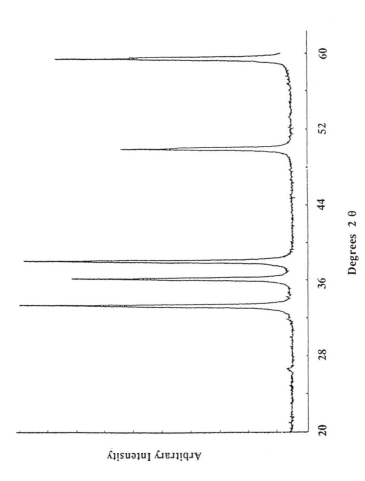

Figure 2. X-ray powder diffraction pattern of **V** fired to 1800°C.

fabricated into complex shapes using plastic forming methods. The use of aluminum imine complexes as feedstock reagents for alazane polymer formation has been demonstrated. The imine complexes can be thermally coverted to aluminum-nitrogen polymers. During the preparation of such polymers, both saturated and unsaturated hydrocarbon by-products are evolved, providing evidence for non-trivial imine decomposition pathways. This synthetic approach is quite versatile due to the commercial availability of a large number of nitriles of various molecular weights and functional group substitution. Consequently, the resulting polymers have a spectrum of properties encompassing curable liquids to thermoplastic solids. When unsaturated nitriles are used to prepare these imine complexes, liquid polymers which thermoset to solid materials via crosslinking through centers of unsaturation can be prepared. All of the alkylaluminum imine-derived polymers prepared above convert to AlN on pyrolysis in inert atmospheres.

Acknowledgments

The author wishes to thank L. J. Barret, and J. A. Mahoney for technical assistance, and Dr. J. M. Schwark and Dr. A. Lukacs III for helpful discussions.

Literature Cited

1. Slack, G. A.; Tanzilli, R. A.; Pohl, R. O.; Vandersande, J. W.; *J. Phys. Chem. Solids* **1987**, *48*, 641.
2. Prohaska, G. W.; Miller, G. R. *Mat. Res. Soc. Symp. Proc.* **1990**, *167*, 215.
3. Sheppard, L. M. *Am. Ceram. Soc. Bull.* **1990**, *69*, 1801.
4. Virkar, A. V.; Jackson, T.B.; Cutler, R. A. *J. Am. Ceram. Soc.* **1989**, *72*, 2031.
5. Pouskouleli, G. *Cer. Int.* **1989**, *15*, 213.
6. Developments in Inorganic Nitrogen Chemistry, C. B. Colburn Ed.; Elsevier Publishing Co.: Amsterdam, 1966; 508-510.
7. Laubengayer, A. W. Inorganic Polymers - An International Symposium; The Chemical Society: London, 1961, 85-88.
8. Janik, J. F.; Duesler, E. N.; Paine, R. T. *Inorg. Chem.* **1987**, *26*, 4341.
9. Clark, T. J.; Johnson, R. E. U. S. Patent 4,687,657, Aug. 18, 1987.
10. Czekaj, C. L.; Hackney, M. L. J.; Hurley, W. J., Jr.; Interrante, L. V.; Sigel, G. A.; Schields, P. J.; Slack, G. A. *J. Am. Ceram. Soc.* **1990**, *73*, 352.
11. Paciorek, K. J. L.; Nakahara, J. H.; Hoferkamp, L. A.; George, D.; Flippen-Anderson, J. L.; Gilardi, R.; Schmidt, W. R. *Chem. Mater.* **1991**, *3*, 82.
12. Jensen, J. A. U. S. Patent 5,229,468, July 20, 1993.
13. Wiberg, E.; in FIAT Review of German Science, Vol. 24, Bahr, G. Editor-in-Chief, Inorganic Chemistry, Part 2, Klemm, W. Ed. 1948, 155.
14. Su, S. R. U. S. Patent 4,783,430, Nov. 8, 1988.
15. Interrante, L. V.; Carpenter, L. E.; Whitmarsh, C.; Lee, W.; Slack, G. A. in Better Ceramics Through Chemistry II, Brinker, C. J.; Clark, D. E.; Ulrich, D. R. Eds.; Materials Research Society: Pittsburgh, PA, 1986, Vol. 73, 359-366.
16. Manasevit, H. M.; Erdman, F. M.; Simpson, W. I. *J. Electrochem. Soc.* **1971**, *118*, 1864.

17. Interrante, L. V.; Lee, W.; McConnel, M.; Lewis, N.; Hall, E. *J. Electrochem. Soc.* **1989**, *136*, 472.
18. Jiang, Z.; Interrante, L. V. *Chem. Mater.* **1990**, *2*, 439.
19. Baker, R. T.; Bolt, J. D.; Reddy, G. S.; Doe, D. C.; Staley, R. H.; Tebbe, F. N.; Vega, A. J. in Better Ceramics Through Chemistry III, Brinker, C. J.; Clark, D. E.; Ulrich, D. R. Eds.; Materials Research Society: Pittsburgh, PA, 1988, Vol. 121, 471-476.
20. Tebbe, F. N.; Bolt, J. D.; Young, R. J.; Van Buskirk, O. R.; Mahler, W.; Reddy, G. S.; Chowdry, U. *Adv. Ceram.* **1989**, *26*, 63-68.
21. Tebbe, F. N. U. S. Patent 4,696,968, Sept. 29, 1987.
22. Hiai, A.; Kazuo, W.; Tanaka, M.; Tanaka, T. U.S. Patent 4,869,925, Sept. 26, 1989.
23. Wiberg, E.; May, A. Z. *Naturforsch.* **1955**, *10B*, 229.
24. Maya, L. *Adv. Cer. Mater.* **1986**, *1*, 150.
25. Ochi, A.; Bowen, H. K.; Rhine, W. E. in Better Ceramics Through Chemistry III, Brinker, C. J.; Clark, D. E.; Ulrich, D. R. Eds.; Materials Research Society: Pittsburgh, PA, 1988, Vol. 121, 663-666.
26. Janik, J. F.; Paine, R. T. *J. Organomet. Chem.* **1993**, *449*, 39.
27. Cesari, M.; Cucinella, S. in The Chemistry of Inorganic Homo- and Heterocycles Haiduc, I.; Sowerby, D. B., Eds.; Academic Press: London, 1987, Vol. 1; 167-90.
28. Sugahara, Y.; Onuma, T.; Tanegashima, O.; Kuroda, K.; Kato, C. *J. Ceram. Soc. Japan* **1992**, *100*, 101
29. Becke-Goehring, M.; Krill, H. *Chem. Ber.* **1961**, *94*, 1059.
30. Schleich, D. M. U. S. Patent 4,767,607, Aug. 30, 1988.
31. Seibold, M. Russel, C. in Better Ceramics Through Chemistry III, Brinker, C. J.; Clark, D. E.; Ulrich, D. R. Eds.; Materials Research Society: Pittsburgh, PA, 1988, Vol. 121, 477-482.
32. Seibold, M. M.; Russel, C. *J. Am. Ceram. Soc.* **1989**, *72*, 1503.
33. Teusel, I.; Russel, C. *J. Mater. Sci.* **1990**, *25*, 3531.
34. Ross, C. B.; Wade, T.; Crooks, R. M.; Smith, D. M. *Chem. Mater.* **1991**, *3*, 768.
35. Ehrlich, R; Perry, D. D. U. S. Patent 3,505,246, Apr. 7, 1970.
36. Ehrlich, R.; Young, A. R.; Lichstein, B. M.; Perry, D. D. *Inorg. Chem.* **1964**, *3*, 628.
37. Lloyd, J. E.; Wade, K *J. Chem. Soc.* **1965**, 2662.
38. Jennings, J. R.; Lloyd, J. E.; Wade, K *J. Chem. Soc.* **1965**, 5083.
39. Zakharkin, L. I.; Khorlina, I. M. *Proc. Acad. Sci. USSR* **1957**, *112*, 879.
40. Zakharkin, L. I.; Khorlina, I. M. *Bull. Acad. Sci. USSR, Eng. Transl.* **1959**, 523.
41. Jensen, J. A. *J. Organomet. Chem.* **1993**, *456*, 161.
42. Jensen, J. A. in Better Ceramics Through Chemistry V, Hampden-Smith, M. J.; Klemperer, W. G.; Brinker, C. J. Eds.; Materials Research Society: Pittsburgh, PA, 1992, Vol. 271, 845-850.
43. Jensen, J. A. *Polymer Preprints* **1993**, *34*, 302.
44. Jensen, J. A. U. S. Patent 5,260,397, Nov. 9, 1993.
45. Jensen, J. A. U. S. Patent 5,276,105, Jan. 4, 1994.
46. Bent, B. E.; Nuzzo, R. G.; Dubois, L. H. *J. Am. Chem. Soc.* **1989**, *111*, 1634.

RECEIVED April 8, 1994

METAL-CONTAINING POLYMERS

Chapter 33

Ring-Opening Polymerization of Strained, Ring-Tilted Metallocenophanes

New Route to Polymers Based on Transition Metals

Daniel A. Foucher, Ralf Ziembinski, Rudy Rulkens, James Nelson, and Ian Manners[1]

Department of Chemistry, University of Toronto, 80 Saint George Street, Toronto M5S 1A1, Ontario, Canada

Recent developments involving the use of ring-opening polymerization (ROP) as a new route to high molecular weight polymers containing skeletal transition metal atoms are described. ROP of strained, ring-tilted silicon-bridged [1]ferrocenophanes provides a route to high molecular weight poly(ferrocenylsilanes). These polymers possess an unusual backbone of alternating ferrocene groups and silicon atoms and the current knowledge of the properties of these interesting materials is reviewed with emphasis on their electrical and electrochemical properties, thermal transition behavior, and morphology. Related [1]ferrocenophanes containing germanium or phosphorus in the bridge also undergo ROP to yield poly(ferrocenylgermanes) and poly(ferrocenylphosphines), respectively. The recent extension of this ROP approach to [2]metallocenophanes to yield new iron and ruthenium containing polymers is also discussed.

Polymers containing transition elements in the main chain are of considerable interest because of their potentially attractive physical and chemical properties. However, until very recently, most routes to these materials have yielded products which are either insoluble, of low molecular weight, or of poorly defined structure (1-16). Ring-opening polymerization (ROP) reactions generally occur via chain-growth processes which are not subject to the stringent stoichiometry and conversion requirements which often impede the preparation of high molecular weight polymers via condensation (step-growth) routes. Indeed, ROP provides a versatile route to organic (17,18) and, increasingly, to inorganic and organosilicon-based macromolecules (19-30). In contrast, ROP represents a virtually unexplored methodology for the preparation of transition metal-based polymers and very few examples of the ROP of cyclic compounds containing skeletal transition elements have been reported (31,32).

As part of our studies of new classes of polymers based on main group elements (28-30) or transition metals (33-44) we recently reported the discovery of a novel, ROP route which provided access to the first examples of high molecular weight poly(ferrocenylsilanes) (33). This involved the thermal ROP of [1]ferrocenophanes **1**

[1]Corresponding author

0097–6156/94/0572–0442$08.00/0

containing a silicon atom in the bridge and the resulting macromolecules **2** possess an unusual main chain comprising ferrocene units and silicon atoms. In this Chapter we provide an overview of recent work in this area and the extension of the ROP route to prepare other classes of transition metal-based macromolecules (33-44).

ROP of Silicon-Bridged [1]Ferrocenophanes

Previous work on [1]ferrocenophanes such as **1** has shown that they possess strained, ring-tilted structures and that stoichiometric ring-opening processes readily occur (45-47). The strain is apparent in the molecular structure of **1** (R = R' = Me) (Figure 1) where the planes of the cyclopentadienyl ligands are tilted with respect to one another by an angle of $20.8(5)°(35)$.

We found that when **1** (R = R' = Me) is heated at 130°C thermal ROP takes place to yield the corresponding poly(ferrocenylsilane) **2** in essentially quantitative yield (33). Gel permeation chromatography (GPC) indicated that **2** (R = R' = Me) possessed an approximate weight average molecular weight (M_w) of 5.2×10^5 and a number average molecular weight (M_n) of 3.4×10^5. Because the polymerizations of [1]silaferrocenophanes are essentially quantitative it is possible to use differential scanning calorimetry (DSC) to obtain an estimate of the strain energy present in the monomers **1**. A DSC thermogram for **1** (R = R' = Me) indicates that this species melts at 78°C and then polymerizes exothermically at 120-170°C. Integration of the latter exotherm for a known amount of the monomer indicated the strain energy to be <u>ca</u> 80 kJmol^{-1} (33). The phenylated [1]silaferrocenophane **1** (R = R' = Ph) was similarly studied and in this case the strain energy was estimated to be <u>ca</u> 60 kJmol^{-1} (33). These values are quite large and lie inbetween those found for cyclopentane (<u>ca</u> 42 kJmol^{-1}) and cyclobutane (<u>ca</u> 118 kJmol^{-1}) and are far greater than the value reported for the cyclotrisiloxane [Me$_2$SiO]$_3$ (19 kJmol^{-1}).

Because of the substantial instrinsic strain energy present we have found that thermal polymerization is quite general for species **1** and even species with bulky substituents such as phenyl (as mentioned above), ferrocenyl, or norbornyl undergo ROP. Representative examples of the poly(ferrocenylsilanes) prepared to date are given in Table 1. These polymers are of high molecular weight which contrasts with the materials previously prepared by condensation routes (48). In addition, virtually all of the poly(ferrocenylsilanes) dissolve in polar organic solvents and polymers such as **2** (R = R' = n-Hex) are even soluble in non-polar solvents such as hexanes. However, the phenylated poly(ferrocenylsilane) **2** (R = R' = Ph) is insoluble although lower molecular weight fractions can be extracted using hot THF (34).

The detailed properties of poly(ferrocenylsilanes) are under active investigation and the published results in this area from their discovery up to mid-1993 have been recently reviewed (38). One of the most interesting characteristics of these polymers involves their use as pyrolytic precursors to magnetic ceramics (36). However, in this Chapter we focus in some detail on the electrical and thermal characteristics of these novel materials.

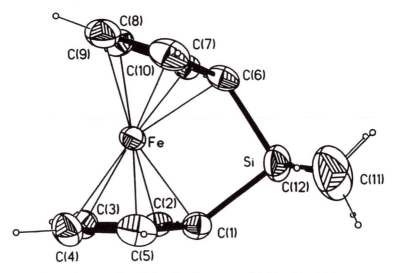

<u>Figure 1</u>   The Molecular Structure of **1** (R = R' = Me).
                 Reproduced with permission from reference 35.

Table 1 Characterization Data for Selected Poly(ferrocenylsilanes) 2

R	R'	$M_w{}^a$	$M_n{}^a$	$\delta(^{29}Si)^b$	$T_g{}^c$	$T_m{}^c$
Me	Me	5.2×10^5	3.4×10^5	-6.4	33	122
Et	Et	7.4×10^5	4.8×10^5	-2.7	22	108
nBu	nBu	1.1×10^6	4.9×10^5	-4.9	3	116,129
nHex	nHex	1.2×10^5	7.6×10^4	-2.3	-26	-
Ph	Ph	5.1×10^4	3.2×10^4	-12.9^e	-	-
Me	H	8.6×10^5	4.2×10^5	-20.0	9	87, 102
Me	TFP^d	2.7×10^6	8.1×10^5	-4.3	59	-
Me	Fc^d	1.6×10^5	7.1×10^4	-9.8	99	-
Me	n-C$_{18}$H$_{37}$	1.4×10^6	5.6×10^5	-5.3	1	16
Me	Vinyl	1.6×10^5	7.7×10^4	-10.3	28	-
Me	Ph	3.0×10^5	1.5×10^5	-7.7	54	-
Me	Nor^d	1.6×10^5	1.1×10^4	-3.4	81	-
Ph	Vinyl	1.1×10^5	7.6×10^4	-12.6	-	-

a   Polymer molecular weights were determined by Gel Permeation Chromatogaphy in THF containing 0.1 % by weight of [NBu$_4$]Br using polystyrene standards.
b ^{29}Si NMR data recorded in C$_6$D$_6$
c T_g and T_m values were determined by Differential Scanning Calorimetry, the latter generally after annealing.
d   Fc = ferrocenyl, TFP = trifluoropropyl, and Nor = norbornyl
e   Determined by solid state ^{29}Si NMR.

Electrical and Electrochemical Properties of Poly(ferrocenylsilanes)

The electrochemical properties of poly(ferrocenylsilanes) are quite intriguing. In contrast to organic polymers with ferrocenyl side groups such as poly(vinylferrocene) which show the presence of a single reversible oxidation wave, poly(ferrocenylsilanes) show two reversible oxidation waves (33, 37-39). This is illustrated by the cyclic voltammograms for 2 (R = R' = Me) dissolved in CH$_2$Cl$_2$ (Figure 2a) and as a (initially) surface confined species which dissolves on oxidation so as to give rise to progressively weaker traces (Figure 2b). In the latter case the different shapes of the two waves are probably a consequence of solvation effects. The presence of two waves for poly(ferrocenylsilanes) in solution has been explained in terms of the presence of cooperative interactions between the iron centers. In contrast, in poly(vinylferrocene) the ferrocenyl groups do not interact with one another. Thus, the initial oxidation of poly(ferrocenylsilanes) is believed to occur at alternating iron sites and the subsequent oxidation of iron centers next to those already oxidized is more difficult and occurs at higher potentials (33,39).

Interestingly, comparative studies of different poly(ferrocenylsilanes) show that the peak separation, ΔE, which is a measure of the interaction between the iron centers, changes with the substituents at silicon and increases in the order (Me < Et < n-Bu). Molecular modelling indicates that the distance between the iron centers in poly(ferrocenylsilanes) is probably over 6 Å which suggests that the interactions probably occur through the bridge rather than through space.

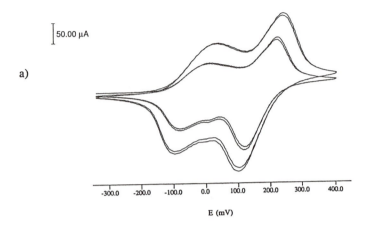

a)

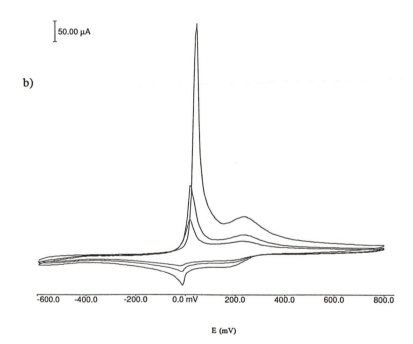

b)

Figure 2   Cyclic Voltammograms for **2** (R = R' = Me)
                  a) 5 x 10⁻⁴ M in 0.1 M [NBu₄][PF₆] in CH₂Cl₂
                  (Scans at 500 and 1000 mVs⁻¹)
                  b) as a Film on at Pt Electrode dipped in 0.1 M
                  [NBu₄][PF₆] in MeCN (Scans at 250 mVs⁻¹).
                  Potentials are relative to the ferrocene/ferrocenium
                  couple at E = 0.00 V.Figure 2a is reproduced with
                  permission from reference 39.

In view of the evidence for interactions between the iron atoms in the polymer main chain and the appreciable σ-delocalization in polysilanes the electrical conductivity of poly(ferrocenylsilanes) has been studied. In the undoped state poly(ferrocenylsilanes) are insulators with conductivities of ca 10^{-14} Scm^{-1}. The lack of delocalization in these undoped materials is supported by UV/vis spectroscopy which shows the polymers possess spectra similar to those for monomeric ferrocenes such as bis(trimethylsilyl)ferrocene (34, 38). Oxidation of poly(ferrocenylsilanes) can be achieved in solution using a variety of reagents such as I_2, o-quinones, and FeCl$_3$ and is accompanied by the growth of a characteristic "ferrocenium-type" visible absorption with λ_{max} = ca 645 nm as the colors of the solution change from amber to blue. Similar behavior is detected for poly(ferrocenylsilane) films which are electrochromic and can be cycled between amber and dark blue-green states. Conductivities for I_2-doped samples of 2 (R = R' = Me) rise into the weak semiconductor range (ca 10^{-7} Scm^{-1}) which is consistent with an electron hopping model (38). An essentially localized structure for this particular polymer is supported by Mössbauer spectroscopy which shows the presence of both an outer (IS = 0.38 mms^{-1}, ΔEq = 2.27 mms^{-1}) and an inner (IS = 0.29 mms^{-1}, ΔEq = 0.69 mms^{-1}) quadrupolar doublet. These were assigned to discrete Fe^{II} and Fe^{III} sites, respectively (Figure 3). This indicates that at least for doped samples of 2 (R = R' = Me) at 25°C delocalization does not occur on the Mössbauer timescale. Nevertheless, work on small molecule [m,m]ferrocenophanes indicates that the factors that influence delocalization in these types of materials are likely to be very complex and so some further understanding of structure-property relations is probably required in order to access more highly conducting materials in the future (38,39).

Thermal Transition Behavior and Morphology of Poly(ferrocenylsilanes)

In order to investigate the thermal properties of poly(ferrocenylsilanes) studies by DSC, thermogravimetric analysis (TGA) and Dynamic Mechanical Analysis (DMA) were carried out. In addition, in order to gain insight into the morphology of these materials Wide Angle X-Ray Scattering (WAXS) studies of poly(ferrocenylsilanes) were also performed (34,38,40).

Studies of the glass transition temperatures of poly(ferrocenylsilanes) by DSC have indicated that the T_gs are generally higher than those of polysilanes which can be attributed to the relative rigidity and bulk of the skeletal ferrocenyl units. In addition, the T_gs decrease with the length of the side group in the series 2 (R = R' = Me, Et, n-Bu, and n-Hex) which can be rationalized by free volume effects (Table 1) (34). Studies of the polymers by TGA have shown that they are thermally stable to ca 350°C but above this temperature they decompose to yield interesting ceramic and depolymerization products (36).

The melting characteristics and morphology of poly(ferrocenylsilanes) has been studied by DSC and WAXS, respectively, and have been found to be very sensitive to sample thermal history (38,40). Samples of the symmetrically substituted poly(ferrocenylsilanes) 2 (R = R' = Me, Et, n-Bu, and n-Hex) have been studied in some detail. WAXS of unannealed samples of the polymers with methyl and ethyl side groups show diffractograms which are indicative of the presence of some order with broad lines superimposed on typical amorphous halos (Figures 4a and 4b). The amorphous halos, which correspond to a d-spacing of ca 6.0 Å in 2 (R = R' = Me) and 6.9 Å in 2 (R = R' = Et), probably arise from the contribution of the first coordination sphere of the iron-iron correlation function. Samples of these polymers show melting transitions (T_ms) at 122°C and 108°C, respectively, when annealed at ca 90 - 100°C and sometimes without annealing which indicates that sample thermal history is important. This behavior is illustrated by the DSC thermograms for 2 (R =

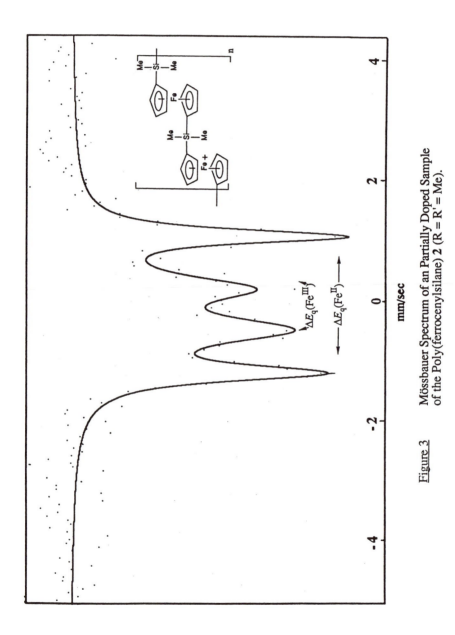

mm/sec

Figure 3   Mössbauer Spectrum of an Partially Doped Sample
of the Poly(ferrocenylsilane) 2 (R = R' = Me).

R' = Et) for an annealed sample (90°C, 2 h) (Figure 5a) and for a subsequent run
where a T_m was not detected (Figure 5b).

A WAXS study of the poly(ferrocenylsilane) with n-Bu side groups, 2 (R = R' =
nBu), shows much more evidence of order with several sharp peaks (Figure 4c) and a
T_m is observed at 129°C for annealed samples and often for samples that were not
deliberately annealed prior to study. Interestingly, annealing of 2 (R = R' = n-Bu) at ca
80 - 90°C for 2 h - 3 days leads to a new endotherm at 116°C in addition to that at
129°C. One of these peaks may be a consequence of side chain crystallization, which is
a well-known phenomenon for poly(di(n-alkyl)silanes) (27), and further studies in this
area are in progress. In contrast to the poly(ferrocenylsilanes) with shorter n-alkyl
groups, the n-hexyl substituted polymer 2 (R = R' = n-Hex) shows only very broad
amorphous halos at d-spacings of 5.3 Å and 13.6 Å (Figure 4d) and a T_m has not yet
been detected for this material.

ROP of Germanium- and Phosphorus-Bridged [1]Ferrocenophanes

The discovery of the facile thermal ring-opening polymerization of
[1]silaferrocenophanes suggested that [1]ferrocenophanes with other elements in the
bridge structure might also polymerize (33). This has been found to be the case and
[1]ferrocenophanes with germanium in the bridge 3 also undergo ROP to yield
poly(ferrocenylgermanes) 4 (Table 2) (38,41,42).

Table 2   Molecular Weight and Glass Transition Data for Selected
Poly(ferrocenylgermanes) 4

R	R'	$M_w{}^a$	$M_n{}^a$	$T_g{}^b$	T_m
Me	Me	2.0×10^6	8.5×10^5	28	-
Et	Et	7.4×10^5	4.8×10^5	12	-
nBu	nBu	8.9×10^5	3.4×10^5	- 7	88, 94
Ph	Ph	1.0×10^6	8.2×10^5	114	-

a   Polymer molecular weights were determined by Gel Permeation Chromatogaphy
    in THF containing 0.1 % by weight of [NBu$_4$]Br using polystyrene standards.
b T_g and T_m values were determined by Differential Scanning Calorimetry.

The properties of these polymers are under investigation and some characterization
data for selected examples is given in Table 2. Interestingly, electrochemical studies
have shown that similar cooperative interactions exist between the iron centers in
poly(ferrocenylgermanes) to those detected for their silicon analogues (39).

In addition, phosphorus-bridged [1]ferrocenophanes such as 5 (R = Ph) and
related species also polymerize thermally to yield poly(ferrocenylphosphines) such as

a)

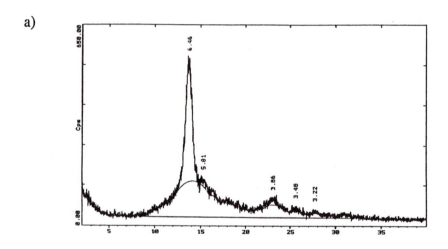

b)

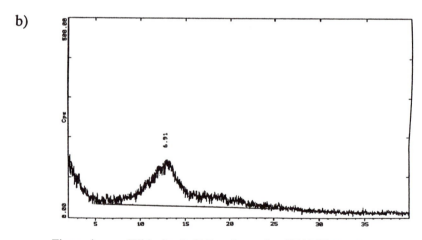

Figure 4   Wide Angle X-Ray Scattering (WAXS) Powder
                  Diffractograms of Poly(ferrocenylsilanes)
                  a) **2** (R = R' = Me) b) **2** (R = R' = Et),
                  c) **2** (R = R' = nBu) and d) **2** (R = R' = nHex).

Continued on next page

c)

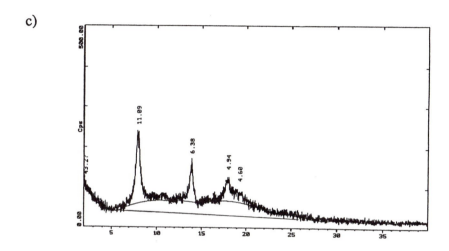

d)

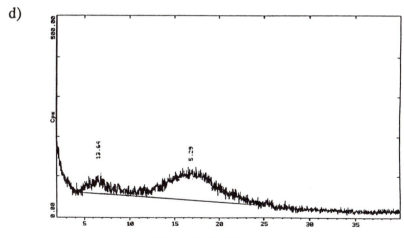

Figure 4. *Continued*

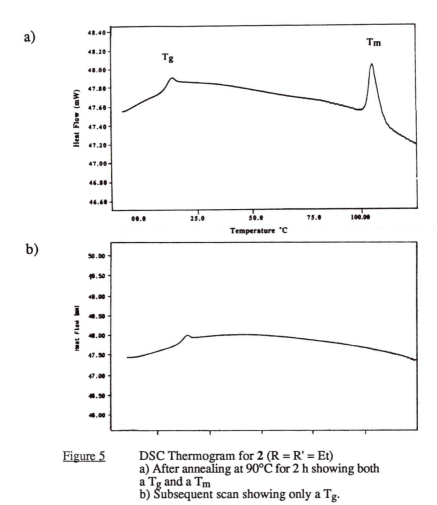

Figure 5   DSC Thermogram for 2 (R = R' = Et)
                a) After annealing at 90°C for 2 h showing both
                a T_g and a T_m
                b) Subsequent scan showing only a T_g.

6 (R = Ph). These materials are of identical structure to the polymers prepared previously by condensation routes (49). It should be noted that stoichiometric ring-opening reactions of species of type **5** are also known and in fact oligomers with 2 - 5 repeat units can be formed with certain anionic reagents. However, previous attempts to induce the ROP of species **5** were unsuccessful (49). Poly(ferrocenylphosphines) **6** also show evidence for interactions between the iron centers in their electrochemical behavior (39).

5　　　　　　　　　　　　　　　**6**

ROP of [2]Metallocenophanes with a Hydrocarbon Bridge

Although [1]ferrocenophanes with a single silicon, germanium, or phosphorus atom were found to readily polymerize, to date, our attempts to extend the ROP methodology to [2]ferrocenophanes which possess *two* silicon atoms in the bridge using either thermal or catalytic initiation have been unsuccessful (35). The reduced propensity for such species to polymerize has been attributed to the lower degree of ring strain present which is reflected by the very small cyclopentadienyl ring tilt-angle of only ca 4 °in compounds such as $Fe(\eta\text{-}C_5H_4)_2(SiMe_2)_2$ (35). However, [2]ferrocenophanes with a hydrocarbon bridge **7**, are significantly more strained than disilane-bridged analogues because of the smaller size of carbon relative to silicon and these compounds can possess ring tilt-angles of ca 21°. We have found that the hydrocarbon bridged species **7** will undergo ROP which has provided access to the first examples of well-characterized poly(ferrocenylethylenes) **8** (44). These polymers possess backbones consisting of alternating ferrocene groups and aliphatic C_2 units and are insoluble if R = H but are readily soluble in solvents such as THF if R is an organic group such as methyl.

7 M = Fe　　　　　　　　　**8**　M = Fe
9 M = Ru　　　　　　　　　**10**　M = Ru

Poly(ferrocenylethylenes) **8** possess ferrocene units which are further separated from one another than in polymers derived from [1]ferrocenophanes such as the poly(ferrocenylsilanes) **2**. As mentioned above, the electrochemistry of the latter polymers is indicative of the presence of substantial cooperative interactions between the iron centers. In contrast, studies of the electrochemistry of **8** (R = Me) showed the presence of only a single reversible oxidation wave which indicated that the ferrocene groups interact to much less significant extent (39).

In order to prepare [2]metallocenophanes that are even more strained than **7** we attempted to synthesize analogous species with a larger ruthenium atom in the place of iron. Such [2]ruthocenophanes would be expected to possess much greater ring tilt-angles and, as ruthenocene is known to possess significantly different electrical properties compared to ferrocene, the ruthenium analogues of poly(ferrocenylethylenes) would be very interesting to investigate.

We have recently synthesized the first examples of [2]ruthenocenophanes and **9** (R = H) possesses a tilt angle of 29.6(5)°, which is the largest known to date for any neutral iron group metallocenophane (40). In addition, we have found that such species with hydrocarbon bridges undergo ROP more readily than their iron analogues to yield poly(ruthenocenylethylenes) **10** (40).

These results indicate that ROP provides a versatile route to a variety of high molecular weight transition metal-based polymers. We are now attempting to extend the ROP route still further whilst concentrating on detailed studies of the properties of the polymers prepared to date, and also their mechanisms of formation. The different possible pathways for the ROP processes are outlined in a forthcoming review article (50).

Acknowledgements

Our work on the ring-opening polymerization of strained rings containing transition elements was initially supported by the Ontario Center of Materials Research (OCMR) and has subsequently been funded by the ACS Petroleum Research Fund (PRF), the Institute of Chemical Science and Technology (ICST), and the Natural Sciences and Engineering Research Council of Canada (NSERC). The research on the detailed properties of the polymers is being carried out in collaboration with the Polymer Materials Science Research Group of Professor G. Julius Vancso at the University of Toronto.

Literature Cited

1. Wright, M. E.; Sigman M.S. *Macromolecules*, **1992**, 25, 6055.
2. Fyfe, H.B.; Mlekuz, M.; Zargarian, D.; Taylor, N.J.; Marder, T.B. *J. Chem. Soc. Chem. Commun.* **1991**, 188.
3. Davies, S.J.; Johnson, B.F.G.; Khan, M.S.; Lewis, J. *J. Chem. Soc., Chem. Commun.*, **1991**, 187.
4. Tenhaeff, S.C. and Tyler, D.R. *J. Chem. Soc. Chem. Commun.* **1989**, 1459.
5. Neuse, E.W.; Bednarik, L. *Macromolecules*, **1979**, *12*, 187.
6. Sturge, K. C. ; Hunter, A. D. ; McDonald, R. ; Santarsiero, B. D. *Organometallics*, **1992**, *11*, 3056.
7. Pollagi, T.P.; Stoner, T.C.; Dallinger, R.F.; Gilbert, T.M.; Hopkins, M.D. *J. Am. Chem. Soc.*, **1991**, *113*, 703.
8. Bayer, R.; Pohlmann, T.; Nuyken, O. *Makromol. Chem. Rapid Commun.* **1993**, *14*, 359.
9. Gonsalves, K.; Zhanru, L.; Rausch, M.V. *J. Am. Chem. Soc.*, **1984**, *106*, 3862.
10. Brandt, P.F.; Rauchfuss, T.B.*J. Am. Chem. Soc.*, **1992**, *114*, 1926.
11. Dembek, A.A.; Fagan, P.J.; Marsi, M. *Macromolecules*, **1993**, 26, 2992.
12. Gilbert, A.M.; Katz, T.J.; Geiger, W.E.; Robben, M.P.; Rheingold, A.L. *J. Am. Chem. Soc.*, **1993**, *115*, 3199.
13. Nugent, H.M.; Rosenblum, M.; Klemarczyk, P. *J. Am. Chem. Soc.*, **1993**, *115*, 3848.
14. Sheats, J.E.; Carraher, C.E.; Pittman, C.U. *Metal Containing Polymer Systems*, Plenum, **1985**.
15. Neuse, E.W.; Rosenburg, H. *J. Macromol. Sci. Revs., Macromol. Chem.* **1970**, C4(1), 1.
16. Manners, I. *J. Chem. Soc. Ann. Rep. Prog. Chem.* (A) **1991**, 77.

17. *Ring Opening Polymerization* Ivin, K.J. and Saegusa, T. Eds, Elsevier, New York, **1984**.
18. *Ring Opening Polymerization* McGrath, J.E.; Ed.; ACS Symp. Ser. 286, Washington D.C. **1985**.
19. Sargeant, S. J.; Zhou, S. Q.; Manuel, G.; Weber, W. P. *Macromolecules*, **1992**, *25*, 2832.
20. West, R.; Hayase, S.; Iwahara, T. *J. Inorg. Organomet. Polym.* **1991**, *1*, 545.
21. Anhaus, J.T.; Clegg, W.; Collingwood, S.P.; Gibson, V.C. *J. Chem. Soc. Chem. Commun.* **1991**, 1720.
22. Manners, I.; Renner, G.; Allcock, H.R.; Nuyken, O. *J. Am. Chem. Soc.,* **1989**, *111*, 5478.
23. Ngo, D.C.; Rutt, S.J., Allcock, H.R. *J. Am. Chem. Soc.,* **1991**, *113*, 5075.
24. Wu, H.J.; Interrante, L.V. *Chem. Mater.*, **1989**, *1*, 564.
25. Allcock, H.R.; Dodge, J. A.; Manners, I.; Riding, G. H.; *J. Am. Chem. Soc.,* **1991**, *113*, 9596.
26. Cypryk, M.; Gupta, Y.; Matyjaszewski, K. *J. Am. Chem. Soc.,* **1991**, *113*, 1046.
27. Mark, J.E.; Allcock, H.R.; West, R. *Inorganic Polymers*, Prentice Hall, **1992**.
28. Liang, M.; Manners, I. *J. Am. Chem. Soc.,* **1991**, *113*, 4044.
29. Ni, Y.; Stammer, A.; Liang, M.; Massey, J.; Vancso, G.J.; Manners, I *Macromolecules* **1992**, *25*, 7119.
30. Manners, I. *Polymer News* **1993**, *18*, 133.
31. Roesky, H.W.; Lücke, M. *Angew. Chem. Int. Ed. Engl.*, **1989**, *28*, 493.
32. Roesky, H.W.; Lücke, M. *J. Chem. Soc., Chem. Commun*, **1989**, 748.
33. Foucher, D.A.; Tang, B.Z., Manners, I. *J. Am. Chem. Soc.*, **1992**, *114*, 6246.
34. Foucher, D.A.; Ziembinski, R.; Tang, B.Z.; Macdonald, P.M.; Massey, J.; Jaeger, R.; Vancso, G.J.; Manners, I. *Macromolecules*, **1993**, *26*, 2878.
35. Finckh, W.; Tang, B.Z.; Foucher, D.A.; Zamble, D.B.; Ziembinski, R.; Lough, A.; Manners, I. *Organometallics*, **1993**, *12*, 823.
36. Tang, B.Z., Petersen, R.; Foucher, D.A.; Lough, A.; Coombs, N.; Sodhi, R.; Manners, I. *J. Chem. Soc. Chem. Commun.* **1993**, 523.
37. Foucher, D.A.; Petersen, R.; Tang, B.Z.; Ziembinski, R.; Coombs, N.; MacDonald, P.M.; Sodhi, R.N.S.; Massey, J.; Vancso, J.G.; Manners, I. *Polymer Prepr. (Am. Chem. Soc., Div. Polym. Chem.)* **1993**, *34(1)*, 328.
38. Manners, I. *J. Inorg. Organomet. Polym.* **1993**, *3*, 185.
39. Foucher, D.A.; Honeyman, C.; Nelson, J.M.; Tang, B.Z.; Manners, I. *Angew Chem Int. Ed. Engl.* **1993**, *32*, 1703.
40. Manners, I. *Advanced Materials* **1994**, *6*, 68.
41. Honeyman, C.; Foucher, D.A.; Mourad, O.; Rulkens, R.; Manners, I. *Polymer Prepr. (Am. Chem. Soc. Div. Polym. Chem.)* **1993**, *34(1)*, 330.
42. Foucher, D.A.; Manners, I. *Makromol. Chem., Rapid Commun.* **1993**, *14*, 63.
43. Ziembinski, R.; Honeyman, C.; Mourad, O.; Foucher, D.A.; Rulkens, R.; Liang, M.; Ni, Y.; Manners, I. *Phosphorus, Sulfur, Silicon and Rel. Elem.*, **1993**, *76*, 219.
44. Nelson, J.M.; Rengel, H.; Manners, I. *J. Am. Chem. Soc.* **1993**, *115*, 7035.
45. Stoeckli-Evans, H.; Osborne, A.G.; Whiteley, R.H.; *Helv. Chim. Acta.* **1976**, *59*, 2402.
46. Osborne, A.G.; Whiteley, R.H.; Meads, R.E.; *J. Organomet. Chem.* **1980**, *193*, 345.
47. Fischer, A.B.; Bruce, J.A.; McKay, D.R.; Maciel, G.E.; Wrighton, M.S.; *Inorg. Chem..* **1982**, *21*, 1766.
48. Rosenberg, H.; U.S. Patent 3,426,053 (1969).
49. Withers, H.P.; Seyferth, D.; Fellmann, J.D.; Garrou, P.; Martin, S. *Organometallics*, **1982**, *1*, 1283.
50. Manners, I. *Adv. Organomet. Chem.* **1995**, 37, in press.

RECEIVED July 14, 1994

Chapter 34

New Organometallic Polymeric Materials

Search for $\chi^{(2)}$ Nonlinear Optical Organometallic Polymers

Michael E. Wright[1], Brooks B. Cochran[1], Edward G. Toplikar[1], Hilary S. Lackritz[2], and John T. Kerney[2]

[1]Department of Chemistry and Biochemistry, Utah State University, Logan, UT 84322−0300
[2]School of Chemical Engineering, Purdue University, West Lafayette, IN 47907−1283

A brief overview of organometallic $\chi^{(2)}$ nonlinear optical (NLO) polymers is presented. The synthesis of new ferrocene NLO-phores is presented along with their copolymerization with methyl methacrylate to generate organometallic NLO-polymers (NLOPs). A new class of organometallic NLO-phores were synthesized by condensation of ferrocene-carboxaldehyde with fluorene compounds. The fluorenyl-ferrocene NLO-phores displayed interesting linear optical properties and reactivity. NLO-spectroscopy was employed to study the orientation and relaxation behavior of the new organometallic NLOPs and we report on these interesting results.

Transition metal containing polymeric materials represents an important and active area of materials research.[1] Ferrocene containing polymers have been utilized in a broad range of applications, ranging from batteries[2] to NLO materials.[3] Ferrocene has long been recognized to stabilize alpha-carbocations[4] and can serve as an excellent electron-donor for an NLO-phore.[5] Our program has focused on the incorporation of ferrocene based NLO-phores into polymeric materials using covalent bonds.[6] To our knowledge, the ferrocenyl NLOPs presented below, are the only $\chi^{(2)}$ organometallic NLOPs reported in the literature. Only the side-chain organometallic NLOPs have been demonstrated to show second harmonic generation (SHG) after alignment by corona poling.[6]

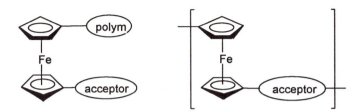

We envisioned that fluorenyl-ferrocenes, as depicted below, would be an interesting new class of NLO-phores. With the placement of electron-withdrawing groups at the 2 and 7-positions the complexes would model the λ-organic NLO-phores designed by Watanabe and coworkers.[7] Subsequent incorporation into a polymer backbone would give rise to an "accordion" NLO polymeric material with alternating apexes fixed.[8] In principle, this should facilitate the alignment process by predisposing part of the polymer backbone into the desired orientation.

In the cartoon below the resonance structure on the right sacrifices one "cyclopentadienyl anion" to generate the fluorenyl carbanion; hence, there should be little net loss in aromaticity. This "no net loss of aromaticity" has recently been exploited by Marder and coworkers in designing NLO-phores with large hyperpolarizabilities.[9]

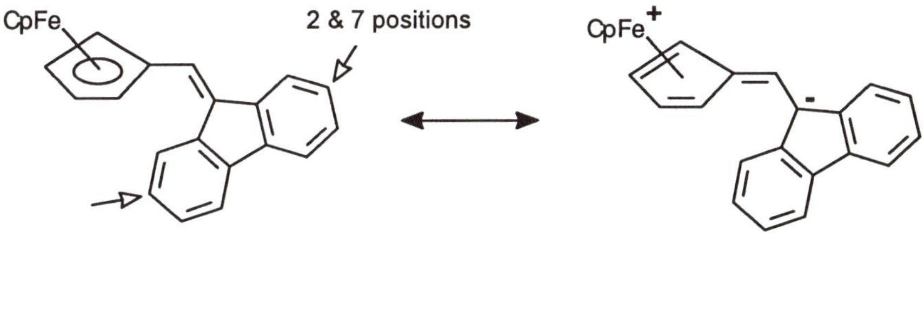

In this paper we will discuss the synthesis and characterization of ferrocene based NLOPs and a study of their orientation and relaxation behavior using NLO-spectroscopy.[10] Included in this work is the synthesis of a new class of fluorenyl-ferrocene NLO-phores and a detailed study of their chemistry and the linear optical properties of the new NLO-phores. The fluorenyl-ferrocene NLO-phore/monomer was successfully incorporated in the backbone of a conjugated organic polymer.

Results & Discussion

Monomer Synthesis & Structural Analysis. We have prepared a variety of vinyl-ferrocene NLO-phores (Scheme 1). The synthetic routes utilize the selective and efficient functionalization of the cyclopentadienyl rings developed in our laboratory.[11] We found through structural analysis for a series of complexes that the orientation of the vinyl moiety is apparently controlled (planar for strong electron acceptors) by the nature of the substituent.[12] We have not observed bond averaging for these organometallic systems as seen in certain organic NLO-phores where strong ground state interaction of acceptor and donor has been proposed.[13] The vinyl-ferrocene NLO-phores display λ_{max} of around 480 nm; thus, one has to be cognizant of resonance enhancement in measurements using the second harmonic generation signal while employing a fundamental laser source of 1064 nm.

Scheme 1

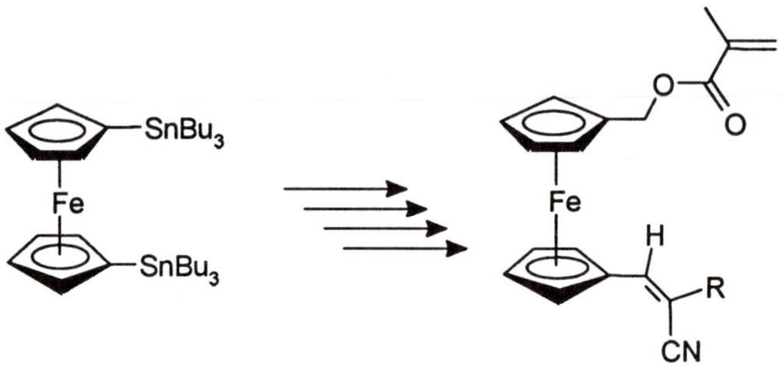

1a, R = CN
1b, R = CO$_2$Et
1c, R = 4-bromophenyl
1d, R = 4-pyridyl

The fluorenyl-ferrocene based NLO-phores have were prepared by condensing 2-bromo-, 2,7-dibromofluorene, and 2-nitrofluorene with ferrocene carboxaldehyde to afford **2a**, **3a**, and **4**, respectively, all in reasonable yield.[14] For 2-nitrofluorene the base used is potassium *tert*-butoxide and for the bromofluorenes, lithium diisopropylamide (LDA). Complexes **2a** and **3a** undergo clean halogen-metal exchange by treatment with *n*-BuLi in tetrahydrofuran (THF) at -78 °C and are then converted to their respective carboxaldehyde[15] and tributylstannyl[16] derivatives by treatment with DMF and Bu₃SnCl, respectively, in moderate yield (Scheme 2). Oxidative cleavage of the aryl-tin bond[17] with I₂ affords the *mono-* and *bis*(iodo) complexes in good yield (Scheme 2).

Scheme 2

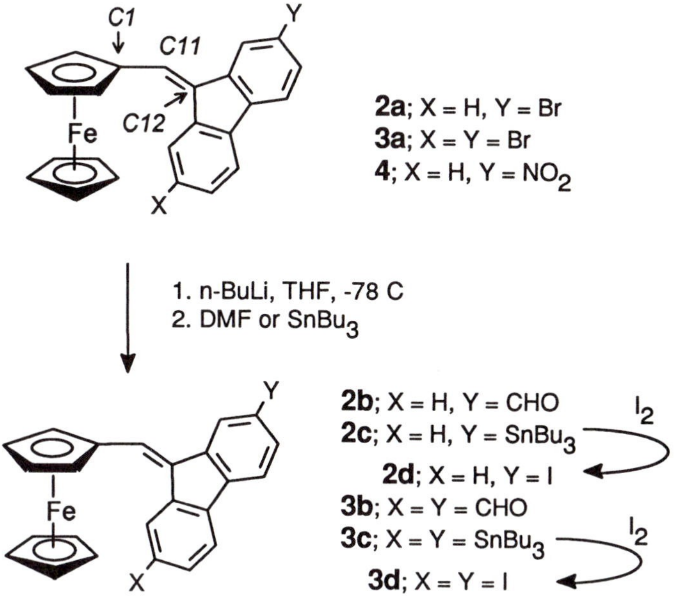

2a; X = H, Y = Br
3a; X = Y = Br
4; X = H, Y = NO₂

1. n-BuLi, THF, -78 C
2. DMF or SnBu₃

2b; X = H, Y = CHO
2c; X = H, Y = SnBu₃
2d; X = H, Y = I
3b; X = Y = CHO
3c; X = Y = SnBu₃
3d; X = Y = I

I₂

A structural study was initiated to determine the orientation of the fluorenyl-ring and possibly reveal any bond distortions (*i.e.* bond alteration or averaging). Single-crystal x-ray molecular structures were completed for complexes **2b** and **3b**. A detailed account of the crystallographic study for both structures is presented elsewhere.[18] The fluorenyl-ring system is 39 (± 1 deg) degrees out of plane with the η⁵-C₅H₄ ring in both complexes. This is different from what we have observed previously in the vinyl-ferrocene systems discussed above where the substituent appeared to control planarity of the system.[12] The exo-double bond (C11-C12) is distorted from planarity by 7-9 degrees (defined by C13-C12-C11-C1). The two structures do not show significant differences in their C1-C11 and C11-C12 bond distances [**2b**, 1.450(6), 1.354(6); **3b**, 1.458(3), 1.340(3); respectively]. We would anticipate that contribution of a dipolar structure to the ground-state configuration would lead to a shorting of the C(1)-C(11) bond and lengthening of the C(11)-C(12) bond.

We tested for solvatochromism in the new fluorenyl-ferrocene NLO-phores. The data is presented in Table I. It is quite evident that the electron-withdrawing groups have an effect on the metal to ligand charge transfer (MLCT) band.[19] These data imply that there is indeed a meaningful interaction between the iron center and the fluorenyl-ring system. The observation of a solvatochromic effect for **4** (hexane => CH_2Cl_2, red-shift of 20 nm) is consistent with a dipolar excited state. It is worth noting that the change in dipole moment ($\mu_{\text{ground-state}} - \mu_{\text{excited state}}$), which is often indicated by a solvatochromic shift, can be used to estimate the hyperpolarizibility of a given material.[20]

Table I. Linear Optical Data For Selected NLO-phores

Compound	Solvent	λ_{max} (nm) [ϵ x 10^3]
3a	CH_2Cl_2	502 [3.01]
3b	CH_2Cl_2	522 [6.50]
4	CH_2Cl_2	512 [6.76]
	EtOH	506
	hexanes	492

Monomer Reactivity and Polymer Synthesis. Monomers **1a-d** were copolymerized with methyl methacrylate (MMA) by free-radical initiation. The monomers were incorporated into the copolymers **5** at the same level as the feed-stock ratio of ferrocenyl monomer/MMA. The copolymers each had an identical λ_{max} as the respective starting monomer, indicating the NLO-phore was indeed intact.

By analogy with our recent success in preparing ferrocenyl polymers via a Knoevenagel polycondensation[21] we treated **3b** with a *bis*(cyanoacetate) comonomer. To our surprise, regardless of the base employed, the reaction afforded a complicated mixture of products, none of which displayed the expected and distinctive cyanocinnamate vinyl proton resonance.[22] A model study using **3b** and ethyl cyanoacetate again gave a complicated mixture of products. The product-mixture is yellow and this is suggestive that some type of reaction is occurring which breaks conjugation between the fluorenyl and Cp rings. Michael-addition at the C(11) vinyl carbon would represent such an event.

The palladium catalyzed cross-coupling of **3a** with phenylacetylene was found to lead to decomposition of the acetylene and recovery of **3a**.[23] In another model study we employed **3c** in the Stille cross-coupling reaction with acid chlorides[24] and found the reaction led to the formation of ketone, but also several additional by-products. With so many side-reactions we deemed the reaction unsuitable for polymerization studies. We did find that the iodo derivatives undergo facile palladium/copper catalyzed alkynylation and were successful in preparing copolymer **6** (Scheme 3).[25] Copolymer **6** was produced in high yield and purity [M_n = 67,000, polydispersity 3.8].

In some polymerization runs the alkyne comonomer underwent slight decomposition, producing imbalanced stoichiometry, which in turn produced a low molecular weight polymer with fluorenyl-iodide end-caps [M_n = 7000, average degree of polymerization = 10]. Copolymer **6** was soluble in common organic solvents and cast very tough, free-standing films.

Scheme 3

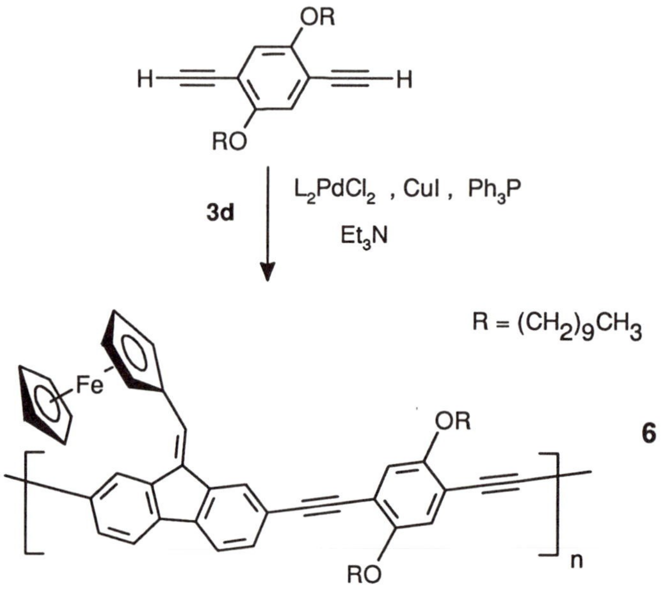

The thermal behavior and stability of copolymer **6** was measured using TGA and DSC. In the DSC scan we observe a small endothermic transition (~80 °C, possibly a melting-point), then an exothermic event (~175 °C), and finally an endothermic event beginning at 275 °C. The middle event is likely some type of organizational process (*e.g.* crystallization).[26] Stopping the DSC scan at 250 °C, cooling the sample, and then reheating the sample showed no endo- or exothermic events below 250 °C. A film of **6** was heated under nitrogen and viewed using a polarizing microscope. We did not observe any apparent phase transitions from 40 to 200 °C. An endothermic event, which began at ~275 °C, was concomitant with a significant weight loss in the TGA. The loss in mass corresponds to removal of two decyl groups per repeating unit. A film of **6** was cast on a sodium chloride plate and infrared spectra were obtained before and after heating the sample to 400 °C for 30 min. The infrared spectrum after heating does not indicate quinone formation (*i.e.* C=O stretch), but is supportive of the ferrocene-fluorenyl remaining intact.

We prepared the model compound **7** by cross-coupling **3d** with 1,4-*bis*(ethynyl)-2,5-*bis*(decoxy)benzene. Complex **7** is isolated as an orange oil and displays spectroscopic data very similar to copolymer **6**. Cyclic voltametry of complex **7** shows two reversible oxidation events, one at +0.11 V (two-electron event) and another at +0.71 V (one-electron, peaks relative to the ferrocene/ferrocenium couple). We interpret this to be oxidation of the ferrocenyl[27] units (uncoupled events) followed by a more difficult oxidation of the hydroquinone moiety. Surprisingly, copolymer **6** (low molecular sample, M_n = ~7,000) shows *only a single* and very reversible oxidation event at +0.13 V. Could an extended array of ferrocenium ions be oriented in such manner as to "protect" the hydroquinone polymer backbone from further oxidation? This is an intriguing result and deserves further investigation.

Orientation and Relaxation Studies For the Organometallic Polymers Using Second Harmonic Generation (SHG). Initial experiments were conducted to determine the optimum conditions for observing SHG for copolymers **5**. Only copolymer **5d** did not show any SHG activity after corona poling at T_g plus 25 °C. Polymers **5a** and **5c** displayed the highest SHG signals when poling temperatures of T_g plus 25 °C are employed. In this temperature region the polymer is mobile enough to allow the NLO-phores to readily orient into the required noncentrosymmetric macroscopic structure, yet the temperature is generally not excessive to degrade the polymer or have electric field effects dominate.[28] It is also important to note that second order susceptibility is inversely proportional to temperature.[29]

The polymers are poled at T_g plus 25 °C for 20 min and then allowed to cool to ambient temperature with the electric field still applied. Upon reaching ambient temperature, the voltage is removed, and decay of the SHG signal is monitored. The decay of $\chi^{(2)}$ over time for **5c** is shown in Figure 1. The decay data following the removal of the electric field was not successfully fit using the Williams-Watt stretched exponential. The WW fit indicated too slow a relaxation process at short times and too fast a relaxation process at long times for both systems. An alternative biexponential fit, $y=\theta_1\exp(-t/\tau_1) + \theta_2\exp((-t/\tau_2)$, was used to fit the data, and is represented as a solid line in the Figures. This equation is based on fitting parameters which do not have direct physical significance for describing relaxation behavior but can be used to illustrate a "short-time" and "long-time" relaxation.

It appears that packing of the polymer chains is the dominant effect in the orientation and relaxation behavior of the organometallic NLO-phore. Packing appears to be dependent not only upon NLO-phore structure, but also the age of the polymer films. Interestingly, the smaller NLO-phore (*i.e.* in **5a**) shows a lower initial SHG signal; however, better long term stability in comparison to the larger NLO-phore in **5c**. Polymer **5a** exhibiting observable signal for hours as opposed to seconds for **5c**. This marked difference in behavior of $\chi^{(2)}$ for the two systems may be rationalized with molecular packing arguments. The polymer chains which pack more efficiently may provide a more stable microenvironment for the NLO-phores than those chains which are not packed as densely. Since densely packed chains exhibit less segmental mobility relative to unpacked chains, those NLO-phores that have been oriented by the electric field would be more likely to retain their orientation.

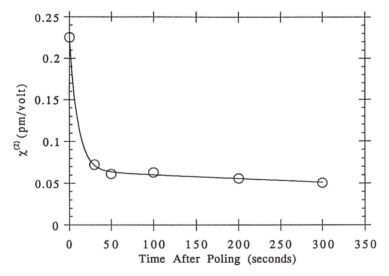

Figure 1. Long term Stability of $\chi^{(2)}$ for copolymer **5c**. Solid line indicates biexponential fit with fitting parameters $\theta_1=0.16$, $\tau_1= 10$ s; $\theta_2=0.07$, $\tau_2= 1231$ s.

Support for the packing argument is found in experiments where the aging of the polymer films is erased by maintaining the films at temperatures well above the T_g for a period of time (at least 1 h) before the poling process is commenced. If the packing argument is correct, then the $\chi^{(2)}$ signal should be less stable and also exhibit a larger initial value. It can be seen in Figure 2 that indeed for both polymers the initial magnitude of the SHG signal is greater and the rate of decay for the signal is increased, particularly so in the case of **5c**.

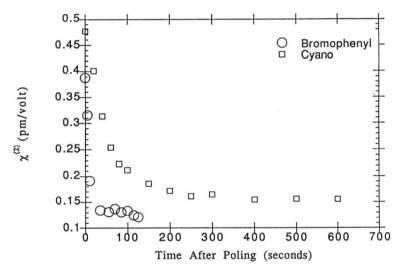

Figure 2. Stability of $\chi^{(2)}$ for copolymers **5c** (O)and **5a** (□) following erasure of thermal history.

Concluding Remarks

Our work in the area of organometallic NLO polymers demonstrates the feasibility of preparing organometallic NLOPs and that NLO-spectroscopy is a useful tool for probing the orientational and relaxation behavior of the organometallic NLO-phores. By varying the structure of the organometallic NLO-phore we were able to determine that packing of the organometallic NLO-phore within the polymer matrix is very important and was shown to in at least one case, dominate size considerations. Physical aging of the polymer films led to a reversible densification of the polymer matrix. The aged polymer films showed a greater resistance to alignment with a concomitant increase in the ability to retain NLO-phore alignment. These data are consistent with a packing argument.

Erasure of the thermal history of aged films by heating prior to poling also supported the idea of tight packing through densification.

Studies are continuing involving the synthesis and evaluation of organometallic NLO-polymers and super-macromolecular organometallic assemblies.

Selected Experimental[30]

SHG Measurements.   A Continuum NY61-10 Q-switched Nd:YAG laser generated p-polarized light at 1.064 μm. The fundamental beam was split so the sample and a y-cut quartz reference could be tested simultaneously, and lenses were used to vary the size and intensity of the laser beam impinging on the sample. Lenses were placed directly behind the sample and reference to focus the frequency doubled light through infrared filters, insuring that only 532 nm light passed into the monochromator and photomultiplier tube (PMT). The PMT signal was sent to a gated integrator and boxcar averager. A Sparc IPC work station was used to collect and store data. The sample was vertically mounted on a temperature-controlled copper block so that the laser beam struck the sample at a 68° angle relative to the normal of the sample. Samples were poled using a corona discharge generated by a tungsten needle across a 1 cm air gap. The corona current was limited to 3 μA to prevent damaging the sample.

SHG Sample Preparation. The polymer samples were dissolved in spectrophotonic grade chloroform (Mallinckrodt) to produce solutions with 10% polymer by weight. Solutions were filtered (5 μm) and then spun cast onto indium tin oxide (ITO) glass substrates. Film thicknesses varied from 2 to 6 μm (±0.5) thick, as measured by diamond stylus profilometry. Films were carefully dried to remove any excess solvent.

$\{\eta^5\text{-}C_5H_4CH_2O_2CC(CH_3)=CH_2\}Fe\{\eta^5\text{-}C_5H_4CH=C(CN)_2\}$ **(1a, R = CN).**
^{1}H NMR (CDCl$_3$) δ 7.68 (s, 1 H, =CH), 6.14 (s, 1 H, =CH$_2$), 5.62 (s, 1 H, =CH$_2$), 4.97 (t, J= 2 Hz, 2 H, Cp), 4.86 (s, 2 H, CH$_2$O), 4.81 (t, J= 2 Hz, 2 H, Cp), 4.40 (t, J= 2 Hz, 2 H, Cp), 4.34 (t, J= 2 Hz, 2 H, Cp), 1.96 (s, 3 H, CH$_3$); ^{13}C NMR (CDCl$_3$) δ 166.9 (CH$_2$O$_2$$\underline{C}$), 162.7 (=CH), 136.0 (=$\underline{C}$Me), 114.7 (C≡N), 114.0 (C≡N), 85.2 (=$\underline{C}$(CN)$_2$), 75.6 (Cp CH), 72.1 (Cp CH), 71.4 (Cp CH), 61.4 (CH$_2$O), 18.3 (CH$_3$); IR (CH$_2$Cl$_2$) $v_{C≡N}$ 2226 cm^{-1}; UV-vis (CH$_2$Cl$_2$) λ_{max} = 526 nm (ε = 2.99 x 10^3). Anal. Calcd for C$_{19}$H$_{16}$FeN$_2$O$_2$: C, 63.36; H, 4.48; N, 7.78. Found: C, 63.10; H, 4.60; N, 7.50.

Preparation of $\{\eta^5\text{-}C_5H_5\}Fe\{\eta^5\text{-}C_5H_4CH=[9\text{-}(2,7\text{-}bis(iodo)fluorenyl)]\}$ (3d). A chilled (0 °C) THF (40 mL) solution of **2d** (1.0 g, 1.1 mmol) was treated with I$_2$ (0.54 g, 2.1 mmol). The solution was allowed to react at 0 °C for 30 min and then diluted with ether (40 mL). The organic layer was washed with 10% Na_2SO_3 (2 x 40 mL), brine (40 mL), and then dried over K_2CO_3. The solvents were removed under reduced pressure and the crude product was crystallized from ether/hexanes to give pure **3d** (0.31 g, 48%); ^1H NMR (CDCl$_3$) δ 8.60 (s, 1 H, Ar CH), 8.08 (s, 1 H, Ar CH), 7.69-7.64 (m, 2 H, Ar CH), 7.47-7.43 (m, 3 H, Ar and vinyl CH's), 4.71 (t, J = 1.9 Hz, 2 H, Cp CH), 4.56 (t, J = 1.8 Hz, 2 H, Cp CH), 4.25 (s, 5 H, Cp CH); ^{13}C NMR (CDCl$_3$) δ 141.5, 138.8, 138.4 (Ar C), 136.5, 136.3, 135.9, 132.9, 130.8, 129.6, 128.7, 121.3 (vinyl and Ar CH's), 80.1 (*ipso*-Cp), 71.0, 70.6 (Cp CH's), 69.7 (Cp CH's); UV-vis (CH$_2$Cl$_2$) λ_{max} = 502 nm (ϵ=3.29 x 10^3). Anal. Calcd for $C_{24}H_{16}FeI_2$: C, 46.94; H, 2.63%. Found: C, 47.50; H, 2.72%.

Preparation of {η^5-C$_5$H$_5$}Fe{η^5-C$_5$H$_4$CH=[9-(2-nitrofluorenyl]} (4).
One equivalent of 2-nitrofluorene (0.99 g, 4.7 mmol} was added to a solution of ferrocenecarboxaldehyde (1.0 g, 4.7 mmol) in THF (100 mL) at room temperature. Two mol-equiv of potassium t-butoxide (1.1 g, 9.4 mmol) were added and the mixture was stirred for 15 min. The mixture was then diluted with ether (100 mL) and run through a flash column (deactivated Al$_2$O$_3$, 1 cm) to remove a heavy black precipitate. The resulting solution was further diluted with ether (50 mL), washed with water (2 x 100 mL), then brine (100 mL), and dried over K$_2$CO$_3$. The crude product was subjected to column chromatography (2 x 15 cm) on deactivated alumina. Elution with EtOAc/hexanes (1/20, v/v) gave two major bands, the first was a dark purple band and this was found to be pure **4** (0.38 g, 20%, mp>300°C). ^1H NMR (CDCl$_3$) δ 9.16 (s, 1 H, Ar CH), 8.64 (s, 1 H, Ar CH), 8.31-8.22 (m, 2 H, Ar CH), 7.88-7.81 (m, 2 H, Ar or vinyl CH), 7.65-7.63 (m, 1 H, Ar or vinyl CH), 7.46-7.33 (m, 2 H, Ar CH), 4.78-4.76 (m, 2 H, Cp CH's), 4.61 (t, J = 1.8 Hz, 1 H, Cp CH's), 4.56 (t, J = 1.8 Hz, 1 H, Cp CH's), 4.25 (s, 5 H, Cp CH's); ^{13}C NMR (CDCl$_3$) δ 146.9, 146.6, 145.8, 143.3, 141.5, 140.5, 138.4, 138.2, 137.1, 135.8 (aromatic C), 131.3, 130.4, 130.1, 128.7, 128.5, 128.1, 127.6, 124.3, 122.9, 122.5, 121.0, 120.9, 119.8, 119.5, 115.1 (aromatic or vinyl CH's), 80.2, 79.9 (*ipso*-Cp) 71.1, 71.0, 70.6 (Cp CH's), 69.7 (Cp CH's). UV-vis (CH$_2$Cl$_2$) λ_{max} = 512 nm (ϵ=6.76 x 10^3).

PMMA Copolymer 5 (R = CN). A Schlenk flask was charged with benzene (3 mL), **1a** (0.34 g, 0.95 mmol), methyl methacrylate (2.02 mL, 18.9 mmol), and AIBN. The mixture was heated to reflux for 8 h and the solvent removed under reduced pressure. The polymer was redissolved (~1.3 g did not dissolve) in chloroform and precipitated into methanol (0.55 g, 25%). ^1H NMR (CDCl$_3$) δ 7.73 (br s, 1 H, =CH), 4.99 (t, J= 2 Hz, 2 H, Cp), 4.84 (t, J= 2 Hz, 2 H, Cp), 4.70 (s, 2 H, CH$_2$), 4.37 (m, 4 H, Cp), 3.58 (s, PMMA OCH$_3$), 1.92 (br s, 2 H, PMMA CH$_2$), 1.79 (br s, 2 H, CH$_2$), 1.00 (s, PMMA CH$_3$), 0.82 (s, 3 H, CH$_3$). Anal. Calcd: N, 1.24. Found: N, 1.21.

Preparation of Polymer 6. A 25 mL Schlenk tube was charged with **3d** (70 mg, 0.11 mmol), 1,4-*bis*(decoxy)-2,5-diethynylbenzene (53 mg, 0.12 mmol), $(PPh_3)_2PdCl_2$ (2 mg, 3 mol-%), CuI (1 mg, 6 mol-%), PPh_3 (2 mg, 6 mol-%), and triethylamine (5 mL) under nitrogen. The mixture was warmed to 50 °C with stirring for 2 h and then diluted with dichloromethane (100 mL). The organic layer was washed with 10% NaCN (100 mL), water (2 x 100 mL), brine (100 mL), and then dried over K_2CO_3. The solvents were removed and the polymer washed with ether (5 mL) and then dried under reduced pressure at 65 °C for 24 h (53 mg, 61%). ^{1}H NMR ($CDCl_3$) δ 8.49-8.46 (br s, 1 H, Ar CH), 8.01-7.97 (br s, 1 H, Ar CH), 7.76-7.72 (m, 2 H, Ar CH), 7.61- 7.54 (m, 3 H, Ar CH), 7.12, 7.09, 7.06, 7.02 (4 br s, 2 H, Ar CH), 4.80-4.78 (m, 2 H, Cp CH), 4.59-4.55 (m, 2 H, Cp CH), 4.26 (br s, 5 H, Cp CH), 4.14-4.03 (m, 4 H, C$\underline{H}_2$O-), 1.95-1.84 (m, 4 H, CH_2's), 1.61-1.51 and 1.31-1.23 (2 m's, 28 H, CH_2's), 0.85-0.83 (m, 6 H, CH_3's). IR (CH_2Cl_2) (alkyne) ν 2206 cm^{-1}; UV-vis (CH_2Cl_2) λ_{max} = 514 nm (ε=5.39 x 10^3). Anal. Calcd for $[C_{58}H_{64}FeO_2]_n$: C, 81.39; H, 7.59%. Found: C, 79.42; H, 7.72%.

Acknowledgment.

This work was supported by the Office of Naval Research (ONR) and the donors to the Petroleum Research Fund, administered by the American Chemical Society. MEW also wishes to express his gratitude to the NSF (CHE-9002379) and Utah State University Research Office for funding the purchase of the single-crystal x-ray diffractometer. HSL would like to acknowledge support from the ONR and NSF (PFF award). A special thanks to Prof. Vernon Parker for helpful discussions and Dr. Kishan Handoo for obtaining the cyclic voltammetry data for compounds **6** and **7**.

References & Notes

1.   Neuse E. W.; Rosenberg H. *Metallocene Polymers*; Marcel Dekker: New York, **1970**. Pittman, C. U., Jr.; Rausch, M. D. *Pure Appl. Chem.* **1986**, *58*, 617. Sheats, J. E., Carraher, C. E., Pittman, C. U., Jr., Eds.; *Metal-Containing Polymer Systems*, Plenum: New York, **1985**. Neuse E. W.; Woodhouse J. R.; Montaudo G.; Puglis C. *Appl. Organomet. Chem.* **1988**, *2*, 53. *Inorganic and Organometallic Polymers: Macromolecule, Containing Silicon, Phosphorus, and other Inorganic Elements*; Zeldin, M., Wynne, K. J., Allcock, H. R., Eds.; ACS Symp. Ser., Washington D. C., **1988**. Pomogailo, A. D.; Uflyand, I. E. *Platinum Metals Review* **1990**, *34(4)*, 184. Sheats, J. E.; Carraher, C. E.; Jr., Pittman, C. U., Jr.; Zeldin, M.; Currell, B. Eds.; *Inorganic and Metal-Containing Polymeric Materials*; Plenum, New York, **1992**. Laine, R. M. *Inorganic and Organometallic Polymers with Special Properties*, Kluwer, Boston, **1992**.

2.   (a) Iwakura, C.; Kawai, T.; Nojima, M.; Yoneyama, H. *J. Electrochem. Soc.* **1987**, *134*, 791. For other examples of typical ferrocene polymers see: (b) Zhan-Ru, L.; Gonsalves, K.; Lenz, R. W.; Rausch, M. D. *J. Polym. Sci. A* **1986**, *24*, 347 and references cited therein. (c) Singh, P,; Rausch, M. D.; Lenz, R. W. *Polym. Bulletin* **1989**, *22*, 247 and references cited therein.

3.   Wright, M. E.; Toplikar, E. G. *Macromolecules* **1992**, *25*, 1838.

4.   For example see: Zou, C.; Wrighton, M. S. *J. Am. Chem. Soc.* **1990**, *112*, 7578. Also see reference 2b in this chapter.

5.   Green, M. L. H.; Marder, S. R.; Thompson, M. E.; Bandy, J. A.; Bloor, D.; Kolinsky, P. V.; Jones, R. J. *Nature* **1987**, *330*, 360. Perry, J. W.; Stiegman, A. E.; Marder, S. E.; Coulter, D. R. In *Organic Materials for Nonlinear Optics*; Hann, R. A., Bloor, D. Eds.; *Spec. Publ. No. 69, The Royal Society of Chemistry: London, England* **1989**. Coe, B. J.; Jones, C. J.; McCleverty, J. A.; Bloor, D.; Kolinsky, P. V.; Jones, R. J. *J. Chem. Soc., Chem. Commun.* **1989**, 1485. Marder, S. R.; Perry, J. W.; Tiemann, B. G.; Schaefer, W. P. *Organometallics* **1991**, *10*, 1896. Calabrese, J. C.; Cheng, L.-T.; Green, J. C.; Marder, S. R.; Tam, W. *J. Am. Chem. Soc.* **1991**, *113*, 7227. Yuan, Z.; Taylor, N. J.; Marder, T. B.; Williams, I. D.; Kurtz, S. K.; Cheng, L.-T. *J. Chem. Soc., Chem. Commun.* **1990**, 1489. Marder, S. R. in *Inorganic Materials*, Bruce, D. W.; O'Hare, D.; Eds.; John Wiley & Sons Ltd, 1992, Chapter 3, pp 116-163. For a theoretical (SCF-LCAO MECI formalism) treatment of organometallic NLO materials, including ferrocene systems, see: Kanis, D. R.; Ratner, M. A.; Marks, T. J. *J. Am. Chem. Soc.* **1990**, *112*, 8203. Kanis, D. R.; Ratner, M. A.; Marks, T. J. *J. Am. Chem. Soc.* **1992**, *114*, 10338.

6.   Wright, M. E.; Sigman, M. S. *Macromolecules* **1992**, *25*, 6055. Wright, M. E.; Svejda, S. A. "Materials for Nonlinear Optics: Chemical Perspectives", Marder, S. R., Sohn, J. E., Stucky, G. D., Eds.; ACS Symp. Series 455, American Chemical Society, Washington DC **1991**, p. 602. Wright, M. E.; Toplikar, E. G. *Contemporary Topics in Polymer Chemistry, Volume 7*, Riffle, J. Ed.; Plenum Publishing Co., New York, **1992**, 285-292.

7.   Watanabe, T.; Kagami, M.; Yamamoto, H.; Kidoguchi, A.; Hayashi, A.; Sato, H.; Miyata, S. *American Chemical Society, Polym. Preprints* **1991**, *32(3)*, 88.

8.   Lindsay, G. A.; Fischer, J. W.; Henry, R. A.; Hoover, J. M.; Kubin, R. F.; Seltzer, M. D.; Stenger-Smith, J. D. *Polym. Preprints* **1991**, *32(2)*, 91. Lindsay, G. A.; Stenger-Smith, J. D.; Henry, R. A.; Hoover, J. M.; Nissan, R. A.; Wynne, K. J. *Macromolecules* **1992**, *25*, 6075 and references cited therein.

9.   Marder, S. R.; Beratan, D. N.; Cheng, L.-T. *Science* **1991**, *252*, 103-106 and references cited therein.

10.   Hampsch, H. L; Yang, J.; Wong, G. K.; Torkelson, J. M.
      Macromolecules **1988**, *21*, 526. Hampsch, H. L.; Yang, J.; Wong, G.
      K.; Torkelson, J. M. *Polym. Commun.* **1989**, *30*, 40. Hampsch, H.
      L.; Torkelson, J. M.; Bethke, S. J.; Grubb, S. G. *J. Appl. Phys.*
      1990, *67*, 1037. Hampsch, H. L.; Yang, J.; Wong, G. K.; Torkelson,
      J. M. *Macromolecules* **1990**, *23*, 3640. Hampsch, H. L.; Yang, J.;
      Wong G. K.; Torkelson, J. M. *Macromolecules* **1990**, *23*, 3648.
      Lackritz, H. S.; Torkelson, J. M. "Polymer Physics of Poled
      Polymers for Second Order Nonlindear Optics" Chapter 8 in
      Molecular Optoelectronics: Materials, Physics, and Devices, Zyss,
      J., Ed. Academic Press, New York **1993**. Loucif-Saibi, R.;
      Nakatani, K.; Delaire, J. A.; Dumont, M.; Sekkat, Z. *Chem.
      Mater.* **1993**, 5, 229. Singer, K. D.; Kuzyk, M. G.; Holland, W. R.;
      Sohn, J. E.; Lalama, S. J.; Comizzoli, R. B.; Katz, H. E.; Schilling,
      M. L. *Appl. Phys. Lett.* **1988**, *53*, 1800 and *Proc. 6th Int. Symp.
      Elect.* **1988**, 647. Eich, M.; Sen, A.; Looser, H.; Bjorklund, G. C.;
      Swalen, J. D.; Twieg, R. J.; Yoon, D. Y. *J. Appl. Phys.* **1989**, *66*,
      2559. Lindsay, G. A.; Henry, R. A.; Hoover, J. M.; Knoesen, A.;
      Mortazani, M. A. *Macromolecules* **1992**, *25*, 4888. Walsh, C. A.;
      Burland, D. M.; Lee, V. Y.; Miller, R. D.; Smith, B. A.; Twieg, R. J.;
      Volksen, W. *Macromolecules* **1993**, *26*, 3720.
11.   Wright, M. E. *Organometallics* **1990**, *9*, 853.
12.   Wright, M. E.; Toplikar, E. G. *Macromolecules* **1992**, *25*, 6050.
13.   A similar structural analysis was reported for a series of donor-
      acceptor organic molecules: Graham, E. M.; Miskowski, V. M.;
      Perry, J. W.; Coulter, D. R.; Stiegman, A. E.; Schaefer, W. P.;
      Marsh, R. E. *J. Am. Chem. Soc.* **1989**, *111*, 8771.
14.   Condensation reactions of ferrocene carboxaldehyde were first
      reported by: Barben, I. K. *J. Chem. Soc.* **1961**, 1827.
15.   The reaction of lithium reagents with DMF to produce aldehydes is
      an established procedure: Wakefield, B. J. *Organolithium Methods*;
      Academic Press: New York, **1988**.
16.   For a general review of organostannane reagents, including
      synthetic routes to the reagents see: Pereyre, M.; Quintard, J.-P.;
      Rahm, A. *Tin in Organic Synthesis*; Butterworths: London, **1987**.
17.   Jung, M. E.; Light, L. A. *Tetrahedron Lett.* **1982**, *23*, 3851. Chen, S.-
      M. L.; Schuab, R. E.; Grudzinskas, C. V. *J. Org. Chem.* **1978**, *43*,
      3450.
18.   Wright, M. E.; Cochran, B. B. *Organometallics* **1993**, *12*, 3873.
19.   Geoffroy, G. L.; Wrighton, M. S. *Organometallic Photochemistry*;
      Academic Press: New York, **1979**, Chapter 1.
20.   DeMartino, R. N.; Choe, E. W.; Khanarian, G.; Haas, D.; Lesile, T.;
      Nelson, G.; Stamatoff, J.; Stuetz, D.; Teng, C. C.; Yoon, H. in
      Nonlinear Optical and Electroactive Polymers, Prasad, P. N.; Ulrich,
      D. R. Eds; Plenum Press: New York, **1988**, pp 169-188. Paley, M. S.;

Harris, J. M.; Looser, H.; Baumert, J. C.; Bjorkland, G. C.; Jundt, D.; Twieg, R. J. *J. Org. Chem.* **1989**, *54*, 3774.

21. Wright, M. E.; Mullick, S. *Macromolecules* **1992**, *25*, 6045. Wright, M. E.; Mullick, S.; Lackritz, H. S.; Liu, L.-Y. *Macromolecules* **1994**, *27*, 0000.

22. The vinyl proton for the β-ferrocenyl(α-cyano acrylates) is typically at δ 8.1 ppm. In the past we have found the chemical shift to be very consistent for both monomeric and polymeric materials.

23. The homocoupling of alkynyl-aryl halide monomers was carried out by: Trumbo, D. L.; Marvel, C. S. *J. Polym. Sci., Polym. Chem. Ed.* **1986**, *24*, 2311. The alkynyl cross-coupling reaction conditions employed were developed by: Takahashi, S.; Kuroyama, Y.; Sonogashira, K.; Hagihara, N. *Synthesis* **1980**, 627.

24. Milstein, D.; Stille, J. K. *J. Am. Chem. Soc.* **1978**, *100*, 3636; *J. Org. Chem.* **1979**, *44*, 1613.

25. The *bis*(alkynyl) monomer was prepared according to the procedure of: Schulz, R. C.; Giesa, R. *Makromol. Chem.* **1990**, *191*, 857.

26. If this is a crystallization process, then the melting point of the material is higher than the observed decomposition point. Other typical processes (T_m, T_g, solvent loss) would be endothermic. In addition, solvent loss is not corroburated by a decrease in sample weight (TGA). The exact nature of this exothermic event is unclear at this time.

27. For an excellent treatment of the electrochemistry for organometallic bridged dimers see: Van Order, N.; Geiger, W. E.; Bitterwolf, T. E.; Rheingold, A. L. *J. Am. Chem. Soc.* **1987**, *109*, 5680 and references cited therein. Also see Kotz, J. C. in *Topics in Organic Electrochemistry*; Fry, A. J.; Britton, W. E. Eds.; Plenum: New York, **1986**, pp 100-109.

28. Hampsch, H. L.; Yang, J.; Wong G. K.; Torkelson, J. M. *Macromolecules* **1990**, 23, 3648. Lackritz, H. S.; Torkelson, J. M. "Polymer Physics of Poled Polymers for Second Order Nonlindear Optics" Chapter 8 in *Molecular Optoelectronics: Materials, Physics, and Devices,* Zyss, J., Ed.; Academic Press: New York, 1993, in press.

29. Shen, Y. R. *Principles of Nonlinear Optics*; John Wiley & Sons: New York, 1984.

30. For experimental details have been published in reference 18 and will appear soon: Wright, M. E.; Toplikar, E. G.; Lackritz, H. S.; Kerney, J. T. *Macromolecules* **1994**, *27*, 0000.

RECEIVED May 11, 1994

Chapter 35

Transition Metallophthalocyanines as Structures for Materials Design

Michael Hanack

Institut für Organische Chemie, Universität Tübingen,
Auf der Morgenstelle 18, 72076 Tübingen, Germany

Bridged macrocyclic transition metal complexes $[MacM(L)]_n$, Mac = phthalocyanine (Pc) and naphthalocyanine (Nc), M = Fe, Ru, Co and L = e.g. pyz, dib, tz, SCN^-, CN^- were synthesized and their electrical properties studied in detail. Regardless of the bridging ligand, stable semiconducting compounds are formed after doping with iodine. In general, these complexes $[MacM(L)]_n$ are insoluble in organic solvents, however, soluble oligomers $[R_xPcM(L)]_n$ were prepared using metallomacrocycles R_4PcM and R_8PcM, R = t-Bu, Et, OR' (R' = C_5H_{11}-$C_{12}H_{25}$), M = Fe, Ru, which are substituted in the peripheral positions. We report here on the systematic investigation of the influence of the bridging ligand on the conductivity of the bridged phthalocyaninato and 2,3-naphthalocyaninato transition metal complexes. Due to the low oxidation potentials of tz and me_2tz and their low lying LUMO the corresponding bridged systems $[MacM(L)]_n$, L = e.g. tz, me_2tz, and others (see Table 1) exhibit *intrinsic* conductivities. The phthalocyanines were characterized by IR, UV, ^1H-NMR and Mößbauer (Fe) spectroscopy.

Metallophthalocyanines and -naphthalocyanines have been used recently for the preparation of materials, which exhibit interesting semiconducting and nonlinear optical properties *(1)*. In this report we concentrate on the electrical properties of phthalocyaninato and naphthalocyaninato transition metal compounds. A necessary condition for achieving good electrical conductivities in

0097–6156/94/0572–0472$08.00/0

phthalocyaninatometal compounds is a special geometric arrangement, namely either planar or stacked.

Metallophthalocyanines and naphthalocyanines are less soluble in common organic solvents. However, soluble phthalocyanines R_4PcM (R = t-Bu, Et, $CH_2OC_{12}H_{25}$, C_5H_{11} - C_8H_{17}; M = Fe, Ru) and R_8PcM (R = C_5H_{11} - $C_{12}H_{25}$, OC_5H_{11} - $OC_{12}H_{25}$; M = Fe, Ru) have been prepared recently. Quite often the solubility of tetra-substituted metallophthalocyanines is higher than the octa-substituted ones. This is due to the larger dipole moment arising from unsymmetrical structures. The synthesis of tetra-substituted phthalocyaninatometal compounds always leads to four structural isomers *(2,3)* and recently, the separation (and characterization by ^{1}H-NMR spectroscopy) of all four isomers was successful in the case of tetrakis(2-ethylhexyloxy)phthalocyaninatonickel(II)   [(2-Et-$C_6H_{12}O)_4$PcNi]   *(4)*. Phthalocyanines substituted in peripheral positions by long chain substituents form liquid crystalline phases *(5,6)*, which are either discotic (from the disc-like shape of the molecules) or columnar meso phases *(7)*.

In this paper we will concentrate on stacked phthalocyaninato (PcM) and naphthalocyaninato (NcM) transition metal complexes. If it is possible to arrange phthalocyaninato transition metal compounds in a stacked fashion and produce charge carriers by oxidation or reduction, good semiconducting properties can be obtained. In addition to their high thermal stability a further advantage in using these metal complexes is their accessibility.

Except for the macrocyclic modification of phthalocyaninatolead(II) *(8)* in its monoclinic arrangement, the stacked arrangement has never been observed. Metallophthalocyanines in general crystallize in the α- or ß-modification which is not favourable for π-orbital overlap and thus for the formation of a conduction band. However, if certain transition metal phthalocyanines, e.g. phthalocyaninatonickel(II) are doped with iodine, stacked structures are obtained which exhibit high conductivities *(9)*.

A few years ago we developed alternative route to obtain a stacked arrangement with macrocyclic transition metal compounds. This leads, as will be pointed out later, to the possibility of achieving intrinsic conductivities in these systems by the so-called "shish kebab" arrangement of macrocyclic transition metal compounds. Here the stacking of the metal macrocycles is achieved by bisaxially connecting the central transition metal atoms with bidentate bridging ligands (L). Such bridged macrocyclic metal compounds $[MacM(L)]_n$ with transition metals, e.g., Fe, Ru, Os, Co, Rh, in various oxidation states have been synthesized and investigated by us in detail *(1,10,11)*. A schematic structure of these type of compounds is shown in Figure 1.

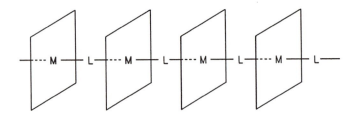

M = Fe^{2+}, Fe^{3+}, Co^{2+}, Co^{3+}, Ru^{2+}, Mn^{2+}, Mn^{3+}, Cr^{3+}

L = ... , CN-⟨ ⟩-NC , CN^-

☐ = Pc^{2-}, R_4Pc^{2-}, R_8Pc^{2-}, $1,2-Nc^{2-}$, $2,3-Nc^{2-}$, TBP^{2-}

Figure 1. Schematic structure of bridged macrocyclic transition metal complexes

Within this new class of materials a large spectrum of compounds was synthesized and studied systematically with respect to their physical properties. The π-electron containing bridging ligands (L) are linear organic molecules e.g. pyrazine (pyz), p-diisocyanobenzene (dib) and substituted p-diisocyanobenzenes, tetrazine (tz) and substituted tetrazines (me_2tz). If the oxidation state of the central metal atom is +3 (Co^{3+}, Fe^{3+}) charged bridging ligands such as cyanide (CN^-), thiocyanate (SCN^-) and others can also be used. In addition to phthalocyanines and tetrabenzoporphyrins other macrocycles used are 1,2- and 2,3-naphthalocyanines and tetranaphthoporphyrins. In general, the complexes $[MacM(L)]_n$ are practically insoluble in organic solvents. However, soluble oligomers $[R_4PcM(L)]_n$ and $[R_8PcM(L)]_n$ have been prepared using substituted metallomacrocycles R_4PcM, in which R = t-Bu, Et, OR' (R' = C_5H_{11} - $C_{12}H_{25}$) and R_8PcM (R = C_5H_{11} - $C_{12}H_{25}$, OC_5H_{11} - $OC_{12}H_{25}$) and M = Fe, Ru. These types of oligomers are completely soluble in most common organic solvents, thereby allowing the determination of the chain lengths by ^{1}H-NMR spectroscopy. The soluble oligomers $[R_4PcFe(dib)]_n$ form Langmuir-Blodgett films *(12)*.

The powder conductivities of most of the bridged phthalocyaninato transition metal complexes $[PcM(L)]_n$ for M = Fe, Ru, Os, Co, Rh; L = pyz, dib, etc. are low in the range of $10^{-6} - 10^{-7}$ S/cm. However, many of these compounds can be doped either chemically (with iodine) or electrochemically *(13,14)* leading to good semiconducting properties ($\sigma_{RT} = 10^{-2} - 10^{-1}$ S/cm) with thermal stabilities up to 120°C. The doping does not destroy the bridged structure of the coordination polymers, because the oxidation process takes place at the macrocycle forming radical cations. The long term stability of these compounds is high, since even after two years of exposure to air and moisture, the conductivities did not change.

The doping process has been studied very carefully by us using not only different macrocycles and central metal atoms, but also using different bridging ligands with a comparatively low oxidation potential. The question addressed was whether the doping process occurs only at the phthalocyanine macrocycle leading to a radical cation or also at the bridging ligand. Fig. 2 shows a phthalocyaninatoiron oligomer in which the bridging ligand is 9,10-diisocyanoanthracene (9,10-dia), a ligand which has a comparatively low oxidation potential. It could be shown that doping with iodine indeed leads to oxidation of the bridging ligand, thereby increasing the powder conductivity from 3×10^{-7} to 8×10^{-3} S/cm *(15)*.

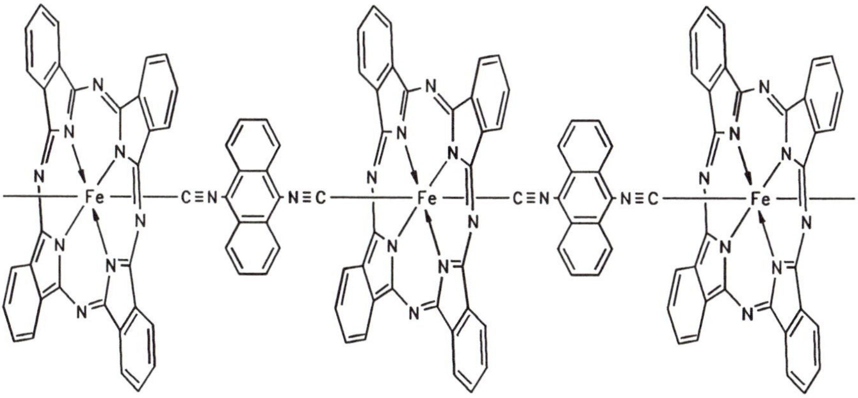

Figure 2. Schematic structure of $[PcFe(9,10\text{-dia})]_n$

Changing the macrocycles in the coordination polymers shown in Fig. 1 from phthalocyanine to one with an extended π-electron system, e.g. 2,3-naphthalocyanine leads to an interesting effect concerning the semiconducting properties of the corresponding bridged systems. A systematic investigation of the oxidation potentials of the metallomacrocycles used for the synthesis of the corresponding bridged systems by cyclic voltammetry in different solvents shows that the first and second oxidation potentials of the metallomacrocycles are very much dependent upon the structure: the phthalo-cyaninatoiron (PcFe) used here has almost the same oxidation potential as the unsymmetrical 1,2-naphthalocyaninatoiron (1,2-NcFe). 2,3-Naphthalocyani-natoiron, however, exhibits a much lower oxidation potential as compared to TBPFe and 2,3-TNPFe. The lower the oxidation potential of the metallo-macrocycle, the higher the conductivity of the corresponding $[MacM(L)]_n$-system, e.g. the bridged compound $[2,3\text{-NcFe(dib)}]_n$ shows a powder con-ductivity of 4×10^{-3} S/cm, which is due to oxygen doping of the macrocycle.

PcFe, PcRu, PcOs and 2,3-NcFe react with tetrazine (tz) and dimethyl-tetrazine (me_2tz) depending on the conditions with formation of the corre-sponding monomers $MacM(L)_2$ (L = tz, me_2tz) and the bridged systems $[MacM(L)]_n$ (L = tz, me_2tz) *(16)*. The tetrazine bridged systems, in contrast to other bridged compounds $[MacM(L)]_n$ with M = Fe, Ru or Os and L e.g pyz or dib, show good semiconducting properties without external oxidative doping (σ_{RT} = 0.05 - 0.3 S/cm) *(16)*.

The powder conductivities of bridged transition metallomacrocycles in the non-doped state are listed in Table I. All the bridged complexes $[MacM(L)]_n$ (L = pyz, tz, dabco), contain cofacially arranged macrocycles, which are separated by approximately the same distance (about 600 pm). While the monomeric complexes $PcM(L)_2$ (L = pyz, dabco, tz; M = Fe, Ru, Os) show insulating behaviour, it can be seen (Table I) that the ligand L has a significant effect on the conductivity of the bridged complexes $[MacM(L)]_n$.

The following conclusion can be drawn from Table I: because dabco is a ligand containing no π-orbitals that are capable of interaction with the metallomacrocycle (see below), the complex $[PcFe(dabco)]_n$ is an insulator. A clear increase in conductivity is observed for the pyrazine-bridged com-pounds $[MacM(pyz)]_n$, which exhibit conductivities in the low semiconduct-ing region. However, by changing the bridging ligand from pyrazine to *s*-tetrazine the conductivity is increased by 3 to 5 orders of magnitude without external oxidative doping (see Table I).

One of the factors responsible for the electrical conductivities in bridged macrocyclic transition metal complexes $[MacM(L)]_n$ is the band gap, which is determined by the energy difference between the LUMO of the bridging ligand and the HOMO of the transition metallomacrocycle. There-fore to achieve semiconducting properties the metallomacrocycle should con-tain a high lying HOMO, while a bridging ligand with a low lying LUMO should be used. A systematic change of the bridging ligand in this respect has been investigated. It was found that not only tetrazine (tz) and dimethyl-tetrazine (me_2tz), but also other bridging ligands, e.g., *p*-diaminotetrazine can be used to prepare the corresponding bridged macrocyclic transition metal compounds. p-Diaminotetrazine $(NH_2)_2tz$ is coordinated to the ruthenium atom in $[PcRu(NH_2)_2tz]_n$ via the nitrogen atoms of the amino groups: its powder conductivity is about 100 times higher than that of $[PcRu(p$-$(NH_2)_2C_6H_4)]_n$, in which the bridging ligand p-phenylenediamine contains no hetero atom in the aromatic ring (Table I). On the other hand, electron withdrawing substituents in the peripheral positions of the macrocycle show the expected effect: e.g. $[(CN)_4PcFe(tz)]_n$ has a conductivity which is at least 3 orders of magnitude less than the conductivities of other tetrazine-bridged compounds investigated so far (Table I) *(17)*.

Table I. DC-Powder Electrical Conductivities of Monomeric and Bridged Macrocyclic Transition Metal Complexes.

Compound	σ_{RT} $[S/cm^{-1}]$[a]
PcFe(dabco)$_2$	10^{-10}
[PcRu(dabco)$_2$ x 1.4 CHCl$_3$]$_n$	10^{-10}
PcFe(pyz)$_2$	3×10^{-12}
[PcFe(pyz)]$_n$	1×10^{-6}
PcFe(tz)$_2$	$< 10^{-9}$
[PcFe(tz)]$_n$	2×10^{-2}
PcRu(tz)$_2$	$< 10^{-11}$
[PcRu(tz)]$_n$	1×10^{-2}
PcOs(tz)$_2$	4×10^{-8}
[PcFe(me$_2$tz)]$_n$	4×10^{-3}
[PcRu(NH$_2$)$_2$tz]$_n$	4×10^{-3}
[PcRup-(NH$_2$)C$_6$H$_4$]$_n$	5×10^{-9}
[PcRuCl$_2$(tz)]$_n$	3×10^{-3}
[PcOs(pyz)]$_n$	1×10^{-6}
[PcOs(tz)]$_n$	1×10^{-2}
[2,3-NcFe(pyz)]$_n$	5×10^{-5}
[2,3-NcFe(tz) x 0.2 CHCl$_3$]$_n$	3×10^{-1}
[(me)$_8$PcFe(pyz)]$_n$	3×10^{-9}
[(me)$_8$PcFe(tz)]$_n$	1×10^{-2}
[PcRu(me$_2$tz)]$_n$	4×10^{-3}
[(CN)$_4$PcFe(pyz)]$_n$	5×10^{-9}
[(CN)$_4$PcFe(tz)]$_n$	1×10^{-6}

[a] Room temperature, pressed pellets, 1 kbar.

The low band gaps of all the tetrazine-bridged coordination polymers with group VIII transition metals $[MacM(L)]_n$ [Mac = Pc; M = Fe, Ru, Os; Mac = 2,3-Nc; M = Fe; L = e.g. tz, me$_2$tz, (NH$_2$)$_2$tz] can be demonstrated by physical properties that are not shown by the corresponding systems $[MacM(L)]_n$ in which L = e.g. pyz or dib: all the tetrazine-bridged coordination polymers with group VIII transition metals $[MacM(tz)]_n$ show broad bands in the UV/Vis/NIR spectra between 1250 and 2500 nm with different maxima, e.g., $[PcFe(tz)]_n$ at 1515 nm (0.75 eV) and $[PcRu(tz)]_n$ at 1053 nm (1.18 eV). The corresponding pyrazine-bridged systems $[PcM(pyz)]_n$ (M = Fe, Ru, Os) exhibit normal UV/Vis spectra with Soret- and Q-bands between 245 and 700 nm, respectively, but show no absorption between 1200 and 2000 nm. The absorption bands in the near infrared correlate well with the electrochemically estimated energy gap between the HOMOs of the different metallomacrocycles and the LUMO of s-tetrazine in all the tetrazine-bridged systems described here. The broad band observed in the absorption spectra of these complexes can be assigned to charge transfer process from the metallomacrocycle to the π^*-orbital of s-tetrazine.

Table II. Mößbauer Data of s-Tetrazine Bridged Complexes.

Complex	δ [mm/s]	Δ_{EQ} [mm/s]
ß-PcFe	0.38	2.58
PcFe(py)$_2$	0.26	2.02
PcFe(pdz)$_2$	0.21	1.82
PcFe(tz)$_2$ [c]	0.15	1.79
$[PcFe(tz)]_n$	0.13	2.23
$[PcFe(bpytz)]_n$	0.24	1.90
PcFe(me$_2$tz)$_2$ x 0.5 me$_2$tz [d]	0.36	2.67
(me)$_8$PcFe	0.38	2.55
(me)$_8$PcFe(py)$_2$	0.26	1.97
$[(me)_8PcFe(tz)]_n$	0.14	2.11
(meO)$_8$PcFe	0.36	2.50
(meO)$_8$PcFe(py)$_2$	0.25	1.95
$[(meO)_8PcFe(tz)]_n$ [c]	0.15	2.19

[a] Measured at room temperature, unless otherwise indicated.
[b] Relative to metallic iron.
[c] Additional doublet, due to PcFe present.
[d] Measured at 77° K.

The Mößbauer data of *s*-tetrazine-coordinated iron complexes are summarized in Table II along with the data for the iron macrocycles and the corresponding pyridine complexes *(18)*. For the tetrazine-bridged complexes $[MacFe(tz)]_n$ the isomer shift δ is similar to the monomeric $PcFe(tz)_2$ whereas a clear increase of Δ_{EQ} in comparison to $[PcFe(pyz)]_n$ is measured. This is a further indication for a low band gap, because this effect can be explained by the thermal activation of electrons of the highest occupied band. As the contribution of such delocalized electrons to the occupation of metal centred d-orbitals is diminished, the increase in Δ_{EQ} is the result of the low band gap of the quasi one-dimensional chain structure of the tetrazine-bridged complexes.

Other examples of macrocyclic transition metal complex exhibiting intrinsic semiconducting properties are the cyano-bridged systems $[MacM(CN)]_n$ (Mac = Pc, TBP, 2,3-Nc; M = Fe, Co). The cyano-bridged phthalocyanine complex $[PcCo(CN)]_n$ can be easily synthesized, e.g., from $PcCoCl_2$ with an excess of NaCN to form the soluble $Na[PcCo(CN)_2]_n$ *(19)*. The crystal structure of this monomer shows the bisaxial coordination of the CN-groups. HCN can be cleaved from the monomer using different methods, e.g. treatment with H_2O at 100°C, to form $[PcCo(CN)]_n$ in quantitative yield *(19)*. It was shown by physical methods that these compounds possess the cyano-bridged structure *(19,20)*. All of the cyano-bridged macrocyclic metal complexes show comparatively high powder conductivities in the range of 10^{-2} - 10^{-1} S/cm.

The bridged transition metal complexes described here are of interest for technical applications due to their comparatively high thermal stability and their good semiconducting properties without external oxidative doping.

Acknowledgment

The reported results are based on the very efficient experimental work of my co-workers Anton Beck, Reinhold Dieing, Michael Dreßen, Carola Feucht, Ronald Großhans, Ahmet Gül, Shigeru Hayashida, Andreas Hirsch, Sabine Kamenzin, Young-Goo Kang, Armin Lange, José Osío Barcina, Jörg Pohmer, Haiil Ryu, Gabriele Schmid, Hanna Schultz, Michael Sommerauer and Elisabeth Witke. My sincere thanks go to all of them.

Literature

(1)   Schultz, H., Lehmann, H., Rein, M., Hanack, M. in *"Structure and Bonding 74"*, Springer-Verlag, Heidelberg 1991, p. 41.
(2)   Hanack, M., Thies, R. *Chem. Ber.* **1988**, *121*, 1225.
(3)   Hanack, M., Meng, D., Beck, A., Sommerauer, M., Subramanian, L.R. *J. Chem. Soc., Chem. Commun.* **1993**, 58.
(4)   Hanack, M., Schmid, G., Sommerauer, M. *Angew. Chem.* **1993**, *105*, 1540.
(5)   Giroud-Godquin, A.-M., Maitlis, P.M. *Angew. Chem.* **1991**, *103*, 370.
(6)   Espinet, P., Esternelas, M.A., Oro, L.A., Serrano, J.L., Sola, E. *Coord. Chem. Rev.* **1992**, *117*, 215.
(7)   Chandraskhar, S., Ranganath, G.S. *Rep. Prog. Phys.* **1990**, *53*, 570.
(8)   Ukei, K. *Acta Cryst.* **1973**, *B29*, 2290.
(9)   Schramm, C.J., Scaringe, R.P., Stojakovic, D.R., Hoffman, B.M., Ibers, J.A., Marks, T.J. *J. Am. Chem. Soc.* **1980**, *102*, 6702.
(10)   Hanack, M., Deger, S., and Lange, A. *Coord. Chem. Rev.* **1988**, *83*, 115.
(11)   Hanack, M., Datz, A., Fay, R., Fischer, K., Keppeler, U., Koch, J., Metz, J., Mezger, M., Schneider, O., and Schulze, H.-J. *"Handbook of Conducting Polymers"*, M. Dekker Inc., New York 1986, Vol. 1.
(12)   Rauschnabel, J. *Diploma Thesis*, Tübingen, **1992**.
(13)   Diel, B.N., Inabe, T., Jakki, N.K., Lyding, W., Schneider, O., Hanack, M., Kannewurf, C.R., Marks, T.J., Schwarz, L.H. *J. Am. Chem. Soc.* **1984**, *106*, 3207.
(14)   Keppeler, U., Schneider, O. Stöffler, W., Hanack, M. *Tetrahedron Lett.* **1984**, *25*, 3679.
(15)   Hanack, M., Ryu, H. *Synth. Met.* **1992**, *46*, 113.
(16)   Hanack, M., Lange, A., Großhans, R. *Synth. Met.* **1991**, *45*, 59.
(17)   Hanack, M., Großhans, R., *Chem. Ber.* **1989**, *122*, 1665.
(18)   Hanack, M., Keppeler U., Lange, A., Hirsch A., and Dieing, R., in: *"Phthalocyanines, Properties and Applications"*, VCH Publishers, Weinheim 1993, Vol. 2, p. 43.
(19)   Metz, J., Hanack, M. *J. Am. Chem. Soc.* **1983**, *105*, 828.
(20)   Hanack, M., Hedtmann-Rein, C., Datz, A., Keppeler, U., Münz, X. *Synth. Met.* **1987**, *19*, 787.

RECEIVED April 14, 1994

Chapter 36

New Class of Photochemically Reactive Polymers Containing Metal–Metal Bonds Along the Polymer Backbone

Synthesis, Characterization, and Reactivity

David R. Tyler, Jeffrey J. Wolcott, Gregory F. Nieckarz, and Steve C. Tenhaeff

Department of Chemistry, University of Oregon, Eugene, OR 97403

A synthetic method was developed for incorporating metal-metal bonds into the backbones of polymers. The method involves the use of difunctional, cyclopentadienyl-substituted metal carbonyl dimers. Polyurethanes, polyureas, polyamides, and polyvinyls were synthesized in a demonstration of the new synthetic methodology. The polymers are photodegradable because the metal-metal bonds homolyze when irradiated with visible light. The photochemical reactions of the polymers in solution are essentially identical to the photochemical reactions of the discrete metal-metal bonded dimers. Typical reactions include metal-metal bond disproportionation and chlorine atom abstraction from carbon tetrachloride. The quantum yields for degradation in solution decrease as the molecular weight of the polymers increase. The polymers are also photochemically degradable in the solid state; thin films of the polymers degrade when irradiated with visible light in the presence of oxygen.

Photochemically reactive polymers are of considerable interest because they are potentially useful as degradable plastics, photoresists, biomedical materials, and precursors of ceramic materials (*1, 2*). Although several types of photodegradable polymers are on the market, most require ultraviolet light for degradation. In order to expand the repertoire of photodegradable polymers to include those that will degrade with visible light, we recently reported the synthesis of a new class of polymers that contain metal-metal bonded organometallic dimers along the polymer backbone (*3-5*). It is well known that metal-metal bonds can be cleaved photochemically with visible light (equation 1), and it was our hypothesis that similar reactivity would occur when the metal-metal bonds were incorporated into polymer backbones (equation 2). In this chapter, the synthesis and characterization of polymers containing $Cp_2Fe_2(CO)_4$ or $Cp_2Mo_2(CO)_6$ units along the backbone is discussed. The photodegradation of these polymers in solution and in the solid state is also covered.

$$L_nM\!-\!ML_n \xrightarrow{\;h\nu\;} 2 \cdot ML_n \tag{1}$$

$$ML_n = CpMo(CO)_3 \ (Cp = \eta^5\text{-}C_5H_5), \ CpW(CO)_3, \ Mn(CO)_5, \ Re(CO)_5, \ CpFe(CO)_2$$

0097–6156/94/0572–0481$08.00/0

$$\text{MM—M} \text{M—M} \text{M—M}$$

$$\downarrow h\nu \qquad\qquad (2)$$

$$\text{M• + •M} \text{M• + •M} \text{M• + •M}$$

Results and Discussion

Synthetic Strategy. Our syntheses of the metal-metal bond-containing polymers are based on the known chemistry of polymeric ferrocenes (*6-12*). Ferrocene is incorporated into polymers by substituting its Cp rings with appropriate functional groups and then reacting with appropriate difunctional organic monomers (e.g., equation 3).

$$(3)$$

Our analogous strategy for synthesizing metal-metal bond-containing polymers uses difunctional, cyclopentadienyl-substituted metal dimers. A sample polymerization reaction is shown in equation 4, which shows the reaction of a metal-metal bonded "diol" with a hexamethylene diisocyanate (HMDI) to form a polyurethane (*3*).

$$(4)$$

This polymerization strategy is quite general, and a number of organometallic polymers have been made from monomers containing functionalized Cp ligands (6-13). However, no similarly constructed polymers containing metal-metal bonded units had been reported prior to our work.

The synthesis of metal-metal bonded dimers with functional groups substituted on the Cp rings is a synthetic challenge. The reason lies in the comparative weakness of the metal-metal bonds. Because they are relatively weak, the metal-metal bonds will not stand up to the harsh conditions typically required for the substitution of metal-coordinated Cp rings. For this reason, it was necessary to synthesize first the substituted Cp molecules and then coordinate these rings to the metals. High yield synthetic routes to various substituted Cp rings are summarized in Schemes I-III. (The vinyl-substituted $Cp_2Fe_2(CO)_4$ dimer is an exception to this general synthetic strategy. The synthesis of $(Cp-CH=CH_2)_2Fe_2(CO)_4$ is discussed in the section following.)

Following the synthesis of these substituted Cp ligands, the difunctional dimers were synthesized by the routes in Schemes I-III. Note that the general route shown for the Mo-containing dimers (involving Fe^{3+} oxidation of the anionic species) was developed by Birdwhistell (14). In all cases, the difunctional dimers were characterized by elemental analysis, NMR and infrared spectroscopy, mass spectrometry, and electronic absorption spectroscopy. There are no particular spectroscopic peculiarities in any of the complexes; the electronic absorption and infrared spectra are virtually identical to those of the unsubstituted dimers. The $(CpCH_2CH_2OH)_2Mo_2(CO)_6$ dimer was further structurally characterized by x-ray crystallography (15).

Scheme I

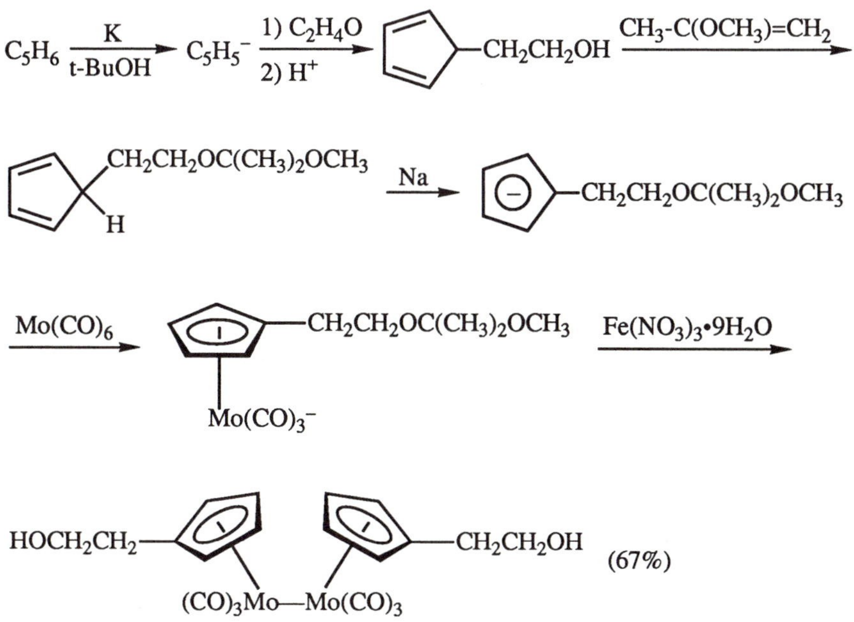

Scheme II

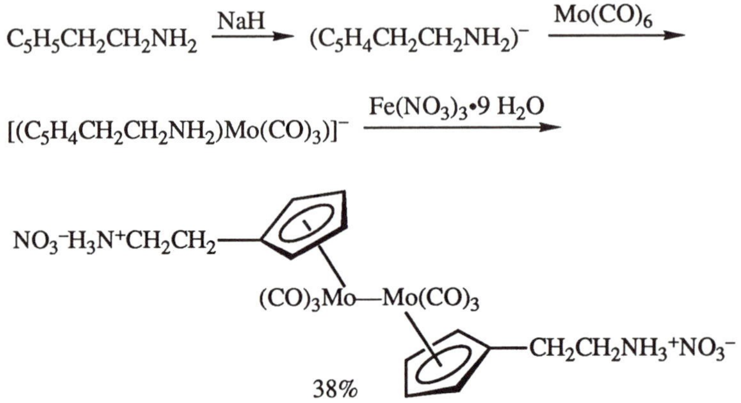

$$C_5H_5CH_2CH_2NH_2 \xrightarrow{\text{NaH}} (C_5H_4CH_2CH_2NH_2)^- \xrightarrow{\text{Mo(CO)}_6}$$

$$[(C_5H_4CH_2CH_2NH_2)Mo(CO)_3]^- \xrightarrow{\text{Fe(NO}_3)_3 \cdot 9\ H_2O}$$

$$NO_3^-H_3N^+CH_2CH_2-$$
$$(CO)_3Mo—Mo(CO)_3$$
$$-CH_2CH_2NH_3^+NO_3^-$$

38%

Scheme III

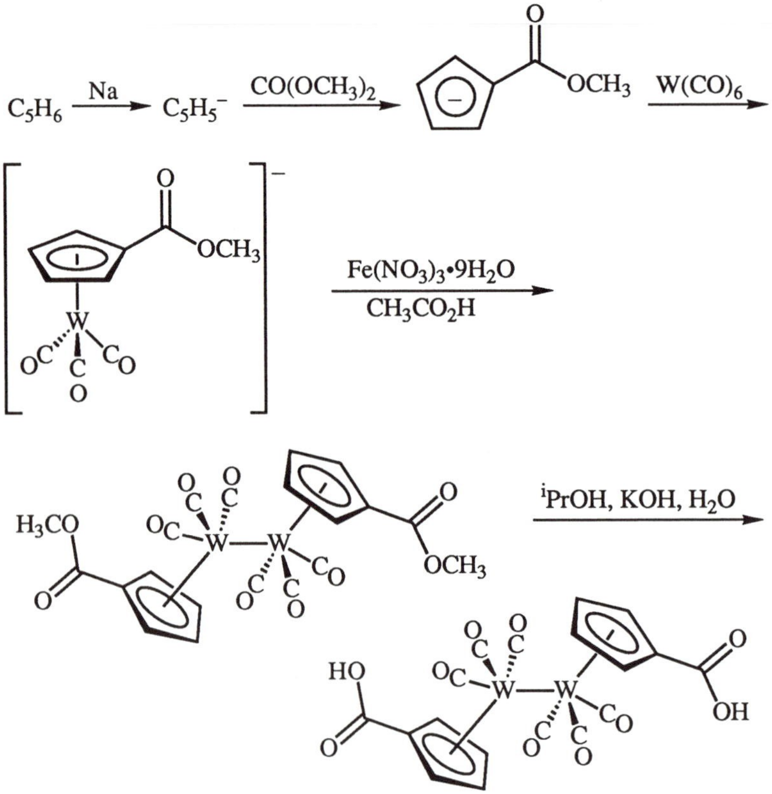

$$C_5H_6 \xrightarrow{\text{Na}} C_5H_5^- \xrightarrow{\text{CO(OCH}_3)_2}$$

$$\xrightarrow{\text{W(CO)}_6}$$

$$\xrightarrow[\text{CH}_3\text{CO}_2\text{H}]{\text{Fe(NO}_3)_3 \cdot 9H_2O}$$

$$\xrightarrow{^i\text{PrOH, KOH, H}_2\text{O}}$$

Synthesis of (Cp-CH=CH$_2$)$_2$Fe$_2$(CO)$_4$. The synthesis of (Cp-CH=CH$_2$)$_2$Fe$_2$(CO)$_4$ is outlined in Scheme IV. It involves the synthesis of the iodocyclopentadiene iron complex as outlined by Herrmann (*16*). The vinyl group is then attached to the Cp by a palladium-catalyzed cross-coupling reaction (*17*), and finally the dimer is made by treatment with sodium naphthalide.

Scheme IV

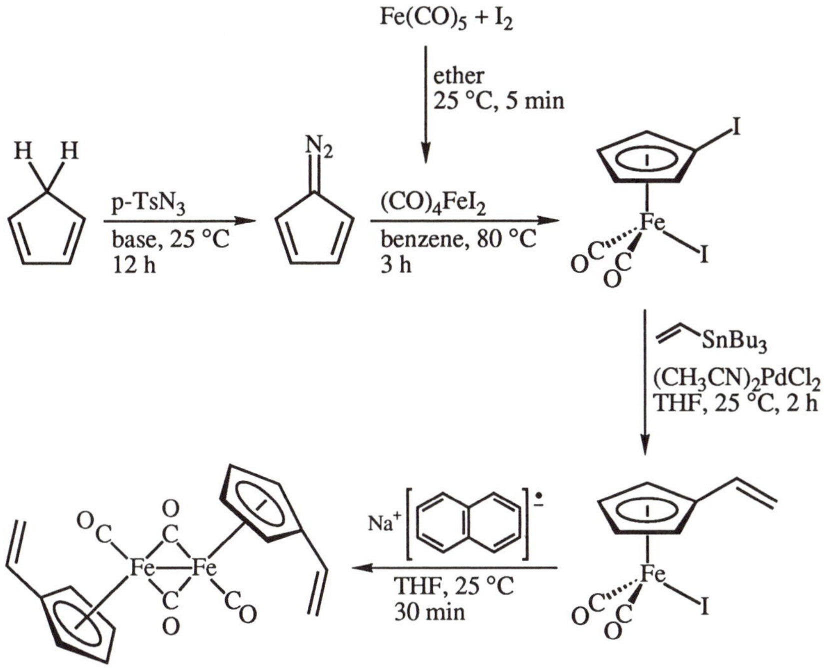

Polymer Synthesis. Just as the comparatively weak metal-metal bonds pose problems for the synthesis of the difunctional dimers, they cause similar problems in the synthesis of the polymers. The relative weakness of the metal-metal bonds makes them more reactive than the bonds found in standard organic polymers; thus, under many standard polymerization reaction conditions, metal-metal bond cleavage would result. For example, metal-metal bonds react with acyl halides to form metal halide complexes. Therefore, the synthesis of polyamides using metal-metal bonded "diamines" and diacyl chlorides would simply lead to metal-metal bond cleavage rather than polymerization. Likewise, metal-metal bonded complexes are incompatible with many Lewis bases because the Lewis bases cleave the metal-metal bonds in disproportionation reactions (*18*). This type of reactivity thus rules out many standard condensation polymerization reactions in which bases are used to neutralize any acids produced. Similarly, it prevented us from using acylchlorides in the synthesis of polyamides. In summary, all of our polymerization strategies are carefully designed to avoid cleaving the metal-metal bond during the polymerization process.

A sample polymerization reaction, showing the synthesis of a polyurethane, was shown above in equation 4. Using similar synthetic strategies, various polyurethanes, polyureas (e.g., equation 5), polyvinyls (e.g., equation 6), and polyamides (e.g., equation 7) were synthesized. Note that the polymer in equation 4 has a metal-metal bond in every repeating unit. We found that it was not necessary to have a metal-metal bond in every repeating unit in order to photochemically degrade the polymers, so our emphasis shifted to copolymers. Sample copolymers are shown in Scheme V.

$$(5)$$

$$(6)$$

(7)

Scheme V

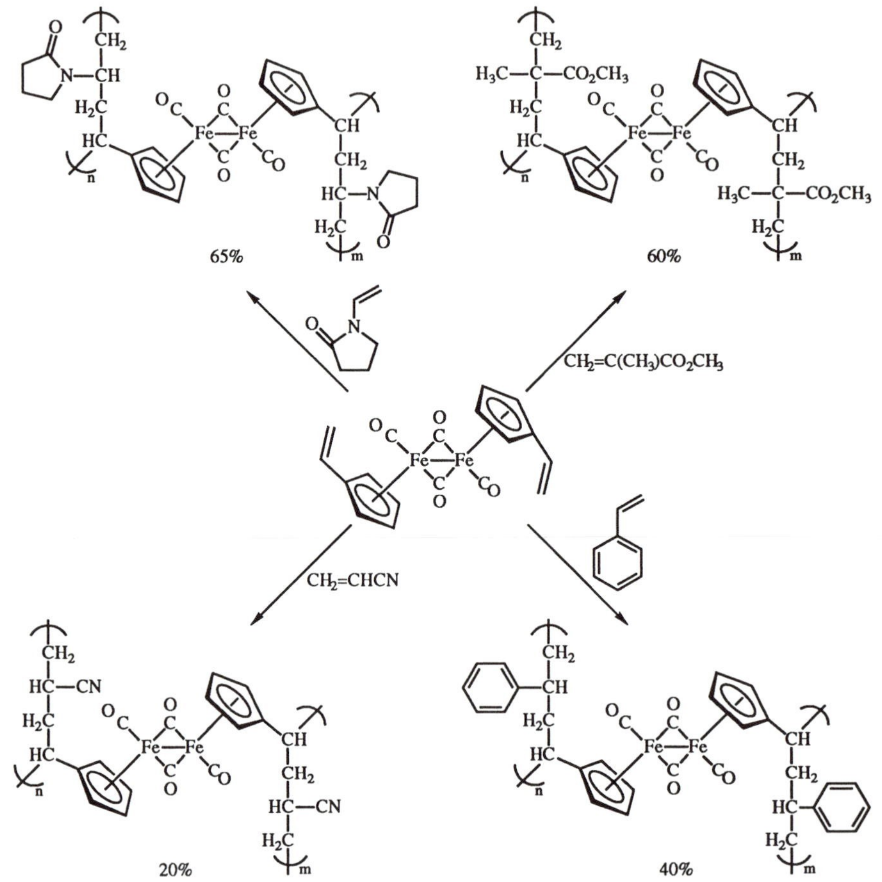

65%

60%

20%

40%

Yet another polymer synthesis strategy is to react the difunctional dimer molecules with prepolymers. Equation 8 shows an example of this technique.

Characterization of the Polymers. The polymers were characterized by comparison of their infrared, electronic, and NMR spectra to model complexes (*3-5*). For example, molecule **1**, a model complex for the polymer in equation 4, was synthesized by reaction of $(CpCH_2CH_2OH)_2Mo_2(CO)_6$ with a *mono*isocyanate (equation 9).

In all cases, the spectroscopic agreement between the polymers and the model complexes was excellent. Sample data showing the close agreement are shown in Table I.

Molecular weights were measured by VPO or GPC. Typical molecular weights (M_n) were between 5,000 and 20,000 (n = 7-25). Thus, in many cases, the polymers are best described as oligomers. However, it is important to note that no effort was made to maximize the molecular weights.

Photochemical Reactions in Solution. Metal-metal bonds react photochemically in solution according to one of three fundamental types of reactions (*4*). (1) They react with radical traps to form monomeric complexes (e.g., equation 10); (2) they react with ligands to form ionic disproportionation products (e.g., equation 11); and (3) they react with oxygen to form metal oxides (equation 12). (The latter reaction is likely a radical trapping reaction in the sense that an oxygen molecule reacts with a photogenerated metal radical.)

Table I. Sample Data Showing a Spectroscopic Comparison of a Polymer to a Model Complex

Compound	^1H NMR δ, ppm	λ, nm (ε)	IR (KBr), cm^{-1} $\nu(C=O)$	other bands
(Model complex: $(CO)_3Mo\!-\!Mo(CO)_3$ dimer with cyclopentadienyl ligands bearing $^{d}CH_2{}^{f}CH_2NHC(=O)NHCH(CH_2)_5CH_3$ side chain; $CH_3(CH_2)_6NHC(=O)NHCH_2CH_2$—)	a: 0.84 (t, $J = 5.7$ Hz, 6 H)[a] b: 1.22 (broad, 12 H) c: 1.30 (m, 4 H) d: 2.41 (t, $J = 5.7$ Hz, 4 H) e: 2.93 (quartet, $J = 5.7$ Hz, 4 H) f: 3.08 (quartet, $J = 5.7$ Hz) g: 5.21-5.58 (4 broad resonances, 8 H) h: 5.81, 5.88 (2 broad resonances, 4 H)	498 (2100)[b] 388 (19000)	2011 (w) 1961 (sh, s) 1940 (s) 1920 (s) 1902 (s)	ν(NH) 3358 (broad & w) ν(C=O), 1627 (m) δ(NH), 1574 (m) ν(C-N), 1261 (w) ν(CH$_3$, sp^3), 2957 (w) ν(CH$_2$, sp^3), 2928 (w)
(Polymer: repeating unit with $(CO)_3Mo\!-\!Mo(CO)_3$ dimer, cyclopentadienyl rings bearing $^{c}CH_2{}^{e}CH_2NHC(=O)NH(CH_2)_6NHC(=O)$—, $\big[\!-\!NHCH_2CH_2$ ... $\big]_n$)	a: 1.21 (broad, 4 H)[a] b: 1.31 (broad, 4 H) c: 2.42 (broad, 4 H) d: 2.94 (m, 4 H) e: 3.08 (broad, 4 H) f: 5.21-5.57 (4 broad resonances, 8 H) g: 5.72, 5.85, 5.90 (broad, 4 H)	497 (1930)[b] 386 (15900)	2011 (m) 1949 (s) 1891 (s)	ν(N-H), 3326 (w) ν(C=O), 1627 (m) δ(NH), 1565 (m) ν(C-N), 1252 (m) ν(CH$_2$, sp^3), 2931 (m) ν(CH$_2$, sp^3), 2857 (m)

[a]DMSO-d$_6$. [b]DMSO.

$$Cp_2Mo_2(CO)_6 + 2\ CCl_4 \xrightarrow{\ h\nu\ } 2\ CpMo(CO)_3Cl + 2\ \bullet CCl_3 \tag{10}$$

$$Cp_2Mo_2(CO)_6 + 2\ PR_3 \xrightarrow{\ h\nu\ } CpMo(CO)_3^- + CpMo(CO)_2(PR_3)_2^+ + CO \tag{11}$$

$$Cp_2Mo_2(CO)_6 \xrightarrow[O_2]{\ h\nu\ } \text{unidentified Mo oxides} \tag{12}$$

In all of the cases examined thus far, the photochemical reactions of the polymers in solution are analogous to the reactions of the discrete metal-metal bonded dimers in solution (*4, 5*). Sample reactions of the polymers showing the three types of reactivity are shown in equations 13-15.

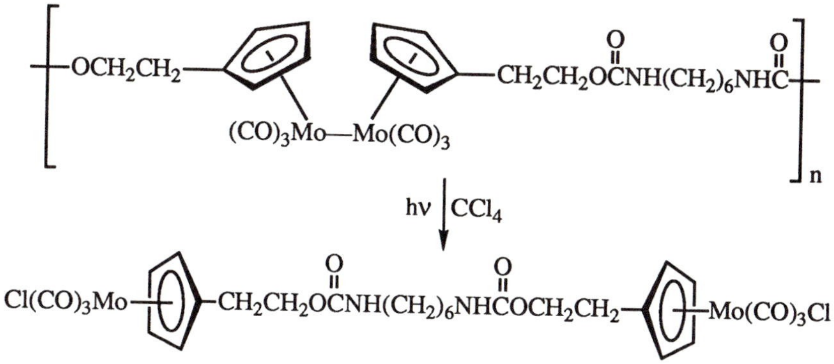

(13)

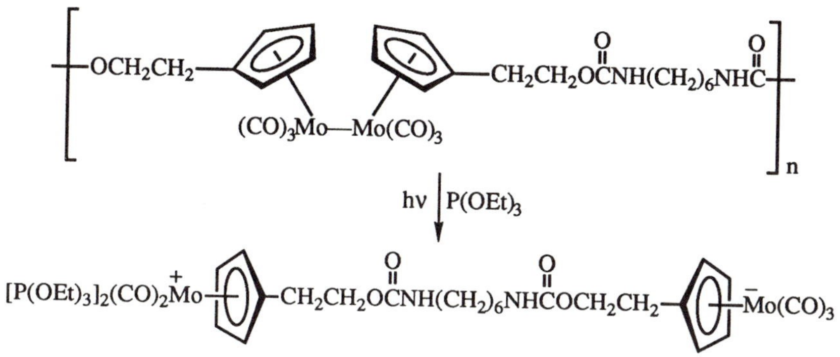

(14)

$$\left[-OCH_2CH_2-\underset{(CO)_3Mo-Mo(CO)_3}{\text{Cp}-\text{Cp}}-CH_2CH_2O\overset{O}{\overset{\|}{C}}NH(CH_2)_6NH\overset{O}{\overset{\|}{C}}-\right]_n$$

$$h\nu \left| O_2 \right. \qquad (15)$$

unidentified metal oxide

A fourth type of dimer reactivity, not mentioned above because it does not cleave the metal-metal bond and thus would not lead to polymer fragmentation, is dissociation of a CO ligand from the dimer. (Generally, this type of reactivity increases in relative efficiency as the radiation energy increases.) In solution, this type of reactivity generally leads to substitution. However, in the case of the $Cp_2Mo_2(CO)_6$ molecule, the reaction in equation 16 occurs (4). (Among the dimers, this reaction to form a triply bonded product is unique to the Mo and W species.)

$$\underset{(CO)_3Mo-Mo(CO)_3}{\text{Cp}-\text{Cp}} \xrightarrow{\text{UV}} \text{Cp}-(CO)_2Mo\equiv Mo(CO)_2-\text{Cp} + 2\,CO \qquad (16)$$

An analogous photoreaction occurs with polymers containing the Mo-Mo unit (equation 17).

$$\left[-OCH_2CH_2-\underset{(CO)_3Mo-Mo(CO)_3}{\text{Cp}-\text{Cp}}-CH_2CH_2O\overset{O}{\overset{\|}{C}}NH(CH_2)_6NH\overset{O}{\overset{\|}{C}}-\right]_n$$

$$\updownarrow \text{UV}, -2\,CO \qquad (17)$$

$$\left[-OCH_2CH_2-\text{Cp}-(CO)_2Mo\equiv Mo(CO)_2-\text{Cp}-CH_2CH_2O\overset{O}{\overset{\|}{C}}NH(CH_2)_6NH\overset{O}{\overset{\|}{C}}-\right]_n$$

In both reactions 16 and 17, addition of CO to the product solution causes the system to back-react to reform the starting materials. Once again, the main point to be made is that the solution photochemistry of the polymers is wholly analogous to the solution photochemistry of the discrete metal-metal bonded dimers.

Quantum Yields. The quantum yields for photolysis of several dimers, model complexes, and polymers in 2 M CCl_4/THF are listed in Table II. Note that polymer **3** has a significantly lower quantum yield for metal-metal bond homolysis than model complex **1** and diol **2**. Likewise, polymer **5** has a lower quantum yield for homolysis than diol **4**. The decreasing quantum yields with increasing side chain

Table II. Disappearance Quantum Yields for the Reactions of Selected Species Containing Metal-Metal Bonds with CCl$_4$ in THF

Compound	Quantum Yield (±10%)
HOCH$_2$CH$_2$—[Cp]—[Cp]—CH$_2$CH$_2$OH, (CO)$_3$Mo—Mo(CO)$_3$ **2**	0.66
CH$_3$(CH$_2$)$_5$NHCOCH$_2$CH$_2$—[Cp]—[Cp]—CH$_2$CH$_2$OCNH(CH$_2$)$_5$CH$_3$, (CO)$_3$Mo—Mo(CO)$_3$ **1**	0.59
—OCH$_2$CH$_2$—[Cp]—[Cp]—CH$_2$CH$_2$OCNH(CH$_2$)$_6$NHC—, (CO)$_3$Mo—Mo(CO)$_3$ **3**	0.44
HOCH$_2$CH$_2$—[Cp]—[Cp]—CH$_2$CH$_2$OH, Fe—Fe (CO)$_4$ bridged **4**	0.22
—OCH$_2$CH$_2$—[Cp]—[Cp]—CH$_2$CH$_2$OCNH(CH$_2$)$_6$NHC—, Fe—Fe (CO)$_4$ bridged **5**	0.14

length is expected because polymers frequently show lower fragmentation quantum yields than their monomeric counterparts. In fact, Guillet studied the Norrish Type I and II quantum yields for a variety of straight-chain ketones (*19*). Although it is simplistic to summarize his numerous conclusions in so short a space as this, he found that the quantum yields dropped with increasing chain length and reached a minimum value, after which a further increase in the chain length did not lower the quantum yield. Two explanations could account for this behavior: either the quantum yield drops because (1) cage escape is less efficient for the radical fragments as they increase in molecular weight and size, or (2) excited state decay is more efficient with increasing chain length. Either explanation would account for the decreasing quantum yield.

Photochemistry in the Solid State. Thin films of the polymers were prepared by dissolving the polymers in a suitable solvent (generally THF), spreading the solution on glass microscope slides, and then allowing the solvent to evaporate. The thin films reacted when they were exposed to *visible* light, whether from the overhead fluorescent lights in the laboratory, from sunlight, or from the filtered output of a high pressure Hg arc lamp. All of the films were irradiated in the presence or the absence of oxygen. For each film and its dark reaction control, the absorbance of the $d\pi \rightarrow \sigma^*$ transition was monitored periodically over a period of several months. Figure 1 is a plot of absorbance at 508 nm vs. time for polymer **3** under the various experimental conditions. As indicated in the Figure, the film of polymer **3**, exposed to sunlight in air, completely decomposed in two months. (The $\sigma \rightarrow \sigma^*$ band at 390 nm was also absent, confirming that the Mo-Mo bond was not intact.) The color of the film on this slide changed from red to green. Thin films stored in the dark in air or irradiated in a glovebox under nitrogen showed only a slight loss of absorbance at 508 nm over a one-year period (Figure 1). From this data, it was concluded that the decomposition of the polymers requires both light and air (oxygen). (The small amount of reaction for those slides stored in the glovebox in the light is probably due to reactions with solvent vapors.) The green decomposition product in the reactions of the Mo-containing dimers has not yet been fully analyzed; however, infrared spectra (KBr pellet) showed the material contained no CO ligands, as indicated by the absence of any stretches in the region 1600-2200 cm^{-1}. Infrared and electronic absorption spectra of the polymers stored in the dark or in a nitrogen atmosphere for one year are virtually identical to the spectra of freshly prepared samples. Recall that, as mentioned previously, oxide complexes form in the solution phase reactions of $Cp_2Mo_2(CO)_6$ with O_2.

The data above suggest that oxygen is necessary for the solid-state photochemical reaction to occur. We propose that oxygen traps the metal radicals produced in the photolysis of the metal-metal bonds, thereby preventing radical recombination. If oxygen diffusion is rate-limiting then the relative rates of oligomer photochemical decomposition in the solid-state would reflect the oxygen diffusion rate. Work is continuing in our laboratory to measure the rate of oxygen diffusion into the polymer films and to determine if there is a correlation between this rate and the quantum yields.

Acknowledgment is made to the donors of the Petroleum Research Fund, administered by the American Chemical Society, for the support of this work. Partial support from the Department of Education Areas of National Need Program for J.J.W. and G.F.N. is acknowledged.

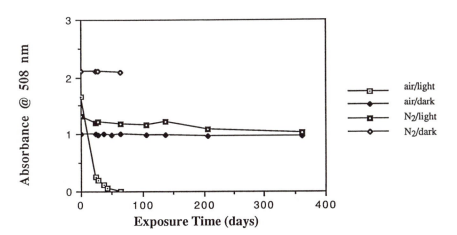

Figure 1. A plot of the absorbance at 508 nm vs time for four thin films of oligomer **3**. As indicated in the figure, one sample was exposed to air and visible light, one sample was kept in the dark in air, and another sample was exposed to light but kept under nitrogen, and the last sample was kept in the dark under nitrogen.

Literature Cited

1. Guillet, J. *Polymer Photophysics and Photochemistry*; Cambridge University Press: Cambridge, 1985.
2. *Inorganic and Organometallic Polymers*; Zeldin, M.; Wynne, K. J.; Allcock, H. R., Eds.; ACS Symposium Series 360; American Chemical Society: Washington, DC, 1988.
3. Tenhaeff, S. C.; Tyler, D. R. *Organometallics* **1991**, *10*, 473-482.
4. Tenhaeff, S. C.; Tyler, D. R. *Organometallics* **1991**, *10*, 1116-1123.
5. Tenhaeff, S. C.; Tyler, D. R. *Organometallics* **1992**, *11*, 1466-1473.
6. Pittman, C. U. Jr.; Rausch, M. D. *Pure Appl. Chem.* **1986**, *58*, 4, 617
7. Gonsalves, K.; Zhan-ru, Lin; Rausch, M. D. *J. Am. Chem. Soc.* **1984**, *106*, 3862.
8. Gonsalves, K. E.; Lenz, R. W.; Rausch, M. D. *Appl. Organomet. Chem.* **1987**, 81.
9. Knobloch, F. W.; Rauscher, W. H. *J. Polym. Sci.* **1961**, *54*, 651.
10. Pittman, C. U. Jr. *J. Polym. Sci., Part A* **1968**, *6*, 1687.
11. Gonsalves, K. E.; Rausch, M. D. *J. Polym. Sci., Part A: Polym. Chem.* **1988**, *26*, 2769.
12. Patterson, W. J.; McManus, S. P. *J. Polym. Sci.: Polym. Chem.* **1974**, *12*, 837.
13. Moran, M.; Pascual, M. C.; Cuadrado, I.; Losada, J. *Organometallics* **1993**, *12*, 811-822.
14. Birdwhistell, R.; Hackett, P.; Manning, A. R. *J. Organomet. Chem.* **1978**, *157*, 239-241.
15. Tenhaeff, S. C.; Tyler, D. R.; Weakley, T. J. R. *Acta Crystallogr.* **1991**, *C47*, 303-305.
16. Herrmann, W. A.; Huber, M. *Chem. Ber.* **1978**, *111*, 3124-3135.
17. Brehm, E. C.; Stille, J. K.; Meyers, A. I. *Organometallics* **1992**, *11*, 938-942.
18. Stiegman, A. E.; Tyler, D. R. *Coord. Chem Rev.* **1985**, *63*, 217-240.
19. Guillet, J. *Polymer Photophysics and Photochemistry*; Cambridge University Press: Cambridge, 1985, p. 273ff.

RECEIVED April 14, 1994

Chapter 37

Structural and Spectroscopic Studies of β-Hematin (the Heme Coordination Polymer in Malaria Pigment)

D. Scott Bohle[1], Brenda J. Conklin[2], David Cox[3], Sara K. Madsen[1], Scott Paulson[1], Peter W. Stephens[4], and Gordon T. Yee[2]

[1]Department of Chemistry, University of Wyoming, Laramie, WY 82071–3838
[2]Department of Chemistry and Biochemistry, University of Colorado at Boulder, Boulder, CO 80309–0215
[3]Department of Physics, Brookhaven National Laboratories, Upton, New York 11973
[4]Department of Physics, State University of New York at Stony Brook, Stony Brook, NY 11794

β-Hematin, the crystalline heme coordination polymer present in malaria pigment, is readily prepared by abstracting HCl from hemin with non-coordinating bases. Spectroscopic results(IR, Raman, and UV-Vis) suggest that polymerization occurs via Fe-O inter-heme linkage of the iron-propionate moieties. X-ray powder diffraction results indicate that β-hematin crystallizes in a triclinic unit cell with two heme groups per cell in the space group $P\bar{1}$.

The polymerization and aggregation of proteins, nucleic acids, carbohydrates, and fats into larger functional structures links biology at the molecular and cellular levels. As a consequence, biopolymers have diverse functional roles ranging from information storage and transport, to the compartmentalization of cellular space and extra-cellular structure. Given these diverse and pervasive roles it is surprising how few biopolymers are recognized as containing inorganic monomers. Apart from the polymeric nucleic acids, which contain a polyphosphate diester backbone, there are few well characterized inorganic biopolymers. It is convenient to define this class of compounds as those that involve inorganic elements directly within the polymeric chain. This definition excludes many biopolymers that contain metals such as the metallothionines, complexes of fulvic and humic acids, and the whole family of metalloproteins. In these cases the metals are ancillary to the polymerized functionality; indeed, it is often possible to exchange one metal or prosthetic group for another in metalloproteins. It is anticipated that inorganic biopolymers are associated with the bulk elements such as phosphorus, calcium, and magnesium, and that as our understanding of the processes involved in their biomineralization develops, new

0097–6156/94/0572–0497$08.00/0

examples of this general class will emerge. The subject of this chapter is the unusual heme containing biopolymer that is produced by the malaria parasite.

Malaria continues to spread despite past accomplishments in drug therapy and vector control. Two main reasons for this are the rise of chloroquine resistant strains of the *plasmodium* genus, and the increasing resistance of mosquitos to pesticides used in their control. Within the mammalian host the predominant portion of the parasites life cycle, shown in Figure 1, is inside red blood cells, where after the initial invasion the contents of the cell are digested and the parasite undergoes asexual schizogony to produce more merozoites. The renowned clinical symptoms of malaria, especially the periodic delirium associated with the fever/chill/sweat cycles, is due to the synchronized rupture of the red blood cells, release of the merozoites, and reinvasion of new red blood cells. Consequences of this include certain anaemias and enlargement of the spleen. In a syndrome termed cerebral malaria, which is due to especially virulent species of malaria, *plasmodium falciparum*, death can result from the occlusion of the cerebral capillaries which leads to a progressive delirium, loss of consciousness, and coma (*1*).

The digestion of red blood cells poses a significant biochemical problem for the growing merozoites in that considerable quantities of heme are liberated as the polypeptide in hemoglobin is digested. Heme is a toxic multifunction regulator which not only regulates its own biosynthesis but also the synthesis of a diverse family of proteins connected by their roles in the transport, regulation, and metabolism of oxygen (*2*). The degradation of heme is performed by the non-heme iron enzyme heme oxygenase, which generates a ferryl intermediate from the reductive cleavage of dioxygen, and breaks the porphyrin ring at the *meso*-methylidyne position between the A and B pyrrole rings to release carbon monoxide, ferric ion, and biliverdin, equation 1 (*3*). The *plasmodium* parasite employs a completely different and unique metabolic pathway to detoxify the released heme, and forms an aggregated insoluble heme-based precipitate termed malaria pigment of hemozoin. It is not understood why the *plasmodium* parasite aggregates the heme in this manner rather than utilizing the more ubiquitous heme oxygenase pathway, but possible rationalizations are that the large quantities of iron and carbon monoxide, which would be released by such metabolism, would pose additional toxicity problems that are avoided by heme aggregation. Furthermore, it has recently been demonstrated that merozoites require extracellular iron, bound to transferrin, and that they are unable to tap the considerable reserves of iron present within β-hematin (*4*).

Upon rupture of the merozoite, step A in Figure 1, the aggregated heme is deposited into the blood which then accumulates in the extensive capillary beds found in the kidney, cerebellum, spleen, and liver. The quantity of accumulated aggregated heme can be so large as to be visible by the naked eye during autopsy and its presence in victims of "marsh miasma" was noted in the eighteenth century, some 150 years before the recognition of the protozoan origin of the disease (*5*).

The mechanism of malaria pigment's biosynthesis is unique and involves a putative heme polymerase enzyme that is associated with the aggregated heme and functions at the low pH of the food vacuole (*6,7*). Recent work demonstrates that heme-aggregation is inhibited by the quinine family of antimalarials

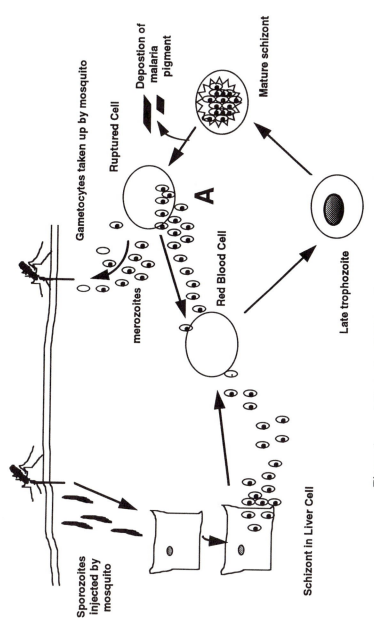

Figure 1.   Life cycle of *Plasmodium* in a mammalian host.

(6). The toxicity of the quinine drugs is attributed to the high concentrations of unaggregated heme that build-up in the merozoite.

The structure and composition of malaria pigment has been the subject of considerable dispute since its identification as being a heme aggregate (8,9). Early suggestions that it was β-hematin were refuted by elemental analyses which suggested the presence of proteinaceous material in these isolates (10). However, more recent purification protocols, which employ a magnetic field to separate the iron containing crystallites from the diamagnetic material, demonstrate that there is in fact no protein component to the pigment (11). When properly purified, malaria pigment consists solely of aggregated heme (11,12). Furthermore, reactivity, spectroscopic, and EXAFS studies have demonstrated that purified malaria pigment and β-hematin are identical (13) and that the most likely mode for aggregation involves inter-heme propionate-iron bond formation to give strands shown in Figure 2.   Our work concerns β-hematin, its synthesis, aggregation, spectroscopy, and crystallography. This chapter will describe the results of our studies to determine the details of the inter-heme interactions in β-hematin.

Synthetic Studies

It was recognized very early in the development of heme chemistry that when hematin, Fe(protoporphyrin-IX)(OH), is treated with mild acids at elevated temperatures for prolonged periods, equation 2, that an insoluble material, termed β-hematin, is produced (14). Although β-hematin is insoluble in water at a pH less than 10, it dissolves in strong alkali to give a product with similar spectroscopic properties to that obtained when either hematin or hemin, Fe(proto-porphyrin-IX)Cl, is treated with strong alkali. These results clearly indicate that β-hematin is an aggregated heme material, and a variety of mechanisms for the inter-heme interaction, including simple electrostatic (15), oxo-bridged dimers (16,17) and carboxylate bridged units have been proposed (18). Discrimination between these proposed interactions is complicated by the fact that the material prepared from the reaction shown in equation 2 is very poorly crystalline, and contains a range of aggregated unknown phases. Furthermore, the workup of these preparations involves numerous alkaline digestions and washes with either a carbonate or phosphate buffer which invariably dissolve large quantities of heme material in the process.

In order to prepare more ordered crystalline samples of β-hematin we initially sought to improve its synthesis by reexamining the reaction of hemin with base. Although many early studies describe the complex reactions between hematins (or hemins) and nitrogenous bases, which give poorly characterized species termed hemochromes (18), these results predate the recognition of the role of the oxo-bridged dimer, 1, in this chemistry (19,20). Moreover, many of these studies were often performed in water or with hydroxide bases. Our strategy for the synthesis of β-hematin was to treat hemin with a non-coordinating base, equation 3 (22). Methanol proved to be a suitable solvent for this preparation in that hemin has moderate solubility in this solvent and all the other reactants and products are either miscible or soluble in methanol.   The key to this development is the recognition that the reaction must be performed *absolutely*

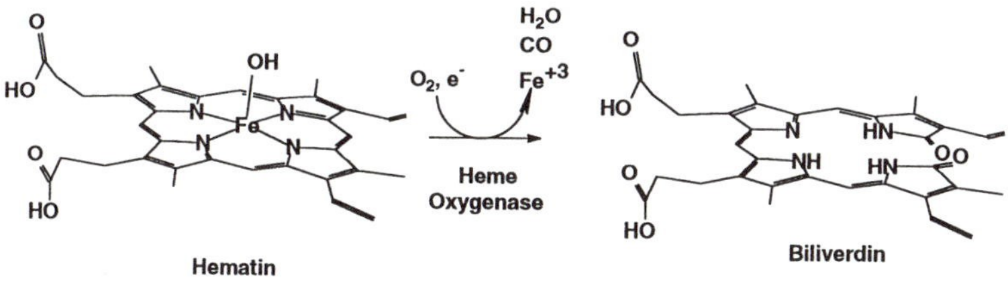

Equation 1

Hematin

Biliverdin

$$n\{Fe(por)(OH)\} \xrightarrow[\text{18 hr, 70 °C}]{\text{H}_2\text{O, CH}_3\text{CH}_2\text{CO}_2\text{H}} \beta\text{-Hematin} + n\{H_2O\} \qquad \text{Equation 2}$$

Hematin

por = protoporphyrin-IX

$$n\{Fe(por)(Cl)\} \xrightarrow[\text{12 hr, 22 °C}]{\text{MeOH, 2,6-lutidine}} \beta\text{-Hematin} + n\{2,6\text{-lutidine·HCl}\} \qquad \text{Equation 3}$$

Hemin

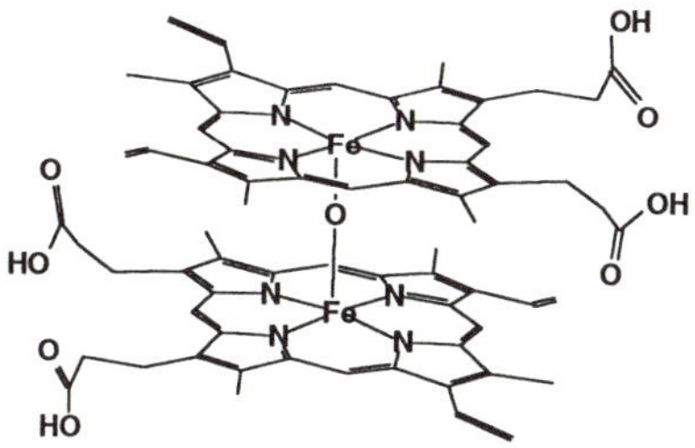

1

water free in order to avoid the formation of the μ-oxohemedimer **1**. In addition to improving the purity and yield of β-hematin, the low temperature conditions employed in equation 3 allows for a greater control over the rate of polymer formation. Thus the phase which results from this preparation has a high fraction of rod shaped needles which typically measure 3 μm in length and 0.4 μm in width as determined by scanning electron microscopy. To our knowledge there are no reports which describe these reactions under rigorously dry conditions. To demonstrate the importance of this experimental detail we have treated hemin with 2,6-lutidine under identical conditions to those described above except that the reaction was performed in the open without predried methanol or 2,6-lutidine. In this experiment we also obtained a dark aggregated methanol insoluble product, but, on treatment with bicarbonate buffer, this completely dissolves and no β-hematin is isolated. The choice of 2,6-lutidine as the base stems from the combination of high basicity with low affinity for transition metal complexes due to the steric interference of two *ortho*-methyl substituents.

The synthesis of β-hematin is improved by employing dimethylsulfoxide, DMSO, as a co-solvent with methanol, equation 4. Although the aggregation process is slower, and requires 9 days at ambient conditions instead of 12 hr, the resulting phase is visibly crystalline (with crystallite dimensions typically between 0.1 mm and 0.08 mm), and the buffered workup described in the experimental section requires only a single wash rather than a more exhaustive treatment. More importantly, this preparation gives a phase which allows for the unit cell determination by X-ray powder diffraction described below. We attribute the result of this synthesis to the better coordinating ability of DMSO as contrasted with methanol. Thus, dissolution of hemin in DMSO/methanol mixtures results in the rapid equilibration of hemin with solvent to give [Fe(protoporphyrin-IX)(DMSO)$_2$]$^+$, **2** in equation 5, which we anticipate will only slowly aggregate upon dissociation of the DMSO ligands.

Spectroscopic and Magnetic Studies

To establish the equivalence of the β-hematin from the two preparations in equations 3 and 4 with prior results we have characterized these compounds by elemental analysis, infrared spectroscopy, Raman spectroscopy, magnetic susceptibility, X-ray powder diffraction, and diffuse reflectance spectroscopy. The analytical results agree within experimental error with those reported by Slayter *et al.* (C, 64.7; H, 5.0; N, 8.7%) for β-hematin prepared by the thermal dehydration reaction in equation 2 (*13*). These results in turn are similar to those obtained for malaria pigment isolated from *P. falciparum* in the late trophozoite stage (*13*).

In Figure 3 the infrared spectra for β-hematin and hemin are contrasted. Notable features of these spectra are a) the differences in the carboxylate stretching regions of the two samples, b) the absence of an intense v(O-H) band between 4000 and 2500 cm^{-1}, and c) the absence of the strong v(Fe-O-Fe)$_{as}$ band *ca.* 900 cm^{-1}.

These features are not only consistent with the carboxylate bridged hypothesis of Slater *et al.* (*7*), Figure 2, but also with the participation of the uncoordinated

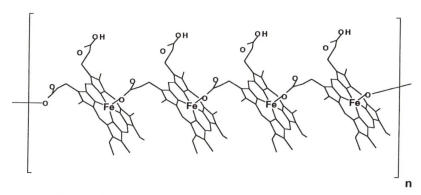

Figure 2. Proposed propionate-iron inter-heme links in β-hematin.

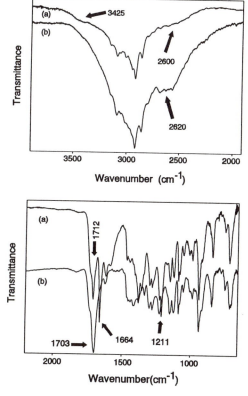

Figure 3. IR spectra for (a) β-hematin, and (b) hemin as a potassium bromide pellet. Top panel shows the ν(C-H) and ν(O-H) region while the bottom panel shows the ν(CO$_2$) region.

$$n\{Fe(por)(Cl)\} \xrightarrow[\text{12 hr, 22 °C}]{\text{MeOH, DMSO, 2,6-lutidine}} \beta\text{-Hematin} + n\{2,6\text{-lutidine·HCl}\} \quad \textbf{Equation 4}$$

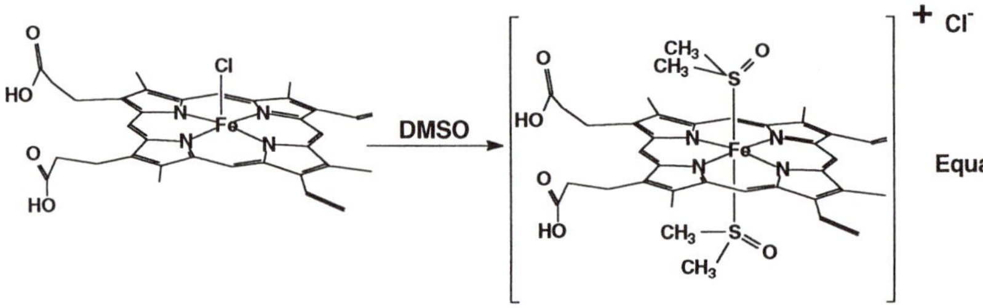

Equation 5

2

propionic side chain in a hydrogen bonded network. The insolubility of β-hematin in 0.1 M $NaHCO_3/Na_2CO_3$, an important feature in the separation described in the experimental section, is also consistent with the propionic acid hydrogen bonding. One possible structure that combines these spectroscopic features is shown in Figure 4, and depicts a possible inter-strand hydrogen bonding.

The resonance Raman spectra shown in Figure 5 also has features consistent with the formulation of β-hematin as containing five coordinate ferric centers with η^1-propionate ligation. Notably absent in Figure 5 (bottom) are bands at 890 and 416 cm^{-1} which correspond to the asymmetric and symmetric Fe-O-Fe bands for the μ-oxo dimer moiety. The shift of ν_{10}, assigned to the band at 1627 cm^{-1}, is consistent with the presence of high spin Fe(III) in a five coordinate environment *(23,24)*. Furthermore, all the bands above 1450 cm^{-1} have values closest to those of hemin, Fe(protoporphyrin-IX)Cl, a high spin five coordinate iron(III) complex. Although a variety of Fe(III)(porphyrin)(η^1-O_2CR) complexes have been prepared *(25,26)* and characterized by diffraction *(27,28)* there are no Raman data available for these complexes to assist in the assignment of the ν(Fe-O) mode. We tentatively assign the weak band at 514 cm^{-1} to this metal-ligand mode, but we note that confirmation of this assignment awaits isotopic substitution studies.

Assignment of these vibrational spectroscopic results provides collaborating evidence for the proposed hydrogen bonding proposed in Figure 4. We suggest that the two strong carboxylate stretching bands, $\nu_{asym}(CO_2)$ at 1712 and 1664 cm^{-1}, are due to the hydrogen bonded and unidentate propionate respectively. Support for the first assignment stems from the similarity of the $\nu(CO_2H)$ band at 1712 cm^{-1} in β-hematin to the band at 1703 cm^{-1} in hemin; The crystal structure of the latter indicates that both propionic acid groups are hydrogen bonded to adjacent porphyrins *(29)*. Although the more intense 1664 cm^{-1} band is higher than the range reported for η^1-coordination, 1645 to 1550 cm^{-1} *(30)*, it has a lower energy than free carboxylic acids, *c.f.* 1725 to 1700 cm^{-1}. A suggested range for the asymmetric stretching bands in carboxylic acid hydrogen bonded dimers is from 1747 to 1692 cm^{-1} *(31,32)*, and encompasses the 1712 cm^{-1} band. Apart from the positions of the bands themselves, the difference in the energy of the asymmetric and symmetric carboxylate stretching modes, $\Delta = \nu_{asym}(CO_2) - \nu_{sym}(CO_2)$, has been used as a diagnostic of the coordination mode for carboxylates, *i.e.* for unidentate carboxylates $\Delta > 200$ cm^{-1}, while for bidentate carboxylates $\Delta < 120$ cm^{-1}. In hydrogen bonded carboxylic acid dimers the magnitude of the difference between the Raman and infrared allowed $\nu_{asym}(CO_2)$ bands, Δ_a, has been used to diagnose dimer formation. In carboxylic acid hydrogen bonded dimers $\Delta_a > 40$ cm^{-1}, but if the hydrogen bonding is to a different functionality, $\Delta_a < 10$ cm^{-1}. If Cerami and Slater's assignment of $\nu(CO_2)_{sym} = 1211$ cm^{-1} for β-hematin is verified by isotopic substitution experiments *(13)*, than the difference, $\Delta = 453$ cm^{-1}, is at the long end of the range associated with unidentate carboxylates, but is nevertheless consistent with our model and assignment. On the other hand the Raman results for β-hematin contains only a single intense band at 1627 cm^{-1}, that is most likely due to a resonantly enhanced ν_{10} skeletal porphyrin mode. There is no indication of a

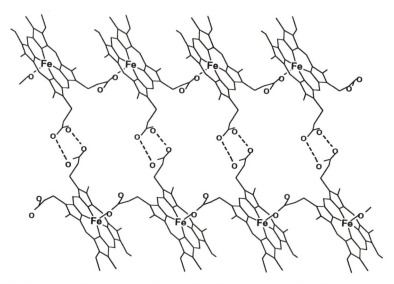

Figure 4.   Proposed hydrogen bonding between the linked heme strands in β-hematin.

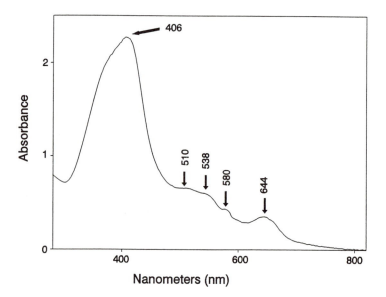

Figure 5.   Raman spectra for β-hematin obtained by the method in equation 4. Laser excitation of 647 nm was used with a sample mounted in a capillary.

band ascribable to a Raman active carboxylate stretch between 1712 cm^{-1} and 1630 cm^{-1}, which, following Lalancette's proposal, would support or refute propionate dimer formation (*33*). However, only very small or negligible enhancement of the carboxylate bands is expected for the non-iron bonded propionates since they are poorly electronically coupled to the porphyrin core by the two intervening saturated methylene groups. Finally Figure 3(top) contains features that are similar, broad bands at 2620 and 2600 cm^{-1}, in both β-hematin and hemin, which support the hydrogen bonding hypothesis. That this inter-heme interaction is different in the two compounds can be ascertained by the relative intensities and energy spread for the two bands, and the small feature at 3425 cm^{-1} that is present in β-hematin and not in hemin.

The UV-vis transmittance spectrum shown in Figure 6 contains a Soret band at 406 nm and Q-bands at 510, 538, 580, and 644 nm, the latter being especially characteristic of hemin aggregates as determined by photoacoustic spectroscopy (*34*) and microspectrophotometry (*35*). Given the number and energies of the bands in the absorption spectrum there exist manifold possibilities for resonance Raman enhancement studies.

The variable temperature magnetic susceptibility of β-hematin down to 4 K as determined with a SQUID magnetometer has been determined and a plot of effective magnetic moment (μ_B) versus temperature is shown in Figure 7. At room temperature effective magnetic moment is 5.73 B.M. and the temperature dependence of the inverse molar magnetic susceptibility χ_m^{-1} is linear from 4 to 300 K, with $C = 4.16$ emu·K·mol^{-1} and a small Curie-Weiss constant of $\theta = -4.65$ K. This result indicates that the iron atoms are only weakly coupled and that the iron is high spin, $S = 5/2$. Not only are these values similar to the magnetic susceptibility of other five coordinate iron (*36*) species such as Fe(TTP)(X) X = halide or pseudohalides, but also they are identical to results published by Brémard *et al.* for malaria pigment (*37,38*). Most notably though, this result demonstrates that the β-hematin resulting from equation 3 and 4 is not contaminated by the μ-oxo dimer, **1**.

Diffraction Results

Native malaria pigment is a highly ordered crystalline substance. This order is shown in the transmission electron micrograph image shown in Figure 8 (*39*), which indicates that there is a distinct preferred axis for growth and that the lattice planes are separated by *ca.* 9 Å.

Characterization of β-hematin by powder X-ray diffraction techniques with both conventional and synchrotron radiation indicates that the phase which results from equations 3 and 4 are sufficiently crystalline to give strong diffraction peaks out to 45° in 2θ with Cu $K_{\alpha 1}$ ($\lambda = 1.540598$ Å) radiation. The pattern obtained with synchrotron radiation ($\lambda = 1.7492$ Å) on beam line X7A at the National Synchrotron Light Source at the Brookhaven National Laboratory is shown in Figure 9.

It has been possible to index this diffraction pattern with the TREOR90 program (*21*) and we have obtained a triclinic unit cell with dimensions $a =$

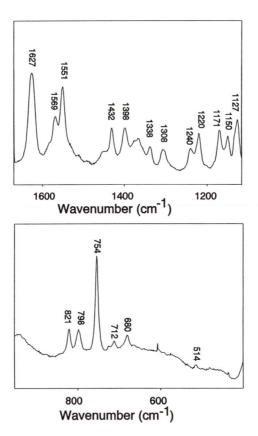

Figure 6.   Transmittance UV-vis spectrum for β-hematin measured as a potassium bromide pellet.

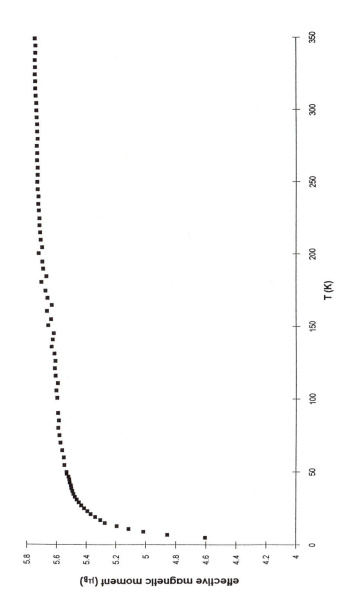

Figure 7.   SQUID Magnetic susceptibility results for β-hematin as a polycrystalline solid. Identical results are also found for a suspension of β-hematin in nujol.

Figure 8.   Transmission electron micrograph image of native malaria
pigment crystals extracted from *Plasmodium berghei* showing the direct
resolution of the lattice planes separated by ~ 8 Å.
(Reproduced with permission from reference 39. Copyright 1974 School of
Tropical Medicine.)

12.19, b = 14.69, c = 8.05 Å, and α = 90.25, β = 96.92, and γ = 98.37°, V = 1416 Å^3, and d_{exper} = 1.393(3) g·ml^{-1}. This cell has a high de Wolff figure of merit, $M(20)$ = 32, and all lines are successfully indexed. The experimentally determined density and volume can be used to calculate the mass of the contents of the unit cell, which in this instance corresponds to two Fe(protoporphyrin-IX) units per cell. Based on the value of the mean of the $|E*E-1|$ for the integrated intensities of the indexed reflections, 1.034, we suggest that the unit cell is centrosymmetric and the space group is uniquely defined as $P\bar{1}$. This space group is coincidentally found for the hemin, Fe(protoporphyrin-IX)Cl, which has a slightly larger unit cell (*29*). Studies are currently underway to use the integrated intensities obtained from high resolution synchrotron radiation powder diffraction pattern to determine the orientation of the hemes within the unit cell of β-hematin. These results will be described elsewhere.

Heme Aggregation

The digestion of hemoglobin and heme aggregation leading to the formation of β-hematin is a highly ordered process which does not involve unassociated free heme (*40*). Our knowledge of heme transport is limited, and perhaps the best understood heme transport protein is hemoplexin which may or may not be present in the food vacuole of *plasmodium* (*41*). Above all else, the inhibition of heme polymerization by the quinine antimalarials is a mechanistic question. In order to understand heme polymerase and the biosynthesis of β-hematin, we need a detailed understanding of heme aggregation in general.

Prior studies have used UV-visible spectroscopy (*42,43*), ultracentrifugation (*44*), and ^{1}H NMR (*45*) to study hematin aggregation. As a result of these studies it is recognized that in alkaline solutions there is a rapid dimerization, to give a μ-oxo bridged species, **1**, (*44*) followed by a slower aggregation process which leads to micelles with an upper limit of 45-50 heme units, distributed around a critical size. Hydrodynamic measurements by ultracentrifugation sedimentation techniques suggest that these micelles are spherically shaped and Blauer proposed the arrangement shown schematically in Figure 10, with the carboxylate groups pointing outward (*44*). By implication, the vinyl groups are oriented inward, and they may therefore be involved in the thermodynamic stabilization of the micelle.

Conclusion

The heme coordination polymer which results from the heme polymerase activity of the *plasmodium* parasite is an insoluble crystalline solid with propionate inter-heme links. How the quinine antimalarials inhibit heme polymerase and how this enzyme is associated with β-hematin remain important, albeit poorly understood, aspects of this biochemistry. Furthermore, heme detoxification by malaria is the only currently recognized example of iron excretion. The homeostasis of iron is usually srictly controlled, and excess is usually stored in ferritin for subsequent

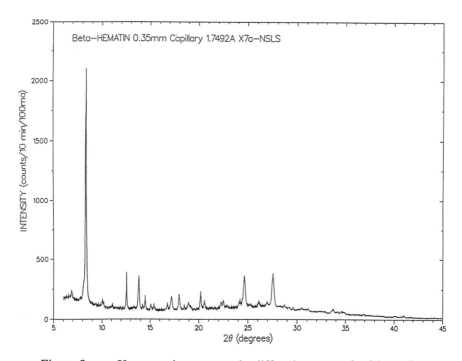

Figure 9.   X-ray synchrotron powder diffraction pattern for β-hematin.

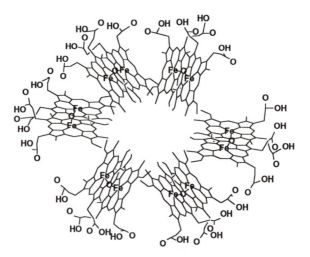

Figure 10.   Schematic depiction of Blauer's proposed micelles formed
from 40-50 units of the oxo bridged dimer 1.

use. β-Hematin represents an intriguing and unique type of inorganic biopolymer. It can be anticipated that other inorganic biopolymers will be discovered as our knowledge of the bulk inorganic elements develops; undoubtedly an understanding of their structures and functions will follow from many of the unique types of observations that have led to our current knowledge of β-hematin.

Experimental

Water and oxygen sensitive reactions and materials were handled and stored in a Vacuum Atmospheres inert atmosphere box with an atmosphere maintained at <5 ppm O_2 and tested for traces of water with $TiCl_4$ (*46*). Bovine hemin (ferriprotoporphyrin-IX chloride) was purchased from Aldrich and dried over phosphorus pentoxide *in vacuo* at room temperature in the dark for 12 hours, and then immediately transferred to the inert atmosphere box and stored in the dark at -32 °C. Methanol (MeOH) and dimethyl sulfoxide (DMSO) were dried and deoxygenated by distillation from Na and CaH_2, respectively, under nitrogen and then further degassed with the inert atmosphere gas before use (*47*). 2,6-Lutidine was dried by distillation under nitrogen from sodium. Infrared spectra were obtained on a Midac corporation FTIR with the sample prepared as a fused potassium bromide pellet. UV-vis spectra were measured on a Hewlett-Packard Diode-Array spectrometer 8452 as a dilute solid sample in a pressed potassium bromide pellet. Resonance Raman Spectra were acquired with a HR320 spectrograph with Photometrics CCD9000 liquid cooled detector. Laser excitation was made by a Spectraphysics 2025 Kr^+ laser operating at 647.1 nm.
X-ray powder diffraction data were acquired with a capillary geometry that was oscillated around θ and with a Ge(111) crystal analyzer. Monochromated synchrotron radiation ($\lambda = 1.7492$ Å) was employed.
Hemin chloride (0.1247g, 0.193 mmol) was dissolved with stirring in 2,6-lutidine (3 mL, 0.030 Mol) over the course of 15 minutes. This mixture is then diluted by addition of 50 mL of MeOH and 50 mL of DMSO and the mixture is allowed to stand without stirring. After nine days it was brought out of the inert atmosphere box and the DMSO/MeOH solvent was decanted off the solid phase. The solid was washed with MeOH until the MeOH washes were colorless and clear. The solid product was dried *in vacuo* at room temperature to give 0.107g, (91 % yield) of a black fluffy powder

Acknowledgments

We gratefully acknowledge support from the National Institutes of Health in the form of a Biomedical Research Support Grant SO7RR07157-17 that was awarded to DSB and the University of Wyoming, the National Institute of Standards and Technology for the use of their SQUID magnetometer (GTY).

Literature Cited

1. Sun, T. *Color Atlas and Textbook of Diagnostic Parasitology*; Igaku-Shoin; New York, NY, 1988; pp 1-5.
2. Padmanaban, G.; Venkateswar, V.; Rangarajan, P.N. *TIBS* **1989**,*14*,492-6.
3. Schacter, B.A. *Semin. Hematol.* **1988**,*25*,349-69.
4. Pollack, S. In *Malaria and the Red Cell 2*; ARL Publishers; New York, NY, 1989; pp 151-61.
5. Wernsdorfer, W. In *Malaria*; Kreier, J.P., Ed.; Academic Press; New York, NY, 1980; Vol. 1, Chapter 1.
6. Slater, A.F.G; Cerami, A. *Nature(London)* **1992**,*355*,167-9.
7. Slater, A.F.G. *Exper. Parasit.* **1992**,*74*,362-5.
8. World Health Organization. *World Malaria Situation, 1988*; Rapport Trimestriel de Sanitares Mondailes; **1990**,*43*,68-79.
9. Wyler, D.J. *N. Engl. J. Med.* **1983**,*308*,875.
10. Yamada, K.A.; Sherman, I.W. *Expt. Parasitol.* **1979**,*48*,61-74.
11. Fairlamb, A.H.; Paul, F.; Warhurst, D.C. *Mol. Biochem. Parasit.* **1984**,*12*,307-12.
12. Fitch, C.D.; Kanjananggulpan, P. *J. Biol. Chem.* **1987**,*262*,15552-55.
13. Slater, A.D.; Swiggard, W.J.; Orton, B.R.; Flitter, W.D.; Goldberg, D.E.; Cerami, A.; Henderson, G.B. *Proc. Natl. Acad. Sci., U.S.A.* **1991**,*88*,325-9.
14. Küster, W. *Handbuch der Biologische Arbeitsmethoden Abt. I, Teil VIII* **1921**,200.
15. Abraham, R.J.; Burbridge, P.A.; Jackson, A.H.; MacDonald, D.B.; *J. Chem. Soc.* **1966**,B620-26.
16. Shack, J.; Clark, W.M. *J. Biol. Chem.* **1947**,*171*,143-87.
17. Fleischer, E.B.; Palmer, J.M.; Srivastava, T.S.; Chatterjee, A. *J. Am. Chem. Soc.* **1971**,*93*,3162-67.
18. Lemberg, R.; Legge, J.W. *Hematin Compounds and Bile Pigments*; Interscience; New York, NY, 1949; pp 174-84.
19. Cohen, I.A. *J. Am. Chem. Soc.* **1969**,*91*,1980-3.
20. Brown, S,B.; Jones, P., Lantzke, I.R. *Nature(London)* **1969**,*223*,960-1.
21. Werner, P.-E. *Z. Kristallogr.* **1964**,*120*,375.
22. Bohle, D.S.; Helms, J.B. *Biochem. Biophys. Res. Commun.* **1993**,*193*,504-8.
23. Choi, S.; Spiro, T.G.; Langry, K.C.; Smith, K.M.; Budd, D.L.; La Mar, G.N. *J. Am. Chem. Soc.* **1982**,*104*,4345-51.
24. Spiro, T.G. In *Iron Porphyrins Part Two*; Lever, A.B.P., Gray, H.B., Eds,; Addison-Wesley Publishing Company; New York, NY, 1983; pp 89-160.
25. Bominaar, E.L.; Ding, X.-Q.; Gismelseed, A.; Bill, E.; Winkler, H.; Trautwein, A.X.; Nasri, H.; Fischer, J.; Weiss, R. *Inorg. Chem.* **1992**,*31*,1845-54.
26. Nasri, H.; Fischer, J.; Weiss, R.; Bill, E.; Trautwein, A. *J. Am. Chem. Soc.* **1987**,*109*,2549-50.

27. Bill, E.; Bominaar, E.L.; Ding, X.-O.; Fischer, J.; Gismelseed, A.; Nasri, H.; Trautwein, A.X.; Weiss, R.; Winkler, H. *Hyperfine Interactions* **1992**,*71*,1295-8.
28. Oumous, H.; Lecomte, C.; Protas, J.; Cocolios, P.; Guilard, R. *Polyhedron* **1984**,*3*,651-9.
29. Koenig, D.F. *Acta Cryst.* **1965**,*18*,663-73.
30. Deacon, G.B.; Phillips, R.J. *Coord. Chem. Rev.* **1980**,*33*,227-50.
31. Hadži, D. *Pure. Appl. Chem.* **1965**,*11*,435-53.
32. Hadži, D. *Chimica* **1972**,*26*,7-13.
33. Vanderhoff, P.A.; Lalancette, R.A.; Thompson, H.W. *J. Org. Chem.* **1990**,*55*,1696-8.
34. Balasubramanian, D.; Rao, Ch. M.; Panijpan, B. *Science* **1984**,*223*,828-30.
35. Morselt, A.F.W.; Glastra, A.; James, *J. Exper. Parasit.* **1973**,*33*,17-22.
36. Mitra, S., In *Iron Porphyrins Part Two*; Lever, A.B.P., Gray, H.B., Eds.; Addison-Wesley Publishing Company; New York, NY, 1983; pp 1-42.
37. Brémard, C.; Merlin, J.C.; Moreau, S.; Girerd, J.J. *Spectroscopy of Biomolecules, Special Publication of Royal Society of Chemistry;* Hester, R.E., Girling, R.B., Eds.; **1991**,329-30.
38. Brémard, C.; Girerd, J.J.; Merlin, J.C.; Moreau, S. *J. Mol. Struc.* **1992**,*267*,117-22.
39. Moore, G.A.; Boothroyd, B. *Ann. Trop. Med. Parasit.* **1974**,*68*,489.
40. Muller-Eberhard, U. *FEBS-Symposium, 58(Transport Proteins)* **1978**,*58*,295-302.
41. Goldberg, D.E.; Slater, A.F.G.; Cerami, A.; Henderson, G.B. *Proc. Natl. Acad. Sci., U.S.A.*, **1990**,*87*,2931-5.
42. Srinivas, V.; Rao, Ch.M. *Biochem. Int.* **1990**,*21*,849-55.
43. Brown, S.B.; Dean, T.C.; Jones, P. *Biochem. J.* **1970**,*117*,733-9.
44. Blauer, G.; Zvilichovsky, B. *Arch. Biochem. Biophys.* **1968**,*127*,749-55.
45. Mazumdar, S.; Mitra, S. *J. Phys. Chem.* **1990**,*94*,561-6.
46. Perrin, D.D.; Armarego, W.L.F. *Purification of Laboratory Chemicals (3rd. Ed.)*; Pergamon Press; New York, NY, 1988; p 217.
47. Shriver, D.F. *The Manipulation of Air-sensitive Compounds*; McGraw-Hill Book Company; New York, NY, 1969; pp 164-89.

RECEIVED June 23, 1994

INDEXES

Author Index

Affiliation Index

Subject Index

A

Production: Susan Antigone & Charlotte McNaughton
Indexing: Deborah H. Steiner
Acquisition: Anne Wilson
Cover design: Alan Kahan

Printed and bound by Maple Press, York, PA

Highlights from ACS Books

Good Laboratory Practice Standards: Applications for Field and Laboratory Studies
Edited by Willa Y. Garner, Maureen S. Barge, and James P. Ussary
ACS Professional Reference Book; 572 pp; clothbound ISBN 0–8412–2192–8

Silent Spring Revisited
Edited by Gino J. Marco, Robert M. Hollingworth, and William Durham
214 pp; clothbound ISBN 0–8412–0980–4; paperback ISBN 0–8412–0981–2

The Microkinetics of Heterogeneous Catalysis
By James A. Dumesic, Dale F. Rudd, Luis M. Aparicio, James E. Rekoske,
and Andrés A. Treviño
ACS Professional Reference Book; 316 pp; clothbound ISBN 0–8412–2214–2

Helping Your Child Learn Science
By Nancy Paulu with Margery Martin; Illustrated by Margaret Scott
58 pp; paperback ISBN 0–8412–2626–1

Handbook of Chemical Property Estimation Methods
By Warren J. Lyman, William F. Reehl, and David H. Rosenblatt
960 pp; clothbound ISBN 0–8412–1761–0

Understanding Chemical Patents: A Guide for the Inventor
By John T. Maynard and Howard M. Peters
184 pp; clothbound ISBN 0–8412–1997–4; paperback ISBN 0–8412–1998–2

Spectroscopy of Polymers
By Jack L. Koenig
ACS Professional Reference Book; 328 pp;
clothbound ISBN 0–8412–1904–4; paperback ISBN 0–8412–1924–9

Harnessing Biotechnology for the 21st Century
Edited by Michael R. Ladisch and Arindam Bose
Conference Proceedings Series; 612 pp;
clothbound ISBN 0–8412–2477–3

From Caveman to Chemist: Circumstances and Achievements
By Hugh W. Salzberg
300 pp; clothbound ISBN 0–8412–1786–6; paperback ISBN 0–8412–1787–4

The Green Flame: Surviving Government Secrecy
By Andrew Dequasie
300 pp; clothbound ISBN 0–8412–1857–9

For further information and a free catalog of ACS books, contact:
American Chemical Society
Distribution Office, Department 225
1155 16th Street, NW, Washington, DC 20036
Telephone 800–227–5558

Bestsellers from ACS Books

The ACS Style Guide: A Manual for Authors and Editors
Edited by Janet S. Dodd
264 pp; clothbound ISBN 0–8412–0917–0; paperback ISBN 0–8412–0943–X

The Basics of Technical Communicating
By B. Edward Cain
ACS Professional Reference Book; 198 pp;
clothbound ISBN 0–8412–1451–4; paperback ISBN 0–8412–1452–2

Chemical Activities (student and teacher editions)
By Christie L. Borgford and Lee R. Summerlin
330 pp; spiralbound ISBN 0–8412–1417–4; teacher ed. ISBN 0–8412–1416–6

Chemical Demonstrations: A Sourcebook for Teachers,
Volumes 1 and 2, Second Edition
Volume 1 by Lee R. Summerlin and James L. Ealy, Jr.;
Vol. 1, 198 pp; spiralbound ISBN 0–8412–1481–6;
Volume 2 by Lee R. Summerlin, Christie L. Borgford, and Julie B. Ealy
Vol. 2, 234 pp; spiralbound ISBN 0–8412–1535–9

Chemistry and Crime: From Sherlock Holmes to Today's Courtroom
Edited by Samuel M. Gerber
135 pp; clothbound ISBN 0–8412–0784–4; paperback ISBN 0–8412–0785–2

Writing the Laboratory Notebook
By Howard M. Kanare
145 pp; clothbound ISBN 0–8412–0906–5; paperback ISBN 0–8412–0933–2

Developing a Chemical Hygiene Plan
By Jay A. Young, Warren K. Kingsley, and George H. Wahl, Jr.
paperback ISBN 0–8412–1876–5

Introduction to Microwave Sample Preparation: Theory and Practice
Edited by H. M. Kingston and Lois B. Jassie
263 pp; clothbound ISBN 0–8412–1450–6

Principles of Environmental Sampling
Edited by Lawrence H. Keith
ACS Professional Reference Book; 458 pp;
clothbound ISBN 0–8412–1173–6; paperback ISBN 0–8412–1437–9

Biotechnology and Materials Science: Chemistry for the Future
Edited by Mary L. Good (Jacqueline K. Barton, Associate Editor)
135 pp; clothbound ISBN 0–8412–1472–7; paperback ISBN 0–8412–1473–5

For further information and a free catalog of ACS books, contact:
American Chemical Society
Distribution Office, Department 225
1155 16th Street, NW, Washington, DC 20036
Telephone 800–227–5558